T0231756

# PHYTOCHEMISTRY

## Volume 2

### Pharmacognosy, Nanomedicine, and Contemporary Issues

# PHYTOCHEMISTRY

## Volume 2

## Pharmacognosy, Nanomedicine, and Contemporary Issues

*Edited by*

**Chukwuebuka Egbuna**
**Shashank Kumar, PhD**
**Jonathan Chinenye Ifemeje, PhD**
**Jaya Vikas Kurhekar, PhD**

Apple Academic Press Inc.
3333 Mistwell Crescent
Oakville, ON L6L 0A2
Canada

Apple Academic Press Inc.
9 Spinnaker Way
Waretown, NJ 08758
USA

© 2019 by Apple Academic Press, Inc.

First issued in paperback 2021

*Exclusive worldwide distribution by CRC Press, a member of Taylor & Francis Group*

No claim to original U.S. Government works

**Phytochemistry, Volume 2: Pharmacognosy, Nanomedicine, and Contemporary Issues**

ISBN 13: 978-1-77463-523-0 set
ISBN 13: 978-1-77463-433-2 (pbk)
ISBN 13: 978-1-77188-760-1 (hbk)

**Phytochemistry, 3-volume set**

ISBN 13: 978-1-77188-762-5 (hbk)

All rights reserved. No part of this work may be reprinted or reproduced or utilized in any form or by any electric, mechanical or other means, now known or hereafter invented, including photocopying and record-ing, or in any information storage or retrieval system, without permission in writing from the publisher or its distributor, except in the case of brief excerpts or quotations for use in reviews or critical articles.

This book contains information obtained from authentic and highly regarded sources. Reprinted material is quoted with permission and sources are indicated. Copyright for individual articles remains with the authors as indicated. A wide variety of references are listed. Reasonable efforts have been made to publish reliable data and information, but the authors, editors, and the publisher cannot assume responsibility for the validity of all materials or the consequences of their use. The authors, editors, and the publisher have attempted to trace the copyright holders of all material reproduced in this publication and apologize to copyright holders if permission to publish in this form has not been obtained. If any copyright material has not been acknowl-edged, please write and let us know so we may rectify in any future reprint.

**Trademark Notice:** Registered trademark of products or corporate names are used only for explanation and identification without intent to infringe.

---

**Library and Archives Canada Cataloguing in Publication**

---

Phytochemistry / edited by Chukwuebuka Egbuna, Shashank Kumar, PhD, Jonathan Chinenye Ifemeje, PhD, Jaya Vikas Kurhekar, PhD.

Includes bibliographical references and indexes.
Contents: Volume 2. Pharmacognosy, nanomedicine, and contemporary issues.
Issued in print and electronic formats.
ISBN 978-1-77188-760-1 (v. 2 : hardcover).--ISBN 978-1-77188-762-5 (set : hardcover).--
ISBN 978-0-429-42619-3 (v. 2 : PDF).--ISBN 978-0-429-42627-8 (set : PDF)

1. Botanical chemistry. I. Egbuna, Chukwuebuka, editor II. Kumar, Shashank, editor III. Ifemeje, Jonathan Chinenye, editor IV. Kurhekar, Jaya Vikas, editor

| QK861.P65 2019 | 572'.2 | C2018-904824-7 | C2018-904875-1 |
|---|---|---|---|

---

CIP data on file with US Library of Congress

---

Apple Academic Press also publishes its books in a variety of electronic formats. Some content that appears in print may not be available in electronic format. For information about Apple Academic Press products, visit our website at **www.appleacademicpress.com** and the CRC Press website at **www.crcpress.com**

# ABOUT THE EDITORS

**Chukwuebuka Egbuna**

Chukwuebuka Egbuna is a chartered chemist, a chemical analyst, and an academic researcher. He is a member of the Institute of Chartered Chemists of Nigeria (ICCON), the Nigerian Society of Biochemistry and Molecular Biology (NSBMB), the Royal Society of Chemistry (RSC), United Kingdom, and the Society of Quality Assurance (SQA), USA. He has been engaged in a number of roles at New Divine Favor Pharmaceutical Industry Limited, Akuzor Nkpor, Anambra State, Nigeria, and Chukwuemeka Odumegwu Ojukwu University (COOU), Nigeria. He has attended series of conferences and workshops and has collaboratively worked and published quite a number of research articles in the domain of phytochemistry. He has edited books with top publishers such as Springer Nature and Elsevier. He is a reviewer and an editorial board member for various journals, including serving as a website administrator for the *Tropical Journal of Applied Natural Sciences* (TJANS), a journal of the faculty of Natural Sciences, COOU. His primary research interests are in phytochemistry, food and medicinal chemistry, analytical chemistry, and nutrition and toxicology. He obtained his BSc and MSc degrees in biochemistry at Chukwuemeka Odumegwu Ojukwu University.

**Shashank Kumar, PhD**

Shashank Kumar, PhD, is working as Assistant Professor at the Center for Biochemistry and Microbial Sciences, Central University of Punjab, Bathinda, India. He obtained his BSc, MSc, and PhD Biochemistry from the Department of Biochemistry, University of Allahabad, India. He worked as Postdoctoral Fellow at the Department of Biochemistry, King George's Medical University, Lucknow, India. Dr. Kumar has about 60 published scientific papers/reviews/editorial articles/book chapters in various national and international peer-reviewed journals and has been cited more than 1200 times. He has edited several books on topics such as the "concepts in cell signaling," "carbohydrate metabolism: theory and practical approach," and so forth. He has expertise in the areas of free radical biology, cancer biology, characterization of plant natural products, xenobiotic metabolism, and microbiology. He is familiar with many biochemical techniques such as spectrophotometry, enzyme-linked immunosorbent assay, electrophoresis, polymerase chain reaction, real-time polymerase chain reaction, flow cytometry, thin-layer chromatography, high-performance liquid chromatography, liquid chromatography–mass spectrometry, cell culture, and microbiological techniques. He has presented his research findings at more than 25 national/international conferences and attended about 30 workshops at different universities and medical colleges throughout the India. Dr. Kumar is a life time member of Italo-Latin American Society of Ethnomedicine, and the Indian Sciences Congress Association, and member of the Asian Council of Science Editors, Dubai, UAE, and Publication Integrity and Ethics, London. He has been awarded the Junior/Senior and Research Associate Fellowships formulated and funded by various Indian agencies, such as Indian Council of Medical Research, University Grants Commission, Council of Scientific and Industrial Research India. Dr. Kumar laboratory has been funded by the University Grant Commission, India, and the Department of Science and Technology, India, for working on effects of various phytochemicals on cancer cell signaling pathway inhibition.

## Jonathan Chinenye Ifemeje, PhD

Jonathan Chinenye Ifemeje, PhD, is an Associate Professor in the Department of Biochemistry, Faculty of Natural Sciences, Chukwuemeka Odumegwu Ojukwu University, Nigeria. He obtained his PhD in applied biochemistry from Nnamdi Azikiwe University, Awka, Nigeria, and his MSc degree in nutrition and toxicology from the University of Port-Harcourt, Nigeria. He has to his credits over 40 publications in both local and international journals. Dr. Ifemeje is currently the Coordinator, Students Industrial Work Experience Scheme (SIWES), COOU, and has served as an external examiner for various institutions. He is the Managing Editor of the *Tropical Journal of Applied Natural Sciences* and is serving as a reviewer and an editorial board member for various journals. He has worked extensively in the area of phytochemistry, nutrition, and toxicology. He is a member of various institutes, including the Institute of Chartered Chemists of Nigeria (ICCON), the Nigerian Society of Biochemistry and Molecular Biology (NSBMB), and the Society of Quality Assurance (SQA).

## Jaya Vikas Kurhekar, PhD

Jaya Vikas Kurhekar, PhD (Microbiology), MBA, (Human Resources), is presently the Head and Associate Professor in Microbiology, teaching undergraduate and postgraduate students in the Department of Microbiology at Dr. Patangrao Kadam Mahavidyalaya, Sangli, India. She is a guide for MPhil and PhD (Microbiology) students at Shivaji University, Kolhapur, Bharati Vidyapeeth University, Pune, and examiner for PhD candidates at Mumbai University, Mumbai, and Cairo University, Egypt. Her areas of research interest are phytochemistry, pharmacognosy, environmental microbiology, medical microbiology, and agricultural microbiology. She has over forty scientific papers published in various international and national journals to her credit. She has also been a reviewer for many international and national journals and worked as an editor for special issues of journals in microbiology. She has been the editor and contributor to a textbook an microbiology and has translations of three books to her credit. She has presented 40 scientific papers at international and national conferences and has been invited as a speaker and chairperson as well. She has significantly contributed to an encyclopedia of scientists in microbiology that is compiled and edited by Maharashtra State Marathi Encyclopedia Creation Board, Mumbai, India. She has successfully undertaken five minor research projects funded by the University Grants Commission, New Delhi, India, and has attended 100 international and national research gatherings. Dr. Kurhekar is a life member of various associations, including the Indian Association of Medical Microbiology, Indian Society of Pharmacognosy, Association of Microbiologists of India, Sangli unit, Society of Environmental Sciences, Microbiologists' Society, Society for Current Sciences, and Swamy Botanical Club.

# CONTENTS

# CONTRIBUTORS

**Maria Aslam**
University Institute of Diet and Nutritional Sciences (UIDNS), University of Lahore (UOL), Pakistan

**Yen San Chan**
Department of Chemical Engineering, Faculty of Engineering and Science, Curtin University CDT 250, 98009, Miri, Sarawak, Malaysia

**Dolli Chauhan**
Centre for Aromatic Plants (CAP), Industrial Estate Selaqui, Dehradun, Uttarakhand 248011, India

**Nirpendra K. Chauhan**
Centre for Aromatic Plants (CAP), Industrial Estate Selaqui, Dehradun, Uttarakhand 248011, India

**Michael K. Danquah**
Department of Civil & Chemical Engineering, University of Tennessee, Chattanooga, TN 37403, United States

**Swagata Das**
Department of Biochemistry and Microbial Sciences, School of Basic and Applied Sciences, Central University of Punjab, Bathinda, Punjab 151001, India

**G. P. Dubey**
Collaborative Program, Institute of Medical Sciences, Banaras Hindu University, Varanasi, India

**Chukwuebuka Egbuna**
Department of Biochemistry, Faculty of Natural Sciences, Chukwuemeka Odumegwu Ojukwu University, Anambra State 431124, Nigeria

**Ashutosh Gupta**
Department of Biochemistry, University of Allahabad, Allahabad Gujarat 211002, India

**S. Zafar Haider**
Centre for Aromatic Plants (CAP), Industrial Estate Selaqui, Dehradun, Uttarakhand 248011, India

**Yiik Siang Hii**
Department of Chemical Engineering, Faculty of Engineering and Science, Curtin University CDT 250, 98009, Miri, Sarawak, Malaysia

**Jonathan C. Ifemeje**
Department of Biochemistry, Faculty of Natural Sciences, Chukwuemeka Odumegwu Ojukwu University, Anambra State 431124, Nigeria

**Stella I. Inya-Agha**
Department of Pharmacognosy and Environmental Medicines, University of Nigeria, Nsukka

**Shaista Jabeen N.**
Department of Zoology, PG, and Research Unit, Dhanabagyam Krishnaswamy Mudaliar College for Women (Autonomous), Sainathapuram, RV Nagar, Vellore, Tamil Nadu 632001, India

**Jaison Jeevanandam**
Department of Chemical Engineering, Faculty of Engineering and Science, Curtin University CDT 250, 98009, Miri, Sarawak, Malaysia

**Saravanan Kaliyaperumal**
PG & Research Department of Zoology, Nehru Memorial College (Autonomous), Puthanampatti, Tiruchirappalli, 621007, India

**Hafsa Kamran**
University Institute of Diet and Nutritional Sciences (UIDNS), University of Lahore (UOL), Pakistan

**Sidra Khalid**
University Institute of Diet and Nutritional Sciences (UIDNS), University of Lahore (UOL), Pakistan

**Ashfaq Ahmad Khan**
Women University of Azad Jammu and Kashmir, Bagh, Pakistan

**Toskë L. Kryeziu**
Department of Clinical Pharmacy, University of Pristina, Kosovo

**Shashank Kumar**
Department of Biochemistry and Microbial Sciences, School of Basic and Applied Sciences, Central University of Punjab, Bathinda, Punjab 151001, India

**Vinesh Kumar**
Department of Sciences, Kids' Science Academy, Roorkee, Uttarakhand, India

**Priyanka Kumari**
Department of Biochemistry and Microbial Sciences, School of Basic and Applied Sciences, Central University of Punjab, Bathinda, Punjab, 151001, India

**Jaya Vikas Kurhekar**
Department of Microbiology, Dr. Patangrao Kadam Mahavidyalaya, Sangli, Maharashtra 416416, India

**Prem Prakash Kushwaha**
Department of Biochemistry and Microbial Sciences, School of Basic and Applied Sciences, Central University of Punjab, Bathinda, Punjab 151001, India

**V. Lakshminarayana**
Department of Botany, Andhra University, Visakhapatnam 530003, India

**Hema Lohani**
Centre for Aromatic Plants (CAP), Industrial Estate Selaqui, Dehradun Uttarakhand 248011, India

**Prareeta Mahapatra**
Department of Biochemistry and Microbial Sciences, School of Basic and Applied Sciences, Central University of Punjab, Bathinda, Punjab 151001, India

**Pragya Mishra**
Centre of Food Technology, University of Allahabad, Allahabad 211002, India

**Raghvendra Raman Mishra**
Medical Laboratory Technology, DDU Kaushal Kendra, Banaras Hindu University, Varanasi, Uttar Pradesh 221005, India

**Sunil Kumar Mishra**
Department of Pharmaceutical, Engineering, and Technology, Indian Institute of Technology, Banaras Hindu University, Varanasi, India

**Azham Mohamad**
Centre of Foundation Studies for Agricultural Science, Universiti Putra Malaysia, 43400 UPM Serdang, Selangor, Malaysia

**Andrew G. Mtewa**
Department of Chemistry, Institute of Technology, Malawi University of Science and Technology, Malawi
Pharmbiotechnology and Traditional Medicine Center of Excellence, Mbarara University of Science
and Technology, Uganda

**Felix Ifeanyi Nwafor**
Department of Pharmacognosy and Environmental Medicines, University of Nigeria, Nsukka,
Enugu State, Nigeria

**Onyeka Kingsley Nwosu**
National Biosafety Management Agency (NBMA), Abuja, Nigeria

**Olumayowa Vincent Oriyomi**
Institute of Ecology and Environmental Studies, Obafemi Awolowo University, Ile-Ife, Osun State,
Nigeria

**Temitope A. Oyedepo**
Nutritional/Toxicology Unit of the Department of Biochemistry, Adeleke University, Ede, Nigeria

**Abhay K. Pandey**
Department of Biochemistry, University of Allahabad, Allahabad 211002, India

**Vijaykumar K. Parmar**
Department of Pharmaceutical Sciences, Sardar Patel University, Vallabh Vidyanagar, India

**Seshu Vardhan Pothabathula**
School of Biotechnology, JNTUK, Kakinada, Andhra Pradesh 533003, India

**Suresh Purohit**
Department of Pharmacology, Institute of Medical Sciences, Banaras Hindu University, Varanasi, India

**G. M. Narasimha Rao**
Department of Botany, Andhra University, Visakhapatnam 530003, India

**Alan Thomas S.**
National Institute of Plant Science Technology, Mahatma Gandhi University, Kottayam,
Kerala  686560 India

**Hameed Shah**
CAS Key Laboratory for Biomedical Effects of Nanomaterials and Nanosafety, National Center for
Nanoscience and Technology, Beijing, China
University of Chinese Academy of Science, Beijing 100049, China

**Nadia Sharif**
Department of Biotechnology, Lahore College for Women University, Lahore, Pakistan

**Yogita Sharma**
Department of Sciences, Kids' Science Academy, Roorkee, Uttarakhand, India

**Parjanya Kumar Shukla**
Department of Pharmaceutical Sciences, Faculty of Health Science, Sam Higginbottom Institute of
Agriculture Technology and Sciences – Deemed University, Allahabad, India

**Rashmi Shukla**
Department of Medicinal Chemistry, Institute of Medical Sciences, Banaras Hindu University,
Varanasi, India

**Pushpendra Singh**
Department of Biochemistry and Microbial Sciences, School of Basic and Applied Sciences,
Central University of Punjab, Bathinda, Punjab 151001, India
National Institute of Pathology, New Delhi, India

**Shibani Sukhi**
Department of Biotechnology, Savitribai Phule Pune University, Pune 411007, India

**Intan Soraya Che Sulaiman**
Centre of Research & Innovation Management, Universiti Pertahanan Nasional Malaysia,
Kem Sungai Besi, 57000, Kuala Lumpur, Malaysia

**Habibu Tijjani**
Natural Product Research Laboratory, Department of Biochemistry, Bauchi State University,
Gadau, Nigeria

**Yamini B. Tripathi**
Department of Medicinal Chemistry, Institute of Medical Sciences, Banaras Hindu University,
Varanasi India

**Frederick O. Ujah**
Department of Chemical Sciences, College of Natural and Applied Sciences, University of Mkar,
Nigeria

**Prabhat Upadhyay**
Department of Pharmacology, Institute of Medical Sciences, Banaras Hindu University, Varanasi, India
Collaborative program, Institute of Medical Sciences, Banaras Hindu University, Varanasi, India

**Ketan Variya**
Ramanbhai Patel College of Pharmacy, Charotar University of Science and Technology (CHARUSAT),
Anand, India
Sun Pharma Advanced Research Company Ltd., Vadodara, India

**Rinki Verma**
Collaborative Program, Institute of Medical Sciences, Banaras Hindu University, Varanasi, India

# ABBREVIATIONS

| | |
|---|---|
| 2PFM | 2-photon fluorescence microscopy |
| 5-OHMeUra | 5-hydroxymethyl uracil |
| 8-OH-G | 8-Hydroxy-guanine |
| AAS | Atomic Absorption Spectroscopy |
| ABTS | 2,2-azino-bis (3-ethylbenzothiazoline-6-sulfonic acid |
| ACH | Acetylcholine |
| ACSM | Alternative Chinese System of Medicine |
| ADH | Alcohol dehydrogenase |
| AFM | Atomic force microscopy |
| AGEs | Advanced glycation end products |
| ALPs | Alkaline phosphatases |
| ALT | Alanine aminotransferase |
| AOSs | Active oxygen species |
| AR | Aldose reductase |
| ARE | Antioxidant response element |
| AST | Aspartate aminotransferase |
| BAM | Brewster angle microscope |
| BDE | Bond dissociation enthalpy |
| BHT | Butylated hydroxytoluene |
| CAT | Catalase |
| CFT | Critical flocculation temperature |
| CG | (-)- Catechin gallate |
| CKD | Chronic kidney disease |
| CLSM | Confocal laser scanning microscopy |
| CNS | Central nervous system |
| COMT | Catechol-O-methyltransferase |
| CRC | Colorectal cancer |
| CSF | Cerebrospinal fluid |
| CTGF | Connective tissue growth factor |
| CTM | Chinese Traditional Medicine |
| CUPRAC | Cupric reducing antioxidant capacity |
| CV % | Coefficient of variation |

| | |
|---|---|
| CVD | Cardiovascular disease |
| DAD | Diode array detector |
| DAG | Diacylglycerol |
| DCM-M | Dichloromethane: methanol |
| DLS | Dynamic light scattering |
| DMA | Dimethylacetamide |
| DM | Diabetes mellitus |
| DMSO | Dimethyl sulfoxide |
| DNA | Deoxyribonucleic acid |
| DN | Diabetic nephropathy |
| DNP | Dinitrophenylhydrazone |
| DOX | Doxorubicin |
| EC | Epicatechin |
| ECG | Epicatechin gallate |
| ECM | Extracellular matrix |
| ED | Erectile dysfunction |
| EDS | energy-dispersive spectroscopy |
| EDX | energy-dispersive X-ray spectroscopy |
| EGC | Epigallocatechin |
| EGCG | Epigallocatechin gallate |
| EGFR | Estimated glomerular filtration rate |
| EL | Ejaculation latency |
| ELISA | Enzyme-linked immunosorbent assay |
| EMT | Epithelial mesenchymal transition |
| ERK | Extracellular-signal-regulated kinases |
| ES | Electrical field stimulation |
| ESRD | End-stage renal disease |
| EVM | Ethnoveterinary medicine |
| FDA | Food and Drug Administration |
| FE-SEM | Field emission scanning electron microscopy |
| FLD | Fatty liver disease |
| FL | Fidelity level |
| FRAP | Ferric reducing antioxidant power |
| FSH | Follicle-stimulating hormone |
| FTIR | Fourier transform infrared spectroscopy |
| GAPs | Good agricultural practices |
| GBM | Glomerular basement thickening |
| GC | (-)-Gallocatechin |

| GCG | (-)-Gallocatechin gallate |
| GCLC | Glutamate-cysteine ligase catalytic subunit |
| GCL | Glutamate-cysteine ligase |
| GCLR | Glutamate-cysteine ligase regulatory subunit |
| GDH | Glutamate dehydrogenase |
| GFR | Glomerular filtration rate |
| GGT | $\gamma$-Glutamyltranspeptidase |
| GI | Gastrointestinal |
| GK | Glucokinase |
| GMPs | Good manufacturing practices |
| GPI | Glycophosphatidylinositol |
| GPS | Geographical positioning system |
| GPX | Glutathione peroxidase |
| GR | Glutathione reductase |
| GSH | Glutathione |
| GSH-Px | Glutathione peroxidase |
| GSR | Glutathione reductase |
| GST | Glutathione S-transferase |
| HACCP | Hazard analysis critical control point |
| HBV | Hepatitis B virus |
| HCV | Hepatitis C virus |
| HDL | High-density lipoprotein |
| HIF-1$\alpha$ | Hypoxia-inducing factor-1$\alpha$ |
| HK | Hepatic hexokinase |
| HO-1 | Heme oxygenase-1 |
| HORAC | Hydroxyl radical averting capacity |
| IIPLC | High-performance liquid chromatography |
| HRs | Heart rates |
| IASM | Indian Alternative Systems of Medicine |
| ICF | Informant consensus factor |
| ICP | Inductively coupled plasma |
| ICSI | Intracytoplasmic sperm injection |
| IDDM | Insulin-dependent diabetes mellitus |
| IF | Intromission frequency |
| IL | Intromission latency |
| IP6 | Inositol hexakisphosphate |
| IPRs | Intellectual Property Rights |
| IR | Infrared |

| | |
|---|---|
| ITAM | Immunoreceptor tyrosine-based activation motif |
| IUCN | International Union for Conservation of Nature |
| JNK | c-Jun NH (2)-terminal kinase |
| KCl | Potassium chloride |
| LDH | Lactate dehydrogenase |
| LD | Laser diffractometry |
| LDL | Low-density lipoprotein |
| LH | Luteinizing hormone |
| LPIC | Lipid peroxidation inhibition capacity |
| LPO | Lipid peroxidation |
| mAb | Monoclonal antibody |
| MAO-B | Monoamine oxidase-B |
| MAP | Mean arterial pressure |
| MAPs | Medicinal and aromatic plants |
| MDA | Malondialdehyde |
| MDH | Malate dehydrogenase |
| MDR1 | Multidrug resistance |
| MF | Mount frequency |
| MIC | Minimum inhibitory concentration |
| MIEN1 | Migration and invasion enhancer 1 |
| ML | Mount latency |
| MPTP | Mitochondrial permeability transition pore |
| MRSA | Methicillin-Resistant Staphylococcus aureus |
| NAA | Neutron activation analysis |
| NAFDAC | National Agency for Food, Drugs, Administration and Control |
| NAFLD | Nonalcoholic fatty liver disease |
| NA | Nalidixic acid |
| NCE | New chemical entity |
| NF-κB | Nuclear factor kappa-B |
| NHDF | Normal human dermal fibroblasts |
| NIDDM | Non-insulin dependent diabetes mellitus |
| NQO1 | NAD(P)H: quinone oxidoreductase 1 |
| Nrf2 | Nuclear erythroid 2-related factor 2 |
| NS5B | Nonstructural 5B protein |
| OD | Optical density |
| OMT | O-methyltransferases |
| ORAC | Oxygen radical absorption capacity |
| PAB | Prooxidant-antioxidant balance |

| | |
|---|---|
| PCA | Principal component analysis |
| PCS | Photon correlation spectroscopy |
| PDAs | Potable display accessories |
| PD | Parkinson's disease |
| PEI | Penile erection index |
| PEL | Post-ejaculatory latency |
| PFK | Phosphofructokinase |
| PFRAP | Potassium ferricyanide reducing power |
| PI3-K | Phosphoinositide 3-kinase |
| PKC | Protein kinase C |
| PLA2 | Phospholipase A2 |
| PPAR-$\alpha$ | Peroxisome proliferator-activated receptor $\alpha$ |
| PUFAs | Polyunsaturated fatty acids |
| R&D | Research and Development |
| RAAS | Renin–angiotensin–aldosterone system |
| RAGE | Receptor for advanced glycation end products |
| RCI | Relative cultural index |
| RDA | Recommended dietary allowance |
| RdRps | RNA-dependent RNA polymerases |
| RNS | Reactive nitrogen species |
| RONS | Reactive oxygen and nitrogen species |
| ROS | Reactive oxygen species |
| RTA | Ricin toxin A |
| RTB | Ricin toxin B |
| SACS | S-allylcysteinesulfoxide |
| SAR | Structure-activity-relationship |
| SCFE | Supercritical fluid extraction |
| SDH | Sorbitol dehydrogenase |
| SEM | Scanning electron microscopy |
| SGLT2 | Sodium glucose cotransporter 2 |
| SMCS | S-methylcysteinesulfoxide |
| SNOM | Scanning near-field optical microscopy |
| SOD | Superoxide dismutase |
| SREBP1 | Sterol regulatory element-binding protein 1 |
| STDs | Sexually transmitted diseases |
| STZ | Streptozotocin |
| T1D | Type 1 diabetes |
| T2D | Type 2 diabetes |

| | |
|---|---|
| TAM | Tumor-associated macrophage |
| TBARS | Thiobarbituric acid reactive substances |
| TCA | Trichloroacetic acid |
| TC | Total cholesterol |
| TEM | Transmission electron microscopy |
| TGF-β | Transforming growth factor β |
| TG | Triglycerides |
| TMP | Traditional medical practitioner |
| TM | Traditional medicine |
| TNF-α | Tumor necrosis factor α |
| TRAP | Total peroxyl radical trapping antioxidant parameter |
| TTM | Traditional Tibetan Medicine |
| USM | Unani System of Medicine |
| UV | Ultraviolet |
| VCAM | Vascular cell adhesion protein 1 |
| VEGF | Vascular endothelial growth factor |
| VLDL | Very low-density lipoprotein |
| WHO | World Health Organization |
| XRD | X-ray diffraction |

# FOREWORD

It is my pleasure to write this foreword and to recommend this book, *Phytochemistry: Volume 2, Pharmacognosy, Nanomedicine, and Contemporary Issues,* to everyone doing research in the field of phytochemistry, pharmacognosy, or related areas. The book's scope is great and one that has incorporated recent developments in the field. The chapters were written by key specialists in the field from diverse academic backgrounds. The volume covers virtually all areas of phytochemistry and pharmacognosy, namely ethnomedicine, ethnobotany, complementary and alternative systems of medicine, ethnoveterinary medicine, herbal medicine, nanomedicine, and so forth.

I commend Chukwuebuka Egbuna for initiating this idea and for his resilient effort in bringing this book to completion. Also, I commend his co-editors and chapter contributors for their excellent work. I am delighted to have known when the book project started and when it was completed.

I am sure the book will receive appreciation from the scientific community. I am also convinced that the book will attract great interest to the industry, especially the pharmaceutical industry.

—**Prof. Reginald Agu**
PhD, FIBD
Visiting Professor, Department of Microbiology
University of Nigeria, Nsukka, Nigeria

# PREFACE

The roadmap to the discovery of new and effective drugs with fewer side effects is one that is very expensive and that which spans a period of 10–12 years from discovery to clinical trials. It is projected that the global drug discovery informatics market of US $1.67 billion will reach US $2.84 billion in 2022 and much higher by 2050 at the compound annual growth rate of 11.2%. The practices leading to the discovery of new drugs cut across several horizons, starting from the indigenous knowledge about the peoples' medicinal plants, *in vitro* research about the phytochemical compositions of the plants, and the *in silico* and *in vivo* studies, including both preclinical and clinical trials. The discovery of drugs encompasses several disciplines, namely ethnomedicine, ethnopharmacology, pharmacognosy, ethnobotany, complementary medicine, and alternative medicine. Several other fields, such as biotechnology, microbiology, zoology, chemistry, plant biochemistry, and marine biochemistry, to mention a few, are involved.

Over the last 50 years, there has been a lot of progress in the discovery of drugs or lead compounds. This volume, *Phytochemistry, Volume 2: Pharmacognosy, Nanomedicine, and Contemporary Issues,* presents a complete coverage, with chapters demonstrating recent advances on the potentials of medicinal plants for the treatment and management of diseases. Since phytomolecules have different mechanisms of action and the fact that their number appears astonishingly high in plants, there is a need for a well-structured book to systematically aid in providing easy understanding for researchers, students, and other users from related areas. This book is comprised of 25 chapters that are grouped into four parts: ethnomedicine and pharmacognosy, medicinal potentials of phytochemicals, nanoparticle biosynthesis and applications, and phytochemicals as friends and foes. The chapters provides information on the recent advances in the discovery of therapeutic drugs from plants. The authors have provided figures, pictures, tables, pathways, and illustrations to aid easy understanding where applicable.

Chapter 1 by Egbuna et al. in Part I of this book presents background information on pharmacognosy, the scope, and the prehistoric uses of medicinal plants. Chapter 2 by Kumar and Sharma discusses the various complementary and alternative systems of medicines. Chapter 3 by Nwafor and Inya-Agha documents the indigenous people's medicinal plants and their ethnobotanical uses. Chapter 4 by Oyedepo presents an overview of herbal medicine and quality control with emphasis on Nigerian medicinal plants. Lakshminarayana and Rao in Chapter 5 documents the Indian ethnoveterinary medicinal plants, their uses, and modes of administration.

In Chapter 6 in Part II, Egbuna discusses antioxidants and phytochemicals. Chapter 7 by Sulaiman et al. is an overview of the roles of phytochemicals in the prevention and treatment of various diseases. Gupta and Pandey in Chapter 8 emphasizes the roles phytochemicals as oxidative stress mitigators. In Chapter 9, Kurhekar presents antimicrobial medicinal plants. The chapter is a combination of review and research on plants that are effective against different strains of microorganisms. Chapter 10 by Tijjani and Egbuna is a systematic study of medicinal plants with antivenom activities. The authors conducted a survey and reached a conclusion on the most reoccurring antivenom plant families, the most used plant parts, and preferred extraction solvents. Ujah in Chapter 11 discusses the medicinal potentials of green tea, *Camellia sinensis*. Chapter 12 by Haider et al. illustrates the antioxidant potentials of cinnamon. Swagata et al. in Chapter 13 details the potentials of phytochemicals as hope for the treatment of hepatic and neuronal disorders. Chapter 14 by Parmar and Variya details the medicinal roles of phytochemicals in the treatment of male infertility. Upadhyay et al. documents medicinal plants and phytochemicals effective against cancer. Chapter 16 by Kushwaha et al. presents an *in silico* study on the potentials of methylated flavonoids as a novel inhibitor of metastasis in the cancer cell. The findings are positive and one that can serve as a lead for the discovery of anticancer drugs. Aslam et al in Chapter 17 provides an overview of diabetes and detail the medicinal plants with active components that are effective in the treatment and management of diabetes mellitus. Shukla et al. in Chapter 18 discusses the roles of phytomolecules in the treatment of diabetic nephropathy with emphasis on medicinal plants. Mishra et al. in Chapter 19 documents phytochemicals and plant sources that can be beneficial for the treatment and management of autoimmune diseases.

Part III details the biosynthesis of nanoparticles and their biomedical applications. In Chapter 20, Pothabathula et al. details the green biosynthesis of metallic nanoparticles. Hii et al. in Chapter 21 presents the cytotoxicity and biomedical applications of metal oxide nanoparticles synthesized from plants. Chapter 22 by Shah et al. details the 'green synthetic approaches and precursors for Carbon Dot Nanoparticles.'

Part IV demonstrates the ability of phytochemicals as friends and how they can be foes as well. Chapter 23 by Egbuna et al. is a review of toxic plants and phytochemicals. Mtewa in Chapter 24 discusses phytochemicals as prooxidants. Chapter 25 by Egbuna is an overview of phytochemical as antinutrient.

This book will to be an invaluable resource for all in the field and related disciplines, and I recommend it to everyone. I extend my heartfelt thanks to the chapter contributors for their great contributions, patience, and cooperation during the editorial process. My sincere thanks goes to the volunteer reviewers and to my co-editors. I will remain grateful to my family for their support and patience during the editorial process of this book. To the management of Apple Academic Press, I extend a special thanks for demonstrating an ability to support authors despite the workload and particularly for a very fast response to emails. To the readers, I appreciate you all and would welcome reviews about the book with an open heart. Thank you.

**—Chukwuebuka Egbuna**
MNSBMB, MICCON, AMRSC
Department of Biochemistry, Faculty of Natural Sciences,
Chukwuemeka Odumegwu Ojukwu University,
Anambra State 431124, Nigeria

# PART I
# Ethnomedicine and Pharmacognosy

# CHAPTER 1

# PHARMACOGNOSY AND PREHISTORIC USES OF MEDICINAL PLANTS

CHUKWUEBUKA EGBUNA[1,*], NADIA SHARIF[2], and
SHAISTA JABEEN N.[3]

*1Department of Biochemistry, Chukwuemeka Odumegwu Ojukwu
University, Anambra State, Nigeria. Tel.: +2347039618485*

*2Department of Biotechnology, Lahore College for Women University,
Lahore, Pakistan*

*3Department of Zoology, PG and Research Unit, Dhanabagyam
Krishnaswamy Mudaliar College for Women (Autonomous),
Sainathapuram, RV Nagar, Vellore, Tamil Nadu 632001, India*

*\*Corresponding author. E-mail: egbuna.cg@coou.edu.ng;
egbunachukwuebuka@gmail.com
\*ORCID: https://orcid.org/0000-0001-8382-0693.*

## ABSTRACT

Pharmacognosy is arguably the oldest modern science which involves the study of drugs in its crude form. The crude drugs could be from a plant, animal, or microbial origin. The term pharmacognosy was coined from two Greek words: "*pharmakon*" meaning drug or medicine, and "*gnosis*" meaning knowledge. Seydler, the German botanist, was the first to use the term "pharmacognosy" but Schmidt, the Australian doctor was credited for begetting the word. Stress areas of pharmacognosy encompasses botany, ethnobotany, ethnomedicine, ethnoveterinary, marine pharmacognosy, zoopharmacognosy, microbiology, herbal medicine, chemistry, biotechnology, phytochemistry, pharmacology, pharmaceutics, clinical pharmacy, and pharmacy practice. Most significantly, plants have been a subject of discussion in pharmacognosy owing to its prolific sources for new bioactive compounds. Plants since antiquity have been utilized for the treatment of diseases and prehistoric evidence from the works of great scientists of all time attests. This chapter provides a foundation for this volume. The definition, scope and prehistoric use of medicinal plants were detailed.

## 1.1   INTRODUCTION

Pharmacognosy is an essential branch of pharmacy which involves the study of medicinal drugs derived from plants or other natural sources. It incorporates the logical investigation of the auxiliary, physical, biochemical, and natural properties of drugs and looks for new actives from plants, animals, and mineral sources. It also investigates a number of medicinal and commercial items, for example, vitamins, compounds, pesticides, allergens and additionally the history, circulation, development, accumulation, readiness, distinguishing proof, assessment, protection, and trade of therapeutic plants. An Austrian doctor J. A. Schmidt (1759–1809) begat the term "pharmacognosy" in his composition "*Lehrbuch der Materia Medica*," distributed in 1811, although C. A. Seydler utilized the term in his book on crude medications "*Analectica Pharmacognostica*" in 1815. Pharmacognosy is the convergence of the two Greek words "*pharmakon*" (medication) and "*gnosis*" (information).

Pharmacognosy has advanced from one being an expressive plant subject to one having a more substance center grasping an expansive range of orders including organic science, phytochemistry, pharmacology, zoology, ethnobotany, sea life science, microbiology, biotechnology, herbal medicine, science, pharmaceutics, clinical drug store, drug store rehearse, and so forth, which makes it today to be an interdisciplinary science (Alamgir, 2017). Pharmacognosy is presently enduring significant change owing to the extensive variety of different procedures, and the current advance in extraction, chromatography, hyphenated systems, screening of common item, biotechnology, and so forth.

Plants produce chemical compounds as part of their normal metabolic activities. These phytochemicals are either primary metabolites such as sugars and fats or secondary metabolites which are produced from primary metabolites. For example, some secondary metabolites are toxins used to deter predation and others are pheromones used to attract insects for pollination. Plant preparations are said to be medicinal or herbal when they are used to promote health beyond basic nutrition. Doubtlessly, plants are prolific sources of new bioactive chemicals such as atropine, ephedrine, morphine, caffeine, salicylic acid, digoxin, taxol, galantamine, vincristine, colchicine, and so forth. (Orhan, 2014).

Nonetheless, periwinkle (*Catharanthus roseus*) and its anticancer specialists, vinblastine and vincristine; St. John's Wort (*Stramonium*) and its synthetic constituents tropane alkaloids, hyoscine, and hyoscyamine; Indian snakeroot (*Rauvolfia* root) and its alkaloids, ajmalicine, reserpine, and rescinnamine; natural products, for example, papaya (*Carica papaya*), kiwifruit (*Actinidia deliciosa* and different species), pineapple (*Ananas comosus*), figs (*Ficus carica*) and their proteases catalyst blend; thyroid organ and its separated hormone, thyroxin; pancreas and its peptide hormone, insulin, and so forth, are similarly critical as topics of pharmacognosy (Eisenberg et al., 2011). The plant kingdom still holds many species of plants containing substances of medicinal value which are yet untapped. Some actives discovered by great scientists were presented in Table 1.1.

## 1.2 SCOPE OF PHARMACOGNOSY

The advancement of pharmacognosy additionally prompts improvement of organic science, scientific categorization, plant biotechnology, plant hereditary qualities, plant pathology, pharmaceutics, pharmacology, phytochemistry, and different branches of science. The major stress areas of pharmacognosy encompass botany, ethnobotany, ethnomedicine, ethnoveterinary, marine pharmacognosy, zoopharmacognosy, microbiology, herbal medicine, chemistry, biotechnology, phytochemistry, pharmacology, pharmaceutics, clinical pharmacy, and pharmacy practice (Table 1.2).

## 1.3 PREHISTORIC USES OF MEDICINAL PLANTS

### 1.3.1 MESOPOTAMIAN CIVILIZATION (3000 BC TO 539 BC)

Some of the major Mesopotamian civilizations include the Sumerian, Assyrian, Akkadian, and Babylonian civilizations. The Sumerians (3000–2400 BC) who inhabited in the lower Mesopotamia a region between the rivers of Tigris and Euphrates (now what is the present-day Iraq and Kuwait) were regarded as the first to have developed the World's first civilization around 4000 BC (Alamgir, 2017). They are the pioneers in using plant-based drugs, wound washing, plasters, and bandaging. The early Babylonians

TABLE 1.1 Some Phytochemicals and their Discoverers.

| Secondary metabolites | Medicinal plants | Name of the scientist | Therapeutics potentials |
|---|---|---|---|
| Alkaloids – morphine Over 12,000 alkaloids are now found | *Papaver somniferum* (opium poppy) | German chemist Friedrich Sertürner (1804). | Anticancer, antimalarial, anti-asthma, vasodilatory, and so forth. |
| Flavonoids Over 4000 flavonoids now found | Paprika and citrus peel are found in almost all plants | The Nobel Prize winner, Dr. Albert Szent-Gyorgyi, and coworkers (1936). Identified flavonoids as vitamin P. Also discovered Vitamin C. | Antioxidant, anti-inflammatory, anti-allergic, antiviral, and anticarcinogenic |
| Glycosides – amygdalin | Rhubarb, cascara, and so forth. | French chemists Pierre Robiquet and Antoine Boutron-Charlard (1830) | Laxative and detoxification |
| Cardiac glycosides | Digitalis (Foxglove plants) and other plants | William Withering (1785) | Dropsy and congestive heart failure |
| Saponins –ginsenoside | Panax species | Garriques (1854) | Antitumor and antimutagenic |
| Terpenes | Conifers and many plants | Leopold Ružička (1953) – Discovery of isoprene unit linkages | Antimicrobial and antifungal |
| Tannins – gallic acid | Gallnuts, sumac, and tea leaves | French Chemist Henri Braconnot (1831) | Antibacterial, binds with proteins and antioxidant |
| Benzoic acid | Plant gums | Nostradamus (1556). Justus von Liebig and Friedrich Wöhler elucidated the structure in 1832 | Antifungal ability |
| Shikimic acid | Plants and microorganism | Johan Fredrik Eylman (1885) | Antioxidant, anti-inflammatory, and antinociceptive activity. Listed as group 3 carcinogen. |

**TABLE 1.2** Branches and Subfields in Pharmacognosy.

| Fields/Subfields | Definition |
| --- | --- |
| Botany | Also called plant science(s), plant biology, or phytology. It is a branch of biology that deals with the study of plant life. |
| Ethnobotany | Study of a region's plants and their practical uses through the traditional knowledge of a local culture and people. See Chapter 3. |
| Ethnomedicine | Deals with a wide range of healthcare systems/structures, practices, beliefs, and therapeutic techniques that arise from indigenous cultural development. Ethnopharmacology inclusive. |
| Ethnopharmacology | Study of the pharmacological qualities of traditional medicinal substances |
| Chemistry | Study of the structure, properties, composition, mechanisms, and reactions of organic compounds. |
| Phytochemistry | Study of plant-derived chemicals, particularly the secondary metabolites. Also, see volume 1 and 3 of this book for comprehensive details. |
| Pharmacology | Branch of biology concerned with the study of drug action. |
| Pharmacy | Science and technique of preparing and dispensing drugs. |
| Pharmaceutics | A discipline of pharmacy that deals with the process of turning a new chemical entity (NCE) or old drugs into a medication to be used safely and effectively by patients. |
| Marine Pharmacognosy | Study of chemicals derived from marine organisms. See volume 3 of this book for detailed information. |
| Microbiology | Study of microorganisms which encompasses virology, mycology, parasitology, and so forth. See Chapter 9 of this volume for more information. |
| Herbal medicine | Also called botanical medicine or phytomedicine, refers to using a plant's seeds, berries, roots, leaves, bark, or flowers for medicinal purposes. See Chapter 4 of this volume for detailed information. |
| Ethnoveterinary medicines | Concerned with traditional knowledge for primary healthcare treatment of domestic animals to order to keep them productive and healthy. See Chapter 5 of this volume for detailed information. |
| Biotechnology | A technology based on biology that deals with the use of living systems and organisms to develop or make products for example, the synthesis of natural bioactive molecules using biotechnology. See volume 3 of this book for more information. |
| Zoopharmacognosy | The science of animal self-medication. |
| Medical anthropology | Involves the study of human health and disease, healthcare systems, and biocultural adaptation. |

(2200–1300 BC) learned to manufacture soaps, leather, vinegar, beer, wine, glass extract natural plant aroma, and animal products (Alamgir, 2017). They are much aware of the use of medicinal plants and plant drugs such as balm of Gilead, colocynth, hellebore, licorice, mustard, myrrh, oleander, opium, opopanax, and storax. Documents showing the use of senna, coriander, saffron, cinnamon, garlic and the preparation of liniments for external applications for sprains, bruises on the skin, and elixirs for internal impairment. In this era diagnosis and treatment separated followed by the preparation of medicines which was handled by assistants called apothecaries.

## 1.3.2   ANCIENT CHINESE

The Chinese Emperor Shen Nung at about 2000–2500 BC explored the hidden treasure of medicinal plants. His document "Pen T-Sao" constitutes about 365 herbal drugs and its procedures of preparations. Most of the drugs he had tried upon himself. Some of the medicinal plants he mainly focused on such as opium, podophyllum, ginseng, rhubarb, cinnamon bark, valerian, and so forth. The Chinese physicians practiced "moxibustion" which means the placement of powdered leaves on an acupuncture point and set alight. These results in yang, that is, the appearance of blisters on that region (see Chapter 2 of this volume). The idea behind this is that they believed that to discard the pain of a disease is by creating a new one. They are not much aware of anatomy at that time. Chinese traditional medication was purely based on Yin and Yang theory. They believed that Yin refers to the dark moist feminine side whereas the Yang refers to the bright dry masculine side (see Chapter 2 of this volume for more information).

## 1.3.3   EGYPTIAN CIVILIZATION (3000 BC–1200 BC)

Simultaneously, the Egyptian civilization emerged in North-east Africa. Their inscriptions on tombs, ceramics, Cyperus papyrus, and so forth, have provided information on the Egyptians medicines and surgery (Dawson, 1927). The Egyptians culture provided a number of documentation which imparts knowledge on the use of medicinal plants to treat illnesses. For example, the medical document, Papyrus Ebers was also known as Ebers Papyrus (1550 BC) which was named after the German Egyptologist, Georg Ebers is among the oldest and most important medical papyri of ancient Egypt. It was purchased at Luxor (Thebes) in the winter of 1873–74 by Georg

Ebers. The document is full of incantations and foul applications meant to turn away disease-causing demons, and it also includes 877 prescriptions and 700 drugs derived from plants, animals, and minerals. Among those derived from plants are cumin, Ricinus seeds opium, poppy, castor seed, garlic and Arabic gum, and so forth, animal sources are milk, waxes, livers, and excreta. The minerals include salt, copper, carbonate alum, and stibnite. These papyri – the first systematic classification of medicine – gave "recipes" for the treatment of certain diseases and symptoms. For example, the Ebers Papyrus lists 21 ways to treat coughs, and others deal with at least 15 diseases of the abdomen, 29 of the eyes and 18 of the skin. The Egyptians used wines and beers with milk together as a physical boost in their liquid medicines (Court, 2005). They also used honey in solid pills and waxes in ointments. Their medication includes pills, decoctions, teas, gargles, snuffs, infusions, lotions, plasters, inhalations, fumigation, troches, enema, and so forth.

### 1.3.4 INDIAN HISTORY OF PHARMACOGNOSY

The in-depth knowledge of pharmacology is rooted in Indian historic herbal drugs which has been practiced in India since 5000 years ago. Evidence can be found in the ancient Ayurvedic documentations "Charak Samhita and Sushrutha Samhitha." The earliest herbal medicine used in Ayurvedic system is dated 1200 BC and it consists of about 127 medicinal plants, its formulation, and efficacy upon human health (see Chapter 2 of this volume for more information).

### 1.3.5 HIPPOCRATES AND GREEK MEDICINES (460–377 BC)

In ancient Greek, medicines were regarded as the second to mathematics. Greek medicines acted on three basic sources such as the temple practice of Asclepius or healing by the god, physiological opinion by the philosophers, and finally, practice of the superintendents of the gymnasia (Court, 2005). The Greek physicians were recognized as spiritual healers. The most renowned Greek Physician, Hippocrates (460–377 BC) credited for being the father of medicine, established the doctrine of medicines and used some 300–400 drugs (Jones, 1923). Hippocrates used the Greek word *carcinos*, meaning crab or crayfish, to refer to malignant tumors. It was Celsus who translated the Greek term into the Latin cancer, also meaning crab. The best known of the Hippocratic writings is the Hippocratic Oath which reads:

I swear by Apollo the physician, and Aesculapius, and Health, and All-heal, and all the gods and goddesses, that, according to my ability and judgment, I will keep this Oath and this stipulation – to reckon him who taught me this Art equally dear to me as my parents, to share my substance with him, and relieve his necessities if required; to look upon his offspring in the same footing as my own brothers, and to teach them this art, if they shall wish to learn it, without fee or stipulation; and that by precept, lecture, and every other mode of instruction, I will impart a knowledge of the Art to my own sons, and those of my teachers, and to disciples bound by a stipulation and oath according to the law of medicine, but to none others.

I will follow that system of regimen which, according to my ability and judgment, I consider for the benefit of my patients, and abstain from whatever is deleterious and mischievous.

I will give no deadly medicine to anyone if asked, nor suggest any such counsel, and in like manner I will not give to a woman a pessary to produce abortion. With purity and with holiness I will pass my life and practice my Art.

I will not cut persons laboring under the stone, but will leave this to be done by men who are practitioners of this work. Into whatever houses I enter, I will go into them for the benefit of the sick, and will abstain from every voluntary act of mischief and corruption; and, further from the seduction of females or males, of freemen and slaves.

Whatever, in connection with my professional practice or not, in connection with it, I see or hear, in the life of men, which ought not to be spoken of abroad, I will not divulge, as reckoning that all such should be kept secret.

While I continue to keep this Oath unviolated, may it be granted to me to enjoy life and the practice of the art, respected by all men, in all times! But should I trespass and violate this Oath, may the reverse be my lot!

Hippocrates group medicated on four liquid humors – blood, phlegm, bile, and black bile. They believed that diseases cause an excess of any one of these liquid humor. Their treatment leads to the elimination of such liquid humor from the body by using enemas, purgatives, and emetics to cleanse and purify the impaired body. Thus pharmakon became the remedy for ailments.

Hippocrates believed in medical astrology and insisted his students to study astrology, saying, "He who does not understand astrology is not a doctor but a fool." He believed that each of the astrological signs (along with the Sun, Moon, and planets) is associated with different parts of the human body. He further stated that many plants are referred to in old herbals as being "under the influence of" some planet. This was used as a codification of the plant's properties and used to create mixtures specific to different diseases.

### 1.3.6 THEOPHRASTUS (FATHER OF BOTANY) AND ARISTOTLE

Theophrastus (372–287 BC) a Greek native of Eresos in Lesbos is a pupil of Plato successor of Hippocrates later a student of Aristotle became the head of Lyceum assigned as natural Philosopher and medical person (Paulsen, 2010). He established the field of botany and clearly classified plants as trees, shrubs, and herbs. He gained the knowledge of folk medicines and documented it in his two famous documents "*De causis plantarum* and *De historia plantarum.*" He mainly focused on the morphology, classification, and natural history of plants.

### 1.3.7 AULUS CORNELIUS CELSUS (CA 25 BC–AD 45)

Aulus Cornelius Celsus, a Roman encyclopaedist, flourished in the first century AD in Rome. He was a distinguished author of "*De Medicina*" (eight volumes, see below) dealing with agriculture, military, rhetoric, philosophy, and medicine. It is now considered as one of the most important historical sources for the present-day medical knowledge (Court, 2005). It was discovered by Pope Nicholas V in (1397–1455) and was the first medical works to be published (1478) after the innovation of printing press. He recommended wounds be washed and treated with substance now applied as antiseptics, such as vinegar, thyme oil, and so forth. He enunciated plastic surgery of the face using skin from other parts of the body. He was the pioneer who identified the cardinal signs of inflammation, calor (warmth), dolor (pain), tumor (swelling), and robur (redness and hyperemia) (Cefalu, 2000).

    Book 1 – The History of Medicine
    Book 2 – General Pathology
    Book 3 – Specific Diseases
    Book 4 – Parts of the Body
    Book 5 and 6 – Pharmacology
    Book 7 – Surgery
    Book 8 – Orthopedics

### 1.3.8 PLINY THE ELDER (23–79 AD)

Pliny the Elder (born Gaius Plinius Secundus, AD 23–79) was a Roman officer and the author of the book "Historia Naturalist" an encyclopedia of 37 volumes. In it, 16 volumes which are subjected to Pharmacology. This document assigned as a reference to pharmacologists. Apart from pharmacology, it considered astronomy, mathematics, geography, anthropology, ethnography, human physiology, zoology, botany, horticulture, sculpture, painting, and precious stones. Pliny's pharmacopeia was the seedling towards the field of pharmacotherapy (Tellingen, 2007). In his book on pharmacopeia, he had suggested some 20 species of herbs against cardiovascular diseases some are presented in Table 1.3.

**TABLE 1.3**　Some Medicinal Plants of Pliny's Pharmacopeia.

| Common name | Scientific name | Biological actions |
|---|---|---|
| Garlic | *Allium sativum* | Antiviral, antibacterial, cardiovascular, and antidiuretic |
| Sea onion | *Urginea maritima* | Jaundice, asthma, heart diseases, laxative, and expectorant |
| Laburnum | *Cytisus laburnum* | Anti-smoking |
| Atropine | *Atropa belladonna* | Jaundice and anticholinergic syndrome |
| Myrtle | *Myrtus communits* | Antioxidant activity |
| Laurel | *laurelia* | Antimicrobial and antifungal activity |
| Indian holly | *Ilex aquifolium* | Emetic and central nervous system stimulant |
| White lupine | *Lupinus albus* | Emmenagogues and vermifuge |
| Broom | *Genista sphaerocarpos* | Liver disorder and diuretic |
| Horehound | *Marrubium vulgare* | Wheezing and cough |

## 1.3.9　PEDANIUS DIOSCORIDES (40–80 AD)

A Greek botanist of the first century AD was the most important botanical writer after Theophrastus and he was the most popular Greek physician and pharmacologist in the Roman era (Bender and Thom, 1965). In his document *De Materia Medica*, he described about 600 plants and their medicinal properties. *De Materia Medica* is the prime historical source of information about the medicines used by the Greeks, Romans, and other cultures of antiquity. Later his document became the precursor of pharmacopeias. When people underwent surgery he used the extracts of mandragora to induce anesthesia. He grouped plants into three categories such as aromatic, culinary, and medicinal.

## 1.3.10 AVICENNA: THE "PERSIAN GALEN" (980–1037 AD)

Avicenna (980–1037 AD) was one of the eminent physicians and philosophers during the Arabian era and his work had been documented in Canon Avicenna also termed as Canon of medicine which imparts the knowledge of medicine and Pharmacology (Bender and Thom, 1965). Avicenna was a famous physician, philosopher intellectual, and favorite among Persian rulers. His pharmaceutical work was accepted by the west till the 17th century and still dominant. All his writings were in Arabic. Some of his system of pharmacy and remedies are still in business such as camphor, saffron, rhubarb, mastix, and aloe. He introduced the systems in experimentation and understood the physiology of human and diagnosed patients with experimental based medicines (Paulsen, 2010).

## 1.4 MODERN ERA OF PHARMACOGNOSY

Since initiation in 1811, pharmacognosy has advanced impressively amid the previous two hundred years. Recently, it has increased much significance in light of the inclusive progress of natural items as lead atoms for new medications and additionally, the expanded utilization of correlative restorative items in industrialized countries. At the start of the 21st century, accentuation has been put on (i) examination, (ii) natural testing, and (iii) coordinated effort of pharmacognostical explore. The multidisciplinary qualities of pharmacognosy are ending up increasingly unmistakable the same number of new zones of research and concentrate; for example, molecular pharmacognosy, neuropharmacognosy, and mechanical pharmacognosy are rising in current pharmacognosy with time. In the cutting edge period, the customary herbalism has been authoritatively viewed as a strategy for elective prescription in numerous parts of the world, particularly in some created nations (e.g., USA and UK). The Traditional Chinese Medicine has been used by the Chinese in healing facilities. The World Health Organization assessed that 80% of individuals overall depend on homegrown meds for some piece of their essential human services. In Germany, around 600–700 plant-based drugs are accessible and are endorsed by almost 70% of German doctors. Numerous elective doctors in the 21st-century join herbalism in present-day medication because of the assorted capacities plants have and their low number of symptoms (Alamgir, 2017).

In the 19th century, microscopy was presented in pharmacognosy for the quality control of crude medications, and for a long time, pharmacognosy stayed bound with the magnifying lens based strategies. In the 20th century, the revelation of essential medications from the set of all animals and microorganisms, especially hormones and vitamins, has turned into a vital wellspring of medications. In the end of the 20th century, chromatographic spectrometric techniques were presented in pharmacognostical investigation. In vitro framework bioassay was included at end of the 20th century, and amid this period (1983–1994), countless and antitumor standards from regular sources were found (Cragg et al., 1997; Alamgir, 2017).

## 1.5  CONCLUSION

Pharmacognosy has advanced throughout the years and is now proper to address the difficulties of medication disclosure and improvement. The pharmacognosy is expansive and incorporates the logical investigation of unrefined medications, therapeutic items (e.g., allergens, vitamins, chemicals, antimicrobials, pesticides, and allergenic concentrates), and excipients (e.g., shading, flavoring, emulsifying and suspending specialists, fasteners, diluents, solidifiers, building or filler operators, sweeteners, disintegrants, analgesic guides, and cements). It manages the examination issues in the zones of phytochemistry, microbial science, biosynthesis, biotransformation, chemotaxonomy, and other organic and compound sciences. It also concentrates on harmful, stimulating, and teratogenic plants crude materials for the generation of oral contraceptives, aphrodisiacs, and so forth, and in addition, flavors, refreshments, and toppings.

## KEYWORDS

- pharmacognosy
- medicinal plants
- *De Materia Medica*
- Hippocrates
- Dioscorides

## REFERENCES

Alamgir, A. Origin, Definition, Scope and Area, Subject Matter, Importance, and History of Development of Pharmacognosy. *Therapeutic Use of Medicinal Plants and Their Extracts*; Springer, 2017; Vol. 1, pp 19–60.

Bender, G. A. and Thom, R. The stories and paintings in stories, a History of Pharmacy in Picture. In *The Great Moments in Pharmacy*. Parke Davis & Company. Northwood Institute Press: Detroit, MI, 1965.

Cefalu, W. T. Inflammation, Insulin Resistance, and Type 2 Diabetes: Back to the Future. *Diabetes* **2000,** *58*(2), 307–308.

Court, W. E. *Making Medicines: A Brief History of Pharmacy and Pharmaceuticals*; Anderson, S. Ed.; Pharmaceutical Press: London, 2005.

Cragg, G. M.; Newman, D. J.; Snader, K. M. Natural Products in Drug Discovery and Development. *J. Nat. Prod.*, **1997,** *60*(1), 52–60.

Dawson, W. R. *The Beginning of Medicine: Medicine and Surgery in Ancient Egypt;* Science Reviews 2000 Ltd: 1927, 22(86), pp 275–284. http://www.jstor.org/stable/43430010 (accessed January 22, 2018).

Eisenberg, D. M.; Harris, E. S.; Littlefield, B. A.; Cao, S.; Craycroft, J. A.; Scholten, R.; Bayliss, P.; Fu, Y.; Wang, W.; Qiao, Y. Developing a Library of Authenticated Traditional Chinese Medicinal (TCM) Plants for Systematic Biological Evaluation – Rationale, Methods and Preliminary Results from a Sino-American Collaboration. *Fitoterapia*, **2011,** *82*(1), 17–33.

Jones, W. H. S. *Hippocrate with an English translation;* Harvard University Press: Cambridge, First printed 1923: Reprinted 1929, 1948; Vol. 1.

Orhan, I. E. Pharmacognosy, Science of Natural Products in Drug Discovery. *Bioimpacts*, **2014,** *4*(3), 109–110. PMC. Web. 7 Mar. **2018.**

Paulsen, B.S. Highlights through the history of plant medicine. In: *Bioactive compounds in plants – benefits and risks for man and animals*. Bernhoft, A., Ed.; The Norwegian Academy of Science and Letters: Oslo, 2010; p 18–29.

Tellingen, C. V. Pliny's Pharmacopoeia or the Roman Treat. *Neth. Heart J.* **2007,** *15*(3), 118–120.

Urk, H. V.; Duin, J.; Sutcliffe, J. *A History of Medicne from Prehistory to the Year 2020*. Barnes & Noble Books, 1992; ISBN 0–88029–927–4.

# CHAPTER 2

# COMPLEMENTARY AND ALTERNATIVE SYSTEMS OF MEDICINES

VINESH KUMAR* and YOGITA SHARMA

*Department of Sciences, Kids' Science Academy, Roorkee, Uttarakhand, India, Mobile: +91 9997228095*

*Corresponding author. E-mail: vkresearch47@gmail.com*
*ORCID: https://orcid.org/0000-0002-7560-7355*

## ABSTRACT

Herbal compounds have been used as a medicine to cure diseases and enhance the quality of health and life. From the dawn of human existence, countless generations of mankind have patiently experimented with and discovered a wide variety of herbs, which cure various diseases in plants and animals. The primitive man started to distinguish between nutritional and pharmacologically active plants for his survival. Herbal medicines are an essential part of traditional systems of medicine in almost all cultures. The efforts of various investigators have led to the development of various traditional systems of medicine, such as aromatherapy, Chinese, Indian, Unani, Tibetan, and Siddha systems of alternative medicines. All the civilized nations have developed their own material medica, and compiling details about plants used for therapeutic purposes. This chapter discusses the various alternative systems of medicines such as Chinese, Indian, Unani, Tibetan, Siddha systems, and so forth.

## 2.1 INTRODUCTION

The traditional or alternative medicines are the various types of preventive and therapeutic healthcare practices which include herbal medicines, acupuncture, massages, naturopathy, aromatherapy, magnet therapy, special teas, and spiritual healing. Traditional medicines have been used in many countries like India, China, and Sri Lanka for thousands of years. In India, various treatments like Ayurveda, Unani, Yoga, Siddha, and medication have been used since ancient time to cure illness. China also has a well-developed system of alternative medicine which includes herbal therapy, acupuncture, and Tuina (massage therapy) since ancient time to keep the body healthy, prevent, and treat illnesses. The alternative systems of medicine discussed in this chapter are:

1. Alternative Chinese System of Medicine (ACSM)
2. Traditional Tibetan Medicine (TTM)
3. Unani System of Medicine (USM)
4. Indian Alternative Systems of Medicine (IASM)

## 2.2 THE ALTERNATIVE CHINESE SYSTEM OF MEDICINE

The Alternative Chinese System of Medicine (ACSM) is also called the Tradition Chinese Medicine (TCM) which was developed about 3000 years ago (Erick, 2010). It has unique characteristics when compared with the other systems. The Chinese alternative system of medicine is a multifarious and natural way to cure diseases. Its aim is to pursue synchronization and equilibrium between mind and the organism's body. This system enables the natural ability of the body to regulate and normalize itself.

The TCM explains that health as the state of harmony. In this, the individual body's internal physiological network is in harmony with external environmental networks. The abnormal interactions between these networks are responsible for diseases. The TCM stresses to maintain harmony or homeostasis between nature and individual body.

About 2200 years ago, Chinese inhabitants observed that pressing certain body parts with warm stones could give relief for some sicknesses (Erick, 2010). They also observed that stabbing some body parts with bone needles could cure the pain of other body parts. That was the beginning of acupuncture in China. In the Chinese system of alternative medicine, it focuses on how feelings can affect the human body. They believe that both positive and negative emotions can affect the balance of the body. According to the traditional system, anything which is in excess can harm the body. TCM aims to treat the root cause of symptoms as well as the branch problems which are specific and unique to each individual.

It emphasizes on the self-curing power of body from diseases and keeping healthy. Without this philosophy, TCM would not be able to promote health and healing diseases. Different infectious, chronic, sensible and psychosomatic disorders can be cured by TCM. It has its own power for healing those problems which allopathic medicine cannot solve.

TCM believed that to prevent the body from diseases is to treat in time before it becomes a serious one. It stresses the harmony between body and mind. To diagnose a disease, TCM supports mental healing and believes that heart should be cured first.

The aim of Chinese traditional system of medicine is to diagnose the root source of disease as well as the associated cause which are specific and exclusive to each individual. This can be explained by the example of a plant in which roots are under the soil and associated branches up to the ground. For best growth, a plant needs good soil, water, carbon dioxide and sunlight. If you only provide soil and not sunlight and water to the plant, think what will happen. Right, you will not have much success to grow the plant. Similarly, if you will provide in excess, the plant again would not be able to grow in the best way. You should have a proper balance in all for the best growth of the plant. Therefore, we can say that this system believes that treating a symptom is not the best method to restore the body and mind. It focuses on to reduce the negative stress on mind and the whole body to bring it into a much better state of harmony.

The idea expressed above is supported by the book of Taoism, Laozi, or Daodejing. According to this book, both man and nature are originated from the primary thing called Tao. Therefore, TCM believed in the natural laws led by Tao which should be followed and not destroyed under any circumstances. These books could guide and help researcher and doctors to understand the concept of TCM.

## 2.2.1   THEORIES OF TCM

### 2.2.1.1   BOTH MAN AND NATURE ARE DEVELOPED FROM THE SAME THING

TCM like other traditional system of medicines has its own specific culture and philosophical knowledge. The scholars think about the book of changes because theories of TCM believe that the men, earth, and heaven are movable and changeable. Similarly, the change in one's health and disease is the result of conditions surrounding him that includes time, society, place and oneself mental condition.

How these factors affect the health of an individual person can be explained as all the things are originated from the same thing that exists spiritually or materially, irrespective if they are in the present movement. TCM represents the complete method of interpreting body's physiology and changes occur in the body.

## 2.2.1.2   THE CONCEPT OF YIN AND YANG

The yin and yang are two topographic terms that describe the shady and sunny sides of the hill. Yin represents the cold, dark and passive properties while yang represents the warm, bright and active characteristics. To achieve a healthy state a harmony between the forces of yin and yang, the body should be maintained. The whole TCM used the concept of yin and yang to diagnose the disease.

## 2.2.1.3   THE THEORY OF FIVE ELEMENTS

This theory is also the fundamental theory of TCM. The five elements of this theory are wood, fire, earth, metal, and water with their characteristics. This theory classifies things by their characteristics. According to this theory, all the parts of the body have affinities with particular elements of nature. The lung has an affinity with metal, liver with wood, kidney with water, heart with fire and spleen with earth.

## 2.2.1.4   THE BODY IS IN THE UNITY OF MIND

TCM theories showed that the living body is unity with mind and they are inseparable just as man is in unity with nature. These TCM theories see the body as a complex system of organs, channels, body fluid, blood, five sense organs, limbs, vessel, muscle, skin, and bone. These parts differ from each other in terms of name and functions but all are dominated by the same mind. TCM theory is the theory of vitality and mind and emphasized the functions of the mind. TCM theories classify physical, mental, spiritual, and body–mind aspects on the basis of mind and show the different universal way people observe disease and health.

## 2.2.1.5   IN HOLISTIC MEDICAL PATTERN

According to this concept, before treating the patient doctors should consider the social, mental, psychological, and ethical conditions. This help to educate patients about their lifestyle and mental state. TCM has been practiced for thousands of years and has gained rich experiences. TCM practitioners used emotional inter-checking to treat their patients with acupuncture, herbal therapy, massage, and so forth, and direct patients to do self-care activities like meditation, taiji, and qigong. Therefore, TCM has developed a whole set of practices for the prevention, health education, health promotion, and cure of diseases for the modern society can learn and give credence to it.

## 2.2.2   CAUSES OF DISEASES

The health is the state of the body in which it is in dynamic equilibrium with its environment. According to TCM, disease occurs due to the disharmony of body's internal system with external environment that is, disharmony between yin and yang. The causes of this imbalance are pathogenic factors including exogenous and endogenous factors. Exogenous factors are wind (cold), summer (heat), dampness, dryness and fire whereas the endogenous factors are emotional, improper diet, pathogens, and so forth.

## 2.2.3   PRINCIPLES OF DIAGNOSIS

The TCM has two steps to maintain health: identification of disease and diagnosis of it. They reestablish the harmony between the yin and yang of individual's body. The methods used for diagnosis are inspection, auscultation, and olfaction. Inspection describes the mental state, behavior, excretion, secretion, and observation of pulses and tongue. Auscultation and olfaction include checking listening, hearing, breathing, coughing, smelling body odors, and so forth, while interrogation includes family history, health state, and progress of the illness.

## 2.2.4   ADVANTAGES OF TCM

The Chinese Alternative System of Medicine helps in the following ways:

1. It provides both physical and mental conditions to improve health and well-being simultaneously.
2. It provides an overall sense of peace and the sense of well-being.
3. It increases the immune system of the body.
4. It treats both root cause as well as the associated symptoms of the disease.
5. It allows the body to calm and relax for the better response and results.
6. It is able to treat multiple illnesses.
7. It is a complete system involving identification, dealing, prognosis, and anticipation.

## 2.2.5  THERAPIES OF CHINESE SYSTEM OF MEDICINE

The Chinese alternative system of medicine includes the following practices:

- Herbal therapy.
- Acupuncture.
- Tuina Chinese Medical Massage.
- Qi-Gong.

### 2.2.5.1  HERBAL THERAPY

Herbal therapy is the fundamental of TCM. The Ben Cao Gang Mu (Materia Medica) and Shennong Bencaojing (Classic of Herbal Medicine) are the fundamental literature of TCM. Generally, herbal therapy prescription consists of a mixture of 2–40 medicinal herbs. Herbal therapy experts believe that each medicinal herb has unique chemical composition, unique strengths, and deficiencies. In a prescription, every ingredient should be carefully balanced in terms of quality and quantity. This increases the effectiveness and reduces the harmful effects at the same time (Wei and Zheng, 2008). Herbal therapy uses almost every part of root, stem, bark, leafs, flowers, buds, and fruits of the plant, and some ingredients of animals.

Examples: The *Taxus chinensis* has been used in TCM to remove toxins from the body and relieve a cough. The active chemical constituents are polyoxygenated diterpene which has anticancer effects (Siowet al., 2005). Artemisia has been used to treat fever for more than 2000 years in TCM. The studies showed that it has been effective in the treatment of malaria (Siow et al., 2005).

### 2.2.5.1.1   Classification of Chinese Herbs

In a prescription, each herb is present in a different amount. According to Lee (2000), herbs are classified into four groups:

1. **The Emperor Herbs:** These are the major herbs which consist of bioactive substances for the pathophysiological condition.
2. **The Minister Herbs:** These herbs are used to provide support or enhancing the action of main herbs. They also relieve secondary symptoms of the disease.
3. **The Assistant Herbs:** These herbs help to modify the action of the emperor herbs. They neutralize adverse effects of the herbal mixture.
4. **The Messenger Herbs:** These herbs are used to direct the action of other herbs to the specific part of the body.

### 2.2.5.1.2   Methods of Preparation of Medicinal Herbs

There are different methods for the preparation of medicinal herbs. The most common methods are:

1. **Infusions**: This method is used to extract chemical compounds or flavors from the herbs. In this method, dried or fresh herbs are placed in hot water or oil for 10–15 min and strained in a quart jar then stored in a covered ceramic pot for 4–10 h.
2. **Decoctions**: This method is used for extraction of chemical compounds from hard roots, barks, seeds, and dries berries. The herb should be a grind, crush or even to powdered form before decoction. This increases the surface areas and chemical constituents of herbs easily extracted. In this method, the herb material (ground) is boiled for a longer period of time. Green (2000) suggested that the decoction must start with cold water.
3. **Tincture**: The meaning of tincture is liquid extracts. These can be taken orally under the tongue. The action of tincture is usually fast as they directly enter the bloodstream. Some nutritive tinctures take several weeks of repetitive use for best results. For tincture preparation, the herbs are extracted with alcohol and nonalcoholic solvents like vinegar or glycerine.
   Some herb materials, like dried roots, berries, seeds, and dried barks need a powerful solvent to extract the active chemical

constituents. The alcohol is a nontoxic and potent solvent for this purpose. It is also a powerful preservative and increases the life of tincture for many years. The vegetable glycerine is also a good solvent for the tincture preparation because it has a sweet taste and nonalcoholic. The less potent solvents like vinegar, water, and vegetable glycerine are not able to extract the active constituents effectively. The nonalcoholic tinctures can easily become contaminated with molds and not safe for use.

4. **Preparation of tincture:** One should take fresh or dried herb without any contamination or dirt and place it in a wide mouth dry jar. The solvent of pure quality should be poured over the herb. If the herb material absorbs the solvent then more solvents can be added as needed in the jar. The jar should be covered tightly and placed in dark place for 4–6 weeks. During this period of time, on every 2–3 days the jar is tilted and the level of solvent checked. It is believed that the potency of tincture increases with time.

5. **Macerations**: This process involves a slow soaking of herb material in a suitable solvent like water or alcohol. The herb material is simply merged in cold water and soaked for 24 h. The herb material could be simply suspended in the solvent and left to sit for few weeks to few months depending on the purpose and potency of maceration. The main purpose of using this method is to extract active constituents of herb that might be lost by heating or might be decomposed in a strong solvent like alcohol.

### 2.2.5.1.3   Limitations of Herbal Therapy

The chemical constituents of the medicinal herbs change with a change in the cultivation area, time, and species. The variation in chemical constituents and their concentration affects the quality of the herb.

TCM prescription usually consists of 2–40 medicinal herbs. The nature and concentration of chemical constituents might be changed during the methods of preparation. This variation in product composition and concentration might lead to underdose or overdose. According to Chiu et al. (2009), severe adverse effects have been reported during the use of ACSM. Therefore, identification of biologically active constituents of herbal drugs becomes an important topic for research and development of new herbal-based products. For this purpose, various scientific methods and instruments have been developed.

Therefore, identification and isolation of bioactive constituents have become a serious issue for research and development of the novel herbal product. Several scientific methods have been established to identify the herbs (Liu et al., 2008; Ganzera, 2009; Jiang et al., 2010).

### 2.2.5.2 ACUPUNCTURE: AS CHINESE ALTERNATIVE SYSTEM OF MEDICINE

Acupuncture is another way to prevent, diagnose, and cure diseases and to improve health in the ACSM for more than 2500 years. The acupuncture word is derived from two Latin words, acus which means needle while pungere means to stick. Nowadays it is a balancing medical exercise which involves stimulating specific points on the body. It is defined as the needling and moxibustion (burning of herb) to stimulate the points of acupuncture (Liao and Ng, 1997).

This is carried out with the help of a needle which penetrates the skin to relieve pain or to diagnose various health illnesses. It also improves the body's functions and stimulates self-healing process. The acupuncture experts explained that it modifies the flow of energy throughout the body. It corrects imbalance of vital energy, Qi with inserting the needle at acupuncture points.

i. **Basic principles of acupuncture:** The acupuncture practitioners believe that the body is a complex system where everything within it is interconnected each affects others. The illness begins when the unbalance occurs within this biological system. The body shows the symptoms when it is not able to rebalance. These symptoms lead to chronic or acute diseases of all types.

There are 14 major meridians called energy channels that flow in the body. The energy called Chi circulates along the energy channels to all parts of the body including all cells and organs. This energy is the vital force which keeps us alive. The smooth flow of the energy is responsible for balanced and vibrant health. When blockages occur somewhere along one or more meridians, illness, and pain develop. The acupuncture points, act as a pass-through but when the energy can get slowing down or stop to the point may cause pain and illness. Into the blocked areas, a thin needle is inserted to open the gate. This again allows the energy to flow through the energy channels smoothly. Acupuncture allows the body to reregulate the energy flow and regulate itself to maintain the maximum level of health.

According to Deadman *et al.* (2007), more than 400 acupuncture points have been reported.

ii.    **Applications of acupuncture:** The studies showed that it stimulates the nerve fibers present in the muscle and stimulates the release of vasoactive compounds locally. This enhances healing process by improving the local blood flow (Sandberg and Lindberg, 2004; Cheuket al., 2007). Johnson and Burchief (2004), proves the effectiveness of acupuncture for the treatment of trigeminal neuralgia and neuropathic pain (Kimura et al., 2006).

The acupuncture activates some afferent nerve fibers of the spinal cord and also gives analgesic effects (Cheng, 2009). The study showed that acupuncture has been used to cure lower back pain (Johnson and Burchief, 2004). According to Mc Carney et al. (2004), it can be effective to control chronic asthma. Casimiro et al. (2005) noted that acupuncture is used for the treatment of rheumatoid arthritis. Many other studies also showed that acupuncture has been used to treat migraine prophylaxis, vascular dementia, epilepsy, nausea, vomiting, depression, schizophrenia, smoking cessation, and insomnia (Smith and Hay-Phillipa, 2004;Rathbone and Xia, 2005; Ezzoet al., 2006; Weinaet al., 2007; Cheuk and Wong, 2008; Linde et al., 2009; White et al., 2011).

## 2.2.6   *THE CHINESE MEDICINE THERAPEUTIC MASSAGE (TUINA)*

The meaning of Tuina is a pinch and pull. It is used as a treatment to resolve specific patterns of disharmony, pleasure, and relaxation. In China, it is in practice for more than 4000 years. This is also focused on the smooth flow of energy in different parts of the body to retain a good health. The practitioners of Chinese massage recognized roots with the limited flow and use hand pressure to restore energy circulation in the body. They balance the vital energy qi by restoring the energy pathways jing luo of blockages which is responsible for pain and disease. They use a variety of hand techniques (shou fa), which stimulates (yang) or sedate (yin) of the patient to boost health and lifestyle.

### 2.2.6.1   *BENEFITS OF CHINESE MASSAGE*

This massage eliminates stress and reducing pain. It balances and regulates the flow of vital energy all parts of the body. This restores the lost energy occurs by disharmony between body and nature. The Chinese massage is used to renew body energy and strength, treat lower back pain, improve cardiovascular health, open frozen shoulders, and accelerate the recovery of injury and surgery. Others are reducing symptoms of rheumatoid arthritis, enhancing the immunity of body, improvement of the functioning of the body organs, better sleep, respiration, and circulation (Arellano, 2017; Braverman, 2017; Chinese Massage, 2017; Liu, 2017). Therefore, it reduces stress and induces relaxation which affects energetic, emotional, mental and physical layer, and increases vitality and overall health.

### 2.2.6.2   METHODS OF TUINA

Tuina uses same healing techniques, energy points, energy channels or meridians, and principles as used in acupuncture. Instead of needles, it applies deep pressure by hand to stimulate energy blockages. There are eight basic techniques used by practitioners. These techniques can be combined in a number of ways as to the condition and needs of the patient. These include pushing (Tui), dragging, or pulling (Na), pressing (An), strong pinching pressure (Tao), kneading (Nie), nipping and pinching (Nien), and rubbing (Mao) and tapping (pai).

### 2.2.6.3   THE CHINESE MEDICINE THERAPY (QI-GONG)

Qigong is a Chinese medicine therapy has been documented since 2500 years ago. It is a form of calm practice which includes breathing techniques, movements and stretching of the body, and meditation. It provides strength and circulates the life energy Qi. This increases the circulation of body fluid. Qi-gong exercise includes both external and internal movements. In China, internal movements are called neigong (internal power).

The benefits of qigong therapy are to prevent chronic health problems, poor circulation of fluid, nerve pain, joint pain, and other general illnesses. It can be practiced by one who wants to become physically and mentally healthy. It has many advantages and can be used with other therapies. It can be practiced by standing, seated, or supine postures. It is beneficial for young and old persons.

## 2.3  TRADITIONAL TIBETAN SYSTEM OF MEDICINE (TTM)

TTM is one of the oldest alternative systems of medicine practiced for more than 2000 years. This includes dietary and behavioral changes, lifestyle, herbs, and therapies to treat the basic cause of illness. According to Lobsang and Dakpa (2001), it usually involves physical therapy, balancing in diet, herbal remedies, or combination of one or more therapies. It delivers symptomatic relief and diagnoses the root cause of illness.

## 2.3.1   BASIC PRINCIPLES INVOLVES IN TTM

TTM considers attachment, hatred, and delusion as three mental poisons which are responsible for all human diseases. It believes that human body is composed of five basic cosmic elements such as water, air, fire, earth, and space. These elements are present in every part of the body including all cells. The principle is based that illnesses occurs by the disequilibrium of these elements. These elements are active forces which are characterized by their energetic utilities (Dunkenberger, 2000).

According to TTM, there are three basic principles of energies that is, wind (rlung), bile (mKhrispa), and phlegm (badkan) which are directly associated with the five cosmic elements. Wind is air element. Its equilibrium deals with our nervous system, thoughts, and circulation of blood. The wind-related problems occur by stress, donating too much blood, insufficient diet, and extreme dieting. Bile is related to the health and functioning of liver and control metabolism. Exercises, sports, and excessive sunbathing disturb bile. Phlegm is associated with body's joint health, mental stability, and accurate digestion. The disturbance in phlegm occurs by taking heavy food, swimming in spring and winter, and relaxing after taking food (Kumar, 2013).

## 2.3.2   PROCEDURES FOR DISEASE DIAGNOSIS

The TTM system uses natural and spiritual approach to medicine. The procedure used by TTM includes visual, touch (pulse), and interrogation (Kumar, 2013).

i.   **Visual:** The visual inspection includes the study of urine, five sensory organs, and stool. Visual diagnosis includes the texture of the blood which is present in sputum, urine and stool, and complexion of the skin.

ii.  **Touch:** This involves examination of the energy meridians, body organs, and pulse of the patients by feeling them with the touch. The urine and pulse analysis form the most characteristic and vital part of the diagnosis.

iii. **Interrogation:** Questioning practices are the most useful method for finding the cause and site of illness. The main aims of this are to observe the sign and symptoms, site of illness, and to find the causative factor. Many questions asked to the patient about the history of disease, symptoms, and other which can help for clinical assessment and right diagnosis of the illness.

### 2.3.3 TREATMENT OF ILLNESS

TTM has three approaches for treating the illness and to enhance body health. These three approaches are the dietary approach, behavioral approach, and external approach.

1. **Dietary approach:** According to the Tibetan medical system, every substance on earth has medicinal properties. There are six principals of tastes that are used in TTM. These are sour, salty, hot, astringent bitter, and sweet. These six tastes are present in our food and they have different nature, fire to ripen, earth as a basis, air to cause movement, water to moisten, and space for growth. Any two dominating elements out of five produces astringent tastes. Each taste has a different influence on the functioning of the body. This dietary approach believes that illness can be treated by giving a proper diet of desire taste (Kumar, 2013).

2. **Behavioral approach:** According to this approach, the mental, physical, and emotional behavior in the inadequate and excessive manner can be the cause of illness. This approach emphasizes on routine, seasonal, and incidental behavior. Routine behavior deals with the proper use of speech, body, and mind. Seasonal behavior deals with the energy transformation that takes place in the body with changes in the environmental conditions. The behavior of any individual must harmonize with these changes. The incidental behavior suggests avoiding blocking the impulse of hunger, vomiting, evacuation of mucus, thirst, semen, saliva, gas, and urine, as these might cause disharmony. If these fluids expelled or suppressed forcefully, different types of illness could arise with the immediate disturbance of wind energy.

3.  **External approach:** This approached is used when the diet and routine behavior are unable to recover the illness conditions then herbal medicines are prescribed. This includes Tibetan medicines, moxibustion therapy, and oil therapy.

## 2.3.4  TIBETAN MEDICINES

Tibetan medicines can be used in different forms like pills, syrups, decoctions, powders, and these are prescribed in small doses. TTM prescription constitutes several ingredients, generally 3–20.

These medicines have the combination of hot, salty, hot, and sweet, which helps to eliminate the problems associated with wind energy. The combinations of sweet, bitter, and astringent taste help to eliminate the problems related to bile. Medicines, with a combination of sour, hot, and salty taste help to eliminate problems related to phlegm (Dunkenberger, 2000).

## 2.4  UNANI SYSTEM OF MEDICINE (USM)

It is the traditional system of medicine originated in 5th and 4th Century by Hippocrates in Greece. The history of USM is traced old to ancient Babylon and Egypt. They had adopted the use of medicinal plants to treat diseases. They had also started surgery as a method of treatment. This system was adopted in Arab and Persian lands. It is presently practiced in China, Egypt, Iraq, Iran, India, Sri Lanka, Pakistan, and far East countries. USM delivers promotive, preventive, curative, and rehabilitative medicine. USM is based on holistic and scientific principles to diagnose and treat illness. This system considers that the entire universe including environment, disease, human being, drugs, nature, and so on are to be intrinsically explained by four qualities dry: wet and hot: cold.

## 2.4.1  BASIC PRINCIPLE OF USM

These qualities are revealed in all the basic concepts of USM such as humor, elements, and temperament which are used for correlating and describing health and illness. The USM relies on the theory of four element water, air, earth, and fire and theory of four senses of humor, that is, phlegm, blood, bile, and black bile. Any disharmony in the humoral balance in the body causes disease.

The USM has treatment for diseases related to all organs and systems of the human body. The USM treatments are highly effective and acceptable for liver, reproductive system, immunological, lifestyle disorders, chronic ailments, and disease of the skin.

## 2.4.2  DIAGNOSIS AND TREATMENT OF ILLNESS

The USM diagnoses and treats the patient by looking into their overall mental, physical, and spiritual aspects. In USM, a disease is diagnosed by the examination of stool, pulse, and urine. The therapies used in USM are regimental therapy, diet therapy, pharmacotherapy, and surgery. Regimental therapy along with diet therapy is considered the best way for improving health and treatment of disease. This system also emphasized on psychiatric treatment and the management of various diseases. The Regimental therapy is a backbone of USM which works on the principle of modifying six essential factors of life. Pain management and skin disorders can also be managed with least medication by Regimental therapy. Aromatics, nutraceuticals, cosmeceuticals, and corresponding therapies are also vital parts of USM.

Ahmad et al. (2011) showed the efficacy of a combination of Unani drugs in a patient of trichomonal vaginitis. Another example is the Unani coded drug UNIM-200(G) in type II diabetes. The Unani drugs might be used to treat diabetes mellitus, bronchial asthma, gastrointestinal ailments, cardiovascular protection, eczema, arthritis, and coloasma (Praveenet al., 2008; Zaidi and Yamada, 2009; Ahmad et al., 2011).

## 2.5  INDIAN ALTERNATIVE SYSTEM OF MEDICINE

In India, various therapies like Ayurveda, Siddha, Unani, Yoga, and meditation have been used since ancient time to cure illness. In this chapter, Siddha, Ayurvedic, and Unani alternative systems have been described.

## 2.5.1  SIDDHA SYSTEM OF MEDICINE

In India, Siddha is the oldest system of medicine and considered as the mother medicine of ancient Tamils in South India. The meaning of Siddha means attainments and Siddhars were holy persons who attained results in medicine. Eighteen Siddhars had contributed towards the expansion of this medical system. This system is basically therapeutic in nature. Some of their works are still referenced books of surgery and medicine among the Siddha medical practitioners (Piet, 1952). The Siddha medicines are prepared based on the principles of Panchabuthas (metals of lead, copper, gold, iron, and zinc). Gold and lead are powerful for the maintenance of the body. Iron, zinc, and copper are used for the preservation of heat in the body (Narayansamy, 1975).

### 2.5.1.1   BASIC PRINCIPLE OF SIDDHA THERAPY

The concept of Siddha is closely similar to Ayurveda with specialization in *iatrochemistry*. According to this system, the body is the duplication of the universe and so are the drugs and food regardless of their source. Siddha system considers that all matter in the universe including human body is composed of five basic elements air, water, earth, fire, and sky. The food and drug which the human body takes are made of these five elements. The amount of these elements, present in the drugs and food is varied and responsible for certain activities and therapeutic results.

Siddha system also considers the body as a composite of three senses of humor, seven basic tissues, and the waste products of the body such as sweat, faces, and urine. The food is the basic building material of human body which gets processed into body tissues, humor, and waste products. The balance of humor is measured as health and its imbalance causes disease. This system also deals with the concept of salvation in life. The promoters of this system consider attainment of this state as conceivable by meditation and medicine (Walter et al., 2009).

### 2.5.1.2   SIDDHA MATERIA MEDICA

Siddha system has established a rich and unique wealth of drug knowledge, in which the use of metals and minerals is very much encouraged. There are 25 types of water-soluble inorganic substances in Siddha materia medica. It also mentioned 64 types of mineral drugs which emits vapor when put in the fire but do not dissolve in water. Thirty-two of these drugs are natural and remaining synthetic. There are seven drugs which emit vapor on heating and do not dissolve in water in Siddha materia medica. The Siddha system has classified metals and alloys in separate classes. These metals include iron, lead, copper, tin, gold, and silver.

These are burnt by special methods and used in medicine. There is a class of drugs which exhibit sublimation on heating. This includes mercury and its various forms like red oxide of mercury, a red sulfide of mercury, and mercuric chloride, and so forth.

Sulfur plays a vital place in Siddha materia medica for use in therapeutics and in the maintenance of health. The above classification reveals detailed information and learning of minerals that this system has developed for treatment.

### 2.5.1.3   CHEMISTRY IN SIDDHA

This system chemistry had been found well established into a science auxiliary to medicine and alchemy. It was found beneficial in the preparation of medicine. Siddhars were also conscious of several alchemical processes in which they are divided into several processes like fermentation, distillation, calcination, sublimation, distillation, separation conjunction or combination, fusion, congelation, cibation, extraction purification, incineration of metals, liquefaction, and so on.

Siddhars were even polypharmacists and engaged in dissolving, boiling, precipitating, and coagulating chemical substances. This system is capable of treating all types of sickness except emergency cases. In general, Siddha system is capable of treating all types of diseases including skin problems, diseases of the liver, gastrointestinal tract, urinary tract infections, anemia, diarrhea, fever, arthritis, and allergic disorders.

### 2.5.1.4   CAUSE OF DISEASE

According to this system, illness occurs when the normal equilibrium of three senses of humor pitta, vata, and kapha is disturbed. The factors which affect this equilibrium are diet, physical activities environment, and climatic conditions. According to this system, diet and lifestyle play a critical role not only in health but also in curing diseases. This theory of the Siddha medicine is known as pathya and apathya that is, essentially a list of do's and don'ts (Subbarayappa, 1971; Kandaswamy, 1979; Sharma, 1992; Thottam, 2000).

### 2.5.1.5   DIAGNOSIS AND TREATMENT

The diagnosis of illness involves recognizing its causes. Identification of causative factors is done on the examination of the tongue, the color of the body, pulse, urine, eyes, and study of voice, tongue, and condition of the digestive system. This system has detail procedure of urine checkup which includes color, smell, density, and quantity and oil drop spreading pattern.

This system of medicine emphasizes that medical treatment is not concerned with sickness alone but has to take into account the environment, patient, the meteorological consideration, habits, age, sex, mental frame, diet, habitat, appetite, physiological constitution, physical condition, and so forth. This reveals that the treatment has to be individualistic, so as to ensure that mistakes in diagnosis and treatment are minimal. This system believes in the principle of "Food itself is a kind of medicine." This system is powerful in treating chronic illness of the liver, bleeding piles, skin diseases, anemia, prostate enlargement, rheumatic problems, and peptic ulcer.

## 2.5.1.6   BASICS CONCEPTS OF SIDDHA MEDICINE

The basic concepts of the Siddha medicine are almost alike to Ayurveda. There is only one difference which appears to be that the Siddha medicine recognizes the prevalence of *vatham* in childhood, *pitham* in adulthood and *kapam* in old, while in Ayurveda, it is totally opposite: *vatham* is dominant is old age, *kapam* is in childhood and *pitham* in adults. According to their mode of uses, the Siddha medicine could be categorized into two classes, such as internal and external medicines.

This system believes that various psychological and physiological functions of the body are attributed to the association of seven elements (Kandaswamy, 1979):

1.  **Saram (digestive juice):** It means oxygen is responsible for nourishment, growth, and development.
2.  **Cheneer (blood):** It is responsible for imparting color, nourishing muscles, and improving intellect.
3.  **Ooun (muscle):** It is responsible for the shape of the body.
4.  **Kollzuppu (fatty tissue):** These are responsible for lubricating joints and oil balance.
5.  **Enbu (bone):** This is responsible for body posture and structure and movement.
6.  **Moolai (bone marrow):** It is responsible for the production of RBC,
7.  **Sukila (semen):** This is responsible for reproduction.

## 2.5.1.7   DIAGNOSIS AND TREATMENTS IN SIDDHA

In diagnosis, examination of eight things is essential which is generally known as astasthana pariksa (Liuet al., 2008; Jiang et al., 2010). These are:

1. **Na (tongue):** White in kapha, black in vatha, ulcerated in anemia, and yellow or red in pitha.
2. **Varna (color):** Yellow or red in pitha, pale in kapha, and dark in vatha.
3. **Svara (voice):** Normal in vatha, low pitched in kapha, high pitched in pitha, and slurred in alcoholism.
4. **Kan (eyes):** Yellowish or red in pitha, muddy conjunctiva, and pale in kapha.
5. **Sparisam (touch):** Warm in pitha, dry in vatha, sweating in different parts of the body, and chill in kapha.
6. **Mala (stool):** Yellow in pitha, dark red in the ulcer, black stools indicate vatha, and shiny in terminal illness and pale in kapha.
7. **Neer (urine):** Early morning urine is examined, reddish yellow excessive heat, straw color indicates indigestion, rose in blood pressure, looks like meat, and saffron color in jaundice.

## 2.5.2   AYURVEDIC SYSTEM OF MEDICINE

Ayurveda is about 5000 years old system of natural healing which has its origins in the Vedic civilization of India. It is the science of life (Ayur = life, Veda = knowledge or science). It reminds us that health is composed of a dynamic integration between environment, mind, body, and spirit. It is mainly related to health and diseases (Bratman and Stevan, 1997). India has more than 45,000 diverse plant species. Out of these, about 15,000–20,000 species have medicinal values but only about 7000–7500 plant species have been used in traditional Ayurveda system of India (Rao, 2011).

According to old facts, the traditional texts like Charaka Samhita and Sushruta Samhita of Ayurveda were written around 1000 BC. *The Materia Medica of Ayurveda* includes 600 species of medicinal plants along with therapeutics and processing. Herbs like ginger, garlic, turmeric, fenugreek, and holy basil are an integral part of Ayurvedic medicine. The many species of medicinal plants have been studied for various biological activities and pharmaceutical applications (Vinesh and Devendra, 2013a,b).

## 2.5.2.1  BASIC PRINCIPLES

It is a remarkable individualized system of medicine. In this system of medicine, things seen as medicine can also be a poison, which might be beneficial to you but could harm someone else, and *vice versa*. It all depends on the background, individual, external, and internal environment. It is basically opposed to one-size-fits-all therapies. While there are some practices that considered to be helpful for everyone, Ayurveda generally focused on *you*, the individual concerned.

Ayurveda identifies five elements; fire, water, earth, air, and ether (space) as the essential building blocks of nature. Each matter and all parts of the body even each cell, contains all five of these elements. Out of these, one or two elements are typically predominant over the others.

Ayurveda recognizes that human is the part of nature. It describes three basic energies or doshas; vata (wind), pitta (fire), and kapha (Earth), which manage our inner and outer environments like structure, movement, and transformation.

1. **Vata:** Is the energy of ether and air associated with creativity and connection, movement, and impulse. This controls breathing, muscle movement in general, sensory perception, nerve impulses, communication, the pulsation of the heart and our taste to experience joy, flexibility, and expansive consciousness. In excess amount, it can cause anxiety, fear, physical and emotional constriction, poor circulation, constipation, emaciation, insomnia, dry skin, cracking joints, twitches, tremors, and other abnormal movements.

2. **Pitta:** Is the energy of water and fire associated with digestion and transformation. This controls courage, charisma, appetite, intelligence, digestion, absorption, assimilation, and ambition. In excess, it can cause excessive heat, heartburn, anger, jealousy, loose stools, migraines, rashes, inflammation, bruising, bleeding disorders, difficulty in sleeping sharp hunger, and overactive metabolism.

3. **Kaph:** Is the energy of earth and water associated with structure and cohesiveness, grounding, and stability. This is responsible for lubrication, regeneration, nourishment, growth, strength, stamina, memory, fluid balance, fat regulation, our ability to feel compassion and contentment. In excess, it can cause depression, a sluggish metabolism, congestion, attachment, greed, resistance to change, heaviness in the mind and body, excessive sleep, lack of motivation, water retention, hardening of the arteries, and the formation of masses and tumors.

## 2.5.2.2 AYURVEDIC MEDICINE THERAPY

Each individual body has a unique proportion of these three energies that shape our physical and mental level. If vata prevails in the body the individuals tend to be light, thin, energetic, enthusiastic, and changeable. If pitta predominates, the individual tends to be intelligent, intense, and goal-oriented and have a strong desire for life. When kapha is dominant, the individual tends to be methodical, and nurturing. Although each individual has all three forces and most people have one or two elements that predominate.

According to Ayurveda, each element has a balanced and imbalance expression. When vata is balanced, the individual is creative and lively, but when there is too much movement in the body, an individual tends to experience constipation, anxiety, dry skin, and insomnia. When pitta is in the harmonic state an individual is disciplined, friendly, warm, a good speaker, and a good leader. When pitta is out of balance, an individual tends to be irritable and compulsive and may suffer from inflammatory condition and indigestion. When kapha is balanced, the individual is supportive, sweet and stable but when kapha is out of balance, an individual may experience weight gain, sinus congestion, and sluggishness.

To treat illness, this system uses various herbs, gems, yoga, mantras, foods, aroma, lifestyle, and surgery. The Rigveda includes sections on disease, symptoms, pathogenesis, and the principle of treatments and the nature of health. Atreya Samhita the book on Ayurveda is one of the oldest books in the world. The Charak Samhita, Ashtanga Hridaya Samhita, Sushruta Samhita text books are still used by the Ayurvedic practitioner. Nowadays, Yoga is also used to treat illness and maintain health with Ayurvedic therapy as described in the Ayurvedic text (Sutra, 1950).

## 2.5.2.3 BENEFITS OF AYURVEDIC MEDICINE THERAPY

The Ayurvedic medicine has many benefits. Some of the benefits are:

1. Ayurvedic medicine lowers stress and anxiety.
2. It lowers blood pressure and cholesterol.
3. It helps with recovery from injuries and illnesses.
4. It promotes a nutrient dense antioxidant rich diet.
5. It can help in reducing body weight.
6. It lowers inflammation.
7. It helps with hormonal balance.

8.  Ayurvedic herbal therapies are generally safe when prescribed by a skilled and experienced Ayurvedic practitioner, however, some herbal medicines can cause side effects.

## 2.6  CONCLUSION

In the various countries of the world, different traditional systems of alternative medicines have been used to prevent, diagnose, and treat diseases. In India, various therapies like Ayurveda, yoga, Siddha, and medication have been used to cure illness. China also has a well-developed system of alternative medicine which includes herbal therapy, acupuncture, and Tuina (massage therapy) since ancient time to keep the body healthy, prevent, and treat illnesses. The USM is based on holistic and scientific principles to diagnose and treat illness. This system considers that entire universe including environment, disease, human being, drugs, nature, and so forth. TTM believes that human body is composed of five basic cosmic elements such as water, air, fire, earth, and space. These elements are present in every part of the body including all cells, in which their imbalance results in diseases. According to Siddha system, the body is the duplication of the universe and so are the drugs and food regardless of their source. Siddha system considers that all matter in the universe including human body is composed of five basic elements such as air, water, earth, fire, and sky. The food and drug which the human body takes are made of these five elements. The amount of these elements presents in the drugs and food varies and is responsible for certain activities and therapeutic results. Ayurveda is called as an individualized system of medicine. All systems believe the body has an affinity for nature with harmony.

## KEYWORDS

- **herbal medicines**
- **alternative medicines**
- **traditional system**
- **Unani**
- **Siddha**

## REFERENCES

Ahmad, W.; Hasan, A.; Abdullah, A.; Tarannum, T. Efficacy of a Combination of Unani Drugs in Patients of Trichomonal Vaginitis. *Indian J. Tradit. Knowl.* **2011,** *10*(4), 727–730.

Arellano, V. What is Chinese Massage? http://www.livestrong.com/article/119168-chinese-massage, (accessed Aug 14, 2017).

Bratman, S.*The Alternative Medicine Sourcebook: A Realistic Evaluation of Alternative Healing Method;* Lowell House: Los Angeles, California, 1997.

Braverman, J. Chinese Massage Benefits. http://www.livestrong.com/article/123084-benefits-Chinese-massage (accessed Jul 18, 2017).

Casimiro, L.; Barnsley, L.; Brosseau, L.; Milne, S.; Welch, V.; Tugwell, P.; Wells, G. A. Acupuncture and Electroacupuncture for the Treatment of Rheumatoid Arthritis. *Cochrane Database Syst. Rev.* **2005,** *19*(4), CD003788. DOI:10.1002/14651858.CD003788.pub2.

Cheng,K. J. Neuroanatomical Basis of Acupuncture Treatment for Some Common Illnesses. *Acupunct. Med.* **2009,** *27,* 61–44. DOI: 10.1136/aim.2009.000455.

Cheuk-Daniel, K.; Yeung, W. F.; Chung, K. F.; Wong, V. Acupuncture for Insomnia. *Cochrane Database Syst. Rev.* **2007,** *18*(3), CD005472. DOI: 10 1002 /14651858 CD005472 pub2.

Cheuk-Daniel, K.;Wong, V. Acupuncture for Epilepsy. *Cochrane Database Syst. Rev.* **2008,** *8*(4), CD005062. DOI: 10 1002 /14651858 CD005062 pub3.

Chinese Massage. http://learn.healthpro.com/chinese-massage (accessed Dec 18, 2017).

Chiu, J.;Yau, T.;Epstein, R. J. Complications of Traditional Chinese/Herbal Medicines (TCM) a Guide for Perplexed Oncologists and Other Cancer Caregivers.*Support Care Cancer,* **2009,** *17,* 231–240.

Deadman, P.; Baker, K.; Al-Khafaji, M. A. Manual of Acupuncture, 2nd Ed.; Journal of. Chinese Medicine Publications: 2007. ISBN 978-0951054659.

Dunkenberger,T. *Tibetan Healing Handbook;* Pilgrim Publishing House: Varanasi, India, 2000.

Erick Hao-Shu Hong. Traditional Chinese Medicine from the Perspective of Western Medicine. Master Thesis, Faculty of Medicine, University of Oslo: Ulleval, 2010.

Ezzo, J.;Richardson, M. A.; Vickers, A.;Allen, C.; Dibble, S.; Issell, B. F.; Lao, L.; Pearl, M.; Ramirez, G.;Roscoe, J. A.;Shen, J.; Shivnan, J. C.; Streitberger, K.; Treish, I.; Zhang, G. Acupuncture-Point Stimulation for Chemotherapy-Induced Nausea Or Vomiting. *Cochrane Database Syst. Rev.* **2006,** *19*(2), CD002285. DOI:10.1002/14651858.CD002285.pub2.

Ganzera, M. Recent Advancements and Applications in the Analysis of Traditional Chinese Medicines. *Planta Med.* **2009,** *75,* 776–783.

Green, J. *The Herbal Medicine Makers Handbook;* Crossing Press: Berkeley, CA, 2000.

Jiang, Y; David, B.; Tu, P.; Barbin, Y. Recent Analytical Approaches in Quality Control of Traditional Chinese Medicines – A Review. *Anal. Chim. Acta.* **2010,** *657*,9–18.

Johnson, M. D.; Burchiel, K. J. Peripheral Stimulation for Treatment of Trigeminal Postherpetic Neuralgia and Trigeminal Posttraumatic Neuropathic Pain: APilot Study. *Neurosurgery,* **2004,** *55,* 135–142.

Kandaswamy, P. N. History of Siddha Medicine. Published by The Government of Tamil Nadu: Madras, India, 1979.

Kimura, K.; Masuda, K.; Wakayama, I. Changes in Skin Blood Flow and Skin Sympathetic Nerve Activity in Response to Manual Acupuncture Stimulation in Humans. *Am. J. Chin. Med.* **2006,** *34,* 189–196.

Kumar, V. Medicinal Plants Used in the Practice of Tibetan Medicine. Virology Group. *International Centre for Genetic Engineering and Biotechnology*; Aruna Asaf Ali Marg, New Delhi 110067, India, 2013.

Lee, K. H. Research and Future Trends in the Pharmaceutical Development of Medicinal Herbs from Chinese Medicine. *Public. Health Nutr.* **2000,** *3,* 515–522.

Liao, S. J.; Ng, LKY. Acupuncture: Ancient Chinese and Modern Western. A Comparative Inquiry. *J. Altern. Complementary Med.* **1997,** 11–23.

Linde, K.; Allais, G.; Brinkhaus, B.; Manheimer, E.; Vickers, A.; White, A. R. Acupuncture for Migraine Prophylaxis.*Cochrane Database Syst. Rev.* **2009,** *21*(1), CD001218. DOI:10.1002/14651858.CD001218.pub2.

Liu, S.; Yi L. Z.; Liang, Y. Z. Traditional Chinese Medicine and Separation Science. *J. Sep. Sci.* **2008,** *31,* 2113–2137.

Liu, Z. Clinical Study of Traditional Chinese Massage Combined with Music Therapy in the Treatment of Cerebral Palsy. *Pediatr. Res.* **2017,** *70,* 336. DOI:10.1038/pr.2011.561.

Lobsang, T.; Dakpa, T. *Fundamentals of Tibetan Medicine;* 4th ed.; Mentsekhang Publication: Dharamsala, India, 2001.

Mc Carney, R. W.; Brinkhaus, B.; Lasserson,T. J.; Linde, K. Acupuncture for Chronic Asthma. *Cochrane Database Syst. Rev.* **2004,** *1,* CD000008. DOI: 10 1002 /14651858 CD000008 pub2.

Narayansamy, V. *Introduction to the Siddha System of Medicine;* Pandit S. S. Anandam Research Institute of Siddha: T. Nagar, Madras, 1975; pp 1–51.

Piet, J. H. *Logical Presentation of the Saiva Siddhanta Philosophy;* Christian Literature Society for India: Madras, 1952.

Praveen, S.; Zafar, S.; Qureshi, M. A.; Bano, H. Randomized Placebo Control Clinical Trial of Herbo Mineral Cream to Evaluate Its Topical Effects on Chloasma, *Hippocratic J. Unani Med.* **2008,** *3*(2), 21–29.

Rao, M. P. Merits of Using Hers in the Whole State (Ayurveda's Concept) Over Isolated Fractions. *Int. J. Res. Ayurveda. Pharm.* **2011,** *2*(2), 80–83.

Rathbone, J.; Xia, J. Acupuncture for Schizophrenia. Acupuncture for Schizophrenia. *Cochrane Database Syst. Rev.* **2005,** *4.* DOI: 10 1002 /14651858 CD005475.

Sandberg, M.; Lindberg, L. G.; Gerdle, B. Peripheral Effects of Needle Stimulation (Acupuncture) on Skin and Muscle Blood Flow in Fibromyalgia. *Eur. J. Pain.* **2004,** *8,*163–171.

Sharma, P. V. Siddha Medicine. *History of Medicine in India;* The Indian National Science Academy: New Delhi, 1992; pp 445–50.

Siow, Y. L.; Gong, Y.; Au-Yeung, K. K.; Woo, C. W.; Choy, P. C. Emerging Issues in Traditional Chinese Medicine. *Can. J. Physiol. Pharmacol.* **2005,** *83,* 321–334.

Smith, C. A.; Hay-Phillipa, P.J. Acupuncture for Depression. *Acupuncture for depression Cochrane Database of Systematic Reviews;* John Wiley & Sons, Ltd: Chichester, UK, 2004; p 3. DOI: 10 1002 /14651858 CD004046 pub2.

Subbarayappa,B. V. Chemical Practices, and Alchemy. *A Concise History of Science in India;* Indian National Science Academy: New Delhi, 1971; pp 315–335.

Ashtanga Hridaya Sutrasthana 2.20 (Dinacharya adhyaya), 1950. https://easyayurveda.com/2012/10/10/ayurvedic-daily-routine-ashtanga-hrudaya-sutra-sthana-chapter-2. (accessed Dec 18, 2017)

Thottam, P. J.*Siddha Medicine: A Handbook of Traditional Remedies;* Penguin Books: New Delhi, 2000.

Vinesh, K.; Devendra, T. Phytochemical Screening and Free-Radical Scavenging Activity of *Bergenia stracheyi. J. Pharmacognosy Phytochem.* **2013a,** *2*(2), 175–180.

Vinesh, K.; Devendra, T. Review on Phytochemical, Ethnomedical and Biological Studies of Medically Useful Genus Bergenia. *Int. J. Curr. Microbiol. Appl. Sci.* **2013b,** *2*(5), 328–334.

Walter, T. M.; Rubia, G.; Sathiya, E. Review of Ethics in Traditional Siddha Medicine as Defined by Siddhar Theraiyar. *Siddha Papers*, **2009**, *02*(03).

Wei, G.; Zheng, X. A Survey of the Studies on the Compatible Law of Ingredients in Chinese Herbal Prescriptions. *J. Tradit. Chin. Med.* **2008**, *28,* 223–227.

Weina, P.; Zhao, H.; Zhishun, L.; Shi, W. Acupuncture for Vascular Dementia. *Cochrane Database Syst. Rev.* **2007**, *2,* CD004987. DOI: 10 1002 /14651858 CD004987 pub2.

White, A. R.; Rampes, H.; Campbell, J. Acupuncture and Related Interventions for Smoking Cessation. *Cochrane. Database Syst. Rev.* **2011**, *19*(1), CD000009. DOI: 10 1002 /14651858 CD000009 pub2.

Zaidi, S. F.; Yamada, K.; Kadowaki, M. Bactericidal Activity of Medicinal Plants, Employed for the Treatment of Gastrointestinal Ailments, Against Helicobacter Pylori. *J. Ethnopharmacology,* **2009**, *121*(2), 286–644.

# CHAPTER 3

# ETHNOBOTANICAL STUDY OF INDIGENOUS PEOPLES' MEDICINAL PLANTS

FELIX IFEANYI NWAFOR* and STELLA I. INYA-AGHA

*Department of Pharmacognosy and Environmental Medicines, University of Nigeria, Nsukka, Nigeria*

*Corresponding author. E-mail: felix.nwafor@unn.edu.ng; Tel.: +2348036062242*
*ORCID: https://orcid.org/0000-0003-1889-6311*

## ABSTRACT

Ethnobotany is an autonomous field of study that draws from various disciplines such as botany, medicine, pharmacy, archeology, ecology, economics, sociology, and cultural and religious studies. It provides indigenous knowledge about useful plants, peoples' culture, and interaction with their environment. In modern science, ethnobotanical studies form the basis of drug research and give the researcher guide to possible biological activities of a plant and how best these can be harnessed in the preparation and administration of plant-based drugs. In this chapter, the authors highlighted the significance of an ethnobotanical study of indigenous medicinal plants, the step-by-step procedures to conducting an ethnobotanical survey, and documented the medicinal plants and traditional medicine practices of the indigenous people of Nsukka, Southeast Nigeria.

## 3.1 INTRODUCTION

Plants are arguably the basis of life: they form the integral part of the ecosystem and every other life form has relied on them for daily survival.

They purify the air we breathe, provide us with food, shelter and medicine. From the days of earlyman, the use of plants as medicine has been recorded (Phillipson, 2001). Majority of indigenous people, especially in developing countries, use plant-based medicines for healthcare delivery. Early documentations of cultures of the ancient Chinese, Indians, and Africans provide reliable evidence of man's dependence on plants and non-plant materials for the treatment and management of wide array of ailments. Many archeological documentaries show that even pre-historic man utilized plants in healing practices, and today the World Health Organization (WHO) estimates that 80% of the world's population uses herbal remedies, in one form or the other, to treat and manage illnesses (Okonkwo, 2012).

Plants used as drugs to treat diseases could also, when abused or as a result of insufficient knowledge of use, be toxic to the user (Sofowora, 2008). They contain chemical substances (known as phytochemicals) that are either beneficial or poisonous to man and livestock (Okwu, 2004). It is also worth noting that many of the drugs now prescribed and sold in pharmacy shops were originally isolated from plants used traditionally among the people of ancient cultures. It is estimated that about 120 of most commonly prescribed modern drugs as well as other pharmaceutical products sold today has at least one active ingredient in them derived from plants (Taylor, 2000). Most notable among them are: quinine–an anti-malaria agent from cinchona tree; paclitaxel and vinblastine–antitumor agents from pacific yew and Madagascar periwinkle, respectively. Some other examples of drugs of plant origin are shown in Table 3.1.

"Ethnobotanical survey" is a term used by scholars who research on plants of a particular region and their practical uses and applications by the local people and cultures. It simply means the act of investigating the use of plants based on the traditional knowledge of primitive societies in a given locality. It encompasses the use of plants for food, clothing, shelter, and medicine. However, in modern science, the search for plant-based drugs/active ingredients to tackle some problems of synthetic drugs: decay, diseases and death, newly emerging infections, drug resistance, and quest for novel bioactive compounds, has prompted scientists from different fields of study to come up with such terms as ethnomedicinal survey (Amujoyegbe et al., 2016), ethnopharmacological survey (Baydoun et al., 2015), ethnobiological survey (Khoirul-Himmi et al., 2014), and ethno-veterinary survey (Dilshad et al., 2008) to suit their respective research interests while some others use simpler titles. Therefore, ethnobotanical surveys are conducted by botanists, pharmacists, medical personnel, and veterinary physicians to study and document the plant diversity (including lower plants and fungi)

**TABLE 3.1** Sample Drugs of Plant Origin.

| Drug | Therapeutic use | Source | Common name | Family | Reference |
|------|-----------------|--------|-------------|--------|-----------|
| Aspirin (salicylic acid) | Pain killer | *Salix* spp. | Willow tree | Salicaceae | Taylor (2000) |
| Atropine | Anticholinergic | *Atropa belladonna* | Belladonna/Deadly nightshade | Solanaceae | Kutama et al. (2015) |
| Caffeine | Stimulants | *Cola* spp. | Kola nut | Malvaceae | Nwafor et al. (2017) |
| Digitoxin | cardiotonic | *Digitalis purpurea* | Foxglove | Plantaginaceae | Kutama et al. (2015) |
| Emodin | Laxative | *Cassia* spp. | Senna | Fabaceae | Taylor (2000) |
| Ephedrine | Respiratory dilation | *Ephedra* spp. | Joint-pine/jointfir | Ephedraceae | Kutama et al. (2015) |
| Hyoscyamine | Anticholinergic | *Hyoscyamus niger* | Stinking nightshade | Solanaceae | Nwafor et al. (2017) |
| Morphine and Codeine | Pain killer | *Papaver somniferum* | Opium poppy | Solanaceae | Kutama et al. (2015) |
| Nicotine | Stimulants | *Cola* spp. | Kola nut | Malvaceae | Kutama et al. (2015) |
| Paclitaxel (taxol) | Anti-ovarian cancer | *Taxus brevifolia* | Pacific yew | Taxaceae | Kutama et al. (2015) |
| Quinine | Anti-malaria | *Cinchona officinalis* | Cinchona tree | Rubiaceae | Kutama et al. (2015) |
| Vinblastine and Vincristine | Anticancer | *Catharanthus roseus* | Madagascar periwinkle | Apocynaceae | Nwafor et al. (2017) |
| Glasiovine | Antidepressant | *Octea glaziovii* | Sweetwood | Lauraceae | Taylor (2000) |
| Gossypol | Male contraceptive | *Gossypium* spp. | Cotton | Malvaceae | Taylor (2000) |
| L-Dopa | Antiparkinsonism | *Mucuna pruriens* | Velvet bean | Fabaceae | Taylor (2000) |
| Deserpidine | Antihypertensive | *Rauvolfia* sp. | Snakeroot | Apocynaceae | Taylor (2000) |
| Theobromine | Vasodilator, diuretic | *Theobroma cacao* | Cocoa tree | Malvaceae | Taylor (2000) |
| Yohimbine | Aphrodisiac | *Pausinystalia yohimbe* | Yohimbe | Rubiaceae | Taylor (2000) |

and non-plant materials (including animals or animal products, minerals and nonliving matters) used by a group of people and culture for their basic needs: food, clothing, and construction and medicine for man and livestock (Khoirul-Himmi et al., 2014; Nwafor et al., 2018). This usually forms the basis of drug research and gives the researcher pre-information of the possible biological activities a plant may possess and best possible dosage form, mode of administration, and side effects of such herbal formulation (Chukwuma et al., 2015).

In most developing countries, the major challenges to the utilization of medicinal plants include the fact that those who are knowledgeable in the utilization of medicinal plants are growing old and senile, with the possibility of deaths, the non-documentation and proper teaching as well as threat of biodiversity loss through deforestation, and overexploitation due to urbanization and industrialization are among the major challenges facing developing countries (Sharma and Kumar, 2013; Nwafor and Ozioko, 2018). This situation is exacerbated by inadequate mentoring, sensitization, and lack of enforcement of relevant government policies.

## 3.2  TRADITIONAL MEDICINE (TM)

Traditional medicine (TM) is the totality of all the ingredients and practices of all sorts, whether empirical or otherwise, which from the beginning of time, has been used successfully by a people or culture to diagnose and cure illnesses and to alleviate suffering. The knowledge of TM is passed on from generation to generation, through verbal communication, observation, and practical experience. It remains a major healing method in Asian and African cultures and involves the use of the enormous wealth of plant resources or plant products, or sometimes non-plant materials (naturally occurring mineral substances or animal matters), solely or in combination, in different forms, and for prevention and treatment of diseases.

A traditional medical practitioner (TMP), sometimes called a traditional healer, is a person (man or woman) who is knowledgeable in medicinal plants and their various uses, and who is recognized by his people as competent and reliable in tackling issues related to healthcare (Kokwaro, 1996). He or She is usually a descendant of a traditional healer, who learned the practice from their forefathers or by divine calling. Interested "ordinary" persons including herb sellers and hunters who patronized or came in contact with a traditional healer, in one way or the other in the past, may also through learning and practice, assume the position of a TMP

(Iwu, 1993). Irrespective of the source of the healing ability, the sole aim of every traditional healer is to successfully diagnose the cause of illness and proffer solutions accordingly. Diagnosis can be through spiritual divination or by physical examination and visual observation of the patient (change in eye color, color of urine, excreta or vomited food substances, feverish conditions, and rashes on the body). A traditional healer can assume any of the following:

- **Diviner:** This is a traditional healer who uses the art of divinity and fortune-telling to diagnose the cause of illness and proffers solution (usually through spiritual means).
- **Herbalist:** A traditional healer who specializes in the use of herbs without any spiritual attachment is termed an herbalist.
- **Traditional Bonesetter:** A traditional healer who uses herbs and other non-herbal materials to heal fractured bones and muscles problems.
- **The Traditional Birth Attendant:** This is a traditional herbalist (usually a woman) who takes care of expectant mothers throughout the gestation period and the newborn child.
- **The Traditional Surgeon:** This is a traditional medical expert who uses sharp objects or special types of knives in performing surgical procedures. Their practices include circumcision of the newborns, excision, and issuing of tribal marks.

## 3.3  METHODOLOGIES AND APPROACHES TO ETHNOBOTANICAL STUDIES

The success of every ethnobotanical study depends on the initial design of the project and choosing the appropriate methodology: sampling, collection of voucher specimen, field documentation, and appropriate qualitative and quantitative procedures in data collection. The most commonly adopted approaches are highlighted below:

### 3.3.1  SOURCES OF ETHNOBOTANICAL INFORMATION

Depending on the scope and nature of study, a researcher has varieties of avenues where ethnobotanical study can be undertaken. For the reason of convenience, the most common avenues documented in literatures are given below:

### 3.3.1.1   HOUSEHOLD DWELLERS

This is the most common way through which ethnobotanists obtain ethno-botanical information. They often move from one household to another, interviewing the local dwellers, extracting from them as much knowledge of medicinal plants, and their uses as possible through the use of question-naires, single or group interviews.

### 3.3.1.2   SURVEY OF HERB MARKETS

Herbal markets are rich in ethnobotanical information in that they harbor wild diversity of plants and animals used in ethnomedicine and create active interaction between medicinal herbs and human cultures. Varieties of medicinal plants, wild and cultivated food plants, medicinal animal parts, and other products of biological origin are sold in local markets with their specific values. Therefore, market survey of medicinal plants is one the methods employed to study both the qualitative and quantitative aspects of ethnobotany. It is used to quantify certain aspects like the number of herb sellers (vendors) and herbal products sold, at what prices, and to what extent they are patronized. It is used to study the ethno-economical values of plant products as well as their availability, accessibility, and sustainability within a particular region (Tinitana et al., 2016).

### 3.3.1.3   SURVEY OF HERBAL CLINICS

A researcher may be interested to obtain knowledge of medicinal plants used in herbal clinics for the management and treatment of diseases, as well as, the traditional medicines practices and the patients' perspective of diseases, and herbal remedy. In this study, a researcher would have first-hand experience on how herbal remedies are prepared and administered, feedbacks from patients, proper understanding of the practices by the traditional healer, and so forth.

### 3.3.1.4   HERBARIA AND RESERVED AREAS

A herbarium is a repository of dried and preserved plant materials, arranged in a systematic manner, for teaching and research purposes. Many herbaria around the world do not just collect and preserve plants for the future taxo-nomic purpose but also document information on the use of collected plants by local people including their medicinal benefits. Most notable among them are herbaria associated with Medicinal Plants Research and Development Institutions and Academic Institutions. Such information gives the researcher an idea of the knowledge and use of medicinal plants, and other forms of traditional medicine practices in a given cultural group (Souza and Hawkins, 2017). Reserved areas such as Natural Parks, Game Reserves, Botanical Gardens, Shrines, and Grooms are also rich in ethnobotanical information.

### 3.3.1.5   INTERNET RESOURCES

The advent of modern technologies has made things easier for researchers from developing countries. Ethnobotanical information on the knowledge and successful use of medicinal plants to manage and treat diseases by indigenous people can also be obtained through surfing the internet. Internet search engines are used to search for published journal articles, e-books and documentaries of ethnobotanical surveys, and databases of medicinal plants. Survey of indigenous medicinal plants and their uses can also be conducted through websites, blogs, social media, and so forth.

### 3.3.2   METHODS OF OBTAINING ETHNOBOTANICAL INFORMATION

Obtaining ethnobotanical information majorly depend on interviewing the research respondents (informants) and systematic recording of the events. Depending on the chosen method certain research tools such as field notebooks, electronic voice recorder, geographical positioning system (GPS) device, digital camera, pre-constructed data sheets (questionnaires), and laptops or potable display accessories (PDAs) may be needed (Hoffman and Gallaher, 2007).

### 3.3.2.1   INDIVIDUAL INTERVIEW

This is where the researcher chooses to interrogate the individual respondents one at a time for ethnobotanical information. This method is often preferred by most researchers because it helps to minimize the "group effect" (where a respondent's answers and response affect those of the fellow respondents) witnessed in field studies (Hoffman and Gallaher, 2007).

### 3.3.2.2   GROUP INTERVIEW

This is conducted on a group of people with ethnobotanical knowledge with a view to obtaining their perspectives and uses of medicinal plants. They could be a group of traditional healers or just a group of ordinary villagers empanelled for this purpose (Hoffman and Gallaher, 2007).

### 3.3.2.3   USE OF OPEN-ENDED AND SEMI-STRUCTURED QUESTIONNAIRES

This is an interview that is guided by a preprepared outline of questions and/or hypothesis in form of a questionnaire, but always gives the researcher freedom for unforeseen circumstances or events. An open or semi-structured questionnaire may be issued to the respondents a few days prior to the day of interview or carried along by the researchers during the survey. A questionnaire guides the respondent as to what information required of about a particular plant species or a given disease condition. A sample ethnobotanical data form (questionnaire) is presented as Appendix.

### 3.3.2.4   FREE-LISTING

This is an event that records information about all the medicinal plants and their uses that a research respondent can give at a time. This technique helps to determine the order of importance, cultural use, popularity of a given plant specimen, or disease prevalence. This is because, by free listing, an informant tends to remember the more frequently used medicinal plants before the others and/or more prevalent disease condition in the community before the others (Hoffman and Gallaher, 2007).

Additional methods include direct observation, preference ranking (where the respondent is asked to list the items in the order of personal preference, perceived cultural importance, popularity, or any other criterion), triadic, pile sorting (grouping together of more similar items to create contrast with other item), and paired comparison (where items are presented to the respondent two at a time to tell which is more or which is less in use or efficiency). Detailed information on these analytical methods can be found in Hoffman and Gallaher (2007).

## 3.4    COLLECTION AND PRESERVATION OF MEDICINAL PLANTS

Every ethnobotanical study features collection and preservation of plant specimens encountered in the field survey. Therefore, it is important that the researcher goes with necessary collection tools including adequate protective wears and collection bags. For future research and reference purposes, collected plants are preserved (usually in form of voucher specimens using herbarium techniques) and stored in safe cabinets together with data sheets containing information on their usage. When on ethnobotanical survey a good number (not less than four) of plant samples should be collected to accommodate the vagaries of whether. This should include a twig with flowers and fruits where possible per sample as these characters are important for identification. However, specimens without the reproductive parts (flowers and fruits) can be collected but should be replaced as soon as fertile samples become available. Diseased or ant-infested plants parts should be avoided at all cost. For herbaceous plants, the whole plant is collected to show the underground parts (Carter et al., 2007; Inya-Agha et al, 2017).

For ethnomedicinal purposes, information regarding the local use of the plants is worth recording. Such information include the local name, time (maturity stage, and so forth.) of collection, therapeutic use (s), plant part (s) used, method of preparation, mode of administration, dosage, whether the plant is singly used or as polyherbal, and any other precaution/things to abstain from during preparation and usage. Efforts should be made to provide, as much as possible, every information in detail. For example, symptoms and diagnoses of ailments claimed to be treated by a particular crude drug should be properly narrated. Methods of preparation such as infusion and decoction as well as the solvent/medium of extraction–water, local gin, palm wine, lime, and so forth, have been reported in many ethnomedicinal studies (See Appendix). Inya-Agha et al (2017) noted the following:

Additional information worth noting include the locality, GPS location, habitat (including landform, slope, dominant plant species, vegetation, for example rainforest, savanna woodland, grassland, and so forth.), Soil type and geology, and also a descriptive information about the site (for example, whether the collection site was a disturbed site such as a roadside, mining site, construction site, industrial site, or burnt area).

Information about the individual plants is collected at the site, including height, form, bark type, and color (for trees and large shrubs), leaf texture, presence of rhizomes, presence and color of sap in cut stems, color of new growth, and flower color. Also, the relative abundance of the species is often recorded, particularly for threatened species.

Special caution should be taken when collecting toxic and poisonous plants to avoid inadvertent poisoning. Also, plants that contain controlled substances such as cannabinol, heroine among others, and plants that produce allergic reactions such as Mucuna pruriens should also be collected with care because of their social implications.

The next task after collection is preservation of the plant samples. The first step is drying which is often done with the help of a plant press (of wood or metal type). It is expected that every researcher embarking on ethnobotanical survey should have provision for proper drying of the collected plants sample to avoid defacement and aid further identification. However, where, for any reason, collection and drying are not convenient, the researcher should, at worst, have field photographs of the plants–stored in digital format–to serve the same purpose. Medicinal seeds may be dried and stored in airtight containers.

## 3.5   ANALYZING ETHNOBOTANICAL INFORMATION

Data obtained from a successful ethnobotanical survey remain useless until they are properly analyzed. Statistical approaches through relative cultural index (RCI) values are taken to test for hypothesis and compare significant difference. This can be done with simple statistical methods such as frequency tally, analysis of variance (ANOVA), and so forth, or through multivariate methods such as principal component analysis (PCA), discriminate function analysis, and so forth, to analysis complex relationships within certain RCI variables. Among the many RCI values studied are the use value (UV), informant consensus factor (ICF), and fidelity level (FL). More information on the statistical analysis of quantitative indices in ethnobotany is detailed by Hoffman and Gallaher (2007).

## 3.6   SURVEY OF MEDICINAL PLANTS USED BY THE INDIGENOUS PEOPLE OF NSUKKA, SOUTHEAST NIGERIA

### 3.6.1   STUDY AREA

Nsukka in a wider context represents a political zoning system called "Nsukka Senatorial Zone" in Enugu State that houses six local government areas including Igbo-Etiti, Igbo-Eze North, Igbo-Eze South, Nsukka, Udenu and Uzo-Uwani (Fig. 3.1). Its population was 1,377,001 as of

2007 with a vegetation characteristically derived savanna and patches of tropical rainforest. Relative humidity ranges from 70 to 80%. The wet season extends from March to October/November, the dry season is from November to March and the annual rainfall ranges from 1845 to 2000 mm. Annual temperature is between 25 and 27°C (Nwite and Obi, 2008). Nsukka falls on the Northern part of Igbo nation, therefore, its inhabitants communicate mainly in Igbo language and sometimes in English. The indigenous people of Nsukka are majorly Christians and Traditionalists; farmers, traders, and mid-class civil servants. They believe in traditional medicine practices and have been reported to handle several ailments with herbal remedies.

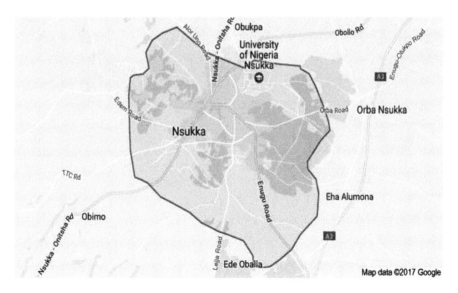

**FIGURE 3.1**    Map showing the study area.

**Source:** Adapted from Google Maps (2017).

## 3.6.2   *DATA COLLECTION AND ANALYSIS*

Ethnobotanical survey data were obtained through series of interview session on the indigenous people of Nsukka on their knowledge and use of medicinal plants and through survey of herbal markets and herbal clinics located within the study region. This was achieved through semi-structured questionnaires (see Appendix), informal personal and group interviews (as shown in Fig. 3.2 [Plates 1 and 2]), and field trips in randomly selected thirteen (13) communities including Alor Uno, Ede Oballa, Eha-Alumona, Obukpa, Enugu-Ezike, Orba, Nguru, Obolo-Afor, Obollo-Eke, Edem, Obollo-Etiti, Opi, and Nsukka. The respondents were properly sensitized on the purpose of the research and anyone who refused to partake in the study was dropped. The medicinal plants collected from the survey were properly identified using Hutchinson and Dalziel's "Flora of Tropical West Africa" and other relevant literatures. Data collected were analyzed on a Microsoft Excel spreadsheet and summarized using descriptive tables.

**PLATE 1** Field photograph with an herbalist residing at Obollo-Etiti, Nsukka Lead author, Felix Nwafor (left), herbalist (center), a plant taxonomist, and Mr. Alfred Ozioko (right).

**PLATE 2** Field photograph of Afor Obollo-Afor herb market. Herb seller (left), lead author, Felix Nwafor (center), and research assistant (right).

**FIGURE 3.2** Lead author in doing field work.

### *3.6.3 RESULTS AND DISCUSSION*

#### *3.6.3.1 DEMOGRAPHIC INFORMATION OF THE RESPONDENTS*

A total of 72 respondents including traditional healers, herb sellers, hunters, and other individuals who are knowledgeable on the use of medicinal plants were interviewed (Table 3.2). Among them were 48 men and 24 women, and fall between the ages of 30 and 75. Most of the respondents are herbalists and herb sellers who combine the occupation with farming while few others who obtained formal education are civil servants. This agrees with previous authors who recorded that the practice of traditional medicine is gender sensitive and mostly lies within the aged and uneducated (Ajibesin, 2012).

**TABLE 3.2** Some Demographic Information of the Respondents.

| Number of respondents | Age (years) | Sex | | Major occupation | | |
|---|---|---|---|---|---|---|
| | | Male | Female | Herbalist | Herb seller | Others |
| 14 | 30–40 | 10 | 4 | 4 | 8 | 2 |
| 16 | 41–50 | 9 | 7 | 8 | 6 | 2 |
| 14 | 51–60 | 8 | 6 | 7 | 4 | 3 |
| 17 | 61–70 | 10 | 7 | 10 | 4 | 3 |
| 11 | 70 & above | 11 | - | 8 | 1 | 2 |
| Total: 72 | | 48 | 24 | 37 | 23 | 12 |

## 3.6.3.2 PLANT DIVERSITY AND DOMESTICATION NEEDS

The list of the medicinal plant species encountered in the study, together with their families, local (Igbo) names, habits, sources, plant parts used, medicinal uses, method of preparation, and mode of administration is given in Table 3.3. A total of 93 species of medicinal plants belonging to 39 families were documented. The plant family with more members of medicinal species is the Fabaceae (17) and this was followed by the Asteraceae (7), Annonaceae (6), Apocynaceae (6) Euphorbiaceae (4), Malvaceae (4), Phyllanthaceae (4), and Rubiaceae (4). Among the plant species, *Ageratum conyzoides* was the most cited (42 times), and followed by *Uvaria chamae* (36), *Annona senegalensis* (34), *Nauclea latifolia* (30), *Aspilia africana* (29), and *Acanthus montanus* (24). The medicinal plants collected from different habitats, with 28 cultivated, while 65 are collected wild. The fact that just a few of these plants are cultivated means that the practice of domesticating medicinal plants has not been established in the area. Several studies on ethnobotanical survey also show that most local people collect their medicinal plants majorly from the wild rather than cultivated.

## 3.6.3.3 PLANT PARTS USED IN PREPARATION OF HERBAL REMEDY

The plant parts recorded to be used by the local Nsukka people in preparation of herbal remedies include leaves, aerial part, seed, root, stem bark, bulb, tuber, fruit, whole stem, and rhizome. Figure 3.3 shows that the most frequently used are the leaves (40.4%), followed by roots (18.2%), aerial parts (14.1%) and stem bark (11.1%), while whole plant (1%), tuber (1%), and rhizome (1%) were the least cited. Similar result was documented by

**TABLE 3.3** Data Obtained from the Ethnobotanical Survey of Medicinal Plants Used by the Indigenous People of Nsukka, Southeast Nigeria.

| Family/Botanical name | Local name | Habit | S | PU | MP | MA | Medicinal uses |
|---|---|---|---|---|---|---|---|
| ACANTHACEAE | | | | | | | |
| Acanthus montanus | Agamsoso and agamebu | Herb | W | L | D | O | Malaria, hepatitis, and typhoid |
| AMARATHACEAE | | | | | | | |
| Amaranthus spinosus | Inine ogwu | Herb | W | Ap | D | O | Skin diseases and malaria |
| AMARYLLIDACEAE | | | | | | | |
| Crinum jagus | Yabasi-ma | Herb | W | Bb | D | O | Respiratory infection, antipoison, and dysentery |
| Crinum ornatum | Yabasi ma | Herb | W | Bb | D | O | Respiratory infection and antipoison |
| ANACARDIACEAE | | | | | | | |
| Mangifera indica | Mangolo | Tree | W; C | Sb; L | D | O | Malaria, typhoid, STDs |
| Spondias mombin | Echikara | Tree | W; C | L | D | O | Fever, pains, infertility in women, and wound healing |
| ANNONACEAE | | | | | | | |
| Annona muricata | Showanshop | Shrub | C | L | D | O | Cancer and fever |
| Annona senegalensis | Uburu ocha | Shrub | W | L | D | O | Anti-malaria, STDs, and internal heat |
| Cleistopholis patens | Ashanda | Tree | W | Sb | D | O | Malaria, fever, and cough |
| Dennettia tripetala | Ose mmimmi | Tree | C | F | Ch | O | Cold and cough |
| Uvaria chamae | Mmimi ohia | Shrub | W | L | D | O | Fever, STDs, and malaria |
| Xylopia aethiopica | Uda | Shrub | C | F | D | O | Postnatal care and fever |
| APOCYNACEAE | | | | | | | |
| Alstonia boonei | Egbu | Tree | W | Sb | D | O | Rheumatism, body pains, and malaria |
| Landolphia dulcis | Akwari | Shrub | W | Ap; F | D | O | General healing |

**TABLE 3.3** *(Continued)*

| Family/Botanical name | Local name | Habit | S | PU | MP | MA | Medicinal uses |
|---|---|---|---|---|---|---|---|
| *Landolphia owariensis* | Utu | Shrub | W | L | D | O | General healing |
| *Rauvolfia vomitoria* | Akanta | Shrub | W | Ap | D | O | Body pains, HBP, and madness |
| *Strophanthus hispidus* | Anu mmi | Shrub | W | R; Sb | D | O | STDs and anti-malaria |
| *Voacanga Africana* | Nga rubber | Tree | W | R | D | O | STDs and Fibroid |
| ARACEAE | | | | | | | |
| *Elaeis guineensis* | Nkwu | Tree | W; C | F | I | O | Infertility in women |
| ARECACEAE | | | | | | | |
| *Anchomanes difformis* | Edemuo | Herb | W | Tb | P; D | T; O | Swellings and Malaria |
| ASCLEPIADACEAE | | | | | | | |
| *Cryptolepis sanguinolenta* | Akpa oku | Herb | W | R | D; I | O | Malaria, Stomach upset, and Aphrodisiac |
| *Gongronema latifolium* | Utazi | Herb | C | L | Ch; I | O | Cough and Diabetes |
| ASTERACEAE | | | | | | | |
| *Ageratum conyzoides* | Agadi isiocha | Herb | W | L; Ap | I | O; T | Wound healing, Ulcer, and Hemorrhage |
| *Aspilia Africana* | Aramjila | Herb | W | L | Ch; I | O; T | Stomach upset, diarrhea, and Wound healing |
| *Chromolaena odorata* | Obiara ohuru | Herb | W | L | I | T; O | Wound healing, ulcer, and diabetes |
| *Emilia praetermissa* | Nti oke | Herb | W | Ap | I | T | Wound healing |
| *Emilia sonchifolia* | Nti oke | Herb | W | L | I | T | Wound healing |
| *Vernonia amygdalina* | Onugbu | Shrub | C | L | I | O | Diabetes and Pile |
| *Vernonia migeodii* | Onubu egu | Herb | W | Ap | D | O | Anti-malaria |
| BIGNONIACEAE | | | | | | | |
| *Newbouldia laevis* | Ogirishi | Tree | W; C | R | D | O | Diabetes |

**TABLE 3.3** *(Continued)*

| Family/Botanical name | Local name | Habit | S | PU | MP | MA | Medicinal uses |
|---|---|---|---|---|---|---|---|
| **CAPPARIDACEAE** | | | | | | | |
| *Buchholzia coriacea* | Wonderful kola | Tree | W; C | S | D | O | STDs and Fibroid |
| **CERCROPIODACEAE** | | | | | | | |
| *Myrianthus arboreus* | Ujuju | Tree | W | Sb | D | O | STDs |
| **COMBRETACEAE** | | | | | | | |
| *Terminalia catappa* | Furut | Tree | W | L | D | O | Diabetes |
| **COMMELINACEAE** | | | | | | | |
| *Commelina Africana* | Mbo agu | Herb | W | Ap | I | O | Infertility in women |
| **CONNARACEAE** | | | | | | | |
| *Byrsocarpus coccineus* | Inri abusi | Shrub | W | R | Ch | O | Antipoison, snake bite, and STDs |
| *Cnestis ferruginea* | Amu nkita | Shrub | W | R | D | O | STDs, laxative, and pile |
| **CUCURBITACEAE** | | | | | | | |
| *Momordica charantia* | Akpana udene | Herb | W | Wp | D | O | Diabetes |
| **EUPHORBIACEAE** | | | | | | | |
| *Euphorbia hirta* | | Herb | W | Ap | I | T | Skin diseases |
| *Hymenocardia acida* | Ago-ozalla | Shrub | W | L | D | O | Pain killer |
| *Ricinodendron heudelotii* | Okwe | Tree | W | Sb | D | O | Miscarriage |
| *Ricinus communis* | Ugba | Shrub | C | L | D | T | Skin rashes |
| **FABACEAE** | | | | | | | |
| *Abrus precatorius* | Anya nnunu | Herb | W; C | Ap | Ch | O | Stomach upset |
| *Albizia adianthifolia* | Ngwu | Tree | W | R | D | O | Rheumatism, STDs, and arthritis |

**TABLE 3.3** (Continued)

| Family/Botanical name | Local name | Habit | S | PU | MP | MA | Medicinal uses |
|---|---|---|---|---|---|---|---|
| *Albizia zygia* | Ngwu | Tree | W | Sb | D | O | Malaria and diabetes |
| *Carica papaya* | Okwuru ezi | Tree | C | L; S | D | O | Malaria and male fertility |
| *Cassia siamea* | Ojima | Tree | W; C | L | D | O | Typhoid and malaria |
| *Cassia sieberiana* | Ojima | Tree | W; C | R | D | O | Typhoid and STDs |
| *Desmodium ramosissimum* | - | Herb | W | L | I | O | Diarrhea |
| *Desmodium velutinum* | Ikeagwuani | Shrub | W | R | I | O | Diarrhea |
| *Dialium guineense* | Nnu nwa egu | Tree | W; C | L | D | O | STDs and wound healing |
| *Erythrina senegalensis* | Echichi | Tree | W; C | R | Ch | O | Cough |
| *Millettia aboensis* | Mkpukpumanya | Tree | W | R | D | O | Laxative and STDs |
| *Parkia biglobosa* | Ugba | Tree | W; C | Sb; L | D | O | Toothache and body pains |
| *Senna alata* | Edema | Shrub | W | L | I | O | Skin infections |
| *Senna tora* | Nsigbu muo | Herb | W | Ap | D | O | Typhoid |
| *Tetrapleura tetraptera* | Isha-isha and ihiorihio | Tree | W | F | D | O | Postnatal care and fever |
| *Vigna sesquipedalis* | Akidi | Herb | C | S | D | O | Infertility in women |
| *Senna occidentalis* | Nsigbu muo | Herb | W | Ap | D | O | Typhoid |
| HYPERICACEAE | | | | | | | |
| *Harungana madagascariensis* | Oturu | Shrub | W | L; Sb; R | D | O | Toothache and dysentery |
| ICACINACEAE | | | | | | | |
| *Icacina trichantha* | Urubia | Shrub | W | Tb | D; I | O | Malaria and cough |

**TABLE 3.3** *(Continued)*

| Family/Botanical name | Local name | Habit | S | PU | MP | MA | Medicinal uses |
|---|---|---|---|---|---|---|---|
| **LAMIACEAE** | | | | | | | |
| *Hyptis suaveolens* | Ogwu anwu | Herb | W | Ap | D | O | Typhoid |
| *Ocimum gratissimum* | Ahinji | Herb | C | L | I | O | Diabetes |
| **LAURACEAE** | | | | | | | |
| *Cassytha filiformis* | Ogba na igorigo | Herb | W | Ap | D | O | Malaria and typhoid |
| **LILIACAEE** | | | | | | | |
| *Allium sativum* | Galiki | Herb | C | Bb | I; P | O; P | STDs, skin rashes, and infant problems |
| *Allium cepa* | Yabasi | Herb | C | Bb | P | T | Convulsion and eye problems |
| **LOGANIACEAE** | | | | | | | |
| *Anthocleista djalonensis* | Oto; Uvuru | Tree | W | R | D | O | Laxative, STDs, and malaria |
| **MALVACEAE** | | | | | | | |
| *Cola gigantean* | Ebenebe | Tree | W | L | I | O | Dysentery and diarrhoea |
| *Cola hispida* | Oji enyi | Shrub | W | L | I | O | Infertility in women |
| *Sida acuta* | Udo | Herb | W | L | I | O | Ulcer |
| *Sida linifolia* | Ire agwo | Herb | W | L | P | T | Swellings |
| **MELIACEAE** | | | | | | | |
| *Azadirachta indica* | Dogonyaro | Tree | W | L; Sb | D | O | Malaria and STDs |
| **MORACEAE** | | | | | | | |
| *Ficus exasperata* | Anwurinwa | Tree | W | L | D | T | Fungal infection |
| *Ficus thonningii* | Ogbu | Tree | W; C | L | Ch | O | To boost breast milk |
| **MYRTACEAE** | | | | | | | |
| *Psidium guajava* | Gova | Tree | C | L | D | O | Malaria |

**TABLE 3.3** *(Continued)*

| Family/Botanical name | Local name | Habit | S | PU | MP | MA | Medicinal uses |
|---|---|---|---|---|---|---|---|
| PHYLLANTHACEAE | | | | | | | |
| *Phyllanthus amarus* | Okwo nwa and enyi kwo nwa | Herb | W | Ap | D | O | Malaria and typhoid |
| *Phyllanthus discoideus* | Isi mkpi | Tree | W | R | D | O | STDs |
| *Phyllanthus muellerianus* | - | Shrub | W | L | D | O | Diabetes and malaria |
| *Phyllanthus niruri* | Okwo nwa | Herb | W | Ap | D | O | Malaria and typhoid |
| PIPERACEAE | | | | | | | |
| *Piper guineense* | Uziza | Herb | C | F | D | O | Postnatal care |
| POACEAE | | | | | | | |
| *Cymbopogon citratus* | Lemongrass | Herb | C | L | D | O | Malaria and typhoid |
| *Imperata cylindrical* | Ata | Herb | W | R | Ch | O | To boost breast milk |
| POLYGALACEAE | | | | | | | |
| *Securidaca longepedunculata* | Agha dibia and Ega ocha | Tree | W | R | D | O | Rheumatism, STDs, pains |
| POLYPODIACEAE | | | | | | | |
| *Platycerium bifurcatum* | Obu egbe | Herb | W | L | D | O | Malaria |
| PROTEACEAE | | | | | | | |
| *Protea madiensis* | Okwo okwo | Shrub | W | L | D | O | Malaria |
| RUBIACEAE | | | | | | | |
| *Fadogia cienkowski* | Ogwu agu | Herb | W | Ap | D | O | Malaria |
| *Morinda lucida* | Ogere | Tree | W | L; R | D | O | Malaria |
| *Nauclea diderrichii* | Uvuru | Tree | W | L | D | O | Malaria |

**TABLE 3.3** *(Continued)*

| Family/Botanical name | Local name | Habit | S | PU | MP | MA | Medicinal uses |
|---|---|---|---|---|---|---|---|
| *Nauclea latifolia* | Uvuru agu and Uburu ilu | Tree | W | L; R | D; I | O; T | Malaria; body pains; skin rashes |
| VERBENACEAE | | | | | | | |
| *Vitex doniana* | Uchakiri | Tree | W | Sb | D | O | STDs, internal heat, and malaria |
| VITACEAE | | | | | | | |
| *Leea guineense* | Odudu nwata | Shrub | W | L | I | T | Wound healing |
| ZINGIBERACEAE | | | | | | | |
| *Aframomum melegueta* | Ose oji and ose aya | Herb | C | S | D; P | O; T | Cough, Swelling, Pains, and Afrodisiac |
| *Aframomum daniellii* | - | Herb | W | S; L | C; P | O; T | Pains, STDs, rheumatism, cough |
| *Zingiber officinale* | Jinja | Herb | C | Rh | D; I | O | STDs, cough, and malaria |

S = source (w = wild, C = cultivated); PU = parts used (L = leaf, Ap = aerial part, Bb = bulb, Sb = stem bark, R = root, S = seed, Tb = tuber, Rh = rhizome, F = fruit, and Wp = whole plant); MP = method of preparation (D = decoction, I = infusion, and Ch = Chewing); and MA = mode of administration (O = oral and T = topical).

Hassan-Abdallah (2013) who studied the medicinal plants and their uses by the people in the Region of Randa, Djibouti and the high preference for leaves have been reported by previous authors (Balick and Cox, 1996; Giday et al., 2003), to be attributed to the relatively high abundance and efficacy. The next in the order of usage are roots and stem and this could be explained by the fact that most plants tend to shade their leaves as a result of unfavorable environmental condition (Hassan-Abdallah, 2013). However, frequent and unsustainable harvesting of roots and stem barks could lead to extinction of the concerned species (Abebe and Ayehu, 1993). On a similar note, woody plants (shrubs and trees) were most cited (59.14%) in the study compared with herbs (40.86%) and this is also in agreement with previous authors. This means they are more in abundance in the area and probably because of their availability at all seasons, as well as the fact that they are less affected by grazing animals and harsh environment (Hassan-Abdallah, 2013).

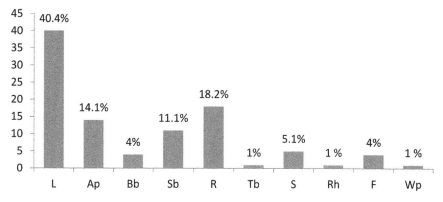

**FIGURE 3.3**   Distribution of plant parts used as encountered in the study.

### 3.6.3.4   METHODS OF PREPARATION AND MODES OF ADMINISTRATION

We were interested to know the various methods of preparing the herbal remedies and how they are administered to the patients. Where there was an alternative method to a standard method we also documented appropriately. For example, when preparing herbal remedy for children by soaking method, an herbalist suggested using ordinary water as choice solvent in lieu of alcohol. Due to the fact that most of them could not express the quantities of the materials to be used in metric systems as grams, kilograms, liters, and so forth, we used terms like "one handful," "two tablespoonfuls," and so forth. The methods of preparation of herbal remedies among the traditional healers were uniform in some ways and they include one or more of the following, boiling, soaking/infusion, pounding, and grinding.

### 3.6.3.5 DISEASE DIVERSITY

The indigenous people of Nsukka have been able to utilize the rich medicinal plant diversity found in their environment to successfully take care of their health related issues. Different categories of diseases were reported in the study (Table 3.3); the most common among them being malaria and sexually transmitted diseases (STDs), probably owing to the high prevalence of the diseases in the area. Others include inflammatory conditions, infertility issues in men and women, hepatitis, typhoid, infant problems, diarrhea, and dysentery among others. It was observed that a particular disease condition could be approached in different ways by the use of different plants, either as single-herb or polyherbal remedy, or simply, by following a different method of preparation.

The most interesting thing to note is that the herbal practitioners were all in agreement with their approaches and most of them gave descriptions that fitted accurately into the general diagnosis of the diseases. Furthermore, most of the plants reported in this study have also been recorded by previous researchers to be useful in the treatment of similar illnesses in other parts of the country (Ajibesin, 2012; Chukwuma et al., 2015).

### 3.6.3.6 CONSERVATION CONCERN

Sustainable utilization and conservation of medicinal plants remain a major concern in most local cultures in developing countries. Most traditional practitioners had mini-medicinal plant gardens within and around their homes where they nurse some medicinal plants, especially, those perceived to be rare species, special or seasonal plants. On personal interview with Mr. Alphonsus Eze of Obukpa community, he argued:

> As traditional practitioners we keep this gardens within our homes to grow and monitor these plants because, for example in dry season, everywhere becomes dry and most herbaceous plants die, so we water them to serve us as we receive patients in all season. If we don't do it this way, you will not be able to satisfy the needs of your patients.

He further stressed:

> There are special plants that you may have encountered in your neighbor's compound. You can ask for the seed or any way you can propagate it and of course you have to monitor it closely to avoid losing it.

Shrines and grooms that exist in the area are characterized by forest thickets rich in plant diversity (usually *Milicia excelsa, Bombax, Ceiba, Acacia*, and *Albizia* species) and similar to most local communities in developing countries. There exist certain taboos that prevent people from indiscriminate harvesting of medicinal herbs and roots from these forests. Though it can be argued that these laws were originally put in place to reverence the inhabiting deities, they have indirectly helped in conservation of the medicinal plants. Some of these notable deities are "*Adoru*" in Aro Uno and "*Kpazaa*" in Obukpa. According to another respondent, Mr. Alfred Ozioko of Orba community:

> If at all collection of medicinal plants must be carried out in such shrines and grooms, the consent of the Chief Priest must be sought, and often certain sacrifices are performed in case of uprooting or complete felling of trees.

Even though certain conservation strategies by the people were encountered in the study, there is still need for strong conservation strategies to be put in place. The local people still enter into any bush at will to harvest plants and sometimes on attempt to get the roots; they succeed in completely destroying the plant. This was confirmed by the experience of the survey of the herbal markets where it was observed that medicinal roots are the most sold (Plate 2). Government policies, if at all they exist in this part of the country, are not properly enforced to curb the situation and if urgent necessary actions are not taken, medicinal plant species will continue to disappear from the surface of the earth. This agrees with the reports of Chukwuma et al. who stressed the needs for urgent intervention by the government in most states in Nigeria where unsustainable harvesting of medicinal plants has become the norm (2015).

## 3.7   CONCLUSION

The study of the local uses of indigenous medicinal plants and traditional practices will continue especially as both individuals and institutions researching in the area of drug discovery have now taken a keen interest in this subject in most developing countries. The case study of the indigenous people of Nsukka, Southeast Nigeria has shown the local people have, right from time immemorial, been making use of the rich plant diversity available to them in management and treatment of diseases. The information obtained from this study, together with existing knowledge of medicinal plants, could contribute to new discovery of therapeutic agents of plant origin.

## 3.8   LIMITATIONS

The researchers encountered a few limitations in the study, which included participants being reluctant to give certain information until stipends were offered to them. In some cases, personal information about them was not revealed to the researchers, otherwise there was full compliance by the informants.

## KEYWORDS

- **ethnobotanical study**
- **methodologies**
- **traditional medicine**
- **medicinal plants**
- **southeast Nigeria**

## REFERENCES

Abebe, D.; Ayehu, A. *Medicinal Plants and Enigmatic Health Practices of Northern Ethiopia.* B.S.P.E: Addis Ababa, Ethiopia, 1993; p 341.

Ajibesin, K. K. Ethnobotanical Survey of Plants Used for Skin Diseases and Related Ailments in Akwa Ibom State, Nigeria. *Ethnobot. Res. Appl.* **2012,** *10,* 463–522.

Amujoyegbe, O. O.; Idu, M.; Agbedahunsi, J. M.; Erhabor, J. O. Ethnomedicinal Survey of Medicinal Plants Used in the Management of Sickle Cell Disorder in Southern Nigeria. *J. Ethnopharmacol.* **2016,** *185,* 347–360.

Balick, M.; Cox, P. *Plants, Culture and People;* Scientific American: New York, 1996.

Baydoun, S.; Lamis, C.; Helena, D.; Nelly, A. Ethnopharmacological Survey of Medicinal Plants Used in Traditional Medicine by the Communities of Mount Hermon, Lebanon. *J. Ethnopharmacol.* **2015**, *173*, 139–156.

Carter, R.; Bryson, C. T.; Darbyshire, S. J. Preparation and Use of Voucher Specimens for Documenting Research in Weed Science. *Weed Tech.* **2007**, *21*, 1101–1108.

Chukwuma, E. C.; Soladoye, M. O.; Feyisola, R. T. Traditional Medicine and the Future of Medicinal Plants in Nigeria. *J. Med. Plants Stud.* **2015**, *3*(4), 23–29.

Dilshad, S. M. R.; Rehman, N.; Iqbal, Z.; Muhammad, G.; Iqba, A.; Ahmad, N. An Inventory of the Ethnoveterinary Practices for Reproductive Disorders in Cattle and Buffaloes, Sargodha District of Pakistan. *J. Ethnopharmacol.* **2008**, *117*, 393–402.

Giday, M.; Asfaw, Z.; Elmqvist, T.; Woldu, Z. An Ethnobotanical Study of Medicinal Plants Used by the Zay People in Ethiopia. *J. Ethnopharmacol.* **2003**, *85*, 43–52.

Hassan-Abdallah, A.; Merito, A.; Hassan, S.; Aboubaker, D.; Djama, M.; Asfawc, Z.; Kelbessa, E. Medicinal Plants and Their Uses by the People in the Region of Randa, Djibouti. *J. Ethnopharmacol.* **2013**, *148*, 701–713.

Hoffman, B.; Gallaher, T. Importance Indices in Ethnobotany. *Ethnobot. Res. Appl.* **2007**, *5*, 201–218.

Inya-Agha, S. I.; Nwafor, F. I.; Ezugwu, C. O.; Ezejiofor, M. Herbarium Techniques. In *Phytoevaluation: Herbarium Techniques and Phytotherapeutics*; Inya-Agha, S. I., Ezea, S. C., Nwafor, F. I., Eds.; Paschal Communication: Enugu State, 2017; pp 8–22.

Iwu, M. M. *Handbook of African Medicinal Plants*; CRC Press: London, 1993.

Khoirul-Himmi, S.; Alie-Humaedi, M.; Astutik, S. Ethnobiological Study of the Plants Used in the Healing Practices of an Indigenous People of Tau Taa Wana in Central Sulawesi. Indonesia *Procedia Environ. Sci.* **2014**, *20*, 841–846.

Kokwaro, J. O. Ethnobotanical Study of East African Medicinal Plants and Traditional Medicine. A symposium on *Science in Africa: Utilizing Africa's Genetic Affluence through Natural Products Research and Development. American Association for the Advancement of Science*; New York, 1996, pp 23–34.

Kutama, A. S.; Dangora, I. I.; Aisha, W.; Auyo, M. I.; Sharif, U.; Umma, M.; Hassan, K. Y. An Overview of Plant Resources and Their Economic Uses in Nigeria. *Global Adv. Res. J. Agric. Sci.* **2015**, *4*(2), 042–067.

Nwafor, F. I.; Chukwube, V. O.; Tchimene, M. K.; Chukwuma, M. O. Plant Resources and Ethnomedicine in Nigeria. In *Phytoevaluation: Herbarium Techniques and Phytotherapeutics;* Inya-Agha, S. I., Ezea, S. C., Nwafor, F. I., Eds.; Paschal Communication: Enugu State, 2017; pp 38–41.

Nwafor, F. I.; Ozioko, A. O. Igbo Indigenous Science: An Ethnobiologist Perspective. In *African Science Education: Gendering Indigenous Knowledge in Nigeria;* Abidogun, J., Ed.; Routledge London, 2018; pp 68–88.

Nwafor, F. I.; Tchimene, M. K.; Onyekere, P. F.; Nweze, N. O.; Orabueze, C. I. Ethnobiological Study of Traditional Medicine Practices for the Treatment of Chronic Leg Ulcer in Southeast Nigeria. *Indian J. Tradit. Knowle.* **2018**, *17*(1), 34–42.

Nwite, J. N.; Obi, M. E. Quantifying the Productivity of Selected Soils in Nsukka and Abakaliki, Southeastern Nigeria Using Productivity Index. *Agro-Sc.* **2008**, *7*(3), 170–178.

Okonkwo, E. E. Traditional Healing Systems among Nsudkka Igbo. *J. Tour. Herit. Stud.* **2012**, *1*(1), 69–81.

Okwu, D. E. Phytochemicals and Vitamin Content of Indigenous Spices of South Eastern Nigeria. *J. Sustain. Agric. Environ.* **2004**, *6*, 30–34.

Phillipson, J. D. Phytochemistry and Medicinal Plants. *Phytochem.* **2001**, *56*, 237–243.

Sharma, M.; Kumar, A. Ethnobotanical Uses of Medicinal Plants: a Review. *Int. J. Life Sci. Pharm. Res.* **2013**, *3*(2), 52–57.

Sofowora, A. Medicinal Plants and Traditional Medicine in Africa, 3rd ed.; Spectrum Books Limited: Ibadan, 2008.

Souza, E. N. F.; Hawkins, J. A. Comparison of Herbarium Label Data and Published Medicinal Use: Herbaria as an Underutilized Source of Ethnobotanical Information. *Econ. Bot.* **2017**, *20*(10), 1–12.

Taylor, L. Plant Based Drugs and Medicines. *Raintree Nutrition Inc.;* Carson City NV, 2000; Available: www.rain-tree.com/plantdrugs.htm. Accessed November 16th, 2017.

Tinitana, F.; Rios, M.; Romero-Benavides, J. C.; Rot, M. C.; Pardo-de-Santayana, M. Medicinal Plants Sold at Traditional Markets in Southern Ecuador. *J. Ethnobiol. Ethnomed.* **2016**, *12*, 29.

# CHAPTER 4

# HERBAL MEDICINE: A CASE STUDY OF NIGERIAN MEDICINAL PLANTS

TEMITOPE A. OYEDEPO

*Nutritional/Toxicology Unit of the Department of Biochemistry, Adeleke University, Ede, Nigeria. Tel.: +2348035838269*

*E-mail: topkay99@gmail.com; topeoyedepo@adelekeuniversity.edu.ng
ORCID: https://orcid.org/0000-0002-6593-1135.*

## ABSTRACT

There have been many scientific investigations that have established the importance and the contribution of many plant families used as medicinal plants. The medicinal values of these plants are due to some phytochemicals that create a distinct effect on the normal functioning of a living organism. However, as much as the natural medicine is generating a lot of attention from both practical and scientific viewpoints, the mechanism of action of local herbal medicines and related products from nature is actually multifaceted than a mechanistic explanation of a single bioactive factor. Since many of these traditional herbs contribute significantly to the development of new drugs, it is necessary to identify the active compounds and standardize such bioactive extracts. The identified bioactive extract should equally go through the safety studies. This chapter presents a review that deals with the impact and present scenario of phytomedicine in the society with the purpose of highlighting the importance of phytochemistry knowledge in herbal formulations and regulations.

## 4.1 INTRODUCTION

Herbal medicine is a major therapy in the traditional system of medicine which has been used for thousands of years. Medicinal plants have significantly contributed to the medical care of about 80% of Africans as phytomedicine are largely used in the treatment of diseases. The use of the indigenous plant is still in existence today because of their numerous biomedical advantages as well as position in cultural beliefs in many parts of the world. Therefore, herbal medicine has made a significant contribution towards maintaining human health (Okigbo and Mmeka, 2006).

Incidentally, a good number of these indigenous medicinal plants are consumed as food and many are used as food spices. The medicinal values of these plants are due to some phytochemicals that create a distinct effect on the normal functioning of a living organism. Some of the important bioactive constituents of these medicinal plants include alkaloids, tannins, flavonoids, and phenolic compounds (Okwu, 1999; Okwu, 2001). Meanwhile, many methodical scientific researches carried out on these medicinal plants have resulted in the identification of a growing number of active constituents (Chang et al., 2013).

The belief that herbal medicine has no side effect and is not toxic like allopathic medicines has contributed immensely to its popularity. Consequently, the number of natural product markets has increased tremendously and interest in traditional systems of medicine is increasing as well (Agarwal, 2005).

Knowledge about indigenous herbal medicines which are used for the treatment of different diseases in Nigeria is based on the experience passed from one generation to another. Basically, the knowledge transfer is done by oral tradition and through cultural practices which are mainly part of the indigenous knowledge of people in different localities (Sofowora, 1993). A good number of these herbal preparations are well-known by traditional medical practitioners (TMPs), but the elderly members of families in the rural areas also have a good grasp of these practices. However, for financial reasons, the herbal knowledge and practices are jealously guarded by traditional healers with extreme secrecy. According to Obute (2007), many of the traditional herbal practitioners also intentionally conceal the identity of plants used for different ailments so that the patient will continue to depend on them for the herbs. Therefore, in order to confound their customers, TMPs will not encourage the cultivation of the medicinal plants. As a matter of fact, the TMPs virtually collect the medicinal plants used in herbal preparations from the wild. Nevertheless, some elders of the rural societies do readily transfer this knowledge to interested people sometimes on payment of an inducement fee. This has somehow helped in the propagation of herbal medicine knowledge in Nigeria.

## 4.2 PHYTOCHEMICAL STUDIES OF SOME MEDICINAL PLANTS IN NIGERIA

Nigeria is the richest country in West Africa with regard to the medicinal plant resources. A great number of the medicinal plants flourish in the country's flora (Gbile, 1986). This is due to the fact that the country exhibits dynamism in climate and topology which has effects on its vegetation and floristic composition. Reports on various herbs that are used to manage common ailments in Nigeria are scattered in many studies which are discussed in this chapter (Obute, 2007; Okoli et al. 2007; Mensah et al., 2008).

A previous study by Odebiyi and Sofowora (1978) examined a total of 47 plant extracts representing 132 genera and 172 species of plants distributed over 59 families from different parts of Nigeria. Extracts of the plants were screened for the presence of alkaloids, anthraquinones, phlobatannins, saponins, and tannins. The number of positive tests obtained was 32.18% for alkaloids, 44.24% for saponins, and 79.52 for tannins. Phlobatannins and anthraquinones were also confirmed to be present in a few of the plants.

Edeoga et al. (2005) investigated the fundamental scientific basis for the use of some Nigeria medicinal plants by defining and quantifying the phytochemical constituents of each plant. The medicinal plants investigated were *Cleome rutidosperma*, *Emilia coccinea*, *Euphorbia heterophylla*, *Physalis angulata*, *Richardia brasiliensis*, *Scoparia dulcis*, *Sida acuta*, *Spigelia anthelmia*, *Stachytarpheta cayennensis*, and *Tridax procumbens*. Alkaloids, flavonoids, and tannins were present in all the plants with the exception of *S. acuta* and *S. cayennensis* which lacked tannins and flavonoids, respectively.

Mensah et al. (2009) examined the plants which were identified by various herbalists in Esanland, Nigeria for treating hypertension. Results of the study revealed that alkaloids, cardiac glycosides, flavonoids, inulins, saponins, and tannins were present in the plants considered. As a matter of fact, cardiac glycosides were detected in all the species studied. However, alkaloids were absent in *Allium cepa, Allium sativum*, *Ocimum gratissimum*, and *Persea americana*. Saponins are equally present in all but absent in the seeds of *Sceloporus accidentalis*. Out of all the plants studied, only garlic (*A. sativum*) does not contain tannin. Flavonoids were not found in the leaves of climbing pepper (*Piper guineensis*), pawpaw (*Carica papaya*), pear leaves (*P. americana*), scent leaves (*O. gratissimum*), and shining bush plant (*Peperomia pellucida*).

*Curculigo pilosa* is a medicinal plant that is commonly used in southwestern Nigeria for herbal preparations as a purgative and also for the management and treatment of gonorrhea, hernia, and infertility. *C. pilosa* is rich in phytochemicals like alkaloids, cardenolides, saponins, tannins, and traces of anthraquinones. A study carried out by Sofidiya et al. (2011) indicated that the 2,2-diphenyl-1-picrylhydrazyl (DPPH) and ferric reducing antioxidant power (FRAP) activities in *C. pilosa* were higher than what is obtained in butylated hydroxytoluene (BHT) and rutin. Ethanol extracts (500 mg/mL) of *C.pilosa* and undiluted essential oil demonstrated antican-didal activity while the water extract (1000 mg/mL) was inactive against isolates (Gbadamosi and Egunyomi, 2010). The minimum inhibitory concentration (MIC) exhibited by the ethanol extract against the tested isolates ranged between 0.020 and 1.500 mg/mL. This is an indication that isolation and identification of the bioactive components of *C. pilosa* can lead to the production of phytomedicine against candidal activities (Gbadamosi and Egunyomi, 2010).

Okwu and Ukanwa (2010) isolated and elucidated the structure from leaf ethanolic extract of *Alchornea cordifolia*. Arg, 5-methyl 4'-propenoxy anthocyanidines 7-O-β-D-diglucopyranoside was isolated in the study and the structure was identified using NMR spectroscopy in combination with IR and MS spectral data. Antibacterial studies of the isolated compound indicated that it can successfully inhibit *Escherichia coli*, *Klebsiella pneumoniae, Proteus mirabilis*, *Pseudomonas aeruginosa*, and *Staphylococcus aureus*. This result proves that the plant can be used in phytomedicine for disease prevention and treatment of infections.

Olanipekun et al. (2013) also studied the following medicinal plants: *Aframomum melegueta, Chromolaena odorata, Cissampelos owariensis, Pergularia daemia, Parquetina nigrescens, Ocimum bascilicum, O. gratissimum*, Tithonia *diversifolia, Vernonia amygdalina*, and *Zingiber officinale* used for curing different ailments in Ekiti State, Nigeria. The bioactive components that were identified in these plants are alkaloids, cardiac glycosides, flavonoids, phlobatannins, saponin, tannins, and terpenoids.

## 4.3   TRADITIONAL PLANTS OF MEDICINAL IMPORTANCE

Many indigenous West African plants have been employed as local remedies for various human ailments in the traditional medicine of the region and as nutritional sources for countless generations. Research has been carried out worldwide by numerous scientists to verify their effectiveness and results of such studies have led to the production of medicines. However, isolation and characterization of the chemical structures for phytochemicals from valuable indigenous plants have only been undertaken for a minimal fraction.

This is basically because the modern techniques of chromatography and mass spectroscopy have only been exploited by African scientists recently. The importance of this development has greatly enhanced the understanding the mechanisms of action for those isolated compounds as pharmacological agents. A lot of studies are still ongoing in this area.

## 4.3.1 HYPERTENSION

Unlike before, the rate of hypertension and its related complications is rapidly increasing in Nigeria and Africa (Addo et al., 2007). Nigerians have developed indigenous and traditional healing practices to manage endemic diseases like hypertension in their local communities. Herbal medicine has been found to have a significant impact on the management of hypertension in Africa (Hendriks, 2012). The choice of plants that are used in the management of hypertension varies among the various cultural groups and is a factor of the flora available in each locality. Nevertheless, the leaf is the part of a plant that has attracted research interest in the various species that have been studied so far. Most of the plants that are used for taking care of hypertension in Nigeria have general distribution and usage, but some are limited to specific localities.

Edible vegetables like bitter leaf, garlic, waterleaf, and scent leaf are also used for treating hypertension in Nigeria. Out of all medicinal plants employed by traditional medicine practitioners, over 150 species are prescribed as food (Odugbemi, 2006). However, in the study carried out by Mensah et al. (2009), the local herbalists disclosed that they do not usually administer the vegetable plants to their patients because they are slow in action. Therefore, it is only patients whose case is mild that are placed on such vegetables. They treat serious conditions of hypertension with more potent herbs (avocado pear leaves, bitter gourd, mistletoe, and pawpaw). This may be administered as a single component or in combination with other herbs.

Herbal homes in Nigeria have the tradition of using many of the herbal preparations for the treatment of multiple ailments. This could be because they have different medicinal properties. Anslem (2006) and Okoli et al. (2007) have reported the efficacy of avocado leaves, bitter gourd, garlic, mistletoe leaves, pawpaw leaves, and pear leaves, for managing hypertension in Nigeria. Regular use of plants such as *A. sativum, A. cepa, C. papaya, Euphorbia hirta, O. gratissimum, Loranthus spectobulus, P. americana, P. pellucida, Psidium guajava, Phrynobatrachus guineensis, Rauwolfia vomitoria*, *Talinum triangulare*, *S. occidentalis*, and *V. amygdalina* in the management of hypertension and other disease condition has also been reported in other parts of Nigeria (Sofowora, and Odebiyi, 1987; Addae-Mensah and Musanga, 1989).

The studies have shown that these medicinal plants can lower heart rates (HRs), diastolic and systolic blood pressures. Furthermore, some of the medicinal plants have been demonstrated to reverse or improve deranged cardiovascular parameters particularly raised blood pressure and other complications associated with these diseases (Nwanjo, 2005; Taiwo et al., 2010). Some of the various herbal preparations from medicinal plants which have been documented for managing hypertension locally in Nigeria are presented below:

a) **African Mistletoe (*Loranthus micranthus*)**: The plant is a semi-parasitic shrub which is found growing on several tree crops in Nigeria. A study by Obatomi et al. (1996) investigated the effect of aqueous extract of *L. micranthus* in normotensive and hypertensive rats. Results of the study indicated that mean arterial pressure (MAP) was lowered in both normotensive and spontaneous hypertensive rats. The mechanism of antihypertensive action of this plant was reported in a study by Iwalokun et al. (2011). The findings of their research indicated that the antihypertensive activity of *L. micranthus* involves antiatherogenic events, vasorelaxation, cardiac arginase reduction, and nitric oxide elevation (Iwalokun et al., 2011).

b) **Avocado (*P. americana*)**: The leaves of the plant are cut into small pieces. It is thereafter dried and made into tea. This is to be taken by the hypertensive patient. Another preparation involves cutting the cotyledons into pieces after which they are dried and ground into powder. A spoonful of this powder in 200 mL hot water is taken after meals for a relief of hypertension (Odugbemi, 2006). The antihypertensive activity of the aqueous seed extract was investigated by Anaka et al. (2009). The effect of the extracts on MAP and HR of hypertensive and naive rats was studied at doses of 240, 260, and 280 mg/kg. The effects of the extracts on MAP and HR were similar to what is obtained with standard drugs.

c) **Bitter Leaf (*V. amygdalina*)**: This is simply by chewing and swallowing the fresh leaves. Alternatively, the fresh leaves are grounded and stirred in water. The liquid is taken to manage hypertension (Okoli et al., 2007). Aqueous extract of *V. amygdalina* was investigated for its cardiovascular effects in normotensive Sprague-Dawley rats by Taiwo et al. (2010). The interpretation of this is that direct vasorelaxant could be the mechanism for the antihypertensive effect of the plant.

d) **Garlic (*A. sativum*)**: Bulbs of garlic are ground and mixed with honey. This will be taken orally by the patient. A research was carried out by Nwokocha et al. (2011) on the cardiovascular effect of garlic on normotensive and two-kidney one-clip (2K1C)-induced hypertensive rats. The report of the study indicated that intravenous injection of *A. sativum* at a dose of 5–20 mg/kg caused a dose-dependent reduction in MAP and HR in both normotensive and 2K1C rats, with more significant effects observed in normotensive rats. According to the authors, no significant difference was observed on the hypotensive and the negative chronotropic actions of the extract when the rats were initially treated with atropine sulfate (2 mg/kg, i.v.). The authors inferred from the study that *A. sativum* caused hypotension and bradycardia which did not involve the cholinergic pathway in both normotensive and 2K1C rats (Nwokocha et al., 2011). The study further recommended that the mechanism of action of *A. sativum* may involve a peripheral mechanism for hypotension (Nwokocha et al., 2011).

e) These research findings were corroborated by Ashraf et al. (2013) in a clinical trial involving patients with stage 1 essential hypertension. The study reported a dose- and a duration-dependent decrease in systolic and diastolic blood pressure of these patients compared to patients that received standard drug (atenolol) or placebo.

f) **Guava (*P. guajava*)**: The leaves are usually soaked in salt water, washed and then squeezed. The leaf of this plant is used traditionally for its hypotensive effect. The product is prepared with fresh water to give a greenish liquid. The usual dosage is to take one glass two times daily for one week for the purpose of increasing blood level and protect the heart from cardiac arrest (Odugbemi, 2006). Ojewole (2005) carried out a research to investigate the hypotensive effects of *P. guajava* leaf aqueous extract (50–800 mg/kg) in Dahl salt-induced hypertension model in rats. Results of the study indicated a dose-dependent decrease in systemic arterial blood pressures and HRs of hypertensive rats when compared to control rats. The numerous phytochemical other chemical compounds present in the plant may be accountable for the hypoglycemic and hypotensive effects exhibited by the plant's leaf extracts (Ojewole, 2005).

g) **Irish Petticoat (*Acalypha wilkesiana Hoffmannii*)**: The leaves of this plant are eaten as vegetables in the traditional management of hypertension because it is a diuretic plant (Omage and Azeke, 2014). A study by Nworgu et al. (2011) evaluated the cardiovascular effect of the plant leaves. The extract decreased diastolic, systolic, and mean arterial blood pressure at a dose of 20 mg/kg. The extract also reduced the rate and force of contraction of isolated rabbit heart at a dose of 10 mg/mL. The authors suggested that the extract of *A. wilkesiana* Hoffmannii leaves exerted hypotensive actions mainly through inhibition of the force and rate of contraction of the heart (Nworgu et al. 2011).

h) **Miracle Leaf (*Bryophyllum pinnatum*)**: Leaf extracts of this plant is commonly used for traditional management of hypertension in South-Western Nigeria. A study by Ojewole (2002) showed that the plant extracts have a remarkable hypotensive effect. The extracts at doses of 50–800 mg/kg produced a dose-dependent reduction in arterial blood pressures and HRs of normotensive and hypertensive rats. Furthermore, at concentrations of 0.25–5.0 mg/mL, the extracts also produced dose-dependent negative inotropic and chronotropic activities on isolated guinea-pig atria. The extracts equally inhibited contractions stimulated by electrical field stimulation (ES)-provoked and contractions of isolated thoracic aortic strips induced by potassium and receptor-mediated agonist drugs in a nonspecific manner. The mechanism for the hypotensive action of the plant is yet to be elucidated.

i) **Roselle (*H. sabdariffa Linn.*)**: Flowers of the plant is brewed up as a local beverage in Nigeria. Local belief across the country is that the plant has several medicinal effects on many ailments including hypertension. The hypotensive effect of the aqueous extracts of *H. sabdariffa* flowers was investigated in anesthetized rats by Adegunloye et al. (1996). A dose-dependent reduction in MAP was observed in rats administered with the extract. However, the hypotensive effect of the extract was inhibited by atropine, cimetidine, and promethazine. It was also observed that the extract failed to inhibit the induction of hypertension by bilateral carotid occlusion. According to the study, cumulative doses of the extract in isolated aortic rings precontracted with noradrenaline created dose-dependent relaxation of the rings. The authors also documented the fact that the mechanism of antihypertensive action of the *H. sadariffa* calyces was not mediated through inhibition of the sympathetic nervous system. Nevertheless, the involvement of acetylcholine-like and histamine-like mechanisms as well as direct vasorelaxant effects was suggested (Adegunloye et al., 1996).

In the same vein, Mojiminiyi et al. (2007) documented a dose-dependent decrease in the blood pressure and HR of hypertensive and normotensive rats following intravenous injection of *H. sabdariffa* (1–125 mg/kg). Other studies by Onyenekwe et al. (1999) and Bako et al. (2010) further corroborated the findings by other researchers on the ability of the extract to significantly lower systolic and diastolic blood pressure in hypertensive and normotensive rats.

The mechanism(s) of the hypotensive action of the crude methanol extract of the flowers of *H. sabdariffa* was investigated on vascular reactivity in isolated aorta from hypertensive rats by Ajay et al. (2007). The research findings indicated that the relaxant effect of the extract to a certain extent depended on the existence of a functional endothelium while the antihypertensive effect was due to vasodilator action in the isolated aortic rings of hypertensive rats. The effects observed were suggested to be mediated through the endothelium-derived nitric oxide-cGMP-relaxant pathway in addition to inhibition of calcium ($Ca^{2+}$) influx into vascular smooth muscle cells (Ajay et al., 2007).

### 4.3.2  DIABETES MELLITUS (DM)

The World Health Organization (WHO, 1994) has strongly emphasized the balanced use of traditional and natural indigenous medicines for treating diabetes mellitus (DM). DM patients are often treated orally by traditional methods using different varieties of plant extracts (Ajgaonkar, 1979). More than 1200 plants species are used globally in diabetes phytotherapy and experimental studies confirm the hypoglycemic activity for many of these plants (Aladesanmi et al., 2007). Many researches have been carried out on indigenous medicinal plants used and recommended by TMPs for the treatment of DM. The researches often make use of animal models to determine antidiabetic properties of such plants. Almost all of these scientific investigations were done with chemically induced diabetic rats. A few of the research was also done with in vitro cell culture-based bioassays (van de Venter et al., 2008).

Apart from the ability of these plants to correct blood glucose levels, several plants with hypoglycemic activities also have potentials in ameliorating lipid metabolism abnormalities that usually accompanies DM (Coon and Ernst, 2003). A review of several research that has been done on these medicinal plant which showed that the polysaccharides, alkaloids, flavonoids, saponins, sterols, terpenoids, and amino acids and their derivatives are the most encountered bioactive principles that exhibited glycemic control in experimental animals (Bnouham et al., 2006).

An ethnobotanical survey was conducted by Negbenebor et al. (2017) with the purpose of identifying the plants which are used for the management of diabetes in Kano, Nigeria. The research findings presented 34 plant species which are often used traditionally to manage diabetes in Kano. Out of the 34 plant species that were mentioned, 21 species including *Acacia nilotica*, *A. cepa*, *A. sativum*, *Azadirachta indica*, *Ficus thinning*, *Guiera senegalensis*, *Mangifera indica*, *Manihot esculenta*, and *Syzygium guineense* have been documented in various publications about medicinal plants which are used for treating diabetes across the nations of the world.

A few of the medicinal plants which are used to prepare herbs for the treatment of diabetes and have been documented in the literature are discussed here:

a) *Moringa oleifera*: Oyedepo et al. (2013) carried out a study to assess the effect of aqueous leaf extract of *M. oleifera* (Moringaceae) on plasma glucose level in male albino rats. The study also investigated the effect of the extracts on total cholesterol (TC), triglycerides (TG), high-density lipoprotein (HDL), and low-density lipoprotein (LDL). Results of the study indicated that aqueous leaf extract of *M. oleifera* has significant hypoglycemic and antidiabetic potential coupled with the ability to reduce the plasma lipid imbalances associated with DM.

b) **Shaddock (*Citrus maxima*)**: Oyedepo and Babarinde (2013) investigated the effects of the plant's fruit juice on glucose tolerance as well as lipid profile in streptozotocin (STZ) induced diabetic rat. After 8 weeks of treatment, a considerable improvement of glucose, cholesterol, HDL, LDL, and TG, in the group treated with 50% *C. maxima* juice was reported.

c) **Onion (*A. cepa*) and Garlic (*A. sativum* L.)**: Several studies have documented the hypoglycemic effect of these two plants. These plants contained sulfur-containing compounds which may account for their medicinal activities. The antidiabetic bioactive components of *A. cepa* L. and *A. sativum* L. were documented to be S-methylcysteinesulfoxide (SMCS) and S-allylcysteinesulfoxide (SACS), respectively (Eidi et al., 2006). The studies also revealed that the antidiabetic property of SMCS and SACS is through stimulation of insulin secretion as well as by competing for insulin-inactivating sites in the liver. Results from the study also documented SMCS and SACS to increase biological synthesis of cholesterol from acetate in the liver. Consequently, allium products will have a little capability of protection against risk factors associated with DM.

d) **Neem (*A. indica*):** Patil et al. (2013) evaluated the alcoholic root bark extract of Neem in diabetes. Results of the study showed a significant antidiabetic effect at a dosage of 800 mg/kg. A similar result was documented by Dholi et al. (2011) as well as Joshi et al. (2011).

e) **West African Pepper (*Piper guineense*):** Morakinyo et al. (2016) examined the hypoglycemic and hypolipidemic effects of the plant's methanolic leaf extract in alloxan-induced diabetic rats. At concentrations of 200 and 400 mg/kg, the extract significantly reduced the TC, TG, and LDL while HDL increased when compared to the negative control group.

f) ***Gongronema latifolium*:** *G. latifolium* is a perennial edible shrub widely employed in Nigeria for various medicinal and nutritional purposes (Ugochukwu et al., 2005). Ethanolic leaf extract of this plant was established to have antihyperglycemic potency in a study conducted by Ugochukwu et al. (2005) which was corroborated by Iweala et al. (2013). The antihyperglycemic effect was considered to be mediated by the activation of glucose-6-phosphate dehydrogenase (G6PDH) hepatic hexokinase (HK), and phosphofructokinase (PFK) coupled with the inhibition of glucokinase (GK) activity in the liver (Ugochukwu et al., 2005). The combination of the plant with Selenium significantly reduced the weight and aspartate transaminase activity. In addition, the extract increased albumin and protein levels, Superoxide dismutase activity as well as glutathione-S-transferase activity (Iweala et al., 2013).

Many of the research studies on antidiabetic activities of indigenous medicinal plants from Nigeria present have indicated that these plants are of tremendous therapeutic and economic importance, which could go beyond local boundaries of healthcare systems and national markets. Unfortunately, one major limitation in advancing the goals of most research exercise is the general inability or failure of research groups to identify the hypoglycemic bioactive principles of the studied medicinal plants and reveal the mechanism of their therapeutic action.

### 4.3.3  ANTIMICROBIAL ACTIVITIES

Medicinal plants are rich sources through which antimicrobial agents may be obtained. Researchers have screened many of these plants used in traditional medical practices for their antimicrobial activities.

A study was carried out by Aladesanmi et al. (2007) on Nigerian plants which are suggested to have antimicrobial and antioxidant activities going by their ethnomedical uses. The antimicrobial activities of the plants were tested against *Bacillus subtilis*, *Candida albicans*, *Candida pseudotropicalis*, *E. coli* (NCTC 10418), *Pseudomonas aeruginosa*, *S. aureus*, and *Trichophyton rubrum*. *Trichilia heudelotii* leaf extract demonstrated both antibacterial and antifungal activities and it has the strongest antibacterial activity against all the strains tested. *Boerhavia diffusa*, *Markhamia tomentosa*, and *Trichilia heudelotii* leaf extracts inhibited *E.coli*, and *P. aeruginosa* strains. However, *Morchella tomentosa*, *T. heudelotii*, and *Sphenoceutrum jollyamum* root inhibited at least one of the fungi tested. The leaf extracts of *M. tomentosa* and *T. heudelotii* demonstrated reasonably strong antimicrobial activities in addition to impressive antioxidant activities. *Musa acuminata* which has the highest antioxidant activity, however, failed to show any antimicrobial activity. The antimicrobial activities were also found in the trichloromethane partition fractions of *T. heudelotii* and *M. tomentosa* leaves.

Similarly, Kubmarawa et al. (2007) in another study screened the ethanolic extracts of 50 plant species which are frequently used in traditional herbal practices in Nigeria for their antimicrobial activity against *B. subtilis*, *Candida albican*, *E. coli*, *P. aeruginosa* and *S. aureus*. From the research findings, 28 extracts out of the 50 plant extracts tested were able to inhibit the growth of one or more test pathogens. Interestingly, four plant extracts exhibited a broad spectrum of antimicrobial ability. Phytochemical studies of the extracts indicated the presence of tannins, saponins, alkaloids, glycosides, flavonoids, and essential oils.

Medicinal plants which are used for oral medications in Delta state of Nigeria were investigated for anti-candida activity by Erute and Egboduku (2013). Twenty-one medicinal plants which are traditionally used in oral medicine for treating infectious diseases were screened for anti-candida activity by using *C. albicans* and *Candida krusei*. Ethanol, methanol, chloroform, n-hexane, and water extracts of the plants were screened for antimicrobial activity using the paper disc diffusion method. Phytochemical study of the medicinal plants indicated significant concentrations of alkaloids, phlobatannins, terpenoids, flavonoids, and cardiac glycosides. Steroids and tannins were not present in considerable concentrations in all the plants that were screened. Nineteen (i.e., 90.48%) ethanol plant extracts were notably active against *C. albicans* and 80.95% of the plants were active with methanol extracts. The number of medicinal plants that showed significant inhibition with n-hexane, chloroform, and aqueous extracts for this test organism was 71.43% (i.e., 15 plants), 61.90% (i.e., 13 plants), and 33.33% (i.e., 7 plants), respectively. 85.71% of the plants presented considerable antimicrobial activity for ethanol and 80.95, 66.67, and 14.29% for methanol, n-hexane, and chloroform, respectively. There was no significant activity observed with the aqueous extract. According to the report, five of the screened plants recorded inhibition zones ≥ 19.00 mm, a value which is significantly greater than the standard antibiotics tested.

### 4.3.4  MALARIA

Malaria is a common ailment throughout Nigeria. Increased resistance of malaria parasites to the frequently used anti-malaria drugs in Nigeria has been reported (Ayoola et al., 2008). About 20% of malaria patients make use of indigenous herbal preparations to treat malaria in African countries (Willcox and Bodeker, 2004). Therefore, there is the need to intensify research in the area of development of new anti-malaria drugs, especially from medicinal plants.

A review of the medicinal plants used in the southwestern part of Nigeria for treating malaria revealed the rich flora diversity that exists in Nigeria (Odugbemi et al., 2007). According to Adebayo and Krettli (2011), about 98 species of plants are used in traditional medicine to treat malaria and/or fever. Most of these anti-malaria plants are used as monotherapy, and only a few plants are taken together in combined therapies. More than half of the Nigerian medicinal plants are also used for malaria treatment and control.

Ogbuehi and Ebong (2015) identified eleven plant species which are used to treat malaria in Onitsha, Nigeria. The identified plants are: *A. indica, Alstonia boonei, C. papaya, Cymbopogon citratus, M. indica, Morinda lucida, Nauclea latifolia, S. acuta, O. gratissimum, P. guajava*, and *V. Amygdalina*. Even though a good number of these plants have been studied there are still many that have not been scientifically investigated for anti-malaria properties. The traditional medicinal uses of some of the plants are discussed here:

a)  ***African Peach (N. latifolia*** Rubiaceae): The plant is used as an aqueous decoction of the root bark against malaria. The bark and roots of the plant contain more than 1% of an opioid that is clinically identical to tramadol, a standard analgesic.

b) **African Basil (*O. gratissimum*):** Aqueous extract of the leaves in combination with other medicinal plants is used for treating the fever that comes with malaria.

c) **Bitterwood Tree** (*Quassia amara*): The plant has a strong anti-malaria reputation for curative and preventive purposes in Nigeria. Its stem bark is used for treating fever and malaria.

d) **Brimstone Tree (*M. lucida*):** Decotions of aerial parts, root, bark, and leaves of the plant are widely used in Nigeria for treating different types of fever like malaria, yellow fever, and other feverish conditions.

e) **Cheese Wood (*A. boonei*):** The plant is reputed to be very useful where affordable anti-malaria drugs are found ineffective, due to drug-resistant malaria parasites. The bark or leaves are administered as decoction or tea. It is also used for malaria steam therapy.

f) **Fagara zanthoxyloides (Rutaceae):** The root of this plant is gener-ally used as chewing stick in Nigeria. Aqueous extract of the root is locally used for malaria treatment.

g) **Mango Tree (*M. indica*):** Aqueous extracts of the bark, leaves, and stems are used for herbal preparations which are used to treat malaria.

h) **Neem Tree (*A. indica*):** The plant is used in Nigeria as a decoction against fever and malaria. It is used in traditional medical practice in form of medicinal preparation made by boiling leaves, stem bark, and the root of the plant.

Ayoola et al. (2008) reported the antioxidant potentials of four medicinal plants which are used in southwestern Nigeria for the treatment of malaria. The plants studied were *C. papaya*, *P. guajava*, *V. amygdalina*, and *Momordica indica*. Phytochemical study of the plants indicated the presence of flavo-noids, terpenoids, saponins, tannins as well as reducing sugars. Cardiac glycosides and alkaloids were absent in *M. indica*. Similarly, alkaloids and anthraquinones were absent in *P. guajava*. Anthraquinones were similarly not present in *V. amygdalina*. All the plants exhibited potent inhibition of DPPH radical scavenging activity with *P. guajava* being the most potent. It is possible that the free radical scavenging (antioxidant) activities of these plants contributed to their effectiveness in malaria therapy.

Ogbonna et al. (2008) carried out a research on three plants which are locally used to treat malaria in the Southeastern part of Nigeria. The three herbs were obtained through a traditional herbalist who makes use of them in his practice. Root ethanolic extracts of *N. latifolia* and *Salacia nitida* as well as stem bark of *Enantia chlorantha* were examined for anti-malaria activity against chloroquine sensitive *Plasmodium berghei* using a 4-day suppressive test procedure. The extracts had intrinsic anti-malaria properties that were dose-dependent. A comparative analysis of the results indicated that 250 mg/kg body weight of *S. nitida* root produced 71.15% inhibition of parasitaemia and the 500 mg/kg body weight of *E. chlorantha* stem bark, *S. nitida* roots, *N. latifolia* roots, and a combination of the three herbs, produced 75.23, 73.28, 71.15, and 77.46%, respectively, when compared with chloroquine that has 71.15% suppression.

## 4.3.5 DIARRHEA

Nduche and Omosun (2016) documented different plant species that have been found useful for treating diarrhea in Nigeria. These include Neem (*A. indica)*, Miracle Leaf (*B. Pinnatum)*, Bitter kola (*Garcinia Kola)*, Mango tree (*M. indica*), and *Physcalis bransilensis*. Some of the plants used for diarrhea treatment whose bioactive components have been profiled in phytochemical studies include: pawpaw (*C. papaya*), stinging nettle (*Laportea aestuans)*, African basil (*O. gratissimum)*, African oil bean (*Pentaclethra macrophylla)*, long pepper (*Piper carniconnectivum)*, snake tree (*Stereospermum colais)*, and ginger (*Z. officinale).*

Similarly, Agunu et al. (2005) studied five medicinal plants (*A. nilotica, Acanthospermum hispidum, Gmelina arborea, Parkia biglobosa,* and *Vitex doniana*) used for treating diarrhea locally in Kaduna State, Nigeria. Diarrhea was induced in mice by castor oil. The aqueous methanol extracts of all the five medicinal plants investigated have pharmacological activity against diarrhea and this explains their usefulness in herbal medicine for diarrhea treatment. In the study, 100% protections were exhibited by extracts of *A. nilotica* (Gum Arabic tree) and *P. biglobosa* (African locus bean) at 100/200 mg/kg. Nevertheless, *V. doniana* (Black plum) showed a dose-dependent effect while *A. hispidum* (Bristly starbur) has the least anti-diarrheal activity among the five plants that were studied.

A survey of medicinal plants used in the treatment of animal diarrhea was carried out by Offiah et al. (2011) in Plateau State, Nigeria. The study documented 132 medicinal plants which are effective in the treatment of animal diarrhea. From the medicinal plants cited, the researchers were able to scientifically identify 57 plants (i.e., 43.18%) which were classified into 25 plant families. The families of Fabaceae (21%) and Combretaceae (14.04%) have the highest occurrence. The plant parts which are mostly used in anti-diarrhea herbal preparations according to the study are the leaves (43.86%) followed by the stem bark (29.82%). The herbal preparations are usually administered orally.

### 4.3.6 INFERTILITY

Soladoye et al. (2014) carried out a study to investigate plants used locally for treating female infertility in three states of Nigeria (i.e., Oyo, Ogun, and Osun state). Oral interview method was applied to aged couples, herbalists, and herb sellers. Responses from the oral interview were documented. Through the research, 75 plant species which belong to 41 different families of angiosperms were discovered to be of use in the traditional management of infertility. Research findings indicated that *A. sativum* L., *Citrullus colocynthis* (L.) Schrad, *Ficus exasperata* Vahl, and *Xylopia aethiopica* (Dunal) A. Rich are central in the different herbal preparations. This may be an indication of their importance in the treatment of infertility problems.

### 4.3.7 SNAKEBITE

Several thousand of Nigerians are victims of snakebites annually. Those who are mainly affected include farmers, herdsmen, and families living in rural areas. Mortality from snakebite is mainly due to poor healthcare services, difficulty in transporting the victims which will consequently result to delay in antivenom administration. Other factors which sometimes limit the use of antisnake venom include undesirable drug reactions and poor storage conditions. Therefore, it is important to study the medicinal plants which are used widely by TMPs in Nigeria.

There are four families of venomous snakes that are often found in Nigeria. These include Atractaspididae, Colubridae, Elapidae, and Viperidae families. However, three species; carpet viper (*Echis ocellatus*), black-necked spitting cobra (*Naja nigricollis*), and puff adder (*Bitis arietans*), belonging to the Viperidae and Elapidae families, are the most significant snakes that are associated with envenoming in Nigeria.

A wide selection of plants and their active principles have been studied by researchers for pharmacological properties. Some medicinal plants which are used locally have been verified to have antivenom properties in preliminary studies. Therefore, it presents medicinal plants as one of the tactical solutions to this problem. Recent pharmacological studies have confirmed the fact that there are several medicinal plants with potentials to treat snakebite in Nigeria. A study by Ameen et al. (2015) presented the medicinal plants which are commonly used for the treatment of snakebite by Fulani Herdsmen in Taraba State, Nigeria. The study documented 19 different plant species belonging to 15 plant families which are used as remedies for the treatment of snakebites by 42.2% of the Fulani herdsmen that were interviewed. Members of the Asteraceae, Liliaceae, Malvaceae, and Mimosaceae families had the highest number of citation of more than 10% while wild soursop *(Annona senegalensis)* was the most recurrent plant species used by 14% of the respondents. *A. senegalensis* was followed by gum acacia *(Acatia senegal)*, wild sunflower *(Aspilia Africana)*, Roselle *(Hibiscus sabdariffa)* and violet tree *(Securidaca longepedunculata)*.

In separate experiments, leaf extract of Sabara (*G. senegalensis*) was established to detoxify (in vitro) venom from two common northern Nigerian snake species that is, *Echis carinatus* and *N. nigricollis*. There was a significant decline in the mortality of albino mice following intraperitoneal administration of reconstituted venom incubated with the extract when compared with those tested with the venom only (Abubakar et al., 2000).

A research study carried out by Ibrahim et al. (2011) showed that sponge guard *(Luffa aegyptiaca)* and wild tobacco (*Nicotiana rustica)* which is often used locally in Nigeria to treat snakebites inhibited the activities of venom protease produced by *Naja nigricolis*. The effectiveness of the methanol extract of the root bark of *A. senegalensis* was tested against cobra (*Naja nigricotlis nigricotli* Wetch) venom in rats. Results showed that the extract reduced the induced hyperthermia and successfully detoxified the snake venom by 16–33%. However, the extract could not restore the biochemical functions of serum alanine aminotransferase (ALT) and aspartate aminotransferase (AST) (Adzu et al., 2004).

Intramuscular administration of the acetone and methanolic stem bark extracts of *Balanites aegyptiaca* to Wistar albino rats, exhibited an antivenin action against *E. carinatus* viper venom at LD50 (0.194 mg/mL). The two extracts were established to be effective at 75 and 100 mg/mL concentrations (Wufen et al., 2007). Tamarind (*Tamarindus indica*) seed extract suppressed the Phospholipase A2 (PLA2), protease, hyaluronidase, l-amino acid oxidase and 5'-nucleotidase enzyme activities of venom and the suppression is dose-dependent. Furthermore, the degradation of the β-chain of human fibrinogen and indirect hemolysis caused by venom was canceled by the extract. Edema, hemorrhage, and myotoxic effects together with lethality, induced by venom were effectively neutralized when different doses of each extract were preincubated with venom before carrying out the assays (Ushanandini et al., 2006).

Di-n-octyl phthalate which was isolated from the Kapok tree (*Ceiba pentandra*) leaves extract was tested for its antivenom properties against *E. ocellatus*. The isolate was biologically active in inhibiting PLA2 activity in a dose-dependent mode as recorded by Ibrahim et al. (2011). Another research by Abubakar et al. (2006) established the fact that both *Aristolochia albida* and *Indigofera pulchra* could effectively neutralize the anticoagulant, hemolytic and phospholipase activities of *N. nigricollis*'s crude venom. Aristolochic acid was also found to form a complex with PLA2, acting as a noncompetitive inhibitor of the enzyme (Vishwanath et al., 1987).

A study by Asuzu and Harvey (2003) established the fact that the water and methanolic extract of African locust bean (*P. biglobosa)* stem bark significantly protected the chick biventer cervicis muscle preparation from *N. nigricollis* venom-induced inhibition of neurally evoked twitches. This effect was observed after it was added to the bath 3–5 min before or after the venom. The extract also reduced the loss of responses to acetylcholine (ACH), carbachol and potassium chloride (KCl), which are usually barred by *N. nigricollis* venom. It effectively decreased the contractures of the preparation that is stimulated by venom. Extracts of *P. biglobosa* at different concentrations (75, 150, and 300 µg/mL), appreciably sheltered $C_2C_{12}$ murine muscle cells in culture against the cytotoxic effects of *N. nigricollis* and *E. ocellatus* venoms. In addition, *E. ocellatus* completely blocked the hemorrhagic actions of the venom at concentrations of 5–10 µg/1.5 µL.

## 4.4   HERBAL PREPARATIONS IN NIGERIA

African traditional medicine is neither similar to Chinese Traditional Medicine (CTM) nor the Indian systems of medicine (e.g., Ayurveda, Unani, and Siddha) that have information that can be accessed in books as well as online. There are still lots of information on African traditional medicine that are yet to be documented.

More than 90% of traditional medicine and herbal preparations in Nigeria contain medicinal plants. Generally, herbal drugs have certain characteristics that distinguish them from synthetic drugs (Calixto, 2000). These are:

i.   They are usually cheaper than conventional drugs.
ii.  They often used as multipurpose therapeutic agents which make them suitable for treatment of many chronic diseases.
iii. The bioactive components of the drug are not usually known.
iv.  The availability of the drug is not always guaranteed.

v. Drug dosage is not clearly stated.

vi. Quality control and standardization of herbal drugs are possible but quite complicated.

vii. Clinical and toxicological studies to establish their efficacy and safety are not common as it is obtained with synthetic drugs.

The traditional dosage form for medicinal plants is usually in the form of infusions and decoctions to be administered orally. These infusions and decoctions are usually good for extracting water-soluble active ingredients such as alkaloids, glycosides, mucilage, polysaccharides, and tannins (see Volume 1 Chapter 9 for details). However, these herbs have some degree of limitations because of their unpleasant taste, shelf life, and poor solubility of many phytochemicals in water. Therefore, many of the herbal practitioners nowadays prefer to use tinctures and fluid extracts rather than infusions and decoctions.

Alcohol and water mixture as the solvent is an efficient method for extracting different active ingredients from medicinal plants. This is in addition to the fact that alcohol can also serve as a good preservative for the herbal preparations. However, a good number of herbal preparations in Nigeria (and indeed Africa) have not been scientifically investigated. A few of the plants used in preparing the herbal mixtures may also contain toxic substances like mutagens and carcinogens with a long-term undesirable effect which may not be immediately apparent (Awosika, 1993).

Some plant extracts which are used locally in various parts of the world have been found to contain cardiac glycosides which are useful for the treatment of heart failure. Other plant extracts which contain active substances such as physostigmine from the seed of *Physostigma venenosun* (from Nigeria) affect cardiac functions (Lawrence et al., 1997). The first herbal preparation recorded as being used for the managing hypertension was *Rauwolfia serpentina* (Indian snakeroot or devils pepper) which contains an alkaloid named reserpine. Local herbalists have been using the West African species *Rauvolfia vomitoria* (devils pepper) for treating the same disease and other ailments for a long time. This species gives a large amount of reserpine than *Rauvolfia serpentina* but it is yet to be exploited commercially.

## 4.5 QUALITY CONTROL OF HERBAL MEDICINE

Quality control of phytomedicine is the process of checking the condition of a drug through the determination of its identity, purity, content, and other chemical, physical or biological properties. This process is enumerated by three important pharmacopeia definitions:

1. **Identity**: This step is carried out to confirm that the herb is truly what it should be. This is done through macro- and microscopic examination. Alterations to the physical form of a plant may lead to wrong identification. Pharmacognostic tests (e.g., microscopic assays by comparison with authentic reference material) are usually used for the identification process.

2. **Purity**: This is not easy to evaluate because the active constituents of most herbal drugs are not known. The purity of a drug is determined by factors like ash values, contaminants, and heavy metals. Determination of purity also involves the evaluation of aflatoxins in the sample, microbial contamination, pesticide residues, and radioactivity. The instrumental analytical techniques which are employed in the determination of purity include gas chromatography and high-performance liquid chromatography.

3. **Content/Assay**: This is usually a quantitative process which is used to evaluate whether the quantity of active constituents is within the defined limits. An important step in quality control of phytomedicines is the identification and quantification of chemical markers. Results of this quantitative analytical techniques help to highlight the nature of phytochemicals or impurities that are present in herbal drugs.

The quality of herbal supplements is the outcomes of different elements as good manufacturing practice and process control. The hazard analysis critical control point (HACCP) is the main and most utilized instrument system. The correct application of HACCP can ensure the safety of herbal products. In addition to standard quality control, herbal supplements need a set of checks to assure the toxicity and safety of the plants used.

## 4.5.1 PARAMETERS FOR QUALITY CONTROL OF PHYTOMEDICINE

The following parameters should be considered in the quality control of herbal drugs:

a) **Determination of Foreign Matter:** Herbs should be totally free from physical contaminants like sand, stones, chemical residues, molds microbial contaminants, and harmful foreign matter.

b) **Microscopic Evaluation:** In order to execute the preliminary identification of herbs, it is very important to carry out the microscopic evaluation. This was not the case in the traditional system of quality control which is majorly determined by appearance. A preliminary visual examination can be done with a magnifying lens to confirm the species of plant. This step can also be done through microscopic analysis.

c) **Determination of Microbial Contaminants:** Herbal drugs can be contaminated by different species of microorganisms which originates from the soil. This may be as a result of poor methods of harvesting, cleaning, drying, handling, and storage which may also lead to additional contamination with species like *E. coli* or *Salmonella spp*. Carrying out a risk assessment of the microbial load of medicinal plants has been a vital focus in the organization of modern HACCP schemes. Limit values and laboratory procedures for examining microbial contaminations are presented in the pharmacopeias, as well as in the WHO guidelines.

d) **Determination of Ash:** This is determined by burning the plant material and the residual ash is measured as the total plus acid-insoluble ash. Total ash is determined by the total quantity of materials left after burning and it includes ash resulting from the part of the plant itself and acid-insoluble ash. The acid-insoluble ash is the remains obtained after boiling the total ash with dilute hydrochloric acid and then burning the remaining insoluble matter.

e) **Determination of Toxic Metals:** Herbal contamination by heavy metals such as mercury, lead, copper, cadmium, and arsenic may be due to factors like environmental pollution, and can create clinically relevant dangers for the health of the users. A simple determination of toxic metals can be found in many pharmacopeias. Instrumental analyses are encouraged if the metals are present in trace quantities and when quantitative analysis is to be carried out. The instrumental methods commonly used include atomic absorption spectrophotometry (AAS), inductively coupled plasma (ICP), and neutron activation analysis (NAA).

f) **Determination of Pesticide Residues:** Herbal drugs may be contaminated with pesticide residues arising from common agricultural practices like spraying of pesticides, soil treatment during cultivation, administration of fumigants during storage among others. It may be better to test herbal drugs for broad groups in general rather than for individual pesticides. For example, pesticides contain chlorine in the molecule and this can be measured through the determination of total organic chlorine while insecticides containing phosphate can be detected through the determination of total organic phosphorus.

g)  **Determination of Radioactive Contamination:** Radioactive contamination may not be a health risk, going by the quantity of herbal medicine that is normally consumed by an individual. Presently, no limits are proposed for radioactive contamination by the regulatory bodies.

## 4.6   CURRENT REGULATIONS FOR STANDARDIZATION OF HERBAL MEDICINE

Herbal medicines are becoming popular due to the belief that they are safer and milder than pharmaceutical drugs. This is usually true because most popular herbal medicines are reasonably harmless at usual doses. The few adverse effects that may be experienced with herbal medicine are primarily mild and infrequent gastrointestinal or dermatological reactions.

Although serious or fatal consequences of herbal medicines are uncommon, there is the absence of regulatory accountability. This is the best illustrated by the inappropriate advertising of ephedra, a traditional Chinese medicine that contains ephedrine and related alkaloids. This Chinese medicine has been misused as a constituent of "natural" herbal weight loss and recreational stimulant in the United States. Several reports have suggested its connection with deaths, strokes, myocardial infarction, seizures, and other serious outcomes (CDC, 1996). Other medicinal herbs and herbal preparations that have been reported to have significant toxicities include chaparral, comfrey, germander, pennyroyal oil, and pokeweed (Newall et al., 1996).

It is the chief responsibilities of the regulatory agencies to make sure that the consumers get the medication guaranteed with purity, safety, potency, and efficacy. The quality control of unrefined drugs and herbal preparations is of vital importance in order to make them acceptable in the modern system of medicine. However, one of the major problems faced by the herbal drug industry is nonavailability of rigid quality control sketch for herbal material and their formulations.

Evaluation of herbal drugs for safety and efficiency are more difficult than those for standard drugs. Despite the excellent therapeutic advantages possessed by medicinal plants, some of their constituents are potentially toxic, carcinogenic, mutagenic, and teratogenic. Therefore, traditional medicine policy and regulation have been made a fundamental part of the WHO proposed critical determinants of herbal medicine safety.

Standardization of drugs is a way of confirming its identity and determining its quality and purity. A variety of methods which include botanical, biological, chemical, and spectroscopic methods are employed in the determination of active constituents that may be present in herbal drugs in addition to their physical constants. Nevertheless, standardization of herbal medicines is often a very difficult task because they contain diverse secondary metabolites. Consequently, their therapeutic actions may be a factor of age, genetic factors and geographical locality of the plant species (Firenzuoli and Gori, 2007). Another problem is that varieties of herbal products from identical plant species may possess variability in their phytochemical contents. Consequently, there will be a variability in the pharmacological activities of these plants. Standardization of herbal medicines worldwide is also difficult to achieve as a result of different harvesting process and period across the nations of the world. In addition to this, incidents of adulterations due to the presence of microorganisms and pesticides are a recurring decimal in some countries (Fong, 2002).

Notwithstanding, quality enhancement of herbal medicines could be achieved by practicing good agricultural practices (GAPs) at the point of cultivation of medicinal plants, good manufacturing practices (GMPs) during the process of manufacture as well as packaging of finished products. Quality enhancement should also include post-marketing quality assurance surveillance (Fong, 2002).

Furthermore, cutting-edge research should be encouraged to examine the shortcoming of herbal remedies and such research findings should be made open to the public. The existence of a vast array of phytochemicals in herbal medicines has opened up meaningful research activities in the area of identification and structure elucidation of the active constituents, characterization of new chemical constituents, biosynthesis, and chemical reactivity of identified bioactive principles. Accordingly, dynamic phytochemical and pharmacological research activities on medicinal plants and herbal drugs are currently ongoing in many Research Institutes and Universities. Research efforts are focused on isolating and identifying bioactive chemical constituents of various herbal drugs and substantiate the claims of their efficacy and safety.

Several regulatory models are currently available for herbal medicines. Harmonization and improvement which shall merge scientific studies and traditional knowledge in the regulatory processes are very important. In Nigeria, there have been interactions between the Government and traditional medicine practitioners. The Federal Ministry of Health in Nigeria established a committee named "National Investigative Committee on Traditional and Alternative Medicine" in 1984. In addition to that, a committee was also constituted to coordinate the conduct of research that will lead to the development of traditional/alternative medicine. The committee was established by the Federal Ministry of Science and Technology in 1988.

The National Agency for Food, Drugs, Administration and Control (NAFDAC) is the agency that lists most ethnomedicinal preparations in the Nigerian market. However, the agency has not been issuing a full status of registration to these products. This is because of doubtful/unverified safety, efficacy, and quality. NAFDAC will only issue full registration status on herbal medicinal products upon evidence of the satisfactory report of clinical trials to establish safety and efficacy. There have been reports of adverse reactions and herb-drug interactions with the use of certain herbal medicine in Nigeria. To achieve standardization of ethnomedicinal preparations in Nigeria, NAFDAC has established Scientific/Expert Committee on verification of herbal medicine claims particularly on safety, efficacy, and quality.

Beyond the application of scientific methods to guarantee the quality and safety of herbal medicine, there should be the integration of control measures during the course of production, storage, and sale and this should conform to obtainable international rules and regulations. This will be effective if it is officially and legally applied using established government institutions and legislative instruments.

## 4.7 RESEARCH EFFORTS ON HERBAL MEDICINE

New approaches and insights into Chinese herbal medicine through Research and Development (R&D) have resulted in the advancement of numerous traditional remedies and innovative drug discovery systems (Pan et al., 2011) and this has immensely to the mainstream biomedical science (Parekh et al., 2009). The active components of many drugs that originate from plants are secondary metabolites (Dobelis, 1993). Therefore, basic phytochemical studies of plant extracts for their bioactive components are quite important.

Botanists, biochemists, microbiologists, natural product chemists, and pharmacologists are presently carrying out researches to investigate medicinal plants for phytochemicals and lead compounds which could be developed into therapeutic substances for several ailments (Fokunang et al.,

2011). This is confirmed by the surplus publications of scientific research papers in many reputable journals. Most of these research studies are in the areas of bioanalytical methodology, isolation, purification, and characterization of the bioactive components of medicinal plants. Furthermore, research efforts in herbal medicine are now targeted at identifying phytochemicals and exposition of their molecular structures. In recent times, a lot of the research studies have also succeeded in establishing the mechanism of action and potential toxicological properties of various medicinal plants. For example, Tanaka et al. (2006) employed spectroscopic data to identify five phytosterols, namely, lophenol, 24-methyllophenol, 24-ethyllophenol, cycloartanol, and 24-methylene cycloartanol as the major hypoglycemic agent in *Aloe vera* gel using Type II diabetic BKS. Cg-m (+/+) Lepr (db/J) (db/db) mice.

The research outcomes on herbal medicine over the years has resulted in a positive move towards therapeutic standardization of herbal drugs whose efficacies have been documented and established through clinical trials (Alvari et al., 2012). WHO-AFRO is equally putting efforts to supplement the various isolated databases on medicinal plants through the provision of guidelines for documentation of herbal recipes (WHO/AFRO, 2012).

## 4.8 HERBAL DRUGS REPURPOSING

There are many factors hindering the development of phytomedicine in Africa which have to be fully addressed so as to promote the African Health Agenda. One major factor that is limiting advancing the goals of most research exercises in Nigeria is the failure of research outcomes to identify the bioactive components of medicinal plants and clarify their therapeutic mechanism. The absence of state-of-the-art scientific infrastructure and poor research funding in most institutions may be responsible for this limitation. Oftentimes when researchers in Africa attempt to carry out a study within the scarcity of available funds, they come up with experimental design approaches that are often too simple. A critical study of such reports will show that consideration of underlying critical issues is either ignored or taken for granted. A good number of times, experiments are slowed down in their preliminary stage. The height of this problem is when useful experimental findings are published in scientific journals of low/no impact factor. This creates a further hindrance to the projection of moving the course of medicinal plant research forward through international collaboration with other researchers.

Other problems as identified by Okigbo and Mmeka (2006) are as follows:

1.  Difficulty in the development of the drug from its natural source. This is due to the fact that the formulation of phytomedicine particularly in crude form requires a specialized expert area that calls for training and experience
2.  Substitution and adulteration of medicinal plants which are used in herbal formulations through incorrect preparation and dosage.
3.  The absence of standardization procedures and quality control of the herbal drugs which are used in clinical trials (Calixto, 2000) and occult practices (Makhubu, 2006).
4.  The possibility of side effect due to toxicity, over-dosage, interaction with conventional drugs, and many other manufacturing problems such as false identification of plants, lack of standardization, non-compliance with good manufacturing practices, field microbial contamination, poor packaging, chemical used, and the environmental condition (e.g., temperature, light exposure) (Elujoba et al., 2005)
5.  Vague diagnosis and dosage of phytomedicine.
6.  Lack of collaboration between TMPs, Orthodox Medical Practitioners and Research Scientists. This leads to the danger of losing the valuable ethnomedical knowledge that the TMPs can present concerning the plant and other aspects of the traditional medicinal system that are fundamentally part of their lives (Makhubu, 2006).
7.  There are inadequate randomizations in a good number of research studies. Patients may not be properly selected and the numbers of patients used in most trials are insufficient for the attachment of statistical significance.
8.  The communication gap between the TMPs and Research Scientists.
9.  Convincing members of a community to trust phytomedicine is difficult after a long use of Orthodox medicine (Makhubu, 2006).
10. Extinction of many species due to the fact that they are not cultivated and sheltered from indiscriminate harvesting. Also, many of the traditional healers are dying because they are advanced in age.

Some of the problems highlighted above can be solved in the following ways:

a)  The African pharmacognosists, pharmacologists, pharmacists, physicians should learn, obtain, document and make use of traditional medicine to help curtail the extinction of plants, and human resources (Elujoba, 2003).

b) In order to eliminate the communication problem between TMPs and Research Scientists, workshops with TMPs should be conducted. Human resources can be sourced through individual contacts as recorded by Makhubu (2006).

c) Collaborative work with TMPs can be accomplished through staff exchange and regular training.

d) The government, the organized private sector as well as nongovernmental agencies can help by funding researches on phytomedicine.

e) Seminars should be organized to raise awareness of the general public on the medical advantages of medicinal plants.

f) Outdated legislation that adds no values should be abandoned and new legislation should be passed to protect indigenous traditional knowledge.

g) Integration of traditional medicine into the orthodox health scheme should be encouraged (Makhubu, 2006).

### 4.8.1   SUGGESTIONS FOR FUTURE ADVANCEMENT OF PHYTOMEDICINE IN NIGERIA

Due to the increasing demand for phytotherapeutic products, some factors have to be developed so as to meet the world herbal medicine's standard of safety and efficacy. The following issues must be emphasized for the improvement of phytomedicine in Nigeria:

i. The importance of domestication, production, biotechnological studies, and genetic development of medicinal plants should be emphasized. The domestication of plants will help in reducing the problem associated with wild-harvested plants and eliminate the issue of wrong identification and field contamination. Biotechnological studies will increase the quality of raw materials and yield through genetic breeding and selection. It will also ensure the production of herbal medicine with resistance to microbial contaminations.

ii. There should be an emphasis on well-controlled and randomized clinical trials to establish the safety and efficiency of herbal medicine. Efforts should be made to improve the quality, efficacy, and safety of phytomedicine which are used in clinical trials in order to produce standardized drugs.

iii. Research studies on traditional medicines should be completed to develop novel therapeutic methods.

iv. There should be advancement in the regulatory processes and universal harmonization of phytomedicine. The integration of African traditional medicine into the orthodox health system should bring harmony between the traditional and orthodox system of healthcare delivery with no threat to each other.

v. A great emphasis should be placed on collaboration work between TMPs and other research scientists. This can include laboratory training for the TMPs.

vi. Detailed legislation on the ownership of intellectual property including patency right has to be established.

## 4.9  FUTURE INVESTIGATION OF HERBAL DRUGS

In every country of the world, regulatory authorities are seeking for research into new analytical methods for the stricter standardization of phytomedicines. Such methods are expected to be both objective and robust and should address the reproducibility of the content of identified chemical profiles.

The WHO has done so much in documenting how medicinal plants are used by various tribes across the nations of the world (Kaido et al., 1997). For this reason, many developing countries, including Nigeria, have intensified their efforts in documenting the ethnomedical data on medicinal plants.

Indigenous efforts by researchers are necessary so as to access the natural biodiversity beyond the geographical boundaries and it is imperative to thoroughly preserve such reported studies with innovative technology. As discussed so far, it is important that drug discovery from new natural sources be given more attention and exploration with a different strategic method. In addition, researchers need to carry out evaluations and characterization of plants and plant constituents which have been used for ages by TMPs against various diseases and ailments.

Nonetheless, appropriate biological target selection will also be of tremendous help in drug discovery. Research on inadequately studied antiviral potencies and modes of actions of different phytochemicals needs to be explored for newer molecules. Cross-screening research approaches which will connect both serendipitous screening methods and rational drug design needs to be strengthened for development of novel drugs from the phytochemical category (Vanpouille et al., 2015). This can be achieved through collaboration between the medicinal chemists and biomedical researchers.

## 4.10  CONCLUSION

Ethnomedicinal studies are very vital for the proper understanding of the social, cultural, and economic factors which influence ideas and actions concerning health and illness among the people of a particular region. Studies like that may help to supply the basic healthcare services to the larger part of the rural areas in an effective way, provided that such studies are conducted alongside with phytochemical, pharmacological, and clinical studies. Every community in Nigeria has its peculiar way of treating different ailments. Many plants are often found useful for such traditional treatment of common diseases and ailments. Products of a medicinal plant are still the primary source of supply of many important drugs in orthodox medicine today. Many problems associated with regulating and assuring quality in the manufacture of herbal medicines is not meant to refute their potential benefits. Further research studies, to clarify the chemical structure of active components of herbs will make room for synthetic modifications for better pharmacokinetic profiles of those herbal products. Quality control protocols should be established for the production of high-quality herbal products.

## KEYWORDS

- herbal medicine
- medicinal plants
- phytochemicals
- phytomedicine
- quality control

## REFERENCES

Abubakar, M. S.; Sule, M. L.; Pateh, U. U.; Abdurahman, E. M.; Haruna, A. K.; Jahun, B. M. In Vitro Snake Venom Detoxifying Action of Guiera Senegalensis Leaf Extract. *J. Ethnopharmacol.* **2000,** *69*(3), 253–257.

Abubakar, M. S.; Balogun, E.; Abdurahman, E. M.; Nok, A. J.; Shok, M.; Mohammed, A.; Garba, M. Ethnomedical Treatment of Poisonous Snakebites: Plant Extract Neutralized *Naja nigricollis*. Venom. *J. Pharm. Biol.* **2006,** *44*(5), 343–348.

Addae-Mensah, I.; Munenge, R. W. Quercetin 3-Neohesperidose (Rutin) and Other Flavonoids as the Active Hypoglycemic Agents in *Bridelia ferruginea. Fitoterapia* **1989,** *2*(4), 359–362.

Adebayo, J. O.; Krettli, A. U. Potential Antimalarials from Nigerian Plants: a Review. *J. Ethnopharmacol.* **2011,** *133*(2), 289–302.

Adegunloye, B. J.; Omoniyi, J. O.; Owolabi, O. A.; Ajagbonna, O. P.; Sofola, O. A.; Coker, H. A. Mechanisms of the Blood Pressure Lowering Effect of the Calyx Extract of *Hibiscus sabdariffa* in Rats. *Afri. J. Med. Sci.* **1996,** *25*, 235.

Addo, J.; Smeeth, L.; Leon, D.A. Hypertension in Sub-Saharan Africa: a Systematic Review. *Hypertension.* **2007,** *50,* 1012–1018.

Adzu, B.; Abubakar, M. S.; Izebe, K. S.; Akumka, D. D.; Gamaniel, K. S. Effect of Annona Senegalensis Root Bark Extracts on *Naja nigricotlis nigricotlis* Venom in Rats. *J. Ethnopharmacol.* **2004,** *96*, 507–513.

Agarwal, A. Critical Issues in Quality Control of Herbal Products. *Pharma. Times* **2005,** *37*(6), 9–11.

Agunu, A.; Yusuf, S.; Andrew, G. O.; Umarzezi, A.; Abdurahman, E. M. Evaluation of Five Medicinal Plants Used in Diarrhea Treatment in Nigeria. *J Ethnopharmacol.* **2005,** *101*(1–3), 27–30.

Ajay, M.; Chai, H. J.; Mustafa, A. M.; Gilani, A. H.; Mustafa, M. R. Mechanisms of the Antihypertensive Effect of *Hibiscus sabdariffa* L. Calyces. *J. Ethnopharmacol.* **2007,** *109*, 388–393.

Ajgaonkar, S. S. Herbal Drugs in Treatment of Diabetes Mellitus: A Review. *Int. Diabetes Fed. Bull.* **1979,** *24*, 10–19.

Aladesanmi, A. J.; Iwalewa, E. O.; Adebajo, A. C.; Akinkunmi, E. O.; Taiwo, B. J.; Olorunmola, F. O.; Lamikanra, A. Antimicrobial and Antioxidant Activities of Some Nigerian Medicinal Plants. *Afr. J. Trad. CAM* **2007,** *4*(2), 173–184.

Alvari, A.; Mehrnaz, S. O.; Ahmad, F. J.; Abdin, M. Z. Contemporary Overview on Clinical Trials and Future Prospects of Hepato-Protective Herbal Medicines. *Rev. Recent. Clin. Trials* **2012,** *7*, 214–223.

Ameen, S. A.; Salihu, T.; Mbaoji, C. O.; Anoruo-Dibia, C. A.; Adedokun, R. A. M. Medicinal Plants Used to Treat Snake Bite by Fulani Herdsmen in Taraba State, Nigeria. *Int. J. Applied Agri. Apic. Res. (IJAAAR)* **2015,** *11*(1&2), 10–21.

Amer, M. M.; Court, W. E. Leaf Alkaloids of Rauwolfia Vomitoria. *Phytochem.* **1980,** *19*, 1833–1836.

Anaka, O. N.; Ozolua, R. I.; Okpo, S. O. Effect of the Aqueous Seed Extract of *Persea americana* Mill (Lauraceae) on the Blood Pressure of Sprague-Dawley Rats. Afr. *J. Pharm. Pharmacol.* **2009,** *3*(10), 485–490.

Anslem, A. *Nature Power. Christian Approach to Herbal Medicine;* Ayitey Smith., Ed.; Don Bosco Training Centre, School of PTP, Printing and Finishing: Akure, Nigeria, 2006; p 206.

Ashraf, R.; Khan, R. A.; Ashraf, I.; Qureshi, A. A. Effects of Allium Sativum (Garlic) on Systolic and Diastolic Blood Pressure in Patients with Essential Hypertension. *Pak. J. Pharm. Sci.* **2013,** *26*(5), 859–863.

Asuzu, I. U.; Harvey, A. L. The Antisnake Venom Activities of *Parkia biglobosa* (Mimosaceae) Stem Bark Extract. *Toxicon.* **2003,** *42*(7), 763–768.

Awosika, F. Local Medicine Plants and the Health of the Consumers. *J. Clinical Pharm. Herbal Medicine.* **1993,** *7*, 3–4.

Ayoola, G. A.; Coker, H. A. B.; Adesegun, S. A.; Adepoju-Bello, A. A.; Obaweya, K.; Ezennia, E. C.; Atangbayila, T. O. Phytochemical Screening and Antioxidant Activities of Some Selected Medicinal Plants Used for Malaria Therapy in Southwestern Nigeria. *Trop. J. Pharm. Res.* **2008,** *7*(3), 1019–1024.

Bako, I. G.; Mabrouk, M. A.; Majem, I. M.; Buraimoh, A. A.; Abubakar, M. S. Hypotensive Effect of Aqueous Seed Extract of *Hibiscus sabdariffa* Linn (Malvaceae) on Normotensive Cat. Inter. *J. Anim. Sci. Vet. Adv.* **2010**, *2*(1),5–8.

Bnouham, M.; Ziyyat, A.; Mekhf, H.; Tahri, A.; Legssyer, A. Medicinal Plants with Potential Anti-Diabetic Activity-A Review of Ten Years of Herbal Research (1990–2000). *Int. J. Diabetes Metab.* **2006**, *14*, 1–25.

Calixto, J. B. Efficacy, Safety, Quality Control, Marketing and Regulatory Guidelines for Herbal Medicines (Phytotherapeutic Agents). *Braz. J. Med. Biol. Res.* **2000**, *33*(2), 179–189.

CDC (Centers for Disease Control and Prevention). Adverse Events Associated with Ephedrine-Containing Products–Texas, Dec. 1993-Sept. 1995. *Morb. Mortal. Wkly. Rept.* **1996**, *45*, 689–693.

Chang, C. L. T.; Lin, Y.; Bartolome, A. P.; Chen, Y. C.; Chiu, S. C.; Yang, W. C. Herbal Therapies for Type 2 Diabetes Mellitus: Chemistry, Biology, and Potential Application of Selected Plants and Compounds. *Evidence-Based Complement Alternat. Med.* **2013**, *378657*, 1–33.

Coon, T.; Ernst, E. Herbs for Serum Cholesterol Reduction: a Systematic View. *J. Fam. Pract.* **2003**, *52*(6), 468–478.

Dholi, S. K.; Raparla, R.; Mankala, S. K.; Nagappan, K. In Vivo Antidiabetic Evaluation of Neem Leaf Extract in Alloxan Induced Rats. *J. App. Pharm. Sci.* **2011**, *1*(4), 100–105.

Dobelis, I. N. *Magic and Medicine of Plants;* The Readers Digest Association Inc. Pleasant: New York, Montreal, 1993; pp 8–48.

Edeoga, H. O.; Okwu, D. E.; Mbaebie, B. O. Phytochemical Constituents of Some Nigerian Medicinal Plants. *Afr. J. Biotechnol.* **2005**, *4*(7), 685–688.

Eidi, A.; Eidi, M.; Esmaeili, E. Antidiabetic Effect of Garlic (*Allium sativum* L.) in Normal and Streptozotocin-Induced Diabetic Rats. *Phytomedicine* **2006**, *13*, 624–629.

Elujoba, A. A. Medicinal Properties of Plants with Oral Health Implication. Proceedings of the 2nd Dr. David Barmes' Memorial Public Health Symposium, 25th March, 2003, Organized by the Regional Center for Oral Health Research and Training for Africa, Jos (Nigeria), in Collaboration with WHO Regional Office, Brazzaville.

Elujoba, A. A.; Odeleye, O. M; Ogunyemi, C. M. Traditional Medical Development for Medical and Dental Primary Health Care Delivery System in Africa. *Afri. J. Tradit. Complementary Altern.* **2005**, *2*(1), 46–61.

Erute, M. A.; Egboduku, O W. Screening of Some Nigerian Medicinal Plants for Anti-Candida Activity. *Am. J. Drug Discovery Dev.* **2013**, *3*, 60–71.

Firenzuoli, F.; Gori, L. Herbal Medicine Today: Clinical and Research Issues. *Evid. Based Complement Alternat. Med.* **2007**, *4*, 37–40.

Fokunang, C. N.; Ndikum, V.; Tabi, O. Y.; Jiofack, R. B.; Ngameni, B.; Guedje, N. M.; Tembe-Fokunang, E. A.; Tomkins, P.; Barkwan, S.; Kechia, F.; Asongalem, E.; Ngoupayou, J.; Torimiro, N. J.; Gonsu, K. H.; Sielinou, V.; Ngadjui, B. T.; Angwafor, F.; Nkongmeneck, A.; Abena, O. M. Ngogang, J.; Asonganyi, T.; Colizzi, V.; Lohoue, J.; Kamsu, K. Traditional Medicine: Past, Present and Future Research and Development Prospects and Integration in the National Health System of Cameroon. *Afr. J. Tradit. Complement Altern. Med.* **2011**, *8*, 284–295.

Fong, H. H. Integration of Herbal Medicine Into Modern Medical Practices: Issues and Prospects. *Integr. Cancer Ther.* **2002**, *1*, 287–293.

Gbadamosi, I. T.; Egunyomi, A. Phytochemical Screening and in Vitro Anticandidal Activity of Extracts and Essential Oil of Curculigo Pilosa (Schum and Thonn) Engl. Hypoxidaceae. *Afr J. Biotechnol.* **2010**, *9*(8), 1236–1240.

Gbile, Z. O. Ethnobotany, Taxonomy and Conservation of Medicinal Plants. *The State of Medicinal Plant Research in Nigeria;* Sofowora, A., Ed.; University of Ibadan Press: Ibadan, Nigeria, 1986; 13–29

GHP. Ghana Herbal Pharmacopeia; Council for Scientific and Industrial Research (CSIR): Accra, 2007; pp 295.

Hendriks, M. E.; Wit, F. W. N. M.; Roos, M. T. L.; Brewster, L. M.; Akande, T. M.; de Beer, I. H; Mfinanga, S. G.; Kahwa, A. M.; Gert Van Rooy, P. G.; Janssens, W.; Lammers, J.; Kramer, B.; Bonfrer, I.; Gaeb, E.; van der Gaag, J.; Rinke de Wit, T. F.; Lange, J. M. A.; Schultsz, C. Hypertension in Sub-Saharan Africa: Cross-Sectional Surveys in Four Rural and Urban Communities. *PLoS One* **2012,** 7(3): 1–10.

Ibrahim, M. A.; Aliyu, A. B.; Abusufiyanu, A.; Bashir, M.; Sallau, A. B. Inhibition of *Naja nigricollis* (Reindhardt) Venom Protease Activity by *Luffa egyptiaca* (Mill) and *Nicotiana rustica* (Linn) Extracts. *Indian J. Exp. Biol.* **2011,** *49*, 552–554.

Iwalokun, B. A.; Hodonu, S. A.; Nwoke, S.; Ojo, O.; Agomo, P. U. Evaluation of the Possible Mechanisms of Antihypertensive Activity of *Loranthus micranthus*: An African Mistletoe. *Biochem. Res. Inter. Volume.* **2011,** 9, Article ID *159439,* 1–9.

Iweala, E. E. J.; Uhuegbu, F. O.; Adesanoye, O. A. Biochemical Effects of Leaf Extracts of *Gongronema latifolium* and Selenium Supplementation in Alloxan Induced Diabetic Rats. *J. Pharmacogn. Phytother.* **2013,** *5*(5), 91–97.

Joshi, B. N.; Bhat, M.; Kothiwale, S. K.; Tirmale, A. R.; Bhargava, S. Y. Antidiabetic Properties of *Azadirachta indica* and *Bougainvillea spectabilis*: in Vivo Studies in Murine Diabetes Model. *Evidence-Based Complementary* **2011,** Article ID 561625, pp 1–19. http://dx.doi.org/10.1093/ecam/nep033.

Kaido, T. L.; Veale, D. J. H.; Havlik, I.; Rama, D. B. K Preliminary Screening of Plants Used in South Africa as Traditional Herbal Remedies During Pregnancy and Labour. *J. Ethnopharmacol.* **1997,** *55*, 185–191.

Kubmarawa, D.; Ajoku, G. A.; Enwerem, N. M.; Okorie, D. A. Preliminary Phytochemical and Antimicrobial Screening of 50 Medicinal Plants from Nigeria. *Afr. J. Biotechnol.* **2007,** *6*(14), 1690–1696.

Lawrence, D. R.; Bennett, P. N.; Brown, M. J. *Clinical pharmacology;* Churchill Livingstone: USA; 1997, pp.399–411.

Makhubu, L. Traditional Medicine: Swaziland. *Afr. J. Trad. CAM.* **2006,** 5(2), 63–71.

Mensah, J. K.; Okoli, R. I.; Ohaju-Obodo, J. O.; Eifediyi, K. Phytochemical, Nutritional and Medical Properties of Some Leafy Vegetables Consumed by Edo People of Nigeria. *Afr. J. Biotechnol.* **2008,** *7*(14), 2304–2309.

Mensah, J. K.; Okoli, R. I.; Turay, A. A.; Ogie-Odia, E. A. Phytochemical Analysis of Medicinal Plants Used for the Management of Hypertension by Esan People of Edo State Nigeria. *Ethnobotanical Leaflets* **2009,** *13*, 1273–1287.

Mojiminiyi, F. B.; Dikko, M.; Muhammad, B. Y.; Ojobor, P. D.; Ajagbonna, O. P.; Okolo, R. U.; Igbokwe, U. V.; Mojiminiyi, U. E.; Fagbemi, M. A.; Bello, S. O.; Anga, T. J. Antihypertensive Effect of an Aqueous Extract of the Calyx of *Hibiscus sabdariffa*. *Fitoterapia.* **2007,** *78*(4), 292–297.

Morakinyo, A. E.; Babarinde, S. O.; Adelowo, J. M.; Olopade, E.; Oyedepo, T. A. *Hypoglycemic and Hypolipidemic Effect of West African Black Pepper (Piper guineense) Leaves on Alloxan Treated Rats,* Proceedings of the iSTEAMS Multidisciplinary Cross Border Conference, University of Professional Studies, Accra, Ghana, **2016,** *9*, 286-297.

Nduche, M. U.; Omosun, G. The Use of Medicinal Plants in the Treatment of Diarrhea in Nigeria: Ethnomedical Inventory of Abia State. *Sch. J. Agric. Vet. Sci.* **2016,** *3*(3), 270–274.

Negbenebor, H. E.; Shehu, K.; Mairami, F. M.; Adeiza, Z. O.; Nura, S.; Fagwalawa, L. D. Ethnobotanical Survey of Medicinal Plants Used by Hausa People in the Management of Diabetes Mellitus in Kano Metropolis, Northern Nigeria. *Eur. J. Med. Plants* **2017,** *18*(2), 1–10.

Newall, C. A.; Anderson, L. A.; Phillipson, J. D. *Herbal Medicines: a Guide for Health-care Professionals*; Pharmaceutical Press: London, 1996.

Nwanjo, H. U. Efficacy of Aqueous Leaf Extract of *Vernonia amygdalina* on Plasma Lipo-Protein and Oxidative Status in Diabetic Rat Models. Nig. *J. Physiol. Sci.* **2005,** *20*(1–2), s39–42.

Nwokocha, C.R.; Ozolua, R. I.; Owu, D. U.; Nwokocha, M. I.; Ugwu, A. C. Antihypertensive Properties of *Allium sativum* (Garlic) on Normotensive and Two Kidney One Clip Hypertensive Rats. Niger. *J. Physiol. Sci.* **2011,** *26*(2), 213–218.

Nworgu, Z. A. M.; Ameachina, F. C.; Owolabi, J.; Otokiti, I.; Ogudu, U. Cardiovascular Effects of Aqueous Extract of *Acalypha wilkesiana* Hoffmannii Leaves in Rabbits and Rats. Nig. *J. Pharm Sci.* **2011,** *10*(2), 45–50.

Obatomi, D. K.; Aina, V. O.; Temple, V. J. Effects of African Mistletoe Extract on Blood Pressure in Spontaneously Hypertensive Rats. Pharmaceut. *Biol.* **1996,** *34*(2), 124–127.

Obute, G. C. Ethnomedicinal Plant Resources of South Eastern Nigeria. *Afr. J. Interdiscip. Studies* **2007,** *3*(1), 90–94.

Odebiyi, O. O. Sofowora, E. A. Phytochemical Screening of Nigerian Medicinal Plants, Part II. *Iloydia.* **1978,** *41*, 1–25.

Odugbemi, T. T. *Outline and Pictures of Medicinal Plants from Nigeria;* Universityof Lagos Press: Lagos, Nigeria, 2006; pp 283.

Odugbemi, T. O.; Odunayo, R. A.; Aibinu, I. E.; Fabeku, O. P. Medicinal Plants Useful for Malaria Therapy in Okeigbo, Ondo State, Southwest Nigeria. *Afr. J. Trad. CAM.* **2007,** *4*(2), 191–198.

Offiah, N. V.; Makama, S.; Elisha, I. L.; Makoshi, M. S.; Gotep, J. G.; Dawurung, C. J.; Oladipo, O. O.; Lohlum, A. S.; Shamaki, D. *BMC Vet Res*, **2011,** *7*, 36.

Ogbonna, D. N.; Sokari, T. G.; Agomuoh, A. A. Antimalarial Activities of Some Selected Traditional Herbs from South Eastern Nigeria Against *Plasmodium* Species. Research. *J. Parasitol.* **2008,** *3*, 25–31.

Ogbuehi, I. H.; Ebong, O. O. Traditional Medicine Treatment of Malaria in Onitsha, South East Nigeria. *Greener J. Med. Sci.* **2015,** *5*(1), 011–018.

Ojewole, J.A.O. Antihypertensive Properties of *Bryophyllum pinnatum* (Lam) Oken Leaf Extracts. *Am J. Hypertens.*, **2002,** *15*(S3), 34A.

Ojewole, J. A. Hypoglycemic and Hypotensive Effects of *Psidium guajava* Linn. (Myrtaceae) Leaf Aqueous Extract. *Methods Find Exp. Clin. Pharmacol.* **2005,** *27*(10), 689–695.

Okigbo, R. N; Mmeka, E. C. An Appraisal of Phytomedicine in Africa. *Kmitl Sci. Tech. J.* **2006,** *6*(2), 83–94.

Okoli, R. I.; Aigbe, O.; Ohaju, O.; Obodo, J. O.; Mensah, J. K. Medicinal Herbs Used for Managing Some Common Ailments Among Esan People of Edo State, Nigeria. *Pak. J. Nutr.* **2007,** *6*(5), 490–496.

Okwu, D. E. Flavoring Properties of Spices on *Cassava fufu. Afr. J. Roots Tuber Crops* **1999,** *3*(2), 19–21.

Okwu, D. E. Evaluation of the Chemical Composition of Indigenous Spices and Flavoring Agents. *Global J. Pure Appl. Sci.* **2001,** *7*(3), 455–459.

Okwu, D. E.; Ukanwa, N. Isolation, Characterization and Antibacterial Activity Screening of Anthocyanide Glycosides from *Alchornea cordifolia* (Schumach. and Thonn.) *Mull. Arg. Leaves. E-J Chem.* **2010,** *1*, 41–48.

Olanipekun, M. K.; Kayode, J.; Akomolafe, D. S. Ethnobotanical Importance and Phytochemical Analysis of some Medicinal Plants Commonly used as Herbal Remedies in Oye Local Government Area of Ekiti State, Nigeria. IOSR *J. Agri. Vet. Sci. (IOSR-JAVS)* **2013,** *5*(6), 28–31.

Omage, K.; Azeke, A. M. Medicinal Potential of *Acalypha wilkesiana* Leaves. *Adv. Res.* **2014,** *2*(11), 655–665.

Onyenekwe, P. C.; Ajani, E. O.; Ameh, D. A.; Gamaniel, K. S. Antihypertensive Effect of Roselle (*Hibiscus sabdariffa*) Calyx Infusion in Spontaneously Hypertensive Rats and a Comparison of Its Toxicity with that in Wistar Rats. *Cell Biochem. Funct.* **1999,** *17,* 199–206.

Oyedepo, T. A.; Babarinde, S. O. Effects of Shaddock (Citrus Maxima) Fruit Juice on Glucose Tolerance and Lipid Profile in Type-II Diabetic Rats. *Chem. Sci. Trans.* **2013,** *2*(1), 19–24.

Oyedepo, T. A.; Babarinde, S. O., Ajayeoba T. A. Evaluation of Anti-Hyperlipidemic Effect of Aqueous Leaves Extract of *Moringa oleifera* in Alloxan Induced Diabetic Rats. *Int. J. Biochem. Res. Rev.* **2013,** *3*(3), 162–170.

Pan, S. Y., Chen, S. B.; Dong, H. G.; Yu, Z. L.; Dong, J. C.; Long, Z. X.; Fong, W. F.; Han, Y. F.; Ko, K. M. New Perspectives on Chinese Herbal Medicine (Zhong-Yao) Research and Development. *Evid. Based Complement. Alternat. Med.* **2011,** pp 1–11. Article ID 403709. http://dx.doi.org/10.1093/ecam/neq056.

Parekh, H. S.; Liu, G.; Wei, M. Q. A New Dawn for the Use of Traditional Chinese Medicine in Cancer *Therapy. Mol. Cancer* **2009,** *8*, 21.

Patil, P. R.; Patil, S. P.; Mane, A.; Verma, S. Antidiabetic Activity of Alcoholic Extract of Neem (*Azadirachta indica*) Root Bark. *Natl. J. Physiol. Pharm. Pharmacol.* **2013,** *3*(2), 142–146.

Sofidiya, M.O.; Oduwole, B.; Bamgbade, E.; Odukoya, O.; Adenekan, S. Nutritional Composition and Antioxidant Activities of *Curculigo pilosa* (Hypoxidaceae) Rhizome. *Afr. J. Biotechnol.* **2011,** *10*, 75.

Sofowora, A. Medicinal Plants and Traditional Medicine. *WHO Document* **1993,** *30*, 69.

Sofowora, E. A.; Odebiyi, O. O. Phytochemical Screening of Nigeria Medicinal Plants. *Lloydia.* **1987,** *4*, 234.

Soladoye, M. O.; Chukwuma, E. C.; Olatunji, M. S.; Feyisola, R. T. Ethnobotanical Survey of Plants Used in the Traditional Treatment of Female Infertility in Southwestern Nigeria. *Ethnobotany Res. Appl.* **2014,** *12*, 81–90.

Taiwo, I. A.; Odeigah, P. G. C.; Jaja, S. I.; Mojiminiyi, F. B. Cardiovascular Effects of Vernonia Amygdalina in Rats and the Implications for Treatment of Hypertension in Diabetes. Researcher. **2010,** *2*(1), 76–79.

Tanaka, M.; Misawa, E.; Ito, Y.; Habara, N.; Nomaguchi, K.; Yamada, M.; Toida, T.; Hayassawa, H.; Takase, M.; Inagaki, M,; Higuchi, R Identification of Five Phytosterols from *Aloe vera* Gel as Anti-Diabetic Compounds. *Biol. Pharm. Bull.* **2006,** *29*(7), 1418–1422.

Ugochukwu, N. H.; Fafunso, P. J.; Boba, A. T.; Babady, N. E. The Various Medicinal Effects of *Gongronema latifolium. Phytother. Res.* **2005,** *12*:46–52.

Ushanandini, S.; Nagaraju, S.; Harish Kumar, K.; Vedavathi, M.; Machiah, D. K.; Kemparaju, K.; Vishwanath, B. S.; Gowda, T. V.; Girish, K. S. The Anti-Snake Venom Properties of *Tamarindus indica* (Leguminosae) Seed Extract. *Phytother. Res.* **2006,** *20*(10), 851–858.

van de Venter, M.; Roux, S.; Bungu, L. C.; Louw, J.; Crouch, N. R.; Grace, O. M.; Maharaj, V.; Pillay, P.; Sewnarian, P.; Bhagwandin, N.; Folb, P. Antidiabetic Screening and Scoring of 11 Plants Traditionally Used in South Africa. *J. Ethnopharmacol.* **2008,** *119*, 81–86.

Vanpouille, C.; Lisco, A.; Grivel, J. C.; Bassit, L. C.; Kaufmann, R. C.; Sanchez, J.; Schinazi, R. F.; Lederman, M. M.; Rodriguez, B.; Margolis, L. Valacyclovir Decreases Plasma HIV-1 RNA in HSV-2 Seronegative Individuals: a Randomized Placebo-Controlled Crossover Trial. *Clin. Infect. Dis.* **2015,** *60*, 1708–1714.

Willcox, M. L., Bodeker, G. Traditional Herbal Medicines for Malaria. *Br. Med. J.* **2004,** *329*, 1156–1159.

World Health Organization (WHO). *Alma Ata Declaration Primary Health Care;* Health for All Series, No 1; 1994.

WHO/AFRO. *Health Systems in Africa: Community Perceptions and Perspectives*; The Report of a Multi-Country Study, June 2012.

Wufen, B. M.; Adamu, H. M.; Cham, Y. A.; Kela, S. L. Preliminary Studies on the Antivenin Potential and Phytochemical Analysis of the Crude Extracts of *Balanites aegyptica* (Linn.) Delile on Albino Rats. *Nat. Prod. Radiance* **2007,** *6*, 18–21.

# CHAPTER 5

# PLANT SPECIES UTILIZED FOR ETHNOVETERINARY PRACTICES IN INDIA

V. LAKSHMINARAYANA and G. M. NARASIMHA RAO*

*Department of Botany, Andhra University, Visakhapatnam 530003, India, Tel. 91 9440559806*

*Corresponding author. E-mail: gmnrao_algae@hotmail.com; gmnarasimharao.bot@auvsp.edu.in*
*ORCID: https://orcid.org/0000-0001-9925-8017*

## ABSTRACT

Ethnoveterinary practices involve the application of traditional knowledge of medicinal plants for the primary healthcare of domestic animals. Medicinal plants contain various chemical compounds which possess important therapeutic properties that can be used in the treatment of ailments like anthrax, corneal opacity, dysentery, dyspepsia, ephemeral fever, gout, horn cancer, maggot, rheumatism, retained placenta, trypanosomiasis, and tympany, besides problems like anorexia, boils, bone fracture, constipation, insect bite, and snakebite are being treated by the folklore knowledge. Studies on medicinal plant species used in folklore remedies have attracted scientists in finding solutions for the problems of multiple resistances to the existing synthetic antibiotics. Plants which seem unique for certain animal diseases need research and conservation. In this chapter, data were summarized on plant species used for ethnoveterinary practices in different geographical regions of India. Their mode of administration/preparation, parts used, and ailments it cures were also presented.

## 5.1  INTRODUCTION

Plant species produce secondary metabolites known as allelochemicals. These secondary metabolites are organic compounds which do not participate in the normal growth of the plants but are produced as a by-product during the synthesis of primary metabolic products. Animals are utilizing these secondary metabolites to cure some diseases. Now, these compounds are vital for the production of a large number of medicinal, nutritional, and industrial products.

Since the prehistoric period, India is a homeland for a large number of folk and cultural groups. These groups have a distinct way of life, traditions, dialects, and diverse cultural heritage. They are custodians of the indigenous knowledge system, and their belief has a great effect on their gesture and psychology about diseases. They usually collect the medicinal plant at a particular growth period and season, some before flowering and fruiting while others after the plant have shed flowers and fruits and seeds presumably to get maximum yield of active principles in order to avoid destruction of a large number of plants, which allows for self-regeneration and conservation. They also use a number of plant species for treatment of single disease, so as to reduce the impact of exploitation. Unfortunately, we have entered into a phase of mass extinction (Raven, 1987; Myer, 1990) and have altered about half of the habitable earth surface (Daily, 1995) by damaging and destroying several ecosystems.

In India, more than 75% of the total population residing in the rural area depends on the plants for their healthcare needs. The history of medical practices in India can be traced back to the Vedic period. The curative properties of plants have been known and are documented in ancient manuscripts, namely *Rig Ved*, *Garuda Puran*, and *Agni Puran* (Priyadarsan, 1991).

Ethnoveterinary medicine (EVM) is a broad field encompassing people's beliefs, skills, knowledge, and practices related to veterinary healthcare (Mc Corkle, 1986). Medicinal plants traditionally used for the treatment of animal diseases plays an important role in local health modalities. Specifically, phytotherapeutics often represents the primary form of therapy in rural veterinary care as allopathic modalities remain inaccessible, especially in the developing countries of the world (Katerere and Luseba, 2010). However, traditional ethnoveterinary knowledge is still orally transmitted from generation through a family lineage or guru–shishya (teacher–student) parampara (heritage) cutting across families and ethnic communities across the country such as the Kurubas (Karnataka), the Konars (Tamil Nadu), and the Yadavas (Uttar Pradesh),that is, in the form of conventional remedies, poems, drawing stories, folk myths, proverbs, and songs. Due to the oral transmission, this form of local knowledge remains fragile and threatened, so, an urgent need for being recorded and documented.

The utilization of medicinal plants goes way back to ancient people, who discovered a wealth of therapeutic properties of agents in the plant kingdom and exploited their healing potential as a remedy for several animal ailments. In the recent years, ethnopharmacology is playing a significant role in the modern medicine for the undeveloped and developing countries of the globe.

Before the advent of the modern allopathic system of medicine, it seems possible that the healing art was almost the same across the world including India. This system of medicine has given the term *ethnomedicine* (when related to human treatment) and EVM (in the context of animal treatment). In India, ethnoveterinary applications were in vogue since ancestor. Though there is no authentication when and how plants came into usage for curing the ailments of domestic animals. Traditional animal doctors are a substantial component of livestock healthcare systems in developing countries. The Indian subcontinent has a rich ethnoveterinary healthcare knowledge owing to the large agriculture-based livelihoods and rich biodiversity. Due to various rationales such as social, economic, and semi urbanization, this tradition is facing the threat of rapid depletion. On the basis of the global rate of species extinction, it is anticipated that around 800–1000 medicinally important plant species will face various degrees of threat across different biogeographic regions. International Union for the Conservation of Nature (IUCN) has identified and categorized about 200 species.

Several studies were made on ethnoveterinary medicinal plants and were published in *different countries* (Adolph et al., 1996; Gueye, 1999; Okoli et al., 2002; Lucia et al., 2003; Tafara et al., 2004; Fajima and Taiwo, 2005; Njoroge and Bussmann, 2006; Raynner et al., 2007; Farooq et al., 2008; Abu et al., 2009; Khan, 2009), *from India* (Jain, 1971, 1991, 2000; Pal, 1981; Sebastian, 1984; Mc Corkle, 1986; Reddy et al., 1991; Gaur et al., 1992; Sudarsanam et al., 1995; Geetha et al., 1996; Saiprasad and Pullaiah, 1996; Rajan and Sethuraman, 1997; Misra and Das, 1998; Reddy et al., 1998; Reddy and Raju, 1999; Singh and Kumar, 2000; Martin et al., 2001; Kumar et al., 2003; Kumar et al., 2004; Mokat and Deokule, 2004; Bandyopadyay and Mukarjee, 2005; Tiwari and Pande, 2006a; Mini and Sivadasan, 2007; Murthy et al., 2007; Ganesan et al., 2008; Praveen et al., 2010; Murty and Narasimha Rao, 2012; Kumar and Bharati, 2012a; Lakshminarayana and Narasimha Rao, 2013a,b; Sanjit et al., 2013 and Singh et al., 2014).

During the course of evolution, millions of secondary products have been synthesized from time to time by different plant species, perhaps to debunk infection caused by a variety of viruses, microorganisms like bacteria, fungi, and parasites. The presence of secondary metabolites has conferred on plants a selective advantage on the ability to be resistance to many pests and micro-organisms. Bell (1978) has suggested that plants synthesize a greater array of secondary compounds than animals because plants cannot move to escape from the predators and therefore, evolved a chemical defense against such predators. Many of these compounds that come under secondary metabolites can be defined as "those substances that are not required for the survival of the plant. They are removed as by-products during the primary or central metabolic pathways of the cell" (Fowler, 1984).

It is necessary to identify secondary metabolites of the medicinal plants which are traditionally used because they have the following features. They:

- Provide a scientific component to traditional approaches.
- Help in understanding functional similarities among different drugs.
- Provide better comprehension of toxicology.
- Help to establish a correlation between the scientific approach and traditional herbal practice.
- Provide tools to communicate with researchers and other practitioners.

Secondary metabolites have a broad scope of applications in the phar-maceutical, chemical, and food industries (George and Ravishankar, 1996). These have possessed phylogenetically as well as ecological significance. The existence of a common pattern of secondary compounds may indeed provide much clearer evidence of common ancestry than morphological similar attributable either to common lineage or to convergent evolution.

## 5.2   ETHNOVETERINARY STUDIES IN INDIA FROM THE 1900S

In the Indian subcontinent, several investigators conducted ethnobotanical and ethnoveterinary studies and gave a detailed data regarding the livestock management through the herbal care and the use of various parts of the plant species. Jain (1971) mentioned some magico-religious beliefs about plants such as *Abrus precatorius, Annona squamosa, Azadirachta indica, Cucumis sativus, Curcuma caesia, Curcuma longa, Datura metel, Ficus hispida, Piper nigrum, Phyllanthus emblica, Trachyspermum ammi, and Tinospora cordifolia are* among the Adivasis of Orissa. Issar (1981) reported traditionally important medicinal plants such as *F. hispida, Paederia scandens, P. nigrum, Prunus domestica, and Tylophora indica* for the folklore of

Uttaranchal Himalaya for animal treatments. Pal (1981) described the plants *Adiantum caudatum, Allium sativum, Cissampelos pareira, Cordia dichotoma, Cyperus rotundus, Marsilea minuta, Moringa oleifera, Pongamia pinnata, Shorea robusta, and Sterculia foetida are* used in the treatment of cattle and birds among the local inhabitants of Eastern India.

Pal (1992) reported 24 plant species belonging to 24 genera and 22 families which are used in veterinary medicine in West Bengal, Bihar, and Orissa. The results were compared through the consultation of old literature (Puranas) and herbaria information on plants used for veterinary medicine. Houghton (1995) mentioned the role of plants in traditional medicine and current therapy. Sudarsanam et al. (1995) described 106 plants species among which are *A. squamosa* (leaves), *Citrus aurantifolia* (leaves and fruits), *Curcuma caesia* (leaves), *D. metel* (fruits), *P. nigrum* (leaves), *Gymnema sylvestre* (leaves), *Plumbago zeylanica* (leaves) [Fig. 5.1, Plate 1], *Premna tomentosa* (roots and leaves), *T. cordifolia* (whole plant), *and T. indica* (fruit), and so forth, that are exclusively used by herbalists for different diseases affecting domestic animals such as anthrax, dysentery [Fig. 5.2, **Plate F**], dyspepsia, ephemeral fever, epistaxis, horn cancer, impaction, inflammatory diseases, maggot infested sores [**Plate C**], rheumatism, corneal opacity, retained placenta, tympany and bone fracture [**Plate G**], insect bites, and so forth. Geetha et al. (1996) studied on the Ethnoveterinary medicinal plants from Kolli hills, Tamil Nadu. Jain and De (1996) observed the ethnobotany of Purulia, West Bengal. Mc Corkle and Mathias (1996) documented the data regarding the ethnoveterinary research and development.

Saiprasad and Pullaiah (1996) enumerated 41 plant species, *A. precatorius, Albizia lebbeck, A. squamosa, Cardiospermum halicacabum, Costus speciosus* [**Plate 10**], *Cyanotis tuberosa, Elytraria acaulis, Gymnema sylvestre, Plumbago zeylanica* [**Plate 1**], *Premna tomentosa, Tinospora cordifolia,* and *T. indica.* These plants belong to 26 families used in EVM in Kurnool district, Andhra Pradesh. Chenchus, Erukalas, and Sugalis as the main tribes inhabiting the district. Vedavathy et al. (1997) worked on tribal medicine from Chittoor district, Andhra Pradesh, Tirupati. Reddy et al. (1997) studied some important medicinal plants utilize by people of the tribal regions (Chenchus, Erukalas, Yanadis, and Sugalis, etc.) of Cuddapah district, Andhra Pradesh, to treat ephemeral fevers and anthrax in cattle, during their ethno-medico-botanical survey, enumerated 17 interesting crude drugs used by local inhabitants which belongs to 16 genera and representing 14 families of angiosperms. Leaves, stem bark, shoots, flowers, and rhizomes were used for veterinary diseases.

Plate (1) *Plumbago zeylanica* L.     (2) *Ailanthus excelsa* Roxb.     (3) *Trichosanthes tricuspidata* Laur

(4) *Alangium salvifolium* L.F. Wag.     (5) *Aristolochia indica* L.     (6) *Balanites aegyptiaca* (L.) Del.

(7) *Bridelia retusa* L.     (8) *Cissus quadrangularis* L.     (9) *Madhuca Longifolia* (Koen.) Mac Br.

(10) *Costus speciosa* (Koen.) Sm.     (11) *Gloriosa superba* L.     (12) *Kalanchoe pinnata* (Lam.) Pers.

FIGURE 5.1   (See color insert.) Plant species of ethnoveterinary importance.

(13) *Macaranga peltata* (Roxb.) Muell-Arg.

(14) *Pergularia daemia* (Forsk.) Chiov.

(15) *Rubia cordifolia* L.

(16) *Strychnos potatorum* L.f.

*Aerva lanata* (L.) Juss. Ex. Sch.

*Amorphophallus paeoniifolius* (Dennst.) Nicolson

*Annona reticulata* L.

*Barringtonia acutangula* (Retz.) Willd.

*Cleistanthus collinus* (Roxb.) Benth. Ex Hook.f.

*Cochlospermum relisiosum* (L.) Alston

*Elephantophus scaber* L.

*Entada pursaetha* DC.

**FIGURE 5.1** *(Continued)*

*Grewia tiliaefolia* Van.

*Holoptelea integrifolia* (Roxb.)
Planch.

*Manilkara hexandra* (Roxb.)
Dubard.

*Pterolobium hexapetalum* (Roth)
Sant. & Wagh.

*Schleichera oleosa* (Lour) Koen

*Semicarpus anacardium* L.f.

**FIGURE 5.1**   *(Continued)*

Rajan and Sethuraman (1997) documented the traditional veterinary practices in rural areas of Dindigul district, Tamil Nadu, India. Girach et al. (1998) reported 25 plant species belonging to 19 families which are used for folk veterinary medicine in Bhadrak district of Orissa. Misra and Das (1998) reported 20 ethnoveterinary plants that belong to 14 families used against 10 animal diseases in Ganjam district of Orissa. Besides, these plants were used for mulching of cattle. Pandey et al. (1999) worked on *Achyranthes aspera, Argemone mexicana, Aristolochia bracteolate, Celosia argentea, Curculigo orchioides, Grewia damine, Holostemma ada-kodien, Solanum anguivi, Vitex negundo, and Tylophora fasciculata* of Gonda region, Uttar Pradesh, India.

## 5.3   ETHNOVETERINARY STUDIES IN INDIA FROM THE 2000S

Ramadas et al. (2000) documented 75 ethnoveterinary plant species of which 22 are exclusively used to treat wounds infested with maggots and the rest of the plants for the treatment of other diseases. Most of the plant species are well known for their antiseptic, astringent, and healing properties.

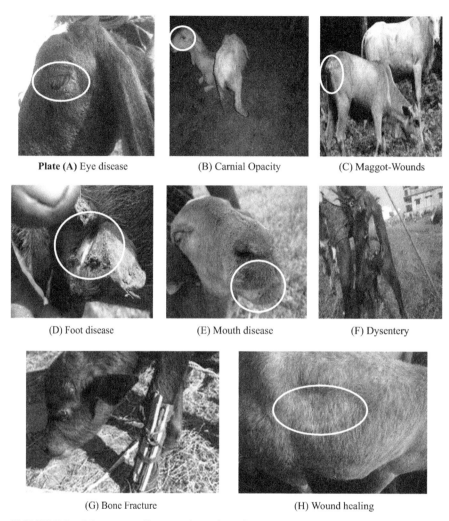

Plate **(A)** Eye disease        (B) Carnial Opacity        (C) Maggot-Wounds

(D) Foot disease        (E) Mouth disease        (F) Dysentery

(G) Bone Fracture        (H) Wound healing

**FIGURE 5.2**    Diseases or ailments of veterinary importance.

**FIGURE 5.3**    (**See color insert.**) Author working in the Field.

Singh and Kumar (2000) reported 14 ethnoveterinary plants that belong to 13 families used by ethnic groups like Gujjars, Gaddi, and other hill communities to treat various animal diseases. The Gaddis were migratory shepherds who practiced herbal therapy for their ailments, animal diseases, and disorders. Bhatt et al. (2001) reported the 32 ethnoveterinary plant species belong to 27 families used for folk veterinary medicine in Gujarat. Mouth diseases [**Plate E**], tympany, fever, swellings mastitis, wounds, foot disease [**Plate D**], ulcers, eczema, anestrous, flatulence, worms, boils, indigestion, septicemia, asthma, easy delivery, diarrhea, prolapse of uterus, and yolk galls were treated with the help of herbal medicine. Ghosh (2002) reported EVM preparations from different plants used by the tribal communities of Bankura and Medinipur districts of West Bengal against 16 common ailments of cattle. Jain (2003) reviewed an ethnoveterinary recipe which includes 836 plant species in EVM belonging to 159 families. Folk medicines among humans include about 2500 plant species; the ratio comes to 1:3. Species used for numerous diseases are fortunately common and familiar plants, for easy availability and teaching these are recommended for cultivation in available land in veterinary institutes. Kumar et al. (2003) reported Ethno-medico-botany of household remedies of Kolayat tehsil in Bikaner district, Rajasthan. Mistry et al. (2003) surveyed on the indigenous knowledge of animal healthcare practices in district Kachchh, Gujarat. Patil and Merat (2003) enumerated 26 plant species such as *Ailanthus excelsa* [**Plate 2**], *Careya arborea, Cissus quadrangularis* [**Plate 8**], *Dalbergia latifolia, Enicostemma axillare, Euphorbia tirucalli, Terminalia bellerica, Tinospora cordifolia, Trichoderma indicum, Uraria picta, and Xanthium strumarium* are used for the ethnoveterinary practices in Satpudas of Nandurbar district of Maharashtra. People inhibiting these areas rear cattle for livelihood, and hence, they mostly on the traditional system of medicine for the healthcare of their domestic animals. Rajagopalan and Harinarayanan (2003) worked on Veterinary wisdom for the Western Ghats.

Bisht et al. (2004) recorded the Ethnoveterinary practices of Kapkot block of Bageshwar district, Uttaranchal. Mandal et al. (2005a, 2005b) published valuable information on ethnomedicine like the information regarding plants uses in various diseases with remedies, the proportion of aids/plant parts, and mode of administration. Mokat and Deokule (2004) reported 36 plant species and parts (*Achyranthes aspera,* Young tender sticks), *Adhatoda vasica* (leaves), *A. squamosa* (leaves), *Asparagus racemosus* (tubers), *C. longa* (tuber and rhizome), *Dillenia indica* (peel), *Garcinia indica* (fruits), *Leea macrophylla* (root), *Mollugo spicata* (bark), *Ricinus communis* (leaves and roots), and *Trichosanthes tricuspidata* (leaves and roots) [**Plate 3**] used by villagers of the forest areas of Ratnagiri district, Maharashtra state as veterinary medicines. Wounds, fever, cough, vermicide, diarrhea, foot and mouth disease, bloat/tympany, skin diseases, increasing milk, scorpion bite, and rinderpest diseases are being cured by using different phytomedicines. Rafiuddin and Vaikos (2004) reviewed 27 plants used for animal treatment from Aurangabad district, Maharashtra. Sadangi and Sahu (2004) gathered the data on 42 plants representing 42 genera and belong to 29 families and these are used to treat 23 types of different veterinary diseases in the Kalahandi district of Orissa, India.

The following researchers worked on ethnoveterinary practices using many plant species for ailment treatment in different chronological years, respectively which is presented in Table 5.1. Thakur and Choudhary (2004); Tiwari and Pande (2004); Misra and Kumar (2004); Lakshmi and Laksh-minarayana (2005); Bandyopadhyay and Mukherjee (2005); Mandal et al. (2005c); Sikarwar and Vivek (2005); Tiwari and Pande (2005); Chitralekha and Jain (2006); Tripathi (2006); Jadeja et al. (2006); Kiruba et al. (2006); Paul and Pal (2006); Somvanshi (2006); Tiwari and Pande (2006); Tiwari and Pande (2006); Reddy et al. (2006); Achuta et al. (2007); Dhandapani et al. (2007); Dhandapani et al. (2007); Praveen et al. (2007); Mankad et al. (2007); Mini and Sivadasan (2007); Pande et al. (2007); Murthy et al. (2007); Reddi et al. (2008); Sanyasi Rao et al. (2008); Ganesan et al. (2008); Raneesh et al. (2008); Shah et al. (2008); Das and Tripathi (2009); Mautushi and Choudhury (2009); Saikia and Borthakur (2010); Nigam et al. (2010); Bharati and Sharma (2010); Mishra et al. (2010); Mokat et al. (2010); Praveen et al. (2010); Satapathy (2010); Thakur et al. (2010); Tiwari and Pande (2010); Murty and Narasimha Rao (2012); Rajakumar and Shivanna (2012); Kumar and Bharati (2012); Mishra (2013); Praveen et al. (2013); Praveen et al. (2013); Kumar and Bharati (2013); Lakshminarayana and Narasimha Rao (2013a); Lakshminarayana and Narasimha Rao (2013b) and Singh et al. (2014).

## 5.4 CONCLUSION

India has a great traditional background in the field of EVM and practices, but in the process of modernization, this knowledge is disappearing slowly. The role of EVM in livestock development is beyond dispute. Advanced research on plants of excessive medicinal values may lead to the new source of drugs which are really beneficial for the healthcare of mankind and other important domestic animals. Documentation and standardization of

**TABLE 5.1** Plants Utilized in Ethnoveterinary Practices and Mode of Administration

| S/ No. | Name of the species | Part(s) used | Ailment | Mode of administration |
|---|---|---|---|---|
| 1. | Abrus precatorius | Stem bark and | Anthrax | Stem bark along with leaves of V. negundo, tubers of Curculigo |
| 2. | Acorus calamus | leaves | | orchioides (each 50 g), 15 g pepper and garlic are pounded and |
| 3. | Adhatoda vasica | | | boiled in water and the decoction is given orally once daily for a |
| 4. | Alstonia scholaris | | | week. |
| 5 | Acacia nilotica | Stem bark, pod | Lactation, indigestion, gas problem | Tender pods are given every morning and evening to enhancing the lactation. The dried powdered bark is boiled in water, filtered and given orally to cattle twice a day. |
| 6 | Alangium salvifolium [Plate 4] | Root, seed | Dog bite | 100 g of root ground with 20 seeds of P. nigrum and a pinch of mustard oil. The paste is applied on the wounds caused by a dog bite. |
| 7 | Aloe vera | Leaf | Mastitis | The leaf pulp is applied on the swollen portion of the udder of cows or buffaloes. |
| 8 | Annona squamosa | | | |
| 9 | Amaranthus spinosus | Whole plant | Galactogogue | Entire plant crunched with cumin seeds and cloves is fed to cows and goats. |
| 10 | Andrographis paniculata | Whole plant | FMD | Powder of the aerial part, jaggery, pinch of rock salt, and water is given once a day for 5 days. |
| 11 | Aegle marmelos | Whole plant and leaves | Insect bite | The whole plant is given as a fodder to cure poisoning in animals due to an insect bite. |
| 12 | Apluda mutica | | | |
| 13 | Argemone Mexicana | Whole plant | Dyspepsia | A glass of whole plant decoction mixed with a powder of 21 seeds of P. nigrum is administered daily once for 2 days. |
| 14 | Aristolochia indica [Plate 5] | | | |
| 15 | Atylosia scarabaeoides | Leaf | Diarrhea | Leaf paste (300 g) is given with fodder to cattle to treat diarrhea. |
| 16 | Azadirachta indica | Leaf | Constipation, anthelmintic | Fruit paste is given to cattle for internal heat. The paste of the leaf mixed with equal quantity of turmeric powder is given once early in the morning for a week. Leaf paste in doses of 100 g twice a day for about 3 days is administered to treat cough, liver diseases and as anthelmintic. |

**TABLE 5.1** *(Continued)*

| S/ No. | Name of the species | Part(s) used | Ailment | Mode of administration |
|---|---|---|---|---|
| 17 | Balanites aegyptiaca [Plate 6] | Fruit | Corneal opacity | Fruit pulp with 10 leaves of Ocimum basilicum, tobacco snuff and turmeric are pounded and kept in a sealed box overnight. It is applied externally to the eye once a day. If both the eyes are infected, the medicine is applied on one of the eyes on the first day and the other eye the next day. |
| 18 | Barleria prionitis | Whole plant | Wounds | The whole plant is crushed and mixed with mustard oil. The preparation is applied on wounds of cattle. |
| 19 | Bridelia retusa [Plate 7] | | | |
| 20 | Brassica nigra | Oil, rhizome | Horn cancer | Pure mustard oil with rhizome paste of Curcuma longa is applied on the mischief parts of cattle horn. |
| 21 | Butea monosperma | | | |
| 22 | Cajanus cajan | Flower and pod | Diarrhea, dysentery | Cooked leaves are fed to cattle and boiled leaves and seeds are mixed with fodder in cattle. Green pods crushed and mixed with cold water are administered twice. |
| 23 | Calotropis procera | | | |
| 24 | Cannabis sativa | Leaf | Rheumatism, eczema | Dried leaves along with a small quantity of salt are given in goats. An aqueous paste of fresh leaves is applied on affected portion. |
| 25 | Carissa spinarum | Root | Wounds | Root is pounded with flowers of Madhuca longifolia [Plate 9] and made into a paste. This is applied to maggot-infested sores. Root paste applied on affected areas of cattle till cure. |
| 26 | Cassia tora | Seed | Galactogogue | Crushed seeds soaked in water overnight and given orally in the morning for 15 days. |
| 27 | Cassytha filiformis | Whole plant | Fractures | It is pounded in human urine and banded over the affected part of the bone. |
| 28 | Ceiba pentandra | Leaf, stem bark | Trypanosomiasis | Leaf and stem bark decoction (15 mL each) is administered twice daily for 3 days. |
| 29 | Chloroxylon swietenia | Leaf, stem bark | Wounds, ulcers, gall | Leaves are ground with turmeric and the paste is applied to the cattle. Stem bark ash mixed with coconut oil is applied locally on yoke gall. |

**TABLE 5.1** *(Continued)*

| S/No. | Name of the species | Part(s) used | Ailment | Mode of administration |
|---|---|---|---|---|
| 30 | Cedrus deodara | Leaf and fruit | Bronchitis, rinderpest | Pickle of the fruits made by adding salt and pounded turmeric is given to animal along with bread of Jowar to cure bronchitis. |
| 31 | Citrus aurantifolia | | | Crushed leaves mixed with and are orally given to animal for curing rinderpest. |
| 32 | Cleome viscosa | Leaf, whole plant | Cuts and injuries | Leaf paste is rubbed on the left horn if the right leg is cut and vice versa. This stops bleeding. The whole plant is crushed. The paste is applied on injuries of bullocks made during the ploughing. |
| 33 | Cocculus hirsutus | Leaf | Diarrhea | Leaf paste is given twice daily for 3 days to cattle. |
| 34 | Coleus amboinicus | Leaf | Tympany and lice | 250 g leaves of juice given orally twice for 4 days to treat tympany. Leaves with of Ximenia Americana, A. indica, and A. squamosa are taken in equal quantities and powdered. The powder is applied on the body of hens daily once for 3 days to eradicate lice in poultry. |
| 35 | Commelina benghalensis | Whole plant | Constipation, sores, helminthiasis | Plant paste is applied to treat yoke sores. Juice of the whole plant is given orally. The whole plant is given as a fodder. |
| 36 | Costus speciosus [Plate 10] | Rhizome | Jaundice | Rhizome used for jaundice. |
| 37 | Cryptolepis buchanani | Leaf | Lactation and galactagogue | Leaf paste in doses of 200 g once a day for 7–10 days for enhancing lactation. Leaf paste in doses of 200 g once a day for 7–10 days in cattle. |
| 38 | Cucumis melo | Seed | Indigestion and tympany | Powder of seeds is soaked in water; water is given orally to cure the gastric trouble. Seeds mixed with black salt are given for curing tympany. |
| 39 | Cuminum cyminum | | | |
| 40 | Curcuma caesia | Rhizome | Gout/Inflammation, dysentery | Fresh rhizome juice mixed with mustard oil is given once daily on empty stomach for 2–3 days in dysentery. It is also applied to cure to gout/inflammation disease. |
| 41 | Curcuma longa | | | |

**TABLE 5.1** (Continued)

| S/ No. | Name of the species | Part(s) used | Ailment | Mode of administration |
|---|---|---|---|---|
| 42 | Cynodon dactylon | Leaf | Mastitis | Fresh and pointed grass and Oryza sativa is used to open the blocked pore of the udder (Mastitis). |
| 43 | Datura metel | Fruit, seed, and Leaf | Diarrhea, dysentery, and insect bite | A paste of roasted unripe fruit in one dose is given orally. Second may be given. Seed powder is used in skin diseases and crushed leaf is applied on insect bite. |
| 44 | Dendrocalamus strictus | Leaf, rhizome and, Root | Anthrax and impaction | Leaves ground with those of ginger, turmeric and a pinch of calcium carbonate is given orally. Leaves ground with tubers of |
| 45 | Derris scandens | | | Gloriosa superba [Plate 11], onion and salt is given orally. |
| 46 | Dillenia pentagyna | Stem bark | Anthrax | Stem bark paste is fed to animal daily twice for 3–4 days to relieve anthrax. |
| 47 | Eclipta prostrata | Leaves | Sore, Ranikhet | Leaf juice is applied to on shoulders caused by carrying heavy loads and also used for swelling of ears in cattle. The alcoholic |
| 48 | Elsholtzia blanda | | | extract has antiviral against Ranikhet disease virus, especially in the hen. |
| 49 | Echinochloa crusgalli | Root, leaf | Maggot wounds, diarrhea | For maggot wounds, grind roots, and salt and wrap the bolus in grass and feed to the animal. 50 g leaves are made into a paste and |
| 50 | Elephantopus scaber | | | mixed with sugar candy which is given orally. |
| 51 | Euphorbia hirta | Leaf and latex | Kill worm | For carbuncle, latex is applied on the affected areas twice a day for 3 days to kill the worms. |
| 52 | Euphorbia neriifolia | | | |
| 53 | Euphorbia nivulia | | | |
| 54 | Fagonia indica | Stem bark, leaf | Constipation, FMD and bronchitis | Stem bark paste is given against constipation. Bark boiled in water for thirty minutes and the lukewarm leachate is applied on |
| 55 | Ficus hispida | | | the effected hoofs during foot and mouth diseases. Shade-dried |
| 56 | Ficus religiosa | | | leaves are powdered and fed two times a day for 7 days to cure bronchitis. |

**TABLE 5.1** (Continued)

| S/ No. | Name of the species | Part(s) used | Ailment | Mode of administration |
|---|---|---|---|---|
| 57 58 59 | G. superba [Plate 11] Glycine max Gomphrena serrata | Whole plant, leaf | Cough, cold | The whole plant with leaves of T. purpurea and seeds of P. nigrum are taken in equal quantities and ground. 50 g of paste with rice washed water is administered daily once for 3 days. |
| 60 61 | Helicteres isora Holarrhena pubescence | Leaf and stem bark | Helminthiasis | 10 g of stem bark paste mixed in a glass of water is administered daily twice for 2 days. Bark powder is applied externally to cure wounds. |
| 62 | Ixora pavetta | Stembark | Ephemeral fever | 30 g stem bark, 10 g garlic, 5 g pepper, 5 g turmeric, and C. spinarum, ground and it is given orally a dose of 50 g daily twice for 7 days. |
| 63 | Kalanchoe pinnata [Plate 12] | Leaf | Insect bite | Leaves massaged with sesame oil are slightly warmed and applied on the body of the cattle affected by bites of insect to get relief from pain. |
| 64 65 66 | Lawsonia inermis Lippia javanica Lippia nodiflora | Whole plant | Lice | For eradication of lice, whole plant used as lice repellent in poultry. |
| 67 | Lycopersicum esculentum | Fruit and leaf | Eye problem | Fruit and leaf juice is administered twice daily for 3 days against eye problem. |
| 68 69 | Macrotyloma uniflorum Mallotus philippensis | Leaf and seed | Galactoggue | 250 g of seeds boiled in water and mixed with husk powder of O. sativa is administered to cattle twice a day during the lactation period. Seed decoction is given orally to cattle for good mulching after delivery. |
| 70 71 72 | Melastoma malabathricum Morinda pubescens Mukia scabrella | Leaves and Stem bark | Rinderpest | Stem bark along with seeds of S. anacardium, camphor, and turmeric are allowed to ferment in water and the infusion is given orally. |
| 73 | Nicotiana tobaccum | Leaves | FMD | Processed leaves are crushed and made into a paste with saw wood and the paste is applied on the hoof of the cattle affected with foot and mouth disease. |

**TABLE 5.1** (Continued)

| S/ No. | Name of the species | Part(s) used | Ailment | Mode of administration |
|---|---|---|---|---|
| 74 | Oxalis corniculata | Root | Enhancing milk | A piece of root is collected on a Lunar eclipse day and burned with seven black peppers in the cattle shed to increase milk production of cows. |
| 75 | Paederia scandens | Flower, leaf and tuber | Anthrax | Leaves along with those of tubers of C. orchioides (each 100 g) and 10 g pepper and garlic pounded and the extract given orally twice daily (morning and evening) for a week. |
| 76 | Pennisetum typhoides | | | |
| 77 | Pergularia daemia [Plate 14] | | | |
| 78 | Phyllanthus niruri | Leaf and seeds | Dysentery | Leaf and seed paste is given in doses of 100 g once a day. |
| 79 | Piper nigrum | | | |
| 80 | Prunus cerasoides | | | |
| 81 | Pongamia pinnata | Leaves and stem bark | Wound | Stem bark along with that of Macaranga peltata [Plate 13] is taken in equal proportions and ground into a paste and is applied on wounds. Half of the above extract is given orally for deworming. |
| 82 | Pterocarpus marsupium | | | |
| 83 | Punica granatum | | | |
| 84 | Ricinus communis | Leaf and seeds | FMD | Powder of the plant materials, jaggery, a pinch of rock salt, and water is given once a day for 5 days |
| 85 | Rubia cordifolia [Plate 15] | | | |
| 86 | Saccharum officinarum | Leaf, stem bark and seeds | Fertility | Decoction of leaf and stem bark is administered once daily for 10 days against to induced fertility. |
| 87 | Salvadora oleoides | | | |
| 88 | Saraca asoca | | | |
| 89 | Semecarpus anacardium | Kernal | Cough | The kernel oil is used for cough. It is applied on the pharyngeal wall with the help of a long stick on alternate days. This treatment is very much use full for goats. |
| 90 | Shorea robusta | Stem bark and seed | Dysentery and worm | Bark paste in doses of 100 g is administered once a day for three days. Seed paste in the same dosage is used to kill worms in the intestine. |
| 91 | Swertia chirata | | | |
| 92 | Strychnos potatorum [Plate 16] | Seed | Eye problem | Seed paste with honey is poured into the eyes for eye infections in doses of few drops thrice a day for three days. |

**TABLE 5.1** (Continued)

| S/No. | Name of the species | Part(s) used | Ailment | Mode of administration |
|---|---|---|---|---|
| 93 94 95 96 97 | Tamarindus indica Tephrosia purpurea Tinospora cordifolia Trachyspermum ammi Tylophora indica | Whole plant leaves and seeds | FMD and infusion | For poultry diseases, infusion of the stem is put in chicken's drinking water. If animals feed on poisonous plants, then whole plant extract is given orally to animals causing vomiting. For foot and mouth, dried aerial pieces are fumigated to reach the mouth area of cattle disease. |
| 98 | Trianthema portulacastrum | Root and leaf | Eye problem and sore | Aqueous extract of root and leaves is used as eye drops for eye disease (Fig. 5.2, Plate A). Leaves ground with musk, saffron, and pepper given orally in sores. |
| 99 | Trigonella foenum–gracum | seed | Haematuria | Infusion of seed flour is given to the animal daily for three days. |
| 100 | Vitex negundo | Leaf | Wound | Dried leaf powder is dusted on the affected parts of the horse to heal wounds. Leaf paste mixed with a pinch of turmeric is applied on the wounds daily twice for three days. |
| 101 | Woodfordia fruticosa | Leaf | Sore | Leaves crushed with those of P. pinnata, Lannea coromandelica, and A. paniculata (each 10 g) and the paste is applied externally for wounds, ulcers, and maggot-infested sores. |
| 102 | Wrightia tinctoria | Leaf | Tympany, pain | 250 mg green leaves of each plant crushed together and juice is extracted. A spoonful of 'jeera' and 'Owa' powder mixture is added in it. This extract is mixed with in one liter of water and given to affected animals with the help of drenching bottles. It expels gas and relieves pain. |
| 103 | Xanthium indicum | Whole plant | Maggot | Whole plant juice in doses of 200 ml is given once a day for swellings on the glands in cattle. A paste of leaves is applied for maggot wounds. |
| 104 105 | Zingiber officinale Ziziphus nummularia | Rhizome and leaves | Dysentery, worm and galactagogue | Leaf and rhizome paste is given in doses of 200 g once a day. |

ethnoveterinary knowledge are also important in the context of Intellectual Property Rights (IPRs) to check the patent claims. Modern medicine plays a very crucial role in treating the chronic ailments of human beings and livestock. Recent developments in veterinary medicine really miraculous for the healthcare of livestock. But still today, people from interior pockets and tribal hamlets of India rely on plants and their products for overcome the health disorders of the man and his pet animals. Ethnoveterinary applications play an important role in the regions where allopathic services could not be reached or available. So, further investigations and exploration work on ethnoveterinary practices are beneficial for the progress and development of new avenues. There is an urgent need for biochemical analysis and pharmaceutical investigations of plant species used by the people of this region to formulate and standardize the medicine for sustainable uses, progress, and development. Some of the valuable plant species may become rare, endangered and threatened in this area due to overexploitation. So, it is suggested that *ex-situ* and *in-situ* conservation is needed to overcome the above said problem.

## KEYWORDS

- **ethnoveterinary**
- **ethnomedicine**
- **veterinary medicine**
- **medicinal plants**
- **secondary products**

## REFERENCES

Abu, A. H.; Ofukwu, R. A.; Mazawaje, D. A. A Study of Traditional Animal Health Care in Nasarawa State, Nigeria. *Am.-Eur. Sustainable Agric.* **2009,** *3*(3), 468–472.

Achuta, N. S.; Singh, K. P.; Kumar, A. Ethnoveterinary Uses of Plants from Achanakmar-Amarkantak Biosphere Reservoir, Madhya Pradesh and Chhattisgarh. *J. Non-Timber For. Prod.* **2007,** *14*(1), 53–55.

Adolph, D.; Blakeway, S.; Linquist, B. J. Ethno-Veterinary knowledge of the Southern Sudan. *A study for the UNICEF Operation lifeline Sudan/Southern Sector Livestock Programme*; Nairobi, Kenya, 1996; pp 1–57.

Bandyopadhyay, S.; Mukherjee S. K. R. Ethnoveterinary Medicine from Koch Bihar District, West Bengal. *Indian J. Tradit. Knowl.* **2005**, *4*(4), 456–461.

Bell, E. A. Toxins in Seeds. In *Biochemical Aspects of Plant–Animal Co-Evolution;* J.B. Harborne., Ed.; Academic Press: London, 1978; pp 143–161.

Bharati, K. A.; Sharma, B. L. Some ethnoveterinary plant records for Sikkim Himalaya. *Indian J. Tradit. Knowl.* **2010**, *9*(2), 344–366.

Bhatt, D. C.; Mitaliya, K. D.; Mehta, S. K. Observations on Ethnoveterinary Herbal Practices in Gujarat. *Ethnobotany* **2001**, *13,* 91–95.

Bisht, N.; Pande, P. C.; Tiwari, L. Ethnoveterinary Practices of Kapkot Block of Bageshwar District, Uttaranchal. *Asian Agri. Hist.* **2004**, *8*(4), 209–314.

Chitralekha, K.; Jain, A. K. Plants Used in Ethnoveterinary Practices in Jhabua District, Madhya Pradesh. *Ethnobotany* **2006**, *18,* 149–152.

Daily, G. C *Science* **1995**, *269,* 350–354.

Das, S. K.; Tripathi, H. Ethnoveterinary practices and socio-cultural values associated with animal husbandary in rural Sunderbans, West Bengal. *Indian J. Tradit. Knowl.* **2009**, *8*(2), 201–205.

Dhandapani, R.; Kavitha, P.; Balu, S. Ethnoveterinary Medicinal Plants of Perambadur District, Tamil Nadu, India. *Adv. Plant. Sci.* **2007**, *20*(2), 563–566.

Fajima, A. K.; Taiwo, A. A. A Review on Herbal Remedies in Animal Parasitic Diseases in Nigeria. *Afr. J. Biotechnol.* **2005**, *4*(4), 303–307.

Farooq, Z.; Iqbal, Z.; Mushtag, S.; Muhammad, G.; Iqbal, M. Z.; Arshad, M. Ethnoveterinary Practices for the Treatment of Parasitic Diseases in Livestock in Cholistan Desert, Pakistan. *J. Ethnopharmacology.* **2008**, *118*(2), 213–219.

Fowler, M. W. Plant-Cell Culture: Natural Products and Industrial Application. *Biotechnol. Gen. Eng. Rev.* **1984**, *2,* 41–44.

Ganesan, S.; Chandrasekaran, M.; Selvaraj, A. Ethnoveterinary Healthcare Practices in Southern Districts of Tamil Nadu. *Indian J. Tradit. Knowl.* **2008**, *7*(2), 347–354.

Gaur, R. D.; Bhatt, K. C.; Tiwari, J. K. An Ethnobotanical Study of Uttar Pradesh, Himalaya in Relation to Veterinary Medicines. *J. Econ. Bot. Soc.* **1992**, *72*:139–144.

Geetha, S.; Lakshmi, G.; Ranjithakani, P. Ethno-veterinary Medicinal Plants of Kolli Hills, Tamil Nadu. *J. Econ. Taxon. Bot. Add. Ser.* **1996**, *12,* 289–291.

George, J.; Ravishankar, G. A. Harnessing High Value Metabolites from Plant Cells. *Indian. J. Pharm. Educ.* **1996**, *30,* 120–129.

Ghosh, A. Ethnoveterinary Medicines from the Tribal Areas of Bankura and Medinipur Districts, West Bengal. *Indian J. Tradit. Knowl.* **2002**, *1*(1), 93–95.

Girach, R. D.; Brahmam, M.; Misra, M. K. Folk Veterinary Medicine of Bhadrak District, Orissa, India. *Ethnobotany* **1998**, *10,* 85–88.

Gueye, E. F. Ethnoveterinary Medicine Against Poultry Diseases in African Villages. *World's poult. Sci. J.* **1999**, *55,* 187–198.

Houghton, P. J. The Role of Plants in Traditional Medicine and Current Therapy. *J. Altern. Complementary Med.* **1995**, *1,* 131–143.

Issar, R. K. Traditionally Important Medicinal Plants and Folklore of Uttaranchal Himalaya for Animal Treatments. *J. Sci. Res. Pl. Med.* **1981**, *2,* 61–66.

Jadeja, B. A.; Odedra, N. K.; Solanki, K. M.; Boraiya, N. M. Indigenous Animal Healthcare Practices in District Porbandar, Gujarat. *Indian J. Tradit. Knowl.* **2006**, *5*(2), 253–258.

Jain, S. K. *Dictionary of Indian folk medicine and Ethnobotany;* Deep Publications: New Delhi, 1991; pp 1–311.

Jain, S. K. Ethnoveterinary Recipes in India – a Botanical Analysis. *Ethnobotany* **2003**, *15,* 23–33.

Jain, S. K. Plants in Indian Ethnoveterinary Medicine Status and Prospects. *Indian J. Vet. Med.* **2000**, *20*(1), 1–11.

Jain, S. K. Some Magico-Religious Beliefs About Plants Among Adivasis of Orissa. *Adivasi* **1971**, *12*(1–4), 39–44.

Jain, S. K.; De, J. N. Observations of Ethnobotany of Purulia, West Bengal. *Bull. Bot. Surv. India* **1996**, *8*, 237–257.

Katerere, D. R.; Luseba, D. *Ethnoveterinary Botanical Medicine: Herbal Medicines for Animal Health;* CRC Press: Boca Raton, 2010.

Khan, F. M. Ethnoveterinary Medicinal Usage of Flora of Greater Cholistan Desert (Pakistan). *Pak. Vet. J.* **2009**, *29*(2), 75–80.

Kiruba, S.; Jeeva, S.; Dhas, S. S. M. Enumeration of Ethnoveterinary Plants of Cape Comorin, Tamil Nadu. *Indian J. Tradit. Knowl.* **2006**, *5*(4), 576–578.

Kumar, R.; Bharati, K. A. Folk Veterinary Medicine in Sitapur District of Uttar Pradesh, India. *Indian J. Nat. Prod. Resour.* **2012b**, *3*(2), 267–277.

Kumar, R.; Bharati, K. A. Folk Veterinary Medicines in Jalaun District of Uttar Pradesh, India. *Indian J. Tradit. Knowl.* **2012a**, *11*(2), 288–295.

Kumar, S.; Goyal, S.; Frazana parveen. Ethnoveterinary plants in Indian arid zone. *Ethnobotany.* **2004**, *16*, 91–95.

Kumar, S.; Goyal, S.; Praveen, F. Ethno-Medico-Botany of Household Remedies of Kolayat Tehsil in Bikaner District, Rajasthan. *Indian J. Tradit. Knowl.* **2003**, *2*(4), 357–365.

Lakshmi, M. K.; Lakshminarayana, K. Observations on Ethnoveterinary Practices in Vizianagaram District, Andhra Pradesh, India. *Recent Trends in Plant Sciences* **2005**, 104–111.

Lakshminarayana, V.; Narasimha Rao, G. M. Ethnoveterinary Practices in Northcoastal Districts of Andhra Pradesh, India. *J. Nat. Rem.* **2013a**, *13*(2), 109–117.

Lakshminarayana, V.; Narasimha Rao, G. M. Traditional Veterinary Medicinal Practices in Srikakulam District of Andhra Pradesh, India. *Asian J. Exp. Biol. Sci.* **2013b**, *4*(3), 476–479.

Lucia, V.; Andrea, P.; Paolo, M. G.; Roberte, V. A Review of Plants Used in Folk Veterinary Medicine in Italy as Basis for a Databank. *J. Ethnopharmacology* **2003**, *89*, 221–244.

Mandal, S.; Shyam; Faizi. Ali Akhtar.; Sinha, A. Ethnomedicine: A Template of Modern Drug Development. In *Ethnomedicine and Human Welfare;* Khan, I. A., Khanam, A., et al., Eds.; Ukaaz Publications: Hyderabad, 2005a; pp 250–261.

Mandal, S.; Shyam; Maurya, K. R.; Sinha, N. K. Ethnomedicine of Weeds. In *Ethnomedicine in Human Welfare;* Khan, I. A., Khanam, A., et al., Eds.; Ukaaz Publications: Hyderabad, 2005b; pp 232–243.

Mandal, S.; Shyam; Maurya, K. R.; Sinha, N. K. Ethnoveterinary Phytomedicine. In: *Role of Biotechnology in Medicinal and Aromatic Plants;* Irfan Ali Khan, Atiya Khanam, et al., Eds.; Ukaaz publications: Hyderabad, India, 2005c, 14, pp 262–317.

Mankad, R. S.; Jedaja, B. A.; Chavda, G. K. Ethnoveterinary Knowledge System of Maldhari Tribe of Girnar Junagadh District, Gujarat, India. *Plant Arch.* **2007**, *7*(2), 843–845.

Martin, M.; Mathias, E.; Mc Corkle, C.M. *Ethnoveterinary Medicine: An Annotated Bibliography of Community Animal Healthcare;* ITDG Publishing: London, UK, 2001.

Mautushi, N.; Dutta, C. M. Ethnoveterinary Practices by Hamar Tribe in Cachar District, Assam. *Ethnobotany* **2009**, *21*, 61–65.

Mc Corkle, C. M. An Introduction to Ethnoveterinary Research and Development. *J. Ethnobiology* **1986**, *6*, 129–149.

Mc Corkle, C. M.; Mathias, E. *Ethnoveterinary Research and Development;* Schillhorn Van Veen, T. W. Ed.; Intermediate Technology publications: London, 1996.

Mini, V.; Sivadasan, M. Plants Used in Ethnoveterinary Medicine by Kurichya Tribe of Wayanad District in Kerala, India. *Ethnobotany* **2007,** *19,* 94–99.

Mishra, D. Cattle Wounds and Ethno-veterinary Medicine: A Study in Polasara Block, Ganjam District, Orissa. India. *Indian J. Tradit. Knowl.* **2013,** *12*(1), 62–65.

Mishra, S.; Sharma, S.; Vasudevan, P.; Bhatt, R. K.; Pandey, S.; Singh, M.; Meena, B. S.; Pandey, S. N. Livestock Feeding and Traditional Healthcare Practices in Bundelkhand Region of Central India. *Indian J. Tradit. Knowl.* **2010,** *9,* (2), 333–337.

Misra, K. K.; Kumar, K. A. Ethnoveterinary Practices Among the Konda Reddi of East Godavari District of Andhra Pradesh. *Stud. Tribes Tribals* **2004,** *2*(1), 37–44.

Misra, M. K.; Das, S. S. Veterinary use of Plants among Tribals of Orissa. *Anc. Sci. Life* **1998,** *17*(3), 214–219.

Mistry, N.; Silori, C. S.; Gupta, L.; Dixit, A. M. Indigenous Knowledge on Animal Healthcare Practices in District Kachchh, Gujarat. *Indian J. Tradit. Knowl.* **2003,** *2*(3), 240–255.

Mokat, D. N.; Deokule, S. S. Plants Used as Veterinary Medicine in Ratnagiri District of Maharashtra. *Ethnobotany* **2004,** *16,* 131–135.

Mokat, D. N.; Mane, A. V.; Deokule, S. S. Plants Used for Veterinary Medicine and Botanical Pesticide in Thane and Ratnagiri District of Maharashtra. *J. Non-Timber For. Prod.* **2010,** *17*(4), 473–476.

Murthy, E. N.; Reddy, Ch. S.; Reddy, K. N.; Raju, V. S. Plants Used in Ethnoveterinary Practices by Koyas of Pakhal Wildlife Sanctuary, Andhra Pradesh, India. *Ethnobotanical Leafl.* **2007,** *11,* 1–5.

Murty, P. P.; Narasimha Rao, G. M. Ethnoveterinary Medicinal Practices in Tribal Regions of Andhra Pradesh. *Bangladesh J. Plant. Taxon.* **2012,** *19*(1), 7–16.

Myer, N. *Global Planet. Change* **1990,** *2,* 175–185.

Nigam, G; Narendra; Sharma, K. Ethnoveterinary Plants of Jhansi District, Uttar Pradesh. *Indian J. Tradit. Knowl.* **2010,** *9*(4), 664–667.

Njoroge, G. N.; Bussmann, R. W. Herbal Usage and Informant Consensus in Ethnoveterinary Management of Cattle Diseases Among the Kikuyus, Central Kenya. *J. Ethnopharmacology* **2006,** *108*(3), 332–339.

Okoli, I. C.; Okoli; C. G.; Ebere, C. S. Indigenous Livestock Production Paradigms Revisited. Survey of Plants of Ethnoveterinary Importance in Southeastern Nigeria. *J. Trop. Ecol.* **2002,** *43*(2), 257–263.

Pal, D. C. Observation on Folklore Plants Used in Veterinary Medicine in Bengal, Bihar and Orissa-II. *J. Econ. Taxon. Bot.* **1992,** *10,* 137–141.

Pal, D. C. Plants Used in Treatment of Cattle and Birds Among Tribals of Eastern India. In *Glimpses of India Ethnobotany;* Jain, S. K., Ed.; Oxford & IBH: New Delhi, 1981; pp 245–257.

Pande, P. C.; Tiwari, L.; Pande, H. C. Ethnoveterinary Plants of Uttaranchal- A Review. *Indian J. Tradit. Knowl.* **2007,** *6*(3), 444–458.

Pandey, P. H.; Varma, B. K.; Narain, S. Ethnoveterinary Plants of Gonda Region, Uttar Pradesh, India. *J. Econ. Taxon. Bot.* **1999,** *23*(1), 199–203.

Patil, S. H.; Manoj M. M. Ethnoveterinary Practices in Satpurdas of Nandurbar District of Maharashtra. *Ethnobotany* **2003,** *15,* 103–106.

Paul, C. R.; Pal, D. C. Traditional Knowledge System About Veterinary Healthcare in and Around of Bankura District, West Bengal, India. In *Herbal Medicine Traditional Practices;* Trivedi, P. C. Ed.; Aavishkar Publishers: Distributors, Jaipur, India, 2006; pp 194–199.

Praveen, G., Jain, A. and Katewa, S.S. Ethnoveterinary medicines used by tribals of Tadgarh-Raoli wildlife sanctuary, Rajasthan. India. Indian J. Tradit. Knowle. **2013b,** *12*(1):56–61.

Praveen, G., Jain, A. and Katewa, S.S. Traditional veterinary medicines used by livestock owners of Rajasthan. India. *Indian J. Tradit. Knowle.* **2013a**, *12*(1):47–55.

Praveen, G.; Jain, A.; Katewa, S. S.; Nag, A.. Animal Healthcare Practices by Livestock Owners at Pushkar Animal Fair, Rajasthan. *Indian J. Tradit. Knowl.* **2010**, *9*(3), 581–584.

Praveen, G.; Nag, A.; Katewa, S. S. Traditional Herbal Veterinary Medicines from Mount Abu, Rajasthan. *Ethnobotany* **2007**, *19*, 120–123.

Priyadarshan, P. E. Herbal Veterinary Medicines in an Ancient Sanskrit Work-The Garuda Purana. *Ethnobotany* **1991**, *3*, 83.

Rafiuddin; N.; Vaikos, N.P. Plants Used in Ethnoveterinary Practices in Aurangabad District, Maharashtra-Ii, India. In *Focus on Sacred Groves and Ethnobotany;* Prism Publications: Mumbai, India, 2004; pp 223–227.

Rajagopalan, C. C.; Hair Narayanan, M. Veterinary Wisdom for the Western Ghats. Amruth., **2003**. *2*(2), 15–17.

Rajakumar, N.; Shivanna, M. B. Traditional Veterinary Healthcare Practices in Shimoga District of Karnataka. India. *Indian J. Tradit. Knowl.* **2012**, *11*(2), 283–287.

Rajan, S.; Sethuraman, M. Traditional Veterinary Practices in Rural Areas of Dindigul District, Tamil Nadu, India. *Indigenous Knowl. Dev. Monit.*, **1997**, *5*, 7–9.

Ramadas, S. R.; Ghotge. N. S.; Ashalata, S.; Mathur, N. P.; Broome, V. G.; Sanyasi Rao. Ethnoveterinary Remedies Used in Common Surgical Conditions in Some Districts of Andhra Pradesh and Maharashtra, India. *Ethnobotany* **2000**, *12*, 100–112.

Raneesh, S.; Hafeel, A.; Hariramamurti, B. A.; Unnikrishnan, P. M. Documentation and Participatory Rapid Assessment of Ethnoveterinary Practices. *Indian J. Tradit. Knowl.* **2008**, *7*(2), 360–364.

Raven, P. H. *We are Killing Our World: The Global Ecosystem Crisis;* Macarthur Foundation: Chicago, 1987.

Raynner R. D. Barboza.; Wedson de M.S. Souto.; Jose da, S. Mourão. The Use of Zoo Therapeutics in Folk Veterinary Medicine in the District of Cubati, Paraiba State, Brazil. *J. Ethnobiology Ethnomed.* **2007**, *3*(32), 1–14.

Reddi, T. V. V. S.; Prasanthi, S.; Rama Rao Naidu, B. V. A.; Nagaiah, K. Folk Veterinary Phytomedicine. In *Role of Biotechnology in Medicinal and Aromatic Plants*; Ali Khan, I., Khanam, A., Eds.; Ukaaz Publications: Hyderabad, India, 2008; Vol. 1.

Reddy, K. N.; Bhanja, M. R.; Raju, V. S. Plants Used in Ethnoveterinary Practices in Warangal District, Andhra Pradesh, India. *Ethnobotany* **1998**, *10*, 75–84.

Reddy, K. N.; Subbaraju, G. V.; Reddy, C. S.; Raju, V. S. Ethnoveterinary Medicine for Treating Livestock in Eastern Ghats of Andhra Pradesh. *Indian J. Tradit. Knowl.* **2006**, *5*(3), 368–372.

Reddy, K. N.; Venkata, Raju R. R. Plants in Ethnoveterinary Practices in Anantapur District, Andhra Pradesh. *J. Econ. Tax. Bot.* **1999**, *23*(2), 347–357.

Reddy, R. D.; Prasad, M. K.; Venkaiah, K. Forest Flora of Andhra Pradesh: Vernacular names. Research and Development Circle. A.P. Forest Department of Hyderabad, 1991.

Reddy, R. V.; Lakshmi, N. V. N.; Venkata Raju, R. R. Ethnomedicine for Ephemeral Fevers and Anthrax in Cattle from the Hills of Cuddapah District, Andhra Pradesh, India. *Ethnobotany* **1997**, *9*, 94–96.

Sadangi, N.; Sahu, R. K. Traditional Veterinary Herbal Practices of Kalahandi District, Orissa, India *J. Nat. Rem.* **2004**, *4*(2), 131–136.

Saikia, B.; Borthakur, S. K. Use of Medicinal Plants in Animal Healthcare-A Case Study from Gohpur, Assam. *Indian J. Tradit. Knowl.* **2010**, *9*(1), 49–51.

Saiprasad Goud, P.; Pullaiah, T. Folk Veterinary Medicine of Kurnool District, Andhra Pradesh, India. *Ethnobotany* **1996**, *8*, 71–74.

Sanjit, M.; Chakravarty, P.; Sanchita, G.; Bandyopadhyay, S.; Chouhan, V. S. Ethno-Veterinary Practices for Ephemeral Fever of Yak: A Participatory Assessment by the *Monpa* Tribe of Arunachal Pradesh. *Indian J. Tradit. Knowl.* **2013,** *12*(1), 36–39.

Sanyasi Rao, M. L.; Varma, Y. N. R.; Kumar, V. Ethnoveterinary Medicinal Plants of the Catchments Area of the River Papagni in the Chittoor and Anantapur Districts of Andhra Pradesh, India. *Ethnobotanical leafl.* **2008,** *12,* 217–226.

Satapathy, K. B. Ethnoveterinary Practices in Jajpur District of Orissa. *Indian J. Tradit. Knowl.* **2010,** *2,* 338–343.

Sebastian, M. K. Plants Used as Veterinary Medicines by Bhils. *Int. J. Trop. Agric.* **1984,** *2,* 307–310.

Shah, R.; Pande, P. C.; Tiwari, L. Traditional Veterinary Herbal Medicines of Western Parts of Almora District, Uttarakhand Himalaya. *Indian J. Tradit. Knowl.* **2008,** *7*(2), 355–359.

Sikarwar, R. L. S; Kumar, V. Ethnoveterinary Knowledge and Practices Prevalent Among the Tribals of Central India. *J. Nat. Rem.* **2005,** *5*(2), 147–152.

Singh, D.; Kachhawaha, S.; Choudhary, M. K.; Meena, M. L.; Tomar, P. K. Ethnoveterinary Knowkedge of *Raikas* of Marwar for Nomadic Pastoralism. *Indian J. Tradit. Knowl.* **2014,** *13*(1), 123–131.

Singh, K. K.; Kumar, K. Observations on Ethnoveterinary Medicine Among the Gaddi Tribe of Kangra Valley, Himachal Pradesh. *Ethnobotany* **2000,** *12,* 42–44.

Somvanshi, R. Veterinary Medicine and Animal Keeping in Ancient India. *Asian Agri-Hist.* **2006,** *10*(2), 133–146.

Sudarsanam, G.; Reddy, M. B.; Nagaraju, N. Veterinary Crude Drugs in Rayalaseema, Andhra Pradesh, India. *Int. J. Pharmacogn.* **1995,** *33*(1), 52–60.

Tafara, M.; Taona, M.; B. Ethnoveterinary Medicine: A Potential Alternative to Orthodox Animal Health Delivery in Zimbabwe. *Int. J. Appl. Res. Vet. Med.* **2004,** *2*(4), 269–273.

Thakur, H. K.; Choudhary, B. L. Folk Herbal Veterinary Medicines of Southern Rajasthan. *Indian J. Tradit. Knowl.* **2004,** *3*(4), 407–418.

Thakur, S. D.; Hanif, S. M.; Chauhan, N.S. Plants in Ethnoveterinary Practice in Kullu District-Himachal Pradesh. *J. Non-Timber For. Prod.* **2010,** *17*(1), 55–58.

Tiwari, L.; Pande, P. C. Ethnoveterinary Medicines in Indian Perspective: Reference to Uttarakhand, Himalaya. *Indian J. Tradit. Knowl.* **2010,** *9*(3), 611–617.

Tiwari, L.; Pande, P. C. Ethnoveterinary Medicines of Holy Doonagiri Hills of Uttaranchal. *J. Econ. Taxon. Bot.* **2006a,** *30,* 51–162.

Tiwari, L.; Pande, P. C. Ethnoveterinary Plants of Uttarkashi district, Uttaranchal, India. *Ethnobotany* **2006b,** *18*(1&2), 139–144.

Tiwari, L.; Pande, P. C. Indigenous Veterinary Practices of Darma Valley of Pithoragarh District, Uttaranchal Himalaya. *Indian J. Tradit. Knowl.* **2006c,** *5*(2), 201–206.

Tiwari, L.; Pande, P. C. Traditional Veterinary Medicinal Plants of Bhilangana Valley of Tehri District, Uttaranchal Himalaya. *Asian Agri. Hist.* **2005,** *9*(3), 253–262.

Tiwari, L.; Pande, P. C. Traditional Veterinary Practices in South-Eastern Part of Chamoli District, Uttaranchal. *Indian J. Tradit. Knowl.* **2004,** *3*(4), 397–406.

Tripathi, H. Approaches in Documenting Ethnoveterinary Practices. *Indian J. Tradit. Knowl.* **2006,** *5*(4), 579–581.

Vedavathy, S.; Mridula, V.; Sudhakar, A. Tribal Medicine of Chittoor. *Anc. Sci. Life* **1997.** 16(4), 307–331.

# PART II
# Medicinal Potentials of Phytochemicals

# CHAPTER 6

# ANTIOXIDANTS AND PHYTOCHEMICALS

CHUKWUEBUKA EGBUNA

*Department of Biochemistry, Faculty of Natural Sciences, Chukwuemeka Odumegwu Ojukwu University, Anambra State 431124, Nigeria, E-mail: egbuna.cg@coou.edu.ng; egbunachukwuebuka@gmail.com; Tel. +2347039618485*

*ORCID: https://orcid.org/0000-0001-8382-0693*

## ABSTRACT

Antioxidants are substances that protect the cells in an organism from damage against the destructive tendencies of reactive species. They act by preventing oxidation of important macromolecules through many mechanisms such as scavenging or chelation of toxic radicals. Antioxidants can be grouped as either enzymatic antioxidants (superoxide dismutase, catalase (CAT), and glutathione peroxidase (GPX)) or nonenzymatic antioxidants (glutathione (GSH), vitamins C, E, ubiquinones, carotenoids, β-carotene, lycopene, polyphenols, uric acid, and mineral antioxidants: zinc, copper, selenium). Antioxidants can also be grouped according to its line of defense. Those in the first line of defense include GSH, superoxide dismutase (SOD), and CAT, while the linebackers or second line of defense are vitamins A, C, D, and E. The third line of defense is the carotenoids, the bioflavonoid, and coenzyme Q10. This chapter discusses antioxidants, their functions, mechanisms, and types.

## 6.1 INTRODUCTION

An antioxidant-rich diet is important for the body to utilize substances that can protect the body against the destructive effects of free radicals. The antioxidants defense systems are known for their neutralizing potentials. A manifold of compounds such as flavonoids, uric acid, vitamin E, vitamin C, glutathione (GSH), and various enzymes (catalase (CAT), superoxide dismutase (SOD), and glutathione peroxidase (GPX)) have been described as antioxidants (Egbuna and Ifemeje, 2017). According to Walks and Doyle (2011), antioxidants usually exist in two forms, their active (reduced) or oxidized form. Each antioxidant can become oxidized themselves while other antioxidants come along and donate electrons to activate them. For example, vitamins A, C, and E protect each other in a circular fashion. Vitamin A regenerates vitamin C, vitamin C regenerates vitamin E, and vitamin E regenerates vitamin A. The antioxidant system has other backup systems, for example, beta-carotene converts to vitamin A in the liver, so it serves as a backup to vitamin A. Bioflavonoid enhance and prolong the antioxidant action of vitamin C and coenzyme $Q_{10}$ works in place of vitamin E when vitamin E levels are low. Antioxidants act in a way which makes it difficult to be understood at a glance. However, scientists are tapping on existing information to unravel the mechanisms of those not fully examined.

## 6.2   FUNCTIONS OF ANTIOXIDANTS

According to McDowell et al. (2007), antioxidants are effective through the listed mechanisms:

1.  Preventive antioxidants: Some antioxidants act by preventing oxidation of vital biomolecules.
2.  Free radical scavengers: These include antioxidants that scavenge free radicals while themselves might be temporarily become radicals but quickly converted back to its reduced form.
3.  Sequestration of metal by chelation: Some antioxidants prevent the body from damages by trapping toxic metallic radicals.
4.  Quenching of active oxygen species (AOSs): This group annuls the oxidative effects of AOSs.

## 6.3   TYPES OF ANTIOXIDANTS

Antioxidants can be grouped in different ways but more generally could be one of the enzymatic antioxidants and nonenzymatic antioxidants. Groupings according to its lines of defense make it very easy to comprehend.

Antioxidants in the first line of defense include glutathione, SOD, and CAT. The linebackers or second line of defense are vitamins A, C, D, and E while the third line of defense is the carotenoids, bioflavonoid, and coenzyme $Q_{10}$ (Walks and Doyle, 2011).

## 6.3.1 ENZYMATIC ANTIOXIDANTS

Enzymatic antioxidants are an indispensable class of antioxidants that are crucial in handling the ravaging effects of reactive oxygen species (ROSs). These enzymes over time has been studied and monitored closely and found to occur in higher concentrations with higher activity in organisms prone to oxidative stress. In one study by Ifemeje et al. (2015), revealed that earthworm samples harvested from active waste dump site had higher activity compared to those found in withdrawing/dormant dumpsite. This is a clear indication that external factor could stimulate the endogenous production of enzymatic antioxidants. These enzymes are often multimeric and require certain types of metals to function properly. This implies that adequate supply of antioxidant minerals in food substances helps in the fight against the destructive effects of ROSs. These enzymes include SOD, CAT, and GPX.

### 6.3.1.1 SUPEROXIDE DISMUTASE (SOD) (EC 1.15.1.1)

Dismutation is a term that refers to a special type of reaction, where two equal but opposite reactions occur on two separate molecules. As the name implies, SOD, takes two molecules of superoxide, strips the extra electron off of one, and places it on the other. So, one ends up with an electron less, forming normal oxygen, and the other ends up with an extra electron.

$$O_2 + e^- \rightarrow .O_2^- \text{ (Formation of destructive superoxide)}$$

$$O_2 + .O_2^- + 2H^+ \rightarrow 2H_2O_2 + O_2 \text{ (A typical reaction catalyzed by SOD)}$$

The one with the extra electron then rapidly picksup two hydrogen ions to form hydrogen peroxide which the cell must detoxify through the help of CAT enzyme because hydrogen peroxide is also a dangerous compound (David, 2007; Ifemeje et al., 2015). SOD is an enzyme that alternately catalyzes the dismutation (or partitioning) of the superoxide ($O_2^-$) radical into either ordinary molecular oxygen ($O_2$) or hydrogen peroxide ($H_2O_2$).

All SODs, irrespective of source, are multimeric metalloproteins that are very efficient at scavenging the superoxide radical. SODs could contain copper and zinc (Cu/ZnSOD), manganese (MnSOD), or iron (FeSOD). With a few exceptions, Cu/ZnSODs are generally found in the cytosol of eukaryotic cells and chloroplasts; the MnSODs are found in the matrix of mitochondria and in prokaryotes; the FeSODs are generally found in prokaryotes and have been reported to exist in some plants (Duke and Salin, 1985). SOD neutralizes superoxide ions by going through successive oxidative and reductive cycles of transition metal ions at its active site (Chaudière and Ferrari-Iliou, 1999). Increased number and enhanced activities of SODs had been reported in plants subjected to a stress condition. Hence, they could serve as an important biomarker. To increase the production of more SOD, food sources rich in trace minerals such as zinc, copper, and manganese needs to be consumed. For example, the intake of oysters, lobster, chicken, cashews, and peas increases the body's zinc level while Mn can be obtained by consuming pumpkin seeds, mussels, and spinach. Rich sources of copper include grains, beans, nuts, potatoes, and so forth.

### 6.3.1.2   CATALASE (CAT) (EC 1.11.1.6)

Catalase is a very important enzyme in protecting the cell from oxidative damage by ROS. It is a tetrameric enzyme consisting of four identical tetrahedrally arranged subunits of 60 kDa that contains a single ferriprotoporphyrin group per subunit, and it has a molecular mass of about 240 kDa (Shinoura et al., 2001). CAT is an enzyme that catalyzes the decomposition of hydrogen peroxide to water and oxygen (Bisht and Sisodia, 2010). It has one of the highest turnover rates for all enzymes: one molecule of CAT can convert approximately 6 million molecules of hydrogen peroxide to water and oxygen each minute (Matés et al., 1999). The following reaction is catalyzed by CAT.

$$2H_2O_2 \rightarrow 2H_2O + O_2$$

Hydrogen peroxide is a harmful by-product of many normal metabolic processes; to prevent damage to cells and tissues, it must be quickly converted into other, less dangerous substances. To this end, CAT is frequently used by cells to rapidly catalyze the decomposition of hydrogen peroxide into less-reactive gaseous oxygen and water molecules. The presence of CAT in a microbial or tissue sample can be tested by adding a volume of hydrogen peroxide and observing the reaction. The formation of bubbles, oxygen, indicates a positive result. Food sources of CAT include carrot, red peppers, cucumbers, spinach, and cabbage.

### 6.3.1.3    GLUTATHIONE PEROXIDASE (GPX) (EC 1.11.1.9)

Glutathione peroxidase (GPX), are a large group of isozymes that use GSH to reduce $H_2O_2$. They play a crucial role in lipid peroxidation process, and therefore, help plant cells from oxidative stress (Gill and Tuteja, 2010). There are four different Se-dependent GPXs present in humans (Chaudière and Ferrari-Iliou, 1999), and are known to add two electrons to reduce peroxides by forming selenols (Se-OH) and the antioxidant properties of these seleno-enzymes allow them to eliminate peroxides as potential substrates for the Fenton reaction. Selenium-dependent GPX acts in association with tripeptide GSH, which is present in high concentrations in cells and catalyzes the conversion of hydrogen peroxide or organic peroxide to water or alcohol while simultaneously oxidizing GSH. It also competes with CAT for hydrogen peroxide as a substrate and is the major source of protection against low-levels of oxidative stress (Chaudière and Ferrari-Iliou, 1999). Food sources rich in selenium need to be consumed.

### 6.3.2    NONENZYMATIC ANTIOXIDANTS

Nonenzymatic antioxidants include low-molecular-weight compounds, such as dietary antioxidants and minerals. They include vitamins (vitamins C and E (Fig. 6.1 (1 and 5), β-carotene (3), uric acid, and GSH, a tripeptide (L-γ-glutamyl-L-cysteinyl-L-glycine) that comprise a thiol (sulfhydryl) group (Esraet al., 2012).

### 6.3.2.1    GLUTATHIONE (GSH)

Glutathione (GSH) is the most significant nonenzymatic oxidant defense molecule. GSH is capable of preventing damage to important cellular components caused by ROS such as free radicals, peroxides, lipid peroxides, and heavy metals (Pompella et al., 2003). It is a tripeptide with a gamma peptide linkage between the carboxyl group of the glutamate side chain and the amine group of cysteine, and the carboxyl of cysteine linked to a glycine. It exists in relatively large amounts (about 7 mm levels) and serves to detoxify peroxides and regenerate a number of important antioxidants (e.g., α-tocopherol and ascorbic acid) (Tarpley et al., 2004). Its vitality is owed to the fact that it can be regenerated from its oxidized form (gluta-thione disulfide (GSSG) back to its reduced form (GSH) by glutathione reductase (GSR) (Couto et al., 2013). In the reduced state, the thiol group of cysteine is able to donate a reducing equivalent ($H^+ + e^-$) to other molecules, such as reactive oxygen species to neutralize them, or to protein cysteines to maintain their reduced forms. Food sources rich in GSH are garlic, cabbage, spinach, walnut, watermelon, and so forth.

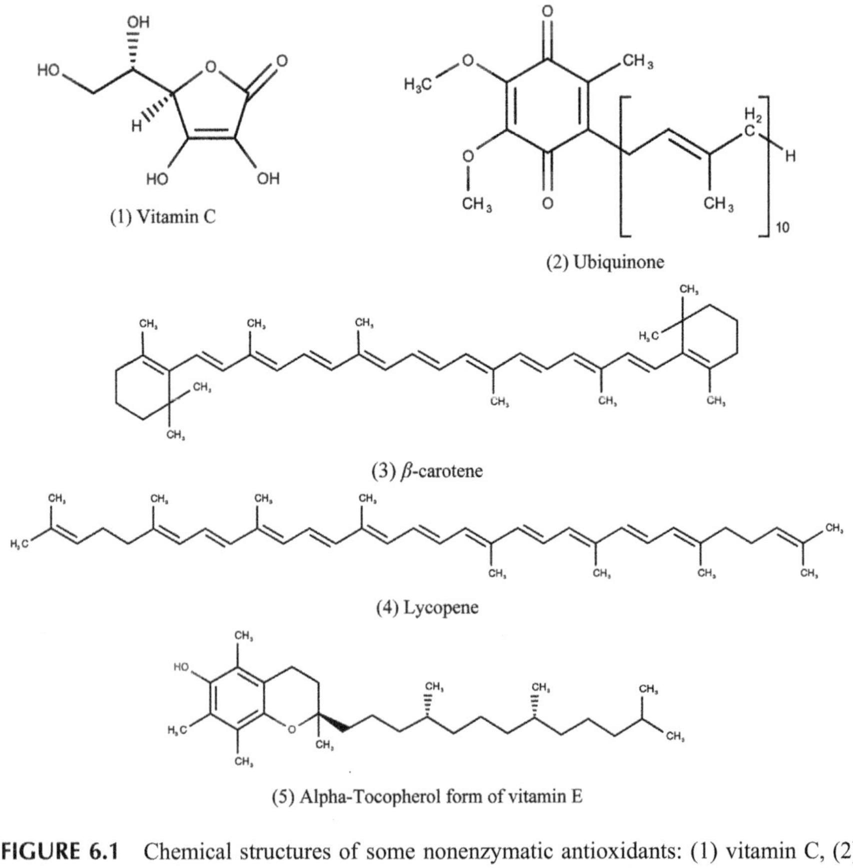

FIGURE 6.1 Chemical structures of some nonenzymatic antioxidants: (1) vitamin C, (2) ubiquinone, (3) β-carotene, (4) lycopene, and (5) alpha-tocopherol form of vitamin E.

## 6.3.2.2 VITAMIN E

Vitamin E (*d*-alpha tocopherol) is a fat-soluble antioxidant. It is stored in body fat and works within the lipid portion of biomembranes to provide an alternative binding site for free radicals, preventing the oxidation of polyunsaturated fatty acids (Chow, 1991). The oxidized α-tocopheroxyl radicals produced in this process may be recycled back to the active reduced form through reduction by other antioxidants, such as ascorbate, retinol, or ubiquinol (Wang et al., 1999). It definitely represents the principal defense against oxidant-induced membrane injury in human tissue because of its role in breaking the lipid peroxidation chain reaction (Burton and Ingold, 1981). It reacts directly with peroxyl and superoxide radicals and singlet oxygen and protects membranes from lipid peroxidation (Bisht and Sisodia, 2010). Vitamin E appears to play a major role as an integral constituent of alveolar surfactant, whose quantity and composition conditions normal lung function (Kolleck et al., 2002). Vitamin E, C, and some other antioxidants work closely together in the antioxidant network at the interface of cytosol and cell membrane, where oxidized vitamin E can be regenerated by vitamin C (Fig. 6.2). Food sources rich in vitamin E are sweet potato, beans, palm oil, nuts, and whole grains.

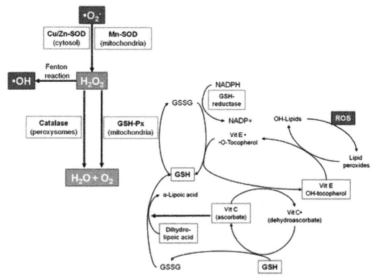

**FIGURE 6.2** The antioxidant network showing the interaction between vitamin E, vitamin C, and thiol. (Reprinted from Kurutas, E. B. The importance of antioxidants which play the role in cellular response against oxidative/nitrosative stress: current state. Nutrition Journal 2016, 15, 71. Open access, CC BY 4.0).

### 6.3.2.3   VITAMIN C

Vitamin C (Fig. 6.1 (1)) is a water-soluble vitamin. It is an excellent reducing agent and scavenges free radical and oxidants. It has been shown that short-term supplementation of vitamin C lasting two to four weeks can significantly reduce the level of free radicals in the body (Naidoo and Lux, 1998). Again, in vitro evidence suggests that vitamin C has a role as a chemical reducing agent both intracellularly and extracellularly (Romieu et al., 2008). Intracellular vitamin C might prevent protein oxidation and regulate gene expression and mRNA translation (Romieu et al., 2008). Extracellular vitamin C protects against oxidants and oxidant-mediated damage (Levine et al., 2006). Generally, fruit sources such as oranges, guava, and red pepper are rich in vitamin C.

### 6.3.2.4   UBIQUINONES

Ubiquinone, also known as Coenzyme $Q_{10}$, ubidecarenone, coenzyme Q, and abbreviated as $CoQ_{10}$, CoQ, or $Q_{10}$, (Fig. 6.1 (2)) is a coenzyme that is ubiquitous in the bodies of most animals. The Q refers to the quinone chemical group and 10 refer to the number of isoprenyl chemical subunits in its tail. This fat-soluble substance, which resembles a vitamin, is present in most eukaryotic cells, primarily in the mitochondria. Although present in food, $CoQ_{10}$ is not considered a vitamin because the body is able to make it from raw materials contained in food (Schachter, 1996). Food substances such as beef, orange, chicken, peanut, and eggs are rich sources of ubiqui-none. It is a component of the electron transport chain and participates in aerobic cellular respiration, which generates energy in the form of ATP and 95% of the human body's energy is generated this way (Ernster and Dallner, 1995; Dutton et al., 2000). There are three redox states of $CoQ_{10}$: fully oxidized (ubiquinone), semiquinone (ubisemiquinone), and fully reduced (ubiquinol). As reviewed by Schachter (1996), the antioxidant nature of $CoQ_{10}$ is derived from its energy carrier function. As an energy carrier, the $CoQ_{10}$ molecule continuously goes through an oxidation-reduction cycle. As it accepts electrons, it becomes reduced. As it gives up electrons, it becomes oxidized. In its reduced form, the $CoQ_{10}$ molecule holds electrons rather loosely, so this CoQ molecule will give up one or both electrons quite easily and, thus, act as an antioxidant. In tight synergy with vitamin E, $CoQ_{10}$ acts to protect cell membranes against oxidative stress (Pincemail and Defraigne, 2016). As an antioxidant, it appears to help correct dietary deficiency of vitamin E in animal models, protects against the toxic effects of adriamycin, protects against low oxygen states which results in large amounts of free radical formation, and reduce oxidative distress that often results from surgery (Schachter, 1996).

## 6.3.2.5 CAROTENOIDS

Carotenoids are natural pigments which are synthesized by plants and are responsible for the bright colors of various fruits and vegetables (Paiva and Russell, 1999). They are present in large quantities in green and yellow leafy vegetables and more than 600 naturally occurring carotenoids have been identified, approximately 50 of which have vitamin A activity (Isler, 1981). These compounds are comparatively nontoxic. $\beta$-Carotene and others carotenoids have antioxidant properties. ROS which is efficiently scavenged by carotenoids are $^1O_2$ and peroxyl radicals (Palozza and Krinsky, 1992). They are the most effective naturally occurring quenchers for $^1O_2$ with quenching rate constants of about $5-12 \times 10^9/\text{mol/s}$ (Diplock et al., 1998). Like vitamin E, carotenoids belong to the group of lipophilic antioxidants present in lipoproteins such as LDL and HDL. It has been shown that they are consumed when isolated LDL is exposed to the process of lipid peroxidation (Diplock et al., 1998). Some of the major sources are carrots ($\alpha$-carotene, $\beta$-carotene), tomatoes (lycopene), citrus fruits ($\beta$-cryptoxanthin), spinach (lutein), and maize (zeaxanthin) (Mangels et al., 1993).

## 6.3.2.6 BETA-CAROTENE

Beta-carotene (Fig. 6.1 (3)) is an effective source of vitamin A in both conventional foods and vitamin supplements, and it's generally safe (CRN, 2013). $\beta$-Carotene is the most abundant form of provitamin A in fruits and vegetables (Olson, 1994; Ross, 1999). The chemical abilities of $\beta$-carotene to quench singlet oxygen and to inhibit peroxyl free-radical reactions are well established (Sies and Stahl, 1995). $\beta$-Carotene also enhances gap-junction communication and, in the rat model, induces hepatic enzymes that detoxify carcinogens (Palozza, 1998). They also protect against cardiovascular disease and cataract prevention (Dietmar and Bamedi, 2001). Most studies have found beneficial associations between the higher intake of nutrients such as $\alpha$-carotene, $\beta$-carotene, $\beta$-cryptoxanthin, and a positive outcome associated with asthma and allergy (Devereux, 2006). In recent studies, protective effects of carotenoids on bladder cancer had been reported (Hung et al., 2006).

### 6.3.2.7   LYCOPENE

Lycopene is a carotenoid pigment, found in tomatoes and other red fruits, like watermelon, papaya, pink grapefruit, and pink guava/Brazilian guava (Lycocard, 2006). Its name is derived from the tomato's species classification, *Solanumly copersicum* (Lycocard, 2006). It is a vibrant red carotenoid that serves as an intermediate for the biosynthesis of other carotenoids which are synthesized by plants and microorganisms but not by animals (Grossman et al., 2004). Lycopene (Fig. 6.1 (4)) is a free radical scavenger like $\beta$-carotene being an acyclic isomer of $\beta$-carotene (DiMascioet al., 1989). The presence of lycopene in the diet has been associated with reduced cancer incidence (Gerster, 1997). Lycopene is insoluble in water and it can be dissolved only in organic solvents and oils. When lycopene is oxidized by its reaction with bleaches or acids, the double bonds between the carbon atoms are broken; cleaving the molecule, breaking the conjugated double bond system, and eliminating the chromophore (Giovannucciet al., 2002). Lycopene obtained from plants tends to exist in an all-trans configuration and in the most thermodynamically stable form (Gerster, 1997).

### 6.3.2.8   POLYPHENOLS

Polyphenols are the most abundant antioxidants in the diet and are widespread constituents of fruits, vegetables, cereals, dry legumes, chocolate, and beverages, such as tea, coffee, or wine (Scalbert et al., 2005). Current evidence strongly supports a contribution of polyphenols to the prevention of cardiovascular diseases, cancers, and osteoporosis and suggests a role in the prevention of neurodegenerative diseases and diabetes (Scalbert et al., 2005). Flavonoids are a large group of polyphenolic antioxidants that occur in several fruits, vegetables, and beverages such as tea, wine, and beer mainly as O-glycosides (Diplock et al., 1998). The term flavonoids summarizes a number of structurally different subgroups including flavanols (catechin, epicatechin), flavonols (quercetin, myricetin, and kaempferol), flavanones (naringenin, taxifolin), flavones (apigenin, hesperetin), isoflavones (genistein), or anthocyanidins (cyanidin, malvidin) (Diplock et al., 1998).

### 6.3.2.9   URIC ACID

Uric acid is a final enzymatic product in the degradation of purine nucleosides and free bases in humans and Great Apes (Sautin and Johnson, 2008).

The ability of urate to scavenge oxygen radicals and protect the erythrocyte membrane from lipid oxidation was originally described by Kellogg and Fridovich (1977). Uric acid (Fig. 6.3) is a powerful scavenger of carbon-centered and peroxyl radicals in the hydrophilic environment but loses an ability to scavenge lipophilic radicals and cannot break the radical chain propagation within lipid membranes (Muraoka and Miura, 2003). Uric acid is a selective antioxidant that removes superoxide by preventing the degradation of SOD and subsequently inhibits its reaction with NO to form peroxynitrite (van der Veen et al., 1997). One of the major sites where the antioxidant effects of uric acid have been proposed is in the central nervous system, particularly in conditions such as multiple sclerosis, Parkinson's disease, and acute stroke (Sautin and Johnson, 2008).

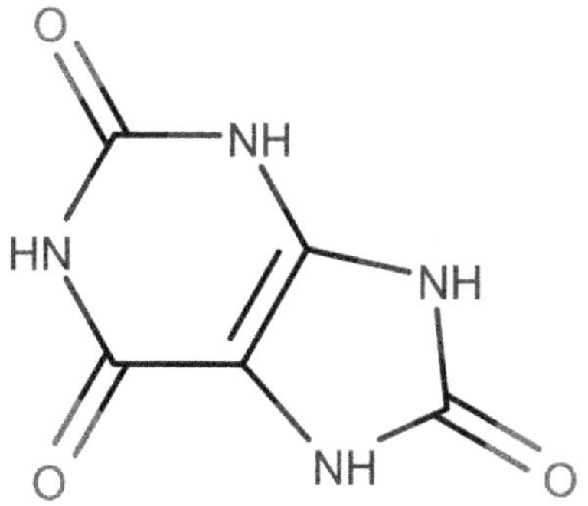

**FIGURE 6.3**   Uric acid.

## 6.3.2.10   MINERAL ANTIOXIDANTS

The body uses these elements to make free radical enzyme scavengers; which neutralize the free radicals. The four most important enzymes that neutralize the free radicals are the SOD enzyme, methionine reductase, CAT, and GPX.

### 1. Zinc

Zn is an important trace metal that performs a number of antioxidant roles in defined chemical systems. Food sources rich in Zn generally include seafoods, lobsters, crabs, oyster, legumes, egg, and so forth. The recommended dietary allowance (RDA) of zinc is the role of zinc in alleviating oxidative stress was evident from chronic administration of Zn and its deficiency. Long-term administration of zinc can be linked to the induction of some other substance that serves as the ultimate antioxidant (Powell, 2010). In this regard, the most studied effectors are the metallothioneins while long-term deprivation of zinc renders an organism more susceptible to injury induced by a variety of oxidative stresses (Powell, 2010). The mechanism by which Zn acts can be summarized thus. Firstly, It competes with iron (Fe) and copper (Cu) ions for binding to cell membranes and proteins and thereby displacing these redox active metals, which catalyze the production of·OH from $H_2O_2$ (Ananda, 2014). Secondly, they bind to sulfhydryl groups of biomolecules protecting them from oxidation (Ananda, 2014).Thirdly, it increases the activation of antioxidant proteins, molecules, and enzymes such as GSH, CAT, and SOD and also reduces the activities of oxidant-promoting enzymes such as inducible nitric acid synthase and NADPH oxidase, and inhibits the generation of lipid peroxidation products (Baoet al., 2013). Fourthly, zinc induces the expression of a metal-binding protein metallothionein, which is very rich in cysteine and is an excellent scavenger of·OH ions (Ananda, 2014).

## 2. Copper

Copper is an essential trace mineral which must be obtained through diet or supplements. The human body contains copper at a level of about 1.4–2.1 mg per kg of body mass (Copper Development Association, 2016). The RDA for copper in normal healthy adults is 0.97 and 3.0 mg/day (NRC/NAS, 1980). Food sources of copper include oyster, shellfish, beans, whole grains, beans, nuts, potatoes, and organ meats. Cu is a zinc-balancing mineral important in many enzymes as well as in the production of hemoglobin, the molecule that transports oxygen. Although normally bound to proteins, Cu may be released and become free to catalyze the formation of highly reactive hydroxyl radicals (Gaetke and Chow, 2003).Data obtained from in vitro and cell culture studies are largely supportive of Cu's capacity to initiate oxidative damage and interfere with important cellular events (Gaetke and Chow, 2003). Interestingly, a deficiency in dietary Cu also increases cellular susceptibility to oxidative damage (Gaetke and Chow, 2003). Therefore, Cu must be balanced otherwise could be deleterious.

## 3. Selenium

Selenium is the only mineral that functions as an antioxidant (Hernandez, 2010). It works as an antioxidant, especially, when combined with vitamin E (Ehrlich, 2015). It works by targeting natural hydrogen peroxide in the body and converting it to water (Hernandez, 2010). Selenium regulates thyroid function and helps the immune system stay strong and healthy. In high concentrations, Se is toxic. In humans, Se is a trace element nutrient that functions as a cofactor for reduction of antioxidant enzymes, such as GPXs (Linus Pauling Institute). Food sources of selenium include garlic and onions, chicken, and seafoods.

## 6.4   CONCLUSION

Antioxidants play a vital role in the protection of the body against the oxidation of important biomolecules such as lipids, proteins, and nucleic acids. The oxidation of these molecules leads to loss of its structure, hence, function, in the absence of antioxidants, the loss of function paves way for oxidative stress which results in various diseases such as cancer, diabetes, neurodegenerative diseases, aging, and so forth. To annul these effects, antioxidants evolved some mechanisms such as chelation, scavenging, sequestration, quenching to mop up oxidants, and protect the body.

## KEYWORDS

- **antioxidants**
- **oxidation**
- **oxidants**
- **oxidative stress**
- **SOD**
- **CAT**
- **GPX**

## REFERENCES

Ananda, S. P. Zinc is an Antioxidant and Anti-Inflammatory Agent: Its Role in Human Health. *Front. Nutr.* **2014,** *1*(14), 1–10.

Bao, B.; Ahmad, A.; Azmi, A.; Li, Y.; Prasad, A. S.; Sarkar, F. H. The Biological Significance of Zinc in Inflammation and Aging. In *Inflammation, Advancing and Nutrition;* Rahman, I., Bagchi, D., Eds.; Elsevier Inc.: New York, NY, 2013; pp 15–27.

Bisht, S.; Sisodia, S. S. Diabetes, Dyslipidemia, Antioxidant and Status of Oxidative Stress. *Int. J. Res. Ayurveda Pharm.* **2010,** *1*(1), 33–42.

Burton, G. W.; Ingold, K.U.Autooxidation of biological molecules. 1. The Antioxidant Activity of Vitamin E and Related Chain-Breaking Phenolic Antioxidants In Vitro. *J. Am. Chem. Soc.* **1981,** *103,* 6472–6477. https://doi.org/10.1021/ja00411a035.

Chaudière, J.; Ferrari-Iliou, R. Intracellular Antioxidants: from Chemical to Biochemical Mechanisms. *Food Chem. Toxicol.* **1999**, *37*, 949–962. https://doi.org/10.1016/S0278–6915 (99)00090–3.

Chow, C. K. Vitamin E and Oxidative Stress. *Free Radical Biol. Med.* **1991**, *11*, 215–232. https://doi.org/10.1016/0891–5849(91)90174–2.

Copper Development Association. http://copperalliance.org.uk/copper-and-society/health (accessed Jun 27, 2016).

Couto, N.; Malys, N.; Gaskell, S.; Barber, J. Partition and Turnover of Glutathione Reductase from Saccharomyces Cerevisiae: A Proteomic Approach. *J. Proteome Res.* **2013**, *12*(6), 2885–94. PMID 23631642, https://doi.org/10.1021/pr4001948.

CRN. *Vitamin and Mineral Safety*, 3rd ed.; Council for Responsible Nutrition (CRN), 2013. www.crnusa.org (Accessed Dec 4, 2017).

David, S.G. Molecule of the Month: Superoxide Dismutase. RCSB PDB and The Scripps Research Institute. 2007. www.pdb.org (Accessed Aug 6, 2017).

Devereux, G.The Increase in the Prevalence of Asthma and Allergy: Food for Thought. *Nat. Rev.Immunol.* **2006**, *6*, 869–847. https://doi.org/10.1038/nri1958.

Dietmar, E. B.; Bamedi, A. Carotenoid Esters in Vegetables and Fruits: A Screening with Emphasis on β-Cryptoxanthin Esters. *J. Agric. Food Chem.* **2001**, *49*, 2064–2067. https://doi.org/10.1021/jf001276t.

DiMascio, P.; Kaiser, S.; Sies, H. Lycopene as the Most Efficient Biological Carotenoid Singlet Oxygen Quencher. *Arch. Biochem. Biophys.* **1989**, *274*(2), 532–538. https://doi.org/10.1016/0003–9861(89)90467–0.

Diplock, A. T.; Charleux, J. L.; Crozier-Willi, G.; Kok, F. J.; Rice-Evans, C.; Roberfroid, M.; Stah, W.; Viña-Ribes, J. Functional Food Science and Defence Against Reactive Oxidative Species. *Br. J.Nutr.***1998**, *80*(1), 77–112. https://doi.org/10.1079/BJN19980106.

Duke, M. V.; Salin, M. L. Purification and Characterization of an Iron-Containing Superoxide Dismutase from a Eukaryote, *Ginko biloba. Arch. Biochem. Biophys.* **1985**, *243*, 305–314. https://doi.org/10.1016/0003–9861(85)90800–8.

Dutton, P. L.; Ohnishi, T.; Darrouzet, E.; Leonard, M. A.; Sharp, R. E.;Cibney, B. R.; Daldal, F.; Moser, C.C. Coenzyme Q Oxidation Reduction Reactions in Mitochondrial Electron Transport. In *Molecular Mechanisms in Health and Disease.* Kagan, V. E., Quinn, P. J., Coenzyme, Q., Eds.; CRC Press: Boca Raton, 2000; pp 65–82.

Egbuna, C.; Ifemeje, J. C. Oxidative Stress and Nutrition.*Trop. J. Appl. Nat. Sci.* **2017**, *2*(1), 110–116. https://doi.org/10.25240/TJANS.2017.2.1.19.

Ehrlich, S. D. Selenium. University of Maryland Medical Centre. 2015. http://umm.edu/health/medical/altmed/supplement/selenium (Accessed July 4, 2016).

Ernster, L., Dallner, G. Biochemical, physiological and medical aspects of ubiquinone function". *Biochimica et Biophysica Acta.* **1995**, *1271*(1), 195–204. doi:10.1016/0925-4439(95)00028-3. PMID 7599208.

Esra, B.; Umit, M. S.; Cansin, S.; Serpil, E.; Omer, K.Oxidative Stress and Antioxidant Defense. *World Allergy Organ. J.* **2012**, *5*(1), 9–19. https://doi.org/10.1097/WOX.0b013e 3182439613.

Gaetke, L. M.; Chow, C. K. Copper Toxicity, Oxidative Stress, and Antioxidant Nutrients. *Toxicology* **2003**, *189*(1–2), 147–63. https://doi.org/$^{10}$.1016/S0300–483X(03)00159–8.

Gerster, H. The Potential Role of Lycopene for Human Health. *J. Am. Coll. Nutr.* **1997**, *16*(2), 109–26.https://doi.org/10.1080/07315724.1997.10718661.

Gill, S. S.; Tuteja, N. Reactive Oxygen Species and Antioxidant Machinery in Abiotic Stress Tolerance in Crop Plants. *Plant Physiol. Biochem.* **2010**, *48*, 909–930. https://doi.org/10.1016/j.plaphy.2010.08.016.

Giovannucci, E.; Willett, W.; Stampfer, M.; Liu, Y.; Rimm, E. A Prospective Study of Tomato Products, Lycopene, and Prostate Cancer Risk. *J. Natl. Cancer Inst.* **2002,** *94*(5), 391–396. https://doi.org/10.1093/jnci/94.5.391.

Grossman, A. R.; Lohr, M.; Im, C.S. Chlamydomonas Reinhardtiiin the Landscape of Pigments. *Annu. Rev. Genet.* **2004,** *38*(1), 119–73. https://doi.org/10.1146/annurev.genet. 38.072902.092328.

Hernandez, A. Which Vitamins & Minerals Are Antioxidants? http://healthyeating.sfgate. com/vitamins-minerals-antioxidants-5006.html (accessed Jun 27, 2016).

Hung, R. J.; Zhang, Z. F.; Rao, J. Y.; Pantuck, A.; Reuter, V. E.; Heber D.; Lu Q.Y. Protective Effects of Plasma Carotenoids on the Risk of Bladder Cancer. *J. Urol.* **2006,** *176,* 1192– 1197. https://doi.org/10.1016/j.juro.2006.04.030.

Ifemeje, J. C.; Udedi, S.C.; Okechukwu, A. U.; Nwaka, A. C.; Lukong, C. B.; Anene, I. N.; Egbuna, C.; Ezeude, I.C. Determination of Total Protein, Superoxide Dismutase, Catalase Activity and Lipid Peroxidation in Soil Macro-fauna (Earthworm) from Onitsha Municipal Open Waste Dump. *J. Sci. Res. Rep.* **2015,** *6*(5): 394–403. https://doi.org/10.9734/JSRR/2015/12552.

Isler, O. Foreword. In *Carotenoids as Colorants and Vitamin A Precursors;* Bauernfeind, J. C., Ed.; Academic Press: NewYork, 1981; p xiii. https://doi.org/10.1016/B978-0-12-082850-0.50005-6.

Kellogg, E. W.; Fridovich, I. Liposome Oxidation and Erythrocyte Lysisby Enzymically Generated Superoxide and Hydrogen Peroxide. *J. Biol. Chem.* **1977,** *252*(19), 6721–6728.

Kolleck, I.; Sinha, P.; Rustow, B. Vitamin E as an Antioxidant of the Lung: Mechanisms of Vitamin E Delivery to Alveolar Type II Cells. *Am. J. Respir Crit. Care Med.* **2002,** *166,* S62–S66. https://doi.org/10.1164/rccm.2206019.

Levine, M.; Katz, A.; Padayatt, S.J. Vitamin C. In *Modern Nutrition in Health and Disease;* Shils, M., Shike, M., Ross, A. C. et al., Eds.; Lippincott Williams & Willkins: Philadelphia, 2006; pp 507–524.

Lycocard. Lycopene and Human Health. http://www.lycocard.com/index.php/lyco_pub/ health/ (accessed July 3, 2016).

Mangels, A. R.; Holden, J. M.; Beecher, G. R.; Forman, R.; Lanza, E. Carotenoid Content of Fruits and Vegetables: An Evaluation of Analytic Data. *J. Am. Diet Assoc.* **1993,** *93,* 284–296. https://doi.org/10.1016/0002-8223(93)91553–3.

Matés, J. M.; Pérez-Gómez, C.; Nú-ezde, C. I. Antioxidant Enzymes and Human Diseases. *Clin. Biochem.* **1999,** *32*(8), 595–603. https://doi.org/10.1016/S0009–9120(99)00075–2.

McDowell, L. R.; Wilkinson, N.; Madison, R.; Felix, T. Vitamins and Minerals Functioning as Antioxidants with Supplementation Considerations. Florida Ruminant Nutrition Symposium. 2007. http://dairy.ifas.ufl.edu/rns/2007/McDowell.pdf (accessed Jun 27, 2016).

Muraoka, S.; Miura, T. Inhibition by Uric Acid of Free Radicals that Damage Biological Molecules. *Pharmacol. Toxicol.* **2003,** *93*(6), 284–289. https://doi.org/10.1111/j.1600–0773. 2003.pto930606.x.

Naidoo, D.; Lux, O. The Effect of Vitamin C and E Supplementation on Lipid and Urate Oxidation Products in Plasma. *Nutr. Res.* **1998,** *18,* 953–961. https://doi.org/10.1016/ S0271–5317(98)00078–5.

NRC/NAS. Copper. In *Recommended Dietary Allowances;* National Research Council/ National Academy of Sciences, Food Nutrition Board: Washington, D.C., 1980; pp 151–154.

Olson, J. A. Vitamin A, Retinoids and Carotenoids. In *Modern Nutrition in Health and Disease;* Shils, M. E., Olson, J. A., Shike, M., Eds.; 8th ed. Philadelphia: Lea and Febiger; 1994; pp 287–307.

Paiva, S. A.; Russell, R.M. Beta-Carotene and Other Carotenoids as Antioxidants. *J. Am. Coll.Nutr.* **1999,** *18*(5), 426–33. https://doi.org/10.1080/07315724.1999.10718880.

Palozza, P. Prooxidant Actions of Carotenoids in Biologic Systems. *Nutr. Rev.* **1998**, *56*(9), 257–265.

Palozza, P.; Krinsky, N. I. Antioxidant Effects of Carotenoids in Vivo and in Vitro: An Overview. *Methods Enzymol.* **1992**, *268*, 127–136. https://doi.org/10.1016/0076-6879(92)13142-k.

Pincemail, J.; Defraigne, J. Coenzyme Q10 or Ubiquinone: a Peculiar Antioxidant. http://www.probiox.com/uk/html/documents/coQ10uk.PDF (accessed Jun 26, 2016).

Pompella, A.; Visvikis, A.; Paolicchi, A.; Tata, V.; Casini, A. F. The Changing Faces of Glutathione, a Cellular Protagonist. *Biochem. Pharmacol.* **2003**, *66*(8), 1499–503. PMID 14555227.

Powell, S. R. The Antioxidant Properties of Zinc. American Society for Nutritional Sciences. **2010**, pp 1447S–1453S. http://jn.nutrition.org/content/130/5/1447S.full.pdf+html (accessed Jun 27, 2016).

Romieu, I.; Castro-Giner, F.; Kunzli, N.; Sunyer, J. Air Pollution, Oxidative Stress and Dietary Supplementation: A Review. *Eur. Respir J.* **2008**, *31*, 179–196.

Ross, A. C. Vitamin A and Retinoids. In *Modern Nutrition in Health and Disease*, 9th ed.; Shils, M.E., Olson, J.A., Shike, M., Ross, A.C., Eds.; Baltimore: Williams & Wilkins,1999; pp 305–327.

Sautin, Y.Y.; Johnson, R. J. Uric Acid: The Oxidant–Antioxidant Paradox. *Nucleosides Nucleotides Nucleic Acids* **2008**, *27*(6), 608–619. doi: 10.1080/15257770802138558.

Scalbert, A.; Manach, C.; Morand, C.; Rémésy, C.; Jiménez, L. Dietary Polyphenols and the Prevention of Diseases. *Crit. Rev. Food Sci.Nutr.* **2005**, *45*(4), 287–306. https://doi.org/10.1080/1040869059096.

Schachter, M. B. Coenzyme Q10. 1996. http://www.mbschachter.com/coenzyme_q10.htm (accessed Jun 26, 2016).

Shinoura, N.; Sakurai, S.; Asai, A.; Kirino, T.; Hamada, H. Cotransduction of Apaf-1 and Caspase-9 Augments Etoposide-Induced Apoptosis in U-373MG Glioma Cells. *Jpn. J. Cancer Res.* **2001**, *92*, 467–474. https://doi.org/10.1111/j.1349-7006.2001.tb01117.x.

Sies, H.; Stahl, W. Vitamins E; C Beta-Carotene, and Other Carotenoids as Antioxidants. *Am. J. Clin. Nutr.* **1995**, *62*, 1315–1321.

Tarpley, M. M.; Wink, D.A.; Grisham, M. B. Methods for Detection of Reactive Metabolites of Oxygen and Nitrogen: In Vitro and In Vivo Considerations. *Am. J. Physiol. Regul. Integr. Comp. Physiol.* **2004**, *286*, R431–R444. https://doi.org/10.1152/ajpregu.00361.2003.

Van der Veen, R. C.; Hinton, D. R.; Incardonna, F.; Hofman, F. M. Extensive Peroxynitrite Activity During Progressive Stages of Central Nervous System Inflammation. *J. Neuroimmunol.* **1997**, *77*, 1–7. https://doi.org/10.1016/S0165-5728(97)00013-1.

Walks, H.; Doyle, H. Medicine for the People. The Miracle of Antioxidants. **2011**. http://www.manataka.org/page2394.html (accessed Jun 27, 2016).

Wang, X.; Quinn P. J.; Quinn, J. Vitamin E and Its Function in Membranes. *Prog. Lipid Res.* **1999**, *38*(4), 309–336. PMID 10793887. https://doi.org/10.1016/S0163-7827(99)00008-9.

# ROLES OF PHYTOCHEMICALS IN THE PREVENTION AND TREATMENT OF VARIOUS DISEASES

INTAN SORAYA CHE SULAIMAN[1,*], SHIBANI SUKHI[2], and AZHAM MOHAMAD[3]

*[1]Centre of Research & Innovation Management, Universiti Pertahanan Nasional Malaysia, Kem Sungai Besi, 57000, Kuala Lumpur, Malaysia, Tel.: +60192097719*

*[2]Department of Biotechnology, Savitribai Phule Pune University, Pune, 411007, India*

*[3]Centre of Foundation Studies for Agricultural Science, Universiti Putra Malaysia, 43400, Serdang, Selangor, Malaysia*

*\*Corresponding author. E-mail: chesoraya007@yahoo.com;*
*\*ORCID: https://orcid.org/0000-0002-4681-6238*

## ABSTRACT

Phytochemicals are plant-derived chemicals that are produced through primary or secondary metabolism. Plant components generally defined as secondary metabolites possess several biological activities which help in plant growth, reproduction and defense against threats integral to their environment. A healthy diet rich in fruits and vegetables has been found to play an important role in the prevention and treatment of various diseases. Sufficient intakes of these antioxidant sources in the human diet are crucial in order to balance the production of reactive species and antioxidative defense in the human body. Hence, many diseases have been reported to be associated with greater levels of radicals in the cells and the oxidation of biological components caused by the reactive species. The anti-inflammatory and antimicrobial properties of phytochemicals are also extensively applied in traditional medicine in Asia, while elucidation of this molecular mechanism of action will help tremendously in modern clinical science.

## 7.1 INTRODUCTION

Phytochemicals are plant-derived chemicals that are produced through primary or secondary metabolism. The biosynthesis of these chemical metabolites is based on their fundamental roles in plants; the primary metabolites are essential for vital activities while the secondary metabolites are biosynthesized to adapt to environmental stressors (Murakami and Ohnishi, 2012). The phytoconstituents that are produced through primary metabolism include proteins, sugars, amino acids, and chlorophyll while the secondary constituents are phenolic metabolites, alkaloids, glycosides, terpenes, fibers, saponins, nitrogen-containing constituents, and many more (Afolabi and Afolabi, 2013; Dillard and German, 2000). Plant components generally defined as secondary metabolites possess various biological activities which help in plant growth, reproduction and defense against threats integral to their environment (Molyneux et al., 2007). Their quality and quantity in plants vary significantly due to several factors, such as cultivar (Vallejo et al., 2002), post-harvest handling and storage conditions (Koh et al., 2009), environmental stress (water availability, temperature, light, nutrient availability) (Dixon and Paiva, 1995), and crop management practices (Chassy et al., 2006). One single plant may possess diverse biological properties that are caused by the activities of the various different secondary metabolites in that plant. Due to the differences in biosynthetic pathways and endogenous functions that are involved in plant defense mechanisms, a variety of secondary metabolites have been produced with specific roles through their individual or synergistic effects (Caris-Veyrat et al., 2004). Three expression modes are found when secondary metabolites in plants respond to environmental challenges (Delgoda and Murray, 2017). Some plant defense mechanisms are constitutive which means they are expressed in plant tissue alone. On the other hand, some may be constitutively expressed but require an activation of the constituents in order to provide some value for the host plant. These modes of expression are similar to prodrugs that require biochemical activation within the body. Others may involve induced expression which provides the plant with an enhanced level of resistance expressed against environmental pressure (Delgoda and Murray, 2017; Pedras and Abdoli, 2017). Thus, a single secondary metabolite may exhibit more than one type of biological activity or may possess multiple functions.

For example, emodin can act as a laxative agent in mammals, as an anti-inflammatory, antimicrobial, and antifeedant agent to phytophagous insects (Izhaki, 2002). Some of these metabolites are of nutritional importance to humans especially in the prevention and treatment of various diseases due to their anti-inflammatory, antiviral, antioxidant, antinociceptive, anticancer and anti-aging properties, and also because of their protective action in the chronic diseases for example, cardiovascular and neurodegenerative diseases and diabetes.

Another example is turmeric which is used for arresting wound infections, clove oil for dental problems, ginger extract for stomach problems, and so forth, are few of the commonly used home remedies. In this era where the fight against multidrug-resistant organisms is on the increase, one can look out for a powerful weapon in these natural products. It has been widely observed that many of these medicines are generally crude extracts of roots, stem, leaves, flowers, or fruits. These can be extracted with water or with certain organic solvents like methanol, ethanol, or ethyl acetate. These extracts show good activity as crude extracts as observed by many researchers. This may be because of the composite effect of the components. Generally, these medicines include extracted oils like clove oil, flavonoid, phenolic compounds, alkaloids, tannins, and so forth. Many of the applications of phytochemicals were without much information about the molecular level mode of action. It is important to illuminate on the molecular mechanisms of action for such phytochemicals. The knowledge of the mode of action will also help to expand the scope of application of these phytochemicals and will supplement the clinical field with effective medication which is devoid of any adverse effects.

## 7.2   ANTIOXIDANT POTENTIALS OF PHYTOCHEMICALS

Plant-based foods provide much energy and a range of nutrients. They are also rich sources of phytochemical antioxidants such as phenolic acids, flavonoids, vitamins, minerals, and fiber. A sufficient intake of these antioxidant sources in the human diet is crucial in order to balance the production of reactive species and antioxidative defense in the human body (Liu, 2013; Mironczuk-Chodakowska et al., 2018). Reactive species refer to reactive oxygen and nitrogen species (RONS) which are two classes of reactive molecules that possess unpaired electrons (Weidinger and Kozlov, 2015). About 1.5–5% of our consumed oxygen is converted into reactive species, namely free radicals (Poljšak et al., 2012). RONS are harmful free radicals that can damage cell structures including proteins and nucleic acids, hence, altering their functions (Birben et al., 2012). The production of RONS in aerobic organisms such as humans is as by-products in the electron transport chain in the mitochondria. In addition, this free radical-generated damage is also contributed by exogenous factors such as the environment, lifestyle, smoking, psychological and emotional stress, hormone changes, pollutant, and poor nutrition (Poljšak et al., 2012; Srivastava and Kumar, 2015). Although the human body possesses natural defenses for its protection against radical scavenging enzymes, sole reliance on its natural defense is not enough. As human biological functions may not progress at the same speed as our lifestyle (Lintner et al., 2009), the imbalanced production of RONS and antioxidative defense will lead to oxidative stress. As a result, oxidative stress may cause oxidative damage to large biomolecules such as DNA, lipids, and proteins. Thus, there will be an increased risk for the pathogenesis of many chronic diseases such as aging and cancer (Ames and Gold, 1991), coronary heart disease, hypertension and metabolic syndrome (Srivastava and Kumar, 2015), cardiovascular diseases (Csányi and Miller Jr., 2014), and diabetes (Zujko et al., 2014). Hence, many diseases have been reported to be associated with greater levels of radicals in the cells (Halliwell, 1991). Antioxidant defense is associated with many biological functions in the human body. Epidemiological studies have shown that an increased consumption of antioxidant metabolites had a good correlation with a reduced incidence of chronic diseases in humans (Birben et al., 2012).

## 7.2.1   POLYPHENOLS

As plants survive in open fields, the environmental factors include massive exposure to ultraviolet radiation, extreme temperatures, insect attacks, pathogen infection, and nutrient deficiency which can significantly increase the amount of antioxidant metabolites such as phenolic compounds (Caris-Veyrat et al., 2004). Approximately 100,000–200,000 secondary metabolites have been identified and phenolic compounds or polyphenols are among the most ubiquitous class of antioxidant metabolites in nature (Pereira et al., 2009). Plants rich in phenolics have been recognized as having high levels of antioxidant activity, due to the stability of phenolics against oxidation (Maestri et al., 2006). In addition, 20% of carbon fixed by photosynthesis was reported to be channeled into the phenylpropanoid pathway which generated a majority of the phenolics in the plant kingdom (Ralston et al., 2005; see

Volume 1, Chapter 2 for more details). Phenolic compounds can be classified into two groups; flavonoids and non-flavonoids. The classification is based on the basic chemical structure and the number of carbons in the molecules that range from simple phenolics (flavonoids, phenolic acids, and phenylpropanoids) to highly polymerized compounds (lignins, tannins, and melanins) (Vermerris and Nicholson, 2007).

## 7.2.1.1   FLAVONOIDS

Flavonoids (1) are a major group of phenolic compounds containing two aromatic rings connected by a three carbon bridge, $C_6$–$C_3$–$C_6$ (Rice-Evans et al., 1996). Figure 7.1 shows the basic structure of flavonoids and some phytochemical antioxidants in plants. Flavonoids occur mostly in the epidermis of leaves and the skin of fruits and play a crucial role in the prevention of damage caused by free radicals (Crozier et al., 2009; Barzegar, 2017). Their presence may influence the color and flavor of the fruits and vegetables. Potatoes, for example, are one of the vegetables abundant in flavonoids such as catechin, kaempferol (2) quercetin (3), rutinose, and anthocyanins (Akyol et al., 2016).

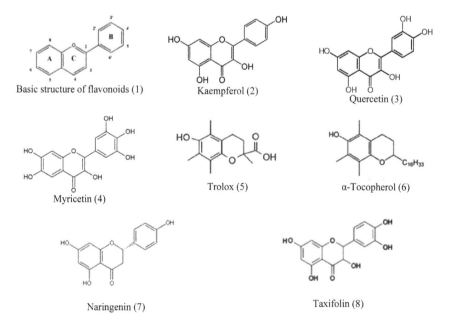

**FIGURE 7.1**   Basic structure of flavonoids and some phytochemical antioxidants in plants.

## 1. Anthocyanins

Anthocyanins are a class of pigmented flavonoids that are responsible for the red-blue color of many fruits and vegetables including red and purple-fleshed potatoes (Burgos et al., 2013). Bornsek et al. (2012) reported the intracellular antioxidant activity of bilberry and blueberry anthocyanins in mammalian cells. The main dietary anthocyanins in bilberries were delphinidin and cyanidin glycosides while malvidin glycosides were predominant in the blueberries. Among the subclasses of dietary flavonoids, flavonols, flavanones, anthocyanidins, flavones, flavan-3-ols, and isoflavones are the most commonly found components throughout the plant kingdom. Flavonols, such as kaempferol, quercetin, isorhamnetin, and myricetin are most widespread as *O*-glycosides (Crozier et al., 2009).

## 2. Kaempferol

Kaempferol (2) is a predominant component of leeks, broccoli, and radish. Kaempferol is isolated from the stems of butterbur (*Petasites japonicus*) and has been reported to attenuate the glutamate-induced oxidative stress in mouse-derived hippocampal neuronal HT22 cells. The neuroprotective effect of kaempferol was due to its ability to decrease levels of $Ca^{2+}$ influx and the overproduction of intracellular ROS, which are upstream factors in apoptosis signaling (Yang et al., 2014). Correspondingly, the ability of kaempferol to reduce intracellular calcium induced by antimycin A protected murine osteoblast-like MC3T3-E1 cells through antioxidant effects and the regulation of mitochondrial function. This demonstrated the potential use of kaempferol in reducing or preventing osteoblast degeneration in osteoporosis or other degenerative disorders (Choi, 2011).

## 3. Quercetin

Studies conducted by Seifi-Jamadi et al. (2017) revealed that a combination of 10 μM quercetin with dimethylacetamide (DMA) was effective in the freezability of goat semen. The synergetic effects of quercetin and DMA helped in preserving the goat sperm motion kinetics and suppressed lipid peroxidation by reducing ROS formation and malondialdehyde concentration. In addition, formulated quercetin lipid nanocapsules protected monocytic cell line THP-1 from oxidative stress by exogenous hydrogen peroxide where exogenous $H_2O_2$ increased the endogenous ROS levels in THP-1 cells (Hatahet et al., 2017). A study conducted by Cao et al. (1997) reported that dietary flavonoids that contained multiple hydroxyl substitutions such as kaempferol (2), quercetin (3), and myricetin (4) which have four, five, and six OH substitutions, respectively, showed stronger antiperoxyl radical activities than trolox (5) and α-tocopherol (6) analogues with a lower number of OH substitutions. A similar relationship between OH substitutions and hydroxyl radical scavenging activity due to the active hydrogen donor ability of hydroxyl substitution was reported by Loganayaki et al. (2013).

## 4. Myricetin

A comparative study conducted by Guitard et al. (2016) revealed that myricetinis a superior natural antioxidant that is more effective than other commercial antioxidants such as α-tocopherol, BHT, BHA, TBHQ, and propyl gallate for the preservation of omega-3 oils against autoxidation. In order to estimate the antioxidant activity of phenol, an important parameter is the bond dissociation enthalpy (BDE) of the ArO-H bond. BDE values can be influenced by the nature, number, and position of the substituents linked to the phenol ring. Myricetin, with the lowest BDE value, was the most effective antioxidant to protect the omega-3 oils from autoxidation. This was due to the ortho-substitution of the OH groups in myricetin that led to a stronger decrease in the BDE (Wright et al., 2001). Previous reports by Decharneux et al. (1992) demonstrated that at least two *ortho*-substituted OH groups (in the B ring) and carbon-3 position must be present in the structure of flavonoids in order to protect lysosomes against oxidative stress. This was due to their scavenging capacity and the ability of flavonoids to inhibit glucose translocation through the lysosomal membrane. However, flavones and monohydroxyflavones with none or lesser number of OH groups showed no activity.

## 5. Highly polymerized tannins

Besides the number of OH substitutions, extended conjugation, arrangement, and molecular weight were also factors that influenced the ability of monomeric phenolics to act as strong antioxidants. Highly polymerized tannins with many phenolic hydroxyl groups were stronger antioxidants due to the ability of condensed tannins to effectively quench free radicals (Hagerman et al., 1998). Flavanones, particularly naringenin (7) and taxifolin (8) are predominant in citrus fruits such as lemons, oranges, tangerines, and grapefruit.

Taxifolin (dihydroquercetin) with five OH substitutions exhibited higher antiperoxyl radical activities than naringenin (Cao et al., 1997). Taxifolin was able to significantly reduce the cholesterol oxidation product-induced neuronal apoptosis by suppressing the Akt and NF-κB activation-mediated cell death (Kim et al., 2017). Taxifolin was also able to stimulate fibril formation and promote the stabilization of collagen fibrils (Tarahovsky et al., 2007).

Naringenin was found to protect human keratinocyte cells against UVB-induced apoptosis and enhanced the removal of cyclobutane pyrimidine dimers (El-Mahdy et al., 2008). Recently, Kumar and Tiku (2016) reported that the oral administration of 50 mg/kg of naringenin protected mice against radiation-induced DNA, chromosomal, and membrane damage. This indicated the ability of naringenin to downregulate radiation-induced apoptotic proteins resulting in radioprotection at the cellular, tissue, and organism levels. Some sources of flavanols for example, epicatechin, catechin, and epigallocatechin have been reported to be mainly found in beverages such as green and black teas (Rice-Evans et al., 1996).

Recent studies on some phytochemical constituents that contributed to the antioxidants properties of plants are tabulated in Table 7.1.

## 7.3 MECHANISMS OF ACTIONS OF PLANT-DERIVED ANTIOXIDANTS

The classification of antioxidants can be made based on various parameters such as catalytic activity, solubility, and molecular size (Nimse and Pal, 2015). In terms of catalytic activity, antioxidants can be divided into two categories, namely enzymatic and nonenzymatic antioxidants. Enzymatic antioxidants include superoxide dismutase (SOD), glutathione reductase, glutathione peroxidase, and catalase which may function with the presence of cofactors such as zinc, copper, manganese, and iron. The human body protects itself from reactive species through enzymatic antioxidants by breaking down and removing free radicals. Nonenzymatic antioxidants which are represented by dietary supplements and synthetic antioxidants work by interrupting free radical chain reactions (Nimse and Pal, 2015). Owing to the multiple methods of generation of free radicals, thus quenching methods may also be different (Laitonjam, 2012). The pharmacological actions of plant polyphenols are multiple. Their mechanisms of action include their function in free radical scavenging, reduction, metal chelation, and gene expression (Soobrattee et al., 2005).

### 7.3.1  FREE RADICAL SCAVENGER

Most antioxidants found in food and the body is free radical scavengers, particularly flavonoids. Flavonoids with strong H-donating activity can

**TABLE 7.1** Summary of Some Phytochemical Antioxidants in Plants

| Antioxidant compounds | Plants | References |
|---|---|---|
| Vitamin C, miraculin, anthocyanin, palmitic acid, oleic acid, and linoleic acid | Miracle fruit (*Synsepalum dulcificum*) | He et al. (2016) |
| Isorhamnetin-3-rutinoside, isorhamnetin-3-glucoside, isorhamnetin, quercetin, kaempferol, ascorbic acid, tocopherols, fatty acids, and carotenoids | Sea buckthorn berries (*Hippophaë rhamnoides* L.) | Guo et al. (2016) |
| Hydroxycinnamic acids, caffeic derivatives, naringenin, myricetin, gallic acid, and ferulic acid | *Antidesma thwaitesianum* Müll. | Dechayont et al. (2017) |
| Phenol, 2-methoxy-4-vinylphenol, 3,5-dimethoxyacetophenone, 1,2-cyclopentanedione, and hexadecanoic acid | Soybeans (*Glycine max* L.) | Alghamdi et al. (2017) |
| Carvacrol, and α-pinene | *Falcaria vulgaris* | Jaberian et al. (2013) |
| Quercetin 3-O-(2'-rhamnosyl)rutinoside, isorhamnetin-O-(rhamnosyl)rutinoside, ferulic acid derivative, hexosyl caffeic acid, and hexosyl coumaric acid | Cactus (*Opuntia microdasys* Lehm.) | Chahdoura et al. (2014) |
| Epicatechin, dihydromyricetin, dihydroquercetin, epicatechin-O-rhamnoside, procyanidin B, and morachalcone | Jackfruit (*Artocarpus heterophyllus* Lam.) | Zhang et al. (2017) |
| Lignans, vitamin E, sterols, folates, ferulic acid, vanillic acid, syringic acid, p-coumaric acid, cyanidin-3-glucoside, peonidin-3-glucoside, delphinidin-3-glucoside, cyanidin, and proanthocyanidin | Barley | Idehen et al. (2017) |
| α-Tocopherol, 4-hydrophenylacetic acid, 4-hydroxybenzoic acid, coumalic acid, quercetin hydrate, stearic acid, palmitic acid, and linoleic acid | Sabah Snake Grass (*Clinacanthus nutans* Lindau) | Che Sulaiman et al. (2015) |
| Gallic acid, catechol, hydroxy benzoic acid, caffeic acid, vanillin, ferulic acid, and salicylic acids | Saptarangi (*Salacia chinensis* L.) | Ghadage et al. (2017) |

scavenge free radical species such as hydroxyl radicals (•OH), superoxides ($O_2$•–), and lipid peroxyl radicals (LOO•) by donating an electron or hydrogen atom to form stable compounds (Terao, 2009). Myricetin and quercetin are among the essential dietary flavonoids that possess powerful antioxidant properties due to their free radical scavenger properties (Maurya et al., 2016). Clinical studies conducted by Maurya et al. (2016) revealed the potential use of quercetin and myricetinas antioxidants for therapeutic strategies of aging and age-related disorders. During oxidative stress, both flavonoids exert antioxidative effects by causing malondialdehyde (MDA) depletion thereby increasing the glutathione and membrane sulfhydryl levels. Their role in targeting ROS metabolism could bypass drug resistance and achieve selectivity of treatment in various pathological conditions. However, Miller et al. (2014) underlined that a good radical scavenger was not always superior in reducing oxidative stress. Many factors such as cytotoxicity, metabolism, and their biomolecular interactions need to be taken into consideration.

## 7.3.2   CHELATING AGENTS

Plant polyphenols, particularly flavonoids have been reported for their ability to chelate iron ($Fe^{3+}$) and copper ($Cu^{2+}$) ions to form metal-flavonoid complexes. Their chelation properties are dependent on the structure and pH of the flavonoids (Mira et al., 2002). The three potential sites targeted for metal chelation with the flavonoids are between the (1) C-3' and C-4' dihydroxyl groups of ring B, (2) C-5 hydroxyl and C-4 carbonyl groups, and (3) C-3 hydroxyl and C-4 carbonyl groups (Fig. 7.2).

Addition of flavonoids in food preparation as chelating agents may retard lipid oxidation (Miller et al., 2014). Although the formation of free radicals are catalyzed by transition metal ions (free metal ions), however, the formation of chelated metals have reduced their effect on oxidation. This is because the electronic structures of the chelated metals are altered which limits their reactivity (Miller et al., 2014). The experimental study conducted by Dowling et al. (2010) reported that the isoflavone metal chelates of biochanin A and genistein were bound to $Cu^{2+}$ and $Fe^{3+}$ through the 4-keto group and C-5 hydroxyl sites of biochanin A and genistein which were confirmed by the presence of chelated water in the complexes. According to Sigurdson et al. (2017), interactions of metal ions and phenolic compounds such as anthocyanin could create many anthocyanin blue hues in nature with a wide range of color intensities.

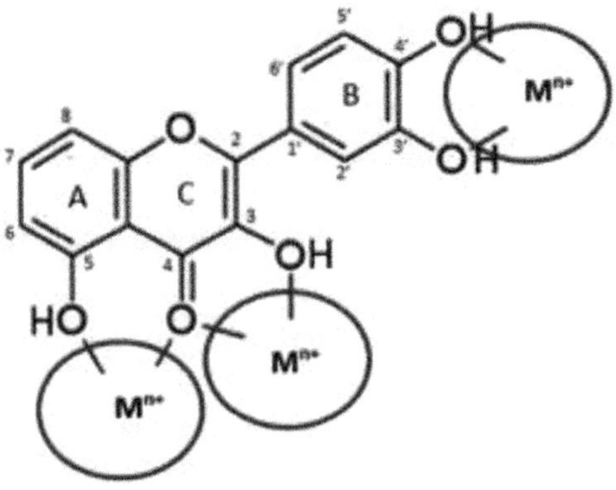

FIGURE 7.2   Potential sites for metal chelation.

## 7.4    ANTI-INFLAMMATORY EFFECTS OF PHYTOCHEMICALS

Inflammation is a basic physiological host response to tissue injury and infection, which is fundamental to our body when facing invading microbes or the diseases. However, if uncontrolled may lead to a negative biological response. There are various anti-inflammatory agents available to control this response, a prolonged use of which can pose some dangers like effect on liver and kidney. Some known phytochemicals can come to our rescue at such point. In fact, there are certain plant products like turmeric which play a dual role in controlling infection as well as inflammation. *Boswellia*, turmeric, ashwagandha, and ginger are a few herbs which offer us a combined remedy to fight inflammation effectively.

It is a well-known fact that the secondary metabolites of plants offer medicinal properties to man. Some of the waste products of plants are also useful for animals. Some of such products which are excreted as resins offer some advantages. In India, these resins are offered during the postpartum period to provide minerals like calcium. One of such plant resin product is available from a tree called *Boswellia serrata*, which is widely used by Ayurveda, Chinese medicine as well as other tropical regions and the Middle East. This product is also found to be effective in osteoarthritis. The active ingredient is boswellic acid has an inhibitory action on pro-inflammatory enzyme like 5-Lipoxygenase along with some other anti-inflammatory effects. An extensive review describing various beneficial effects of *Boswellia* besides anti-inflammatory effect is available (Yadav, 2011). Another such versatile member of this class is *Withania somnifera,* also known as Indian ginseng. The tuberous roots of the plant have been in use traditionally in Ayurveda. It's known as Ashwagandha and is used for various purposes like improving vitality and fertility, boosts immunity (Agarwal et al.,1999) relieves stress and depression, strengthens heart muscles, controls cholesterol, and effectively controls bacterial infections (Pandit et al., 2013). Also known as Ubab in Yemen, a paste from its dried leaves is used for burns and wounds. The effect is due to the plant alkaloids (isopelletierine, anferine) and steroidal lactone (withanolides, withaferins) saponins (sitoindos). *W. somniferra* is also rich in iron. There are certain alkaloids like ashwagandha and somniferrin (Verma and Kumar, 2011) which are also known for their anti-inflammatory properties. Another member in this class is ginger (*Zingiber officinale*). It is commonly used as a spice and flavoring agent in many of the south-east Asian countries. Although it is renowned for its curative properties for various digestive ailments, it is also widely used for the treatment of degenerative diseases like arthritis and rheumatism. The rhizoid of ginger contains volatile as well as nonvolatile components. The volatiles are found to be responsible for flavor and taste while the anti-inflammatory effect was due to components like gingerol (Jolad et al., 2004). Gingerol, shogaol, and other structurally-related substances in ginger inhibit prostaglandin and leukotriene biosynthesis through suppression of 5-lipoxygenase or prostaglandin synthetase. Additionally, they can also inhibit the synthesis of pro-inflammatory cytokines such as IL-1, TNF-α, and IL-8 (Tjendraputra et al., 2001). Turmeric is another important member of this group. Warm milk with a dash of turmeric is a valuable treatment for relief of throat inflammation while an application of warm paste of turmeric with alum gives an effective relief from a sprained ankle. *Curcuma longa* (turmeric) has a long history of use in Ayurvedic medicine as a treatment for inflammatory conditions. Turmeric constituents include the three curcuminoids: curcumin (diferuloylmethane; the primary constituent and the one responsible for its vibrant yellow color), demethoxycurcumin, and bisdemethoxycurcumin, as well as volatile oils (tumerone, atlantone, and zingiberone), sugars, proteins, and resins (Jurenka, 2009).

## 7.5 ANTIMICROBIAL EFFECTS OF PHYTOCHEMICALS

Plants usually contain many biologically active structurally diverse compounds which are useful as drugs, lead structures, or raw materials. Plants are loaded with a wide variety of secondary metabolites such as tannins, alkaloids, terpenoids, flavonoids, and so forth, which possess antimicrobial properties and may serve as a safe antimicrobial for microbial infections. The multidrug-resistant microbial strains are continuously increasing due to the misuse of broad-spectrum antibiotics. Plant-based antimicrobial substances have great therapeutic potential as they have lesser side effects as compared with synthetic drugs and a lesser chance of microbes developing resistance.

There is immense literature with examples of phytochemicals with antimicrobial effects, Cinchona bark as antimalarial, *Tinospora cordifolia* as antibacterial and anti-HIV (Pandey, 2007), *Allium sativum* (garlic) as antibiotic, cinnamon, clove, pomegranate, thyme, and lantana extracts as inhibitors of multidrug-resistant *Pseudomonas aeruginosa* have been reported (Nascimento, 2000). There is an enormous resource of phytochemicals which can work in synergy with clinical antibiotics. The combination of multiple mechanisms of action will help in overcoming multidrug resistance in notorious pathogens like *E. coli* and *P. aeruginosa.*

Plants produce defense molecules called phytoanticipins in return for bacterial invasion (see volume 1 chapter 3). If a plant is attacked by a pathogen they secrete secondary metabolites like alkaloids, glycosteriods, flavonoids, isoflavonoids, and so forth. Their mode of action against bacteria may involve antibacterial action such as membrane disturbance, intercalation of DNA, inhibition of efflux pump, inhibition of synthesis of nucleic acids, and so forth.

A perfect example of phytochemical antibacterial action against resistant bacteria is the essential oil-containing formulation, polytoxinol, which is strongly bactericidal against an extensive range of aerobic bacteria, including antibiotic-resistant bacteria. Polytoxinol is formulated to contain, in addition to constituents from Eucalyptus and Melaleuca species, components long recognized in traditional herbal medicine (Sherry et al., 2001). Berberine commonly found in *Hydrastis canadensis*, Echinacea species, and Berberis species is known to have antibacterial activity(Gibbons and Udo, 2000) and has shown good multidrug resistance inhibitor potential (Ball, 2006; Stermitz et al., 2000). Cinnamaldehyde and eugenol have also been found to inhibit several different MDR pathogenic bacteria such as *E. coli*, *Staphylococcus* spp., *Proteus* spp., *Klebsiella* spp., *Enterobacter* spp., and *Pseudomonas* spp (Suresh et al., 1992; Ali et al., 2005). There is a great amount of interest in the search for phytochemicals with the potential to inhibit bacterial efflux pumps. An effective efflux pump inhibitor could have noteworthy benefits, including the restoration of antibiotic sensitivity in a resistant strain and the reduction in the dose of antibiotic, dropping the adverse toxic effects (Kaatz, 2005). The inhibition of efflux pumps (MexABOprM, MexCD-OprJ, and MexEF-OprN) decreased the level of intrinsic resistance significantly, reversed acquired resistance, and decreased the frequency of emergence of *P. aeruginosa* strains highly resistant to fluoroquinolones. Previously, Stavri et al. (2007) also proposed the use of an efflux pump inhibitor in combination with an antibiotic to delay the emergence of resistance to that antibiotic.

## KEYWORDS

- **antimicrobial**
- **anti-inflammatory**
- **antioxidants**
- **phytochemicals**
- **polyphenols**

## REFERENCES

Afolabi, F.; Afolabi, O. J. Phytochemical Constituents of Some Medicinal Plants in South West, Nigeria. IOSR *J. Appl. Chem.* **2013**, *4*(1), 76–78.

Agarwal, R. et al. Studies on Immunomodulatory Activity of *Withania somnifera* (Ashwagandha) Extracts in Experimental Immune Inflammation. *J. Ethnopharmacology* **1999**, *67*, 27–35.

Akyol, H.; Riciputi, Y.; Capanoglu, E.; Caboni, M. F.; Verardo, V. Phenolic Compounds in the Potato and Its Byproducts: An Overview. *Int. J. Mol. Sci.* **2016**, *17*, 1–19

Alghamdi, S. S.; Khan, M. A.; El-harty, E. H.; Ammar, M. H.; Migdadi, H. M. Comparative Phytochemical Profiling of Different Soybean (*Glycine max* (L.) Merr) Genotypes Using GC-MS. *Saudi J. Biol. Sci.* **2017**, *25*(1), 15–21.

Ali, S. M.; Khan, A. A; Ahmed, I.; Musaddiq, M.; Ahmed, K. S.; Polasa, H.; Rao, L. V.; Habibullah, C. M.; Sechi, L. A.; Ahmed, N. Antimicrobial Activities of Eugenol and Cinnamaldehyde Against the Human Gastric Pathogen *Helicobacter Pylori*. *Ann. Clin. Microbiol. Antimicrob.* **2005**, *4*, 20.

Ames, B. N.; Gold, L. S. Endogenous Mutagens and the Causes of Aging and Cancer. *Mutat. Res.* **1991**, *250*(1–2), 3–16.

Ball, A. R.; Casadei, G.; Samosorn, S.; Bremner, J. B.; Ausubel, F. M.; Moy, T. I.; Lewis, K. Conjugating Berberinetoa Multidrug Efflux Pump Inhibitor Creates an Effective Antimicrobial. *ACS Chem. Biol.* **2006**, *1*, 594–600.

Barzegar, A. The Role of Intramolecular H-bonds Predominant Effects in Myricetin Higher Antioxidant Activity. *Comput. Theor. Chem.* **2017**, *1115*, 239–247

Birben, E.; Sahiner, U. M.; Sackesen, C.; Erzurum, S.; Kalayci, O. Oxidative Stress and Antioxidant Defense. *World Allergy Organ.* **2012**, *5*(1), 9–19.

Bornsek, S. M.; Ziberna, L.; Polak, T.; Vanzo, A.; Ulrih, N. P.; Abram, V.; Tramer, F.; Passamonti, S. Bilberry and Blueberry Anthocyanins Act as Powerful Intracellular Antioxidants in Mammalian Cells. *Food Chem.* **2012**, *134*(4), 1878–1884.

Burgos, G.; Amoros, W.; Muñoa, L.; Sosa, P.; Cayhualla, E.; Sanchez, C.; Díaz, C.; Bonierbale, M. Total Phenolic, Total Anthocyanin and Phenolic Acid Concentrations and Antioxidant Activity of Purple-Fleshed Potatoes as Affected by Boiling. *J. Food Compos. Anal.* **2013**, *30*(1), 6–12.

Cao, G.; Sofic, E.; Prior, R. L. Antioxidant and Prooxidant Behavior of Flavonoids: Structure-Activity Relationships. *Free Radical Biol. Med.* **1997**, *22*(5), 749–760.

Caris-Veyrat, C.; Amiot, M. J.; Tyssandier, V.; Grasselly, D.; Buret, M.; Mikolajczak, M.; Borel, P. Influence of Organic Versus Conventional Agricultural Practice On The Antioxidant Microconstituent Content of Tomatoes and Derived Purees; Consequences on Antioxidant Plasma Status in Humans. *J. Agric. Food Chem.* **2004**, *52*(21), 6503–6509.

Chahdoura, H.; Barreira, J. C. M.; Barros, L.; Santos-Buelga, C.; Ferreira, I. C. F. R.; Achour, L. Phytochemical Characterization and Antioxidant Activity of the Cladodes of *Opuntia macrorhiza* (Engelm.) and *Opuntia microdasys* (Lehm.). *Food Funct.* **2014**, *5*(9), 2129–2136.

Chassy, A. W.; Bui, L.; Renaud, E. N. C.; Horn, M. V.; Mitchell, A. E. Three-year Comparison of the Content of Antioxidant Microconstituents and Several Quality Characteristics in Organic and Conventionally Managed Tomatoes and Bell Peppers. *J. Agric. Food Chem.* **2006**, *54*, 8244–8252.

Che Sulaiman, I. S.; Basri, M.; Chan, K. W.; Ashari, S. E.; Masoumi, H. R. F.; Ismail, M. In Vitro Antioxidant, Cytotoxic and Phytochemical Studies of *Clinacanthus nutans* Lindau Leaf Extracts. *Afr. J. Pharm. Pharmacol.* **2015**, *9*(34), 861–874.

Choi, E. M. Kaempferol Protects MC3T3-E1 Cells through Antioxidant Effect and Regulation of Mitochondrial Function. *Food Chem. Toxicol.* **2011,** *49*(8), 1800–1805.

Crozier, A.; Jaganath, I. B.; Clifford, M. N. Dietary Phenolics: Chemistry, Bioavailability and Effects on Health. *Nat. Prod. Rep.* **2009,** *26,* 1001–1043.

Csányi, G.; Miller, F. J.Jr. Oxidative Stress in Cardiovascular Disease. *Int. J. Mol. Sci.* **2014,** *15,* 6002–6008.

Decharneux, T.; Dubois, F.; Beauloye, C.; DeConinck, S. W.; Wattiaux, R. Effect of Various Flavonoids on Lysosomes Subjected to an Oxidative or an Osmotic Stress. *Biochem. Pharmacol.* **1992,** *44*(7), 1243–1248.

Dechayont, B.; Itharat, A.; Phuaklee, P.; Chunthorng-Orn, J.; Juckmeta, T.; Prommee, N.; Hansakul, P. Antioxidant Activities and Phytochemical Constituents of *Antidesma thwaitesianum* Müll. Arg. Leaf Extracts. *J. Integr. Med.* **2017,** *15*(4), 310–319.

Delgoda, R.; Murray, J. E. Evolutionary Perspectives on the Role of Plant Secondary Metabolites. In *Pharmacognosy;* Elsevier B.V.: New York, 2017; pp 93–100.

Dillard, C. J.; German, J. B. Phytochemicals: Nutraceuticals and Human Health. Review. *J. Sci. Food Agric.* **2000,** *80,* 1744–1756.

Dixon, R. A.; Paiva, N. L. Stress-Induced Phenylpropanoid Metabolism. *Plant Cell,* **1995,** *7,* 1085–1097.

Dowling, S.; Regan, F.; Hughes, H. The Characterisation of Structural and Antioxidant Properties of Isoflavone Metal Chelates. *J. Inorg. Biochem.* **2010,** *104*(10), 1091–1098.

El-Mahdy, M. A.; Zhu, Q.; Wang, Q.; Wani, G.; Patnaik, S.; Zhao, Q.; Wani, A. A. Naringenin Protects Hacat Human Keratinocytes Against UVB- Induced Apoptosis and Enhances the Removal of Cyclobutane Pyrimidine Dimers from the Genome. *Photochem. Photobiol.* **2008,** *84*(2), 307–316.

Ghadage, D. M.; Kshirsagar, P. R.; Pai, S. R.; Chavan, J. J. Extraction Efficiency, Phytochemical Profiles and Antioxidative Properties of Different Parts of Saptarangi (*Salacia Chinensis* L.) – An Important Underutilized Plant. *Biochem. Biophys. Rep.* **2017,** *12,* 79–90.

Gibbons, S.; Udo, E. The Effect of Reserpine, a Modulator of Multidrug Efflux Pumps, on the in Vitro Activity of Tetracycline Against Clinical Isolates of Methicillin Resistant *Staphylococcus aureus* (MRSA) Possessing the Tet(K) Determinant. *Phytother. Res.* **2000,** *14,* 139–114.

Guitard, R.; Paul, J. F.; Nardello-Rataj, V.; Aubry, J. M. Myricetin, Rosmarinic and Carnosic Acids as Superior Natural Antioxidant Alternatives to α-Tocopherol for the Preservation of Omega-3 Oils. *Food Chem.* **2016,** *213,* 284–295.

Guo, R. X.; Guo, X.; Li, T.; Fu, X.; Liu, R. H. Comparative Assessment of Phytochemical Profiles, Antioxidant and Antiproliferative Activities in Sea Buckthorn (*Hippophae rhamnoides* L.) Berries. *Food Chem.* **2016,** *221,* 997–1003.

Hagerman, A. E.; Riedl, K. M.; Jones, G. A.; Sovik, K. N.; Ritchard, N. T.; Hartzfeld, P. W.; Riechel, T. L. High Molecular Weight Plant Polyphenolics (Tannins) as Biological Antioxidants. *J. Agric. Food Chem.* **1998,** *46*(5), 1887–1892.

Halliwell, B. Reactive Oxygen Species in Living Systems: Source, Biochemistry, and Role in Human Disease. *Am. J. Med.* **1991,** *91*(3), 14–22.

Hatahet, T.; Morille, M.; Shamseddin, A.; Aubert-Pouëssel, A.; Devoisselle, J. M.; Bégu, S. Dermal Quercetin Lipid Nanocapsules: Influence of the Formulation on Antioxidant Activity and Cellular Protection Against Hydrogen Peroxide. *Int. J. Pharm.* **2017,** *518*(1–2), 167–176.

He, Z.; Tan, J. S.; Abbasiliasi, S.; Lai, O. M.; Tam, Y. J.; Ariff, A. B. Phytochemicals, Nutritionals and Antioxidant Properties of Miracle Fruit *Synsepalum Dulcificum*. *Ind. Crops Prod.* **2016,** *86,* 87–94.

Idehen, E.; Tang, Y.; Sang, S. Bioactive Phytochemicals in Barley. *J. Food Drug Anal.* **2017,** *25*(1), 148–161.

Izhaki, I. Emodin - A Secondary Metabolite with Multiple Ecological Functions in Higher Plants. *New Phytol.* **2002,** *155*(2), 205–217.

Jaberian, H.; Piri, K.; Nazari, J. Phytochemical Composition and in Vitro Antimicrobial and Antioxidant Activities of Some Medicinal Plants. *Food Chem.* **2013,** *136*(1), 237–244.

Jolad, S. D.; Lantz, R. C.; Solyom, A. M.; Chen, G. J.; Bates, R. B, Timmermann, B. N. Fresh Organically Grown Ginger (Zingiber Officinale): Composition and Effects on LPS-Induced PGE2 Production. *Phytochemistry* **2004,** *65*, 1937–1954.

Jurenka, J.; Thorne, S. Anti-inflammatory Properties of Curcumin, a Major Constituent of *Curcuma longa*: A Review of Preclinical and Clinical Research. *Erratum in Altern. Med. Rev.* **2009,** *14*(3), 277.

Kaatz, G. W. Bacterial Efflux Pump Inhibition. *Curr. Opin. Invest. Drugs,* **2005,** *6*, 191–198.

Kim, A.; Nam, Y. J.; Lee, C. S. Taxifolin Reduces the Cholesterol Oxidation Product-Induced Neuronal Apoptosis by Suppressing the Akt and NF-κB Activation-Mediated Cell Death. *Brain Res. Bull.* **2017,** *134*, 63–71.

Koh, E.; Wimalasiri, K. M. S.; Chassy, A. W.; Mitchell, A. E. Content of Ascorbic Acid, Quercetin, Kaempferol and Total Phenolics in Commercial Broccoli. *J. Food Compos. Anal.* **2009,** *22*, 637–643.

Kumar, S.; Tiku, A. B. Biochemical and Molecular Mechanisms of Radioprotective Effects of Naringenin, a Phytochemical From Citrus Fruits. *J. Agric. Food Chem.* **2016,** *64*(8), 1676–1685.

Laitonjam, W. S. Natural Antioxidants (NAO) of Plants Acting as Scavengers of Free Radicals. In *Studies in Natural Products Chemistry;* Elsevier: Amsterdam, Netherlands, 2012; Vol. 37, pp 259–275.

Lintner, K.; Mas-Chamberlin, C.; Mondon, P.; Peschard, O.; Lamy, L. Cosmeceuticals and Active Ingredients. *Clinics Dermatol.* **2009,** *27*(5), 461–468.

Liu, R. H. Health-Promoting Components of Fruits and Vegetables in the Diet. *Adv. Nutr.*: An International Review Journal, **2013,** *4*(3), 384S–392S.

Loganayaki, N.; Siddhuraju, P.; Manian, S. Antioxidant Activity and Free Radical Scavenging Capacity of Phenolic Extracts from *Helicteres isora* L. and *Ceiba pentandra* L. *J. Food Sci. Technol.* **2013,** *50*(4), 687–695.

Maestri, D. M.; Nepote, V.; Lamarque, A.; Zygadlo, J. Natural Products as Antioxidants. In *Phytochemistry: Advances in Reasearch;* Research Signpost: Trivandrum, Kerala, 2006; pp. 105–135.

Maurya, P. K.; Kumar, P.; Nagotu, S.; Chand, S.; Chandra, P. Multi-Target Detection of Oxidative Stress Biomarkers in Quercetin and Myricetin Treated Human Red Blood Cells. *RSC Adv.* **2016,** *6*(58), 53195–53202.

Miller, D. D.; Li, T.; Liu, R. H. Antioxidants and Phytochemicals. *Ref. Module Biomed. Sci.* **2014,** 1–13.

Mira, L.; Fernandez, M. T.; Santos, M.; Rocha, R.; Florêncio, M. H.; Jennings, K. R. Interactions of Flavonoids with Iron and Copper Ions: A Mechanism for their Antioxidant Activity. *Free Radical Res. 36*(11), **2002,** 1199–1208.

Mironczuk-Chodakowska, I.; Witkowska, M. A.; Zujko, M. E. Endogenous Non-Enzymatic Antioxidants in the Human Body. *Adv. Med. Sci.* **2018,** *63*, 68–78.

Molyneux, R. J.; Lee, S. T.; Gardner, D. R.; Panter, K. E.; James, L. F. Phytochemicals: The Good, the Bad and the Ugly? *Phytochemistry,* **2007,** *68*, 2973–2985.

Murakami, A.; Ohnishi, K. Target Molecules of Food Phytochemicals: Food Science Bound for the Next Dimension. *Food Funct.* **2012,** *3,* 462–476.

Nascimento, G. F.; Locatelli, J.; Freitas, P. C.; Silva, G. L. Antibacterial Activity of Plant Extracts and Phytochemicals on Antibiotic Resistant Bacteria. *Braz. J. Microbiol.* **2000,** *31,* 247–256.

Nimse, S. B.; Pal, D. Free Radicals, Natural Antioxidants, and their Reaction Mechanisms. *RSC Adv.* **2015,** *5*(35), 27986–28006.

Pandey, A. K. Anti-Staphylococcal Activity of a Pan Tropical Aggressive and Obnoxious Weed *Parthenium histerophorus*: An in Vitro Study. *Nat. Acad. Sci. Lett.* **2007,** *30,* 383–386.

Pandit, S.; Chang, K. W.; Jeon, J. G. Effects of *Withania somnifera* on Growth and Virulence Props of *Streptococcus mutans* and *Streptococcus sobrinus* at Sub-MIC Levels. *Anaerobe* **2013,** *19,* 1–8.

Pedras, M. S. C.; Abdoli, A. Pathogen Inactivation of Cruciferous Phytoalexins: Detoxification Reactions, Enzymes and Inhibitors. *RSC Adv.* **2017,** *7*(38), 23633–23646.

Pereira, D. M.; Valentão, P.; Pereira, J. A.; Andrade, P. B. Phenolics: From Chemistry to Biology. *Molecules,* **2009,** *14*(6), 2202–2211.

Poljšak, B.; Dahmane, R. G.; Godić, A. Intrinsic Skin Aging: The Role of Oxidative Stress. *Acta Dermatovenerol. Alp. Pannonica et Adriat.* **2012,** *21*(2), 33–36.

Ralston, L.; Subramanian, S.; Matsuno, M.; Yu, O. Partial Reconstruction of Flavonoid and Isoflavonoid Biosynthesis in Yeast Using Soybean Type I and Type Ii Chalcone Isomerases. *Plant Physiol.* **2005,** *137*(4), 1375–1388.

Rice-Evans, C. A.; Miller, N. J.; Paganga, G. Structure-Antioxidant Activity Relationships of Flavonoids and Phenolic Acids. *Free Radical Biol. Med.* **1996,** *20*(7), 933–956.

Seifi-Jamadi, A.; Ahmad, E.; Ansari, M.; Kohram, H. Antioxidant Effect of Quercetin in an Extender Containing DMA or Glycerol on Freezing Capacity of Goat Semen. *Cryobiology,* **2017,** *75,* 15–20.

Sherry, E.; Boeck, H.; Warnke, P. H. Percutaneous Treatment of Chronic MRSA Osteomyelitis with a Novel Plant-Derived Antiseptic. *BMC Surg.,* **2001,** *1,* 1.

Sigurdson, G. T.; Robbins, R. J.; Collins, T. M.; Giusti, M. M. Effects of Hydroxycinnamic Acids on Blue Color Expression of Cyanidin Derivatives and their Metal Chelates. *Food Chem.* **2017,** *234,* 131–138.

Soobrattee, M. A.; Neergheen, V. S.; Luximon-Ramma, A.; Aruoma, O. I.; Bahorun, T. Phenolics as Potential Antioxidant Therapeutic agents: Mechanism and Actions. *Mutat. Res./Fundam. Mol. Mech. Mutagen.* **2005,** *579*(1–2), 200–213.

Srivastava, K. K.; Kumar, R. Stress, Oxidative Injury and Disease. *Indian J. Clin. Biochem.* **2015,** *30*(1), 3–10.

Stavri, M.; Piddock, L. J.; Gibbons, S. Bacterial Efflux Pump Inhibitors from Natural Sources. *J. Antimicrob.Chemother.* **2007,** *59,* 1247–1260.

Stermitz, F. R.; Tawara-Matsuda, J.; Lorensz, P.; Mueller, P.; Zenewicz, L.; Lewis, K.5'-Methoxyhydnocarpin-D and Pheophorbide A: Berberis Species Components that Potentiate Berberine Growth Inhibition of Resistant Staphylococcus Aureus. *J. Nat. Prod.* **2000,** *63,* 1146–1149.

Suresh, P.; Ingle, V. K.; Vijava, L. V.; Antibacterial Activity of Eugenol in Comparison with Other Antibiotics. *J. Food Sci. Technol.* **1992,** *29,* 254–256.

Tarahovsky, Y. S.; Selezneva, I. I.; Vasilieva, N. A.; Egorochkin, M. A.; Kim, Y. A. Acceleration of Fibril Formation and Thermal Stabilization of Collagen Fibrils in the Presence of Taxifolin (Dihydroquercetin). *Bull. Exp. Biol. Med.* **2007,** *144*(6), 791–794.

Terao, J. Dietary Flavonoids as Antioxidants. *Forum Nutr.* **2009,** *61,* 87–94

Tjendraputra, E.; Tran, V. H.; Liu-Brennan, D.; Roufogalis, B. D.; Duke, C. C. Effect of Ginger Constituents and Synthetic Analogues on Cyclooxygenase-2 Enzyme in Intact Cells. *Bioorg. Chem.* **2001,** *29,* 156–163.

Vallejo, F.; Toms-Barbern, F. A.; Garca-Viguera, C. Potential Bioactive Compounds in Health Promotion from Broccoli Cultivars Grown in Spain. *J. Sci. Food Agric.* **2002,** *82*(11), 1293–1297.

Verma, S. K.; Kumar, A. Therapeutic uses of *Withania somnifera* (Ashwagandha) with a Note on Withanolides and its Pharmacological Actions. *Asian J. Pharmacol. Clin. Res.* **2011,** *4,* 1–4.

Vermerris, W.; Nicholson, R. *Phenolic Compound Biochemistry;* Springer: Netherlands, Dordrecht, 2007.

Weidinger, A.; Kozlov, A. V. Biological Activities of Reactive Oxygen and Nitrogen Species: Oxidative Stress Versus Signal Transduction. *Biomolecules,* **2015,** *5,* 472–484

Wongkrongsak, S.; Tangthong, T.; Pasanphan, W. Electron Beam Induced Water-Soluble Silk Fibroin Nanoparticles as a Natural Antioxidant and Reducing Agent for a Green Synthesis of Gold Nanocolloid. *Radiat. Phys. Chem.* **2014,** *118,* 27–34

Wright, J. S.; Johnson, E. R.; DiLabio, G. A. Predicting the Activity of Phenolic Antioxidants: Theoretical Method, Analysis of Substituent Effects, and Application to Major Families of Antioxidants. *J. Am. Chem. Soc.* **2001,** *123*(6), 1173–1183.

Yadav, V. R.; et al. Boswellic Acid Inhibits Growth and Metastasis of Human Colorectal Cancer in Orthotopic Mouse Model by Downregulating Inflammatory, Proliferative, Invasive and Angiogenic Biomarkers. *Int. J. Cancer,* **2011,** 23.

Yang, E. J.; Kim, G. S.; Jun, M.; Song, K. S. Kaempferol Attenuates the Glutamate-Induced Oxidative Stress in Mouse-Derived Hippocampal Neuronal HT22 Cells. *Food Funct.* **2014,** *5*(7), 1395–1402.

Zhang, L.; Tu, Z.; Xie, X.; Wang, H.; Wang, H.; Wang, Z.; Lu, Y. Jackfruit (*Artocarpus Heterophyllus* Lam.) Peel: A better Source of Antioxidants and A-Glucosidase Inhibitors than Pulp, Flake and Seed, and Phytochemical Profile by HPLC-QTOF-MS/MS. *Food Chem.* **2017,** *234,* 303–313.

Zujko, M. E.; Witkowska, A. M.; Górska, M.; Wilk, J.; Krętowski, A. Reduced Intake of Dietary Antioxidants can Impair Antioxidant Status in Type 2 Diabetes Patients. *Pol. Arch. Med. Wewn.* **2014,** *124*(11), 599–607.

# CHAPTER 8

# PHYTOCHEMICALS AS OXIDATIVE STRESS MITIGATORS

ASHUTOSH GUPTA and ABHAY K. PANDEY*

*Department of Biochemistry, University of Allahabad, Allahabad 211002, India*

*Corresponding author. E-mail: akpandey23@rediffmail.com; Mob.: +91 9839521138
*ORCID: orcid.org/ orcid.org/0000-0002-4774-3085

## ABSTRACT

Free radicals are produced by regular cellular metabolism. These are highly reactive molecules which can damage or alter the cellular structures, namely carbohydrates, lipid, proteins, and nucleic acids. The alteration in the steadiness between free radicals and antioxidants in favor of free radicals is defined as "oxidative stress." Oxidative stress leads to the development of a wide range of pathological conditions and diseases such as cancer, atherosclerosis, hypertension, diabetes, chronic obstructive pulmonary disease, neurological disorders, and asthma. Phytochemicals influencing animal metabolism are being broadly examined for their potential health benefits. Physiological benefits of phytochemicals could be explained by molecular basis of the antioxidant action including their basic mechanisms, targeting signal transduction pathways, especially through the antioxidant response element/nuclear factor (erythroid-derived 2)-like two transcription system. The understanding of the molecular concept of the biological activity of phytoconstituents provides scientific validation to defend their use in foods. The current chapter summarizes the anomalies of oxidative stress, their biomarkers and ameliorative effects of phytochemicals.

## 8.1  INTRODUCTION

Free radicals in the form of reactive oxygen species (ROS) or reactive nitrogen species (RNS) are continuously produced in living beings as byproducts of normal metabolism. Their higher amount is associated with numerous biochemical and pathophysiological alterations including lipid peroxidation, DNA damage, aging process, hepatic damage, cardiovascular diseases and cancer (Lopez-Alarcona and Denicola, 2013). Oxidative stress-related studies have shown that it is a key factor for the occurrence of a wide range of diseases (Gutteridge, 1993). Under normal conditions, ROS/RNS carry out the important functions of chemical defense or detoxification in addition to cell signaling and biosynthetic reactions (Lopez-Alarcona and Denicola, 2013).

Phytochemicals are naturally produced bioactive compounds in plants which provide health assistance in the living being (Hasler et al., 1999). They primarily protect plants from environmental vulnerabilities such as stress, pathogen attack, UV-exposure, pollution, drought in addition to providing color, flavor, and aroma (Mathai, 2000). More than four thousand phytochemicals are known that are classified based on their physical, chemical, and protective characteristics (Meagher and Thomson, 1999). For humans, they are not nutritionally essential but help them to fight and protect against various diseases such as acting as an antimicrobial, antioxidant, anticancer, and immunomodulatory agents.

## 8.2  OXIDATIVE STRESS

Oxidative stress is defined as "a disturbance in the prooxidant (ROS/RNS) to antioxidant balance in favor of the oxidant species, leading to potential damage" (Sies, 1991). Under these circumstances, the antioxidant defense mechanism slows down or degenerates in the aerobic organism leading to a reduction in endogenous systems' potential to combat against the oxidative stress (Persson et al., 2014). ROS triggered oxidative deterioration exerts its adverse effect on the biomolecules which is translated into the impaired physiological functions contributing to disease promoting incidence and reduction in lifespan (Kumar et al., 2013; Maulik et al., 2013). Diseases resulting from oxidative stress include diabetes, hypertension, atherosclerosis, cancer, acute respiratory distress syndrome, asthma, chronic obstructive pulmonary disease, idiopathic pulmonary fibrosis, and neurological disorders (Asami, 1997). Specific biomarkers help in the diagnosis of various pathological conditions. World Health Organization (WHO) has defined biomarkers as "any structure, substance, process or its products that can be measured in the body and influence or predict the frequency of outcome or disease" (2001).

## 8.2.1 EFFECT OF OXIDATIVE STRESS ON DNA

Various mechanisms of free radical-mediated DNA damage have been proposed which comprises mutations, deletions, translocation, single and double-stranded breaks, and degradation of nitrogenous bases. These alterations might lead to aging, carcinogenesis, neurodegenerative, and autoimmune diseases. Monitoring and measurement of many DNA damage markers are estimated out in the tissues and the body fluid (Table 8.1). The formation of 8-hydroxy-guanine (8-OH-G) is one of the best-known examples of oxidative stress-induced DNA damage. It is a prospective biomarker for cancer detection. GC-rich sequences present in transcription factor binding sites of the promoter regions of the genes are susceptible to oxidant attack. The 8-OH-G formation can modify the binding of transcription factors and consequently changes the expression of related genes (Ghosh and Mitchell, 1999). Transcription inhibition potential has also been observed in cyclo-dA (8,5'cyclo-2'-deoxyadenosine) which inhibits the binding of TATA-binding protein (Marietta et al., 2002).

**TABLE 8.1**   Prime Markers of DNA Oxidative Damage.

| S/ No. | Markers | Sample | Methods | References |
|---|---|---|---|---|
| 1. | 8-hydroxy-deoxy-guanosine | Urine, DNA | HPLC-ECD | Shigenaca et al. (1991); Wisean et al. (1995); and Germadnik et al. (1997). |
| 2. | 8-hydroxy-guanine (8-OH-G) | Urine, DNA | GC-MS | Dizdaroglu (1991); Wisean et al. (1995); and Halliwell (1996). |
| 3. | Oxo-2'deoxy-guanosine | Urine, DNA | HPLC-ECD | Shigenaga et al. (1994) and Helbock et al. (1998). |
| 4. | 8-oxo-guanine | Urine, DNA | HPLC-ECD, GC-MS | Shigenaga et al. (1994); Ravanat et al. (1995); Collins et al. (1997); Heath et al. (1997) and Helbock et al. (1998); |
| 5. | 5-(hydroxymethyl) uracil | DNA, synthesized oligonucleotides | GC | Djuric et al. (1991) and Lafrancois et al. (1998). |

Oxidative damage can also break the single-stranded DNA which is easily repaired by the cells. However, ionizing radiation induces double-stranded breaks, this can lead to major risk for cell survival (Caldecott, 2003). Methylation at CpG islands in DNA is an important epigenetic mechanism that may lead to gene silencing. Oxidation of 5-methylcytosine to 5-hydroxymethyl uracil (5-OHMeUra) can occur through deamination/oxidation reactions of thymine or 5-hydroxymethylcytosine intermediates (Cooke et al., 2003). The modification in gene expression, DNA methylation may also induce an effect on chromatin organization (Jones and Wolffe, 1999). Abnormal DNA methylation patterns induced by oxidative attacks also affect DNA repair activity.

## 8.2.2  EFFECT OF OXIDATIVE STRESS ON LIPIDS

Free radicals can induce lipid damage and alter the lipid membrane arrangement, which in turn deactivate the enzymes, membrane-bound receptor, and enhance the permeability of tissue (Girotti, 1985). Lipid autoxidation is accelerated after the action of an acyl side-chain on a fatty acid by any chemical species that eliminated a hydrogen atom from a methylene carbon (Brown and Kelly, 1996). Fatty peroxyl radicals and fatty hydroperoxides are generally defined as early autoxidation intermediates and primary oxidation products, respectively. However, fatty hydroperoxides, are labile species and readily enter into radical reactions which lead to their molecular transformation and decomposition (Janero, 1990). Lipid peroxidation products, namely malondialdehyde and unsaturated aldehydes, are able to inactivate various cellular proteins by forming protein cross-linkages (Siu and Draper, 1982; Hagihara et al., 1984; Esterbauer et al., 1986, Sharma et al., 2017). 4-Hydroxy-2-nonenal causes reduction of intracellular GSH level while peroxide production stimulates epidermal growth factor receptor (Suc et al., 1998), and boost fibronectin synthesis (Tsukagoshi et al., 2002). Thiobarbituric acid reactive substances (TBARS) and isoprostanes, the lipid peroxidation products have been used as indirect biomarkers of oxidative stress, and increased levels were shown in the exhaled breath condensate or bronchoalveolar lavage fluid or lung of chronic obstructive pulmonary disease patients or smokers (Morrison et al., 1999; Nowak et al., 1999; Montuschi et al., 2000). The elevated level of lipid peroxidation can be determined by a number of methods which can be divided between those that measure the primary products of lipid peroxidation such as hydroperoxides, and those that measure the secondary breakdown products of lipid hydroperoxides such as secondary carbonyls (malondialdehyde, 4-hydroxynonenal) and volatile hydrocarbons (ethane and pentane) (Brown and Kelly, 1996).

## 8.2.3   EFFECT OF OXIDATIVE STRESS ON PROTEINS

Free radicals can cause the modification of the peptide bonds, alteration in electrical charge, cross-linking of proteins and specific amino acids oxidation of protein, and hence, leads to increased exposure to proteolysis degradation by specific proteases (Kelly and Mudway, 2003). Protein residues, namely cysteine and methionine are mainly more disposed to oxidation (Dean et al., 1985). Oxidation of methionine residues or sulfhydryl groups of proteins leads to change in their conformations, protein unfolding, and degradation. Additionally, oxidative modification may inhibit the activities of several kinds of enzymes (Fucci et al., 1983; Stadtman 1990). For example, methionine sulfoxide can be oxidized by methionine (Stadtman et al., 2003) and phenylalanine to o-tyrosine (Stadtman and Levine, 2003), carbonyl groups and disulfide bonds can be formed by oxidation of sulfhydryl groups which is introduced into the side chains of proteins. Gamma rays, HOCl, metal-catalyzed oxidation, and ozone can cause the formation of carbonyl groups (Lyras et al., 1997). Changes due to the free radical damage and other kinds of protein damage or anomalies lead to a damaged protein system (Dean et al., 1993). Poor handling of a certain type of oxidized proteins along with the possible alteration in the amount of formation of oxidized proteins may contribute to the observed accumulation and damaging action of oxidized proteins during aging and in several pathologies (Stadtman and Berlett, 1998).

The protein carbonyl content has been widely used as a suitable biomarker of protein damage. Several sensitive methods for the detection and quantitation of protein carbonyl content under the condition of oxidative stress have been developed (Ayala and Cutler, 1996; Berelett and Stadtman, 1997). The presence of such carbonyl group is taken as a probable indicator of protein oxidative modification (Giulivi and Davies, 1994; Levine et al., 1994). For example, protein glycation may add carbonyl groups to amino acid residues. Moreover, there are other carbonyl groups present in tissues that are not produced by oxidative damage, and these are frequently not even proteins such as nucleic acids. Their presence in samples could indicate serious errors and overexpression of oxidative protein damage. The carbonyl content is measured using various techniques (Ayala and Cutler, 1996; Shacter et al., 1996).

## 8.3   ENZYMATIC MARKERS FOR OXIDATIVE STRESS

### 8.3.1   ALANINE AMINOTRANSFERASE (ALT)

The liver cells produce the alanine aminotransferase (ALT) (E. C.2.6.1.2) enzyme. It is an important enzyme that plays a role in the shifting of an amino group from alanine amino acid to α-ketoglutaric acid which produces glutamate and pyruvate (Fig. 8.1). Early biochemical and cytogenetic studies have proposed the presence of two isoforms of ALT in the living organism (Jadhao et al., 2004). Humans express ALT-1 and ALT-2 and these two forms of the enzyme showed 70% similarities. ALT-1 generally expressed in kidney, liver, fat, and heart tissues while ALT-2 present in muscle, fat, brain, and kidney tissues (Yang et al., 2002). Normal levels are in the range of 5–50 U/L. A high amount of ALT is predominantly present in liver and kidney, whereas it is present in small amount in skeletal and cardiac muscles. In case of abnormal hepatic activities, it secreted out of the cell and initially localized in serum and cerebrospinal fluid (CSF), but not in urine (Sakagishi, 1995). Because ALT is located in the cytoplasm, serum ALT level comparatively upsurge during free radical-mediated liver injuries. For that reason, it acts as a potent marker for hepatic injury in preclinical toxicity studies (Yang et al., 2009).

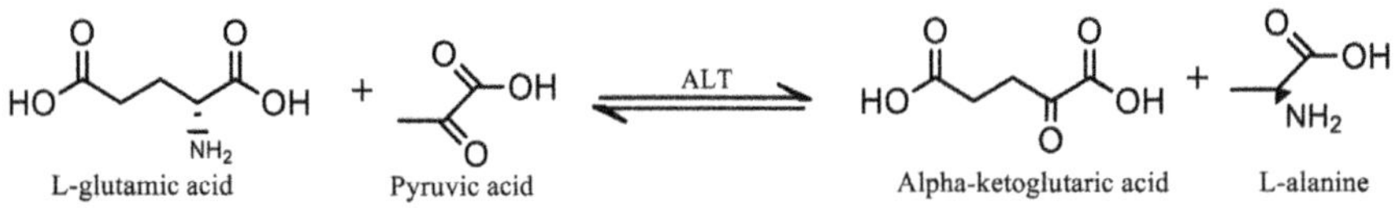

**FIGURE 8.1**   Reaction catalyzed by alanine aminotransferase (ALT).

### 8.3.2   ASPARTATE AMINOTRANSFERASE (AST)

The aspartate aminotransferase (AST) (E. C.2.6.1.1) transfers amino groups from glutamate to oxaloacetate during amino acid catabolism and form α-ketoglutarate and aspartate, respectively (Fig. 8.2). This aspartate is now used as a nitrogen source in the urea cycle. A large fraction of AST is found in the liver. However, it is not a liver-specific enzyme as it is also present in skeletal muscle, red blood cells, renal, and cardiac muscles (Rej, 1989). Normally, the blood contains a small amount of AST, however, during a disease or damaged condition of body tissue or organ, more AST is released into the blood. Normal levels are in the range of 7–40 U/L. The elevated amount of blood AST is directly proportionate to the degree of tissue damage.

The AST to ALT ratio may also help to detect the infected organ whether it is liver or another organ. AST showed similarities with ALT, as both enzymes are associated with liver parenchymal cells, but the difference is that ALT is mainly found in the liver, with clinically insignificant amount found in the heart, skeletal muscle, and kidneys. Whereas AST is present in the liver, heart (cardiac muscle), skeletal muscle, kidneys, and brain.

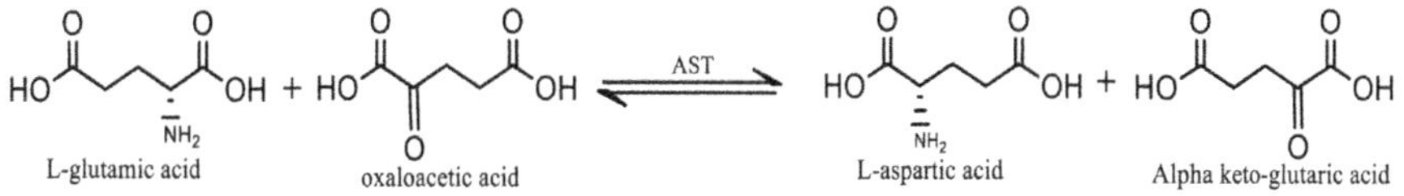

FIGURE 8.2   Reaction catalyzed by aspartate aminotransferase (AST).

## 8.3.3   GLUTAMATE DEHYDROGENASE (GDH)

The glutamate dehydrogenase (GDH) (EC1.4.1.2) is a mitochondrial enzyme which leads the conversion of glutamate to α-ketoglutarate (Fig. 8.3). This conversion is completed in two steps. During the first step, a Schiff base transition state is formed between ammonia and α-ketoglutarate. This Schiff base transition state is essential for the establishment of the alpha carbon atom in glutamate's stereochemistry. The second step involves the Schiff base transition state being protonated, which is achieved by the shifting of a hydride ion from NADPH producing L-glutamate. The GDH is irreplaceable because of its NAD and NADP utilizing the property. The NADP is used in the forward reaction of α-ketoglutarate and free ammonia, which are converted to L-glutamate through a hydride transfer from NADPH to glutamate (Berg et al., 2002). NAD is involved in the reversible reaction which comprises the conversion of L-glutamate to α-ketoglutarate which produces free ammonia through an oxidative deamination reaction (Baker et al., 1997). The investigation of the role of GDH is difficult because numerous isoenzymes function in different directions in vivo. Moreover, as an isoenzyme pattern is sensitive to a broad variety of abiotic factors, GDH function accordingly under different anomalous conditions (Watanabe et al., 2011).

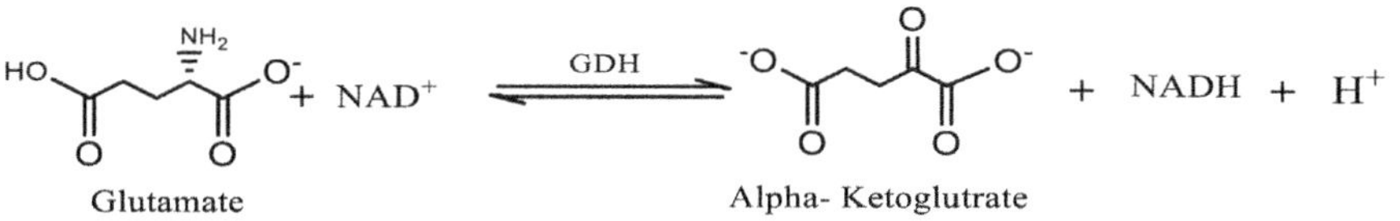

FIGURE 8.3   Reaction catalyzed by GDH.

## 8.3.4  SORBITOL DEHYDROGENASE (SDH)

The sorbitol dehydrogenase (SDH) (EC 1.1.1.14) is a member of the zinc-containing alcohol dehydrogenase family and consists of 357 amino acids. The SDH catalyzes the zinc-dependent inter-conversion of polyols, such as sorbitol and xylitol, to their respective ketoses (Fig. 8.4). It produces fructose from glucose together with aldose reductase without using ATP. SDH has been recognized in various animals as well as human tissues. Although it is widely distributed throughout the body tissues, it is primarily located in the cytoplasm and mitochondria of kidney, liver, and seminal vesicles (El-Kabbani et al., 2004). Normal levels in the plasma are in the range of 1–3 U/L. SDH is a potent marker for hepatic necrosis, but estimation of other important enzymes such as ALT, AST may also be needed along with SDH. Khayrollah et al. also described SDH as an important indicator of hepatic damage and parenchymal diseases (1982). In normal conditions, SDH is present in serum at low concentration, but elevated during the acute liver damage (Dooley et al., 1979).

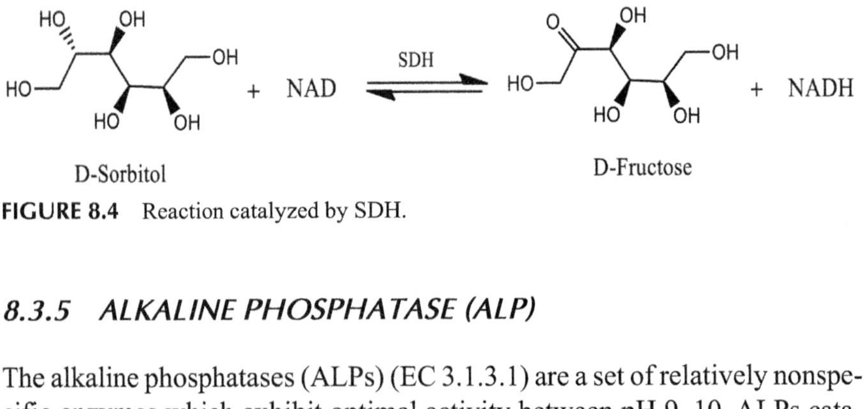

FIGURE 8.4   Reaction catalyzed by SDH.

## 8.3.5  ALKALINE PHOSPHATASE (ALP)

The alkaline phosphatases (ALPs) (EC 3.1.3.1) are a set of relatively nonspecific enzymes which exhibit optimal activity between pH 9–10. ALPs catalyze the hydrolysis of a wide range of natural as well as synthetic substrates (Fig. 8.5). Divalent ions, such as $Mg^{2+}$, $Co^{2+}$, and $Mn^{2+}$ are activators of this enzyme while $Zn^{2+}$ is a constituent metal ion (Ozer et al., 2008). Normal levels are in the range of 20–120U/L. ALP is anchored to cell membranes by glycophosphatidylinositol (GPI) proteins. Breakdown of GPI by phospholipase D, bile acids, and proteases discharge ALP from membranes, causing an increase in ALP levels in blood serum/plasma (Evans, 1996). Increased ALP levels have also been concerning with drug-induced cholestasis (Singh et al., 2011).

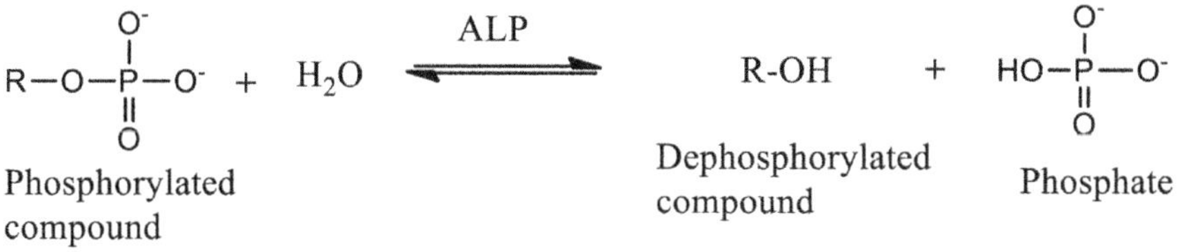

**FIGURE 8.5**   Reaction catalyzed by ALP.

## 8.3.6   Υ-GLUTAMYL TRANSFERASE (GGT)

γ-Glutamyl transferase (GGT) (EC 2.3.2.2) comes under the class of enzyme peptidase, which precisely catalyzes the transfer of γ-glutamyl moiety from peptides and other compounds that contain it to substrate itself, some amino acids or peptides, or even water, in which case a simple hydrolysis takes place (Goldberg, 1980). It beaks the glutathione (GSH) to assist the recapture of cysteine for the synthesis of intracellular GSH. The GGT plays a role in the movement of amino acids across the cellular membrane and leukotriene metabolism (Tate and Meister, 1985). It predominantly participated in amino acid absorption, glutathione metabolism and protection against oxidative damage. A small amount of GGT (5%) is present in the cytoplasm. Solubilization and increased synthesis results in elevated serum GGT level (Albert et al., 1964). Normal levels are in the range of 0–51 U/L. It is a specialized marker of hepatobiliary injury, mainly cholestasis, and biliary effects. It was reported as a key indicator of bile duct lesions in the rat liver (Leonard et al., 1984).

## 8.3.7   LACTATE DEHYDROGENASE (LDH)

The lactate dehydrogenase (LDH) (EC 1.1.1.27) is a tetrameric enzyme that catalyzes the reversible alteration of pyruvate to lactate (Fig. 8.6) (Dawson et al., 1964). This enzyme uses NAD/NADH as a cofactor for this conversion. The LDH is widely distributed intracellular enzyme which is present throughout the body but its high amount is found in those tissues that utilize glucose as an energy source. Normal levels are in the range of 100–220U/L. Isoenzymes of LDH are largely distributed in a tissue-specific manner. LDH-1 and LDH-2 isoenzymes are mainly present in erythrocytes, cardiac muscles, and kidney. Whereas LDH-4 and LDH-5 isoenzymes are predominantly occurring in skeletal and muscle liver. LDH2–4 is found in many other tissues such as the endocrine glands, spleen platelets, and lungs (Milne and Doxey, 1987). As a result of this distribution, an elevated LDH can reveal damage to a number of different tissues such as kidney, liver, skeletal or cardiac muscle. The LDH levels may be increased whenever there is cell necrosis or when there is a neoplastic proliferation of cells, which causes an increase LDH production.

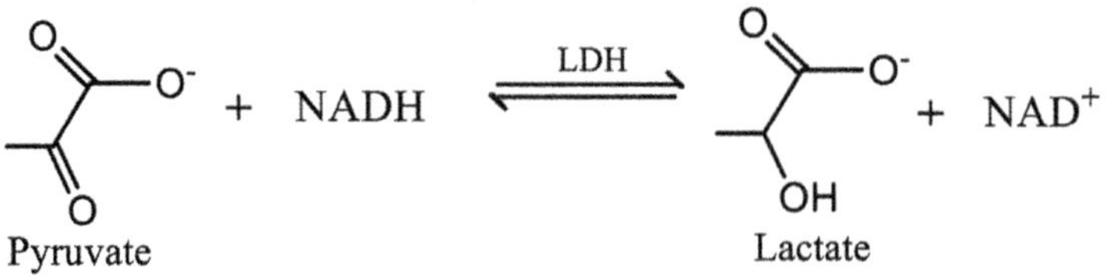

Pyruvate                                              Lactate

**FIGURE 8.6**   Reaction catalyzed by LDH.

### 8.3.8   MALATE DEHYDROGENASE (MDH)

Malate dehydrogenase (MDH) (EC 1.1.1.37) catalyzes the inter-conversion of the malate and oxaloacetate substrate in the presence of NAD/NADH. This reaction performs an important role during the tricarboxylic acid cycle and malate/aspartate shuttle system (Minarik et al., 2002). On the basis of their localization and specificity to the coenzyme such as NADP or NAD, a different type of isoforms of MDH has been reported. The normal value of MDH in healthy human plasma/serum is 23.5–47.7 U/L. Increase in MDH activity was found to correlate with morphological changes (Korsrud et al., 1972). The major activity of MDH was determined in liver followed by heart, brain, and skeletal muscle (Bergmeyer and Gawehn, 1974). The MDH is also a periportal enzyme that is discharged into the serum showing tissue damage. Serum MDH activity is correlated with both liver and heart injury.

### 8.4   PHYTOCHEMICALS AS MEDICINAL AGENTS

Phytochemicals (plant secondary metabolite) have been the most popular source of effective drug leads. Given the advantage of using dietary components with relatively low toxicity, abundance of materials, and cost-effective nutritional therapy, it provides an important strategy for preventing or handling numerous ailments and hence, contributes significantly to the individual's welfare (Cardozo et al., 2013). After a long period of experimental use of herbal formulation, the first separated active compound was alkaloids such as morphine, strychnine, quinine, and so forth. The beginning of 19th century was marked as a discovery of modern medicinal plants and its application, but with time herbal remedies lose their attention due to the remarkable utilization of synthetic drug. On the other hand at the same period application of herbal formulation in Western Europe almost became double (Hamburger and Hostettmann, 1991). However, in the 21st century, the awareness towards the plant-based formulation and perhaps animal origin has grown progressively due to the knowledge of various side effects of synthetic drugs and successful preparation of herbal medicine. Newman et al. suggested that during the period of 1981–2002, approximately 60% of the anticancer medicines and 75% other anti-infectious remedies were approved which were derived from plants (2003). Whereas during this period more than 60% of the total consumed drugs world widely were either plant-based or derived from their bioactive compounds (Gupta et al., 2005). Even though it has been proved that phytochemical may show slower action, but in long-term and mostly in chronic disease it showed better results (Akunyili, 2003). In last few decades, due to the development of new technique and advancement in chemical science and pharmaceutical industries, now it is easy to separate the bioactive compound from different plant parts, and therefore, number of agencies throughout the world are presently exploring therapeutic herbs for potent phytochemicals and lead compounds that could be used in controlling of various diseases (Acharya and Shrivastava, 2008). Some common examples of compounds isolated from plants replace with include extraction of Aspirin (an anti-inflammatory drug) from the bark of the Salix species by Hoffmann and isolation of quinine (antimalarial drug) from Cinchona by Peletier and Caventou. A report of World Health Organization (WHO, 2001) suggested that around 60% of the total population of the world depends on the traditional medicine and near about 80% population of developing countries entirely depends on the traditional medical therapy (Fransworth, 1994).

### 8.4.1   POSSIBLE MECHANISM OF PHYTOCHEMICALS IN ROS-RELATED MOLECULAR TARGETS

The living cells have a number of sensitive mechanisms to fight and counterbalance the damaging effect of free radicals (ROS/RNS) and maintained redox homeostasis (Fig. 8.7). It can be categorized into two types of antioxidants: direct and indirect antioxidants. The direct antioxidants have a short half-life and thus rejuvenate during the redox activities (Jung and Kwak, 2010). Due to their short half-life, they should be taken frequently and comparatively in a high amount to sustain their physiological potensial. Whereas indirect antioxidants act by the amplifying cellular antioxidant capacity such as enhancing specific genes code for antioxidant proteins through the key transcription factor, nuclear factor (erythroid-derived 2)-like 2 (Nrf2) (Fig. 8.7), that is well recognized as a principal regulator of the antioxidant response (Jung and Kwak, 2010; Sahin et al., 2010). Therefore, the effect of indirect antioxidants persists longer as compare to direct anti-oxidants (Mamede et al., 2011).

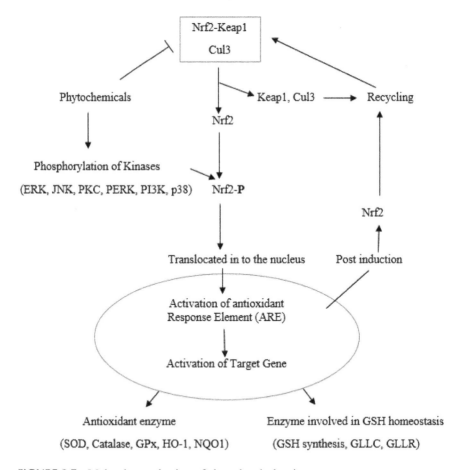

**FIGURE 8.7**   Molecular mechanism of phytochemical action.

Nrf2, a transcription factor that activates a number of antioxidant genes by activating cis-antioxidant response element (ARE) and comprises a conserved sequence which is located in the promoter region (Barve et al., 2008; Shen and Kong, 2009; Hu et al., 2010). The ARE contains glutathione S-transferase (GST), glutathione peroxidase (GSH-Px), heme oxygenase-1 (HO-1), NAD(P)H: quinone oxidoreductase 1 (NQO1), CAT, SOD, glutamate-cysteine ligase (GCL), uridine diphosphate glucuronosyltransferase, and the thioredoxin/peroxiredoxin system. These enzymes system plays a dynamic role in cell protection by enhancing the exclusion of ROS and thus act as an antioxidant agent against oxidative stress (Baird and Dinkova-Kostova, 2011). Under normal physiological condition, Nrf2 presents with a cytoskeleton binding protein Kelch-like ECH associating protein-1 (Keap1) in the cytoplasm, which in turn associates with Cullin 3 (Cul3) to form an E3 ubiquitin ligase complex that constantly degrades the interaction of Nrf2 and Keap1 transcription factors (Cardozo et al., 2013). Now free Nrf2 translocates into the nucleus where it associates with the small musculoaponeurotic fibrosarcoma protein and coactivator proteins which in turn binds to ARE and finally initiate the transcription of cytoprotective genes (Na and Surh, 2008; Baird and Dinkova-Kostova, 2011; Takaya et al., 2012). Various studies suggested that different phytochemicals extracted from vegetables, fruits, spices, and herbs may activate Nrf2 and stimulate expression of antioxidant/phase II detoxifying enzymes (Na and Surh, 2008; Sahin et al., 2010, Saw et al., 2012).

In past few years, ellagic acid, arjunolic acid, gallic acid, and epigallocatechin gallate (EGCG) have become the subject of more exhaustive studies in animal. Additionally, several studies revealed that various phytochemicals can interfere with multiple cell-signaling pathways and they could be used in their natural form in high quantities for mitigation of oxidative stress in the living organism (Sahin et al., 2013).

Figure 8.7 shows phytochemicals increase antioxidant response through the transcriptional activation of nuclear erythroid 2-related factor 2 (Nrf2). Nrf2 under basal conditions is bound to its repressor Kelch-like ECH-associated protein one (Keap1) in the cytoplasm. Keap1 serves as an adaptor protein between Nrf2 and the cullin-3 (Cul3) complex, leading to ubiquitylation of Nrf2 and subsequent degradation by the 26S proteasome. EA interacts with cysteine residues contained in Keap1 decreasing its affinity to Nrf2, releasing it for nuclear translocation. On the other hand, EA could induce Nrf2 phosphorylation at Ser-40 through the activation of kinases, such as extracellular-signal-regulated kinases (ERK), phosphoinositide 3-kinase (PI3-K), c-Jun NH (2)-terminal kinase (JNK), protein kinase C (PKC), protein kinase RNA-like endoplasmic reticulum kinase (PERK), and p38. Once translocated into nucleus, Nrf2 binds to the antioxidant responsive elements (ARE) regulating the expression of genes that encode antioxidant enzymes like superoxide dismutase (SOD), catalase (CAT), glutathione peroxidase (GPx); heme oxygenase-1 (HO-1); NADPH: quinone oxidoreductase 1 (NQO1), as well as enzymes involved in gluta-thione (GSH) homeostasis such as GSH synthetase, glutamate-cysteine ligase catalytic subunit (GCLC), and glutamate-cysteine ligase regulatory subunit (GCLR).

## 8.5   CONCLUSION

Phytochemical evaluation of plants is essential to discover the new sources of therapeutic and pharmacologically active compounds. Nowadays there is growing interest to reveal the pharmacological properties of plant-derived bioactive compounds in term of human welfare. The phytochemical screening and quantitative as well as qualitative estimation of plant bioac-tive compounds show the existence of various secondary metabolites such as flavonoid, alkaloids, saponins, terpenoids, and tannins. These phytochemicals possess a wide range of medicinal and pharmacological activities. Findings suggested that despite the ability of these functional phytochemicals to elevate the expression of phase II (antioxidant) enzymes, it appears that their major role is acting as modifiers of signal transduction pathways to elicit cytoprotective responses through suppressing stress-induced protein activation and enhancing Keap1 dissociation from Nrf2 in response to stressors. Therefore, the suppression of abnormally amplified oxidation signaling and restoration of function as well as activation of antioxidant machinery can provide important strategies for prevention of free radical-induced injuries and augmentation of antioxidant defense in animals. Oxidative stress has a key component for various diseases, yet very few oxidative stress markers have made it into routine clinical use. Stable oxidative alterations, such as lipid oxidation products, DNA/RNA oxidation, 3-nitrotyrosine, and protein carbonyls contribute to some of their positive clinical findings.

## ACKNOWLEDGMENTS

Ashutosh Gupta acknowledges the financial support from UGC-CRET fellowship. Both the authors acknowledge the SAP and DST-FIST facilities of the Department of Biochemistry, University of Allahabad, Allahabad, India.

## KEYWORDS

- **oxidative stress**
- **reactive oxygen species**
- **reactive nitrogen species**
- **DNA damage**
- **bioactive compounds**

## REFERENCES

Acharya, D.; Shrivastava, A. *Indigenous Herbal Medicines: Tribal Formulations and Traditional Herbal Practices;* Aavishkar Publishers Distributor: Jaipur, India, 2008; p 440.

Akunyili, D. N. *The Role of Regulation of Medicinal Plants and Phytomedicine in Socio-economic Development;* AGM/SC of the Nigerian Society of Pharmacognosy: Abuja, 2003; pp 1–7.

Albert, Z.; Szewczuk, A.; Orlowski, M.; Orlowska, J. Histochemical and Biochemical Investigations of Gamma-Glutamyl Transpeptidase in the Tissues of Man and Laboratory Rodents. *Acta Histochem.* **1964,** *18,* 78–89.

American Cancer Society. Phytochemicals. June **2000** http://www.cancer.org/eprise/main/docroot/ETO/content/ETO_5_3X_Phytochemicals.

Asami, S.; Manabe, H.; Miyake, J.; Tsurudome, Y.; Hirano, T.; Yamaguchi, R.; Itoh, H.; Kasai, H. Cigarette Smoking Induces an Increase in Oxidative DNA Damage, 8-Hydroxydeoxyguanosine, in a Central Site of the Human Lung. *Carcinogenesis* **1997,** *18,* 1763–1766.

Ayala, A.; Cutler, R. G. The Utilization of 5-Hydroxyl-2-Amino Valeric Acid as a Specific Marker of Oxidized Arginine and Proline Residues in Proteins. *Free Radical Biol. Med.* **1996,** *21,* 65–80.

Baird, L.; Dinkova-Kostova, A.T. The Cytoprotective Role of the Keap1-Nrf2 Pathway. *Arch. Toxicol.* **2011,** *85,* 241–272.

Baker, P. J.; Waugh, M. L.; Wang, X. G.; Stillman, T. J.; Turnbull, A. P.; Engel, P. C.; Rice, D. W. Determinants of Substrate Specificity in the Superfamily of Amino Acid Dehydrogenases. *Biochemistry* **1997,** *36,* 16109–16115.

Barve, A.; Khor, T. O.; Hao, X.; Keum, Y.; Yang, C. S.; Reddy, B.; Kong, A. T. Murine Prostate Cancer Inhibition by Dietary Phytochemicals-Curcumin and Phenyethylisothiocyanate. *Pharm. Res.* **2008,** *25,* 2181–2189.

Berelett, B. S.; Stadtman, E. R. Protein Oxidation in Aging, Disease and Oxidative Stress. *J. Biol. Chem.* **1997,** *272,* 20313–20316.

Berg, J.M. Tymoczko, J. L. Stryerand, L. *Biochemistry,* 5th ed.; W H Freeman: New York, NY, 2002.

Bergmeyer, H.; Gawehn, K. *Methods of Enzymatic Analysis,* 2nd ed.; Academic Press: New York, 1974; Vol. 2, p 6138.

Brown, R. K.; Kelly, F. J. Peroxides and Other Products. In: *Free radicals. A practical approach.* Punchard, N. A., Kelly, F.J., Eds.; Oxford University Press: Oxford, 1996, pp 119–131.

Caldecott, K. W. Protein-Protein Interactions During Mammalian DNA Single-Strand Break Repair. *Biochem. Soc. Trans.* **2003**, *31*, 247–251.

Cardozo, L. F.; Pedruzzi, L. M.; Stenvinkel, P.; Stockler-Pinto, M. B.; Daleprane, J. B.; Leite, M. Jr.; Mafra, D. Nutritional Strategies to Modulate Inflammation and Oxidative Stress Pathways Via Activation of the Master Antioxidant Switch Nrf2. *Biochimie.* **2013**, *95*, 1525–1533.

Collins, A.; Cadet, J.; Epe, B.; Gedik, C. Problem in the Measurement of 8-Oxoguanine in Human DNA. Report of a Workshop, DNA Oxidation, Aberdeen, UK. *Carcinogenesis* **1997**, *18*, 1833–1836.

Cooke, M, S.; Evans, M. D.; Dizdaroglu, M.; Lunec, J. Oxidative DNA Damage: Mechanisms, Mutation, and Disease. *FASEB J.* **2003**, *17*, 1195–1214.

Dawson, D. M.; Kaplan, N. O.; Goodfriend, T. L. Lactic Dehydrogenases-Functions of 2 Types-Rates of Synthesis of 2 Major Forms Can Be Correlated with Metabolic Differentiation. *Science* **1964**, *143*, 929–233.

Dean, R. T.; Roberts, C. R.; Jessup, W. Fragmentation of Extracellular and Intracellular Polypeptides by Free Radicals. *Prog. Clin. Biol. Res.* **1985**, *180*, 341–350.

Dean, R. T.; Gieseg, S.; Davies, M. J. Reactive Species and Their Accumulation on Radical-Damaged Protein. *Trends Biochem. Sci.* **1993**, *18*, 437–441.

Dizdaroglu, M. Chemical Determination of Free Radical-Induced Damage to DNA. *Free Radical Biol. Med.* **1991**, *10*, 225–242.

Djuric, D.; Luongo, D. A.; Happer, D. A. Quantitation of 5-(Hydroxymethyl) Uracil in DNA by Gas Chromatography with Mass Spectral Detection. *Chem. Res. Toxicol.* **1991**, *4*, 687–691.

Dooley, J. F.; Turnquist, L. J.; Racich, L. Kinetic Determination of Serum Sorbitol Dehydrogenase-Activity with a Centrifugal Analyzer. *Clin. Chem.* **1979**, *25*, 202–269.

El-Kabbani, O.; Darmanin, C.; Chung, R. P. T. Sorbitol Dehydrogenase: Structure, Function and Ligand Design. *Curr. Med. Chem.* **2004**, *11*, 465–476.

Esterbauer, H.; Koller, E.; Slee, R. G.; Koster, J. F. Possible Involvement of the Lipid-Peroxidation Product 4-Hydroxynonenal in the Formation of Fluorescent Chromolipids. *Biochem. J.* **1986**, *239*, 405–409.

Evans, G. O. *Animal Clinical Chemistry: A Primer for Toxicologists*; Taylor & Francis: London, **1996**.

Fransworth, N. R. Ethnopharmacology and Drug Development. In *CIBA Foundation Symposium 185;* Chadwick, D. J., Marsh, J., Eds.; John Wiley and Sons: Chichester, New York, 1994; pp 42–51.

Fucci, L.; Oliver, C. N.; Coon, M. J.; Stadtman, E. R. Inactivation of Key Metabolic Enzymes by Mixed-Function Oxidation Reactions: Possible Implication in Protein Turnover and Ageing. *Proc. Natl. Acad. Sci. USA* **1983**, *80*, 1521–1525.

Germadnik, D.; Pilger, A.; Rudiger, H. W. Assay for the Determination of Urinary 8-Hydroxy-2'-Deoxyguanosine Y High-Performance Liquid Chromatography with Electrochemical Detection. *J. Chromatogr. B Biomed. Sci. Appl.* **1997**, *689*, 399–403.

Ghosh, R.; Mitchell, D. L. Effect of Oxidative DNA Damage in Promoter Elements on Transcription Factor Binding. *Nucleic Acids Res.* **1999**, *27*, 3213–3218.

Girotti, A. W. Mechanisms of Lipid Peroxidation. *J. Free Radic. Biol. Med.* **1985**, *1*, 87–95.

Giulivi, C.; Davies, K. J. A. Dityrosine: a Marker for Oxidatively Modified Proteins and Selective Proteolysis. *Methods Enzymol.* **1994,** *233,* 363–371.

Goldberg, D. M. Structural, Functional, and Clinical Aspects of Gammaglutamyltransferase. *Crit. Rev. Clin. Lab Sci.* **1980,** *12,* 1 58.

Gupta, M.P.; Solis, P.N.; Calderon, A.I.; Guionneau-Sinclair, F.; Correa, M.; Galdames, C.; Guerra, C.; Espinosa, A.; Alvenda, G.I.; Robles, G., Ocampo, R. Medical ethnobotany of the Teribes of Bocas del Toro, Panama. *J. Ethnopharmacol.* **2005,** *96,* 389–401.

Gutteridge, J. M. C. Free Radicals in Disease Processes: a Compilation of Cause and Consequence. *Free Radic. Res. Com* **1993,** *19,* 141–158.

Hagihara, M.; Nishigaki, I.; Maseki, M.; Yagi, K. Age-Dependent Changes in Lipid Peroxide Levels in the Lipoprotein Fractions of Human Serum. *J. Gerontol.* **1984,** *39,* 269–272.

Halliwell, B. Commentary. Oxidative Stress, Nutrition and Health. Experimental Strategies for Optimization of Nutritional Antioxidant Intake in Humans. *Free Rad. Res.* **1996,** *25,* 57–74.

Hamburger, M.; Hostettmann, K Bioactivity in Plants: the Link Between Phytochemistry and Medicine. *Phytochemistry* **1991,** *30,* 3864–3874.

Hasler, C. M.; Blumberg, J. B. Symposium on Phytochemicals: Biochemistry and Physiology. *J. Nutr.* **1999,** *129,* 756S–757S.

Heath, R. L.; Tappel, A. L. A New Sensitive Assay for the Measurement of Hydroperoxidase. *Anal Biochem.* **1997,** *35,* 893–898.

Helbock, H. J.; Beckman, K. B.; Shigenaga, M. K.; Walter, P. B.; Woodall, A. A.; Yeo, H. C.; Ames, B. N. DNA Oxidation Matters: the HPLC-Electrochemical Detection Assay of 8-Oxo-Deoxyguanosine and 8-Oxo-Guanione. *Proc. Natl. Acad. Sci. USA* **1998,** *95,* 288–293.

Hu, R.; Saw, C. L.; Yu, R.; Kong, A. N. Regulation of Nf-E2-Related Factor 2 Signaling for Cancer Chemoprevention: Antioxidant Coupled with Anti-Inflammatory. *Antioxid Redox Signal* **2010,** *13,* 1679–1698.

Human Medical Agents from Plants. In *ACS Symposium series 534;* Kinghorn, A. D., Balandrin, M. F., Eds.; American Chemical Society: San Francisco, USA, 1993.

Jadhao, S. B.; Yang, R.; Lin, Q.; Hu, H.; Anania, F. A.; Shuldiner, A. R.; Gong, D. W. Murine Alanine Aminotransferase: cDNA Cloning, Functional Expression, and Differential Gene Regulation in Mouse Fatty Liver. *Hepatology* **2004,** *39,* 1297–1302.

Janero, D. R. Malondialdehyde and Tiobarbituric Acid-Reactivity as Diagnostic Indices of Lipid Peroxidation Peroxidative Tissue Injury. *Free Rad. Biol. Med.* **1990,** *9,* 515–540.

Jones, P.L.; Wolffe, A. P. Relationships Between Chromatin Organization and DNA Methylation in Determining Gene Expression. *Semin. Cancer Biol.* **1999,** *9,* 339–347.

Jung, K. A.; Kwak, M. K. The Nrf2 System as a Potential Target for the Development of Indirect Antioxidant. *Molecules* **2010,** *15,* 7266–7291.

Kelly, F. J.; Mudway, I. S. Protein Oxidation at the Air-Lung Interface. *Amino Acids* **2003,** *25,* 375–396.

Khayrollah, A. A.; Altamer, Y. Y.; Taka, M.; Skursky, L. Serum Alcohol-Dehydrogenase Activity in Liver-Diseases. *Ann. Clin. Biochem.* **1982,** *19,* 35–42.

Kong, J. M.; Goh, N. K.; Chia, L. S.; Chia, T. F. Recent Advances in Traditional Plants Drugs and Orchids. *Acta. Pharmacologia. Scinc.* **2003,** *24,* 7–21.

Korsrud, G. O.; Grice, H. C.; McLaughlan, J. M. Sensitivity of Several Serum Enzymes in Detecting Carbon Tetrachloride-Induced Liver Damage in Rats. *Toxicol. App. Pharmacol.* **1972,** *22,* 474–483.

Kumar, S.; Gupta, A.; Pandey, A. K. Calotropis Procera Root Extract has the Capability to Combat Free Radical Mediated Damage. *ISRN Pharmacol.* **2013,** Article ID 691372. http://dx.doi.org/10.1155/2013/691372.

Lafrancois, C. J.; Yu, K.; Sower, L. C. Quantification of 5-(Hydroxymethyl) Uracil in DNA by Gas Chromatography/Mass Spectrometry: Problem and Solutions. *Chem. Res. Toxicol.* **1998,** *11,* 786–793.

Leonard, T. B.; Neptun, D. A.; Popp, J. A. Serum Gamma Glutamyl Transferase as a Specific Indicator of Bile-Duct Lesions in the Rat-Liver. *Am J. Pathol.* **1984,** *116,* 262–269.

Levine, R. L.; Williams, J. A.; Stadtman, E. R.; Shacter, E. Crbonyl Assays for Determination of Oxidatively Modified Proteins. *Methods Enzymol.* **1994,** *233,* 346–357.

Lopez-Alarcona, C.; Denicola, A. Evaluating the Antioxidant Capacity of Natural Products: A Review on Chemical and Cellular-Based Assays. *Anal Chim. Acta.* **2013,** *763,* 1–10.

Lyras, L.; Cairns, N. J.; Jenner, A.; Jenner, P.; Halliwell, B. An Assessment of Oxidative Damage to Proteins, Lipids, and DNA in Brain from Patients with Alzheimer's Disease. *J. Neurochem.* **1997,** *68,* 2061–2069.

Mamede, A. C.; Tavares, S. D.; Abrantes, A. M.; Trindade, J.; Maia, J. M.; Botelho, M. F. The Role of Vitamins in Cancer: A Review. *Nutr. Cancer.* **2011,** *63,* 479–494.

Marietta, C.; Gulam, H.; Brooks, P. J. A Single 8, 50-Cyclo-20-Deoxyadenosine Lesion in a Tata Box Prevents Binding of the Tata Binding Protein and Strongly Reduces Transcription in Vivo. **2002,** *1,* 967–975.

Mathai, K. Nutrition in the Adult Years. in Krause'S Food. Nutrition, and Diet Therapy; Mahan, L.K. Escott-Stump, S **2000,** *271,* 274–275.

Maulik, N.; McFadden, D., Otani, H.; Thirunavukkarasu, M.; Parinandi, N. L. Antioxidants in Longevity and Medicine. *Oxid. Med. Cell. Longev.* **2013,** Article ID 820679, 1–3. http://dx.doi.org/10.1155/2013/820679.

Meagher E; Thomson C. Vitamin and Mineral Therapy. In *Medical Nutrition and Disease;*. 2nd ed Morrison, G., Hark, L., Eds.; Blackwell Science Inc: Malden, Massachusetts, 1999; pp 33–58.

Milne, E. M.; Doxey, D. L. Lactate-Dehydrogenase and Its Isoenzymes in the Tissues and Sera of Clinically Normal Dogs. *Res. Vet. Sci.* **1987,** *43,* 2224.

Minarik, P.; Tomaskova, N.; Kollarova, M.; Antalik, M Malate Dehydrogenases: Structure and Function. *Gen. Physiol. Biophys.* **2002,** *21,* 257–265.

Montuschi, P.; Collins, J. V.; Ciabattoni, G.; Lazzeri, N.; Corradi, M.; Kharitonov, S. A. Barnes, P. J. Exhaled 8-Isoprostane as an in Vivo Biomarker of Lung Oxidative Stress in Patients with COPD and Healthy Smokers. *Am. J. Respir. Crit. Care Med.* **2000,** *162,* 1175–1177.

Na, H. K.; Surh, Y. J. Modulation of Nrf2-Mediated Antioxidant and Detoxifying Enzyme Induction by the Green Tea Polyphenol EGCG. *Food Chem. Toxicol.* **2008,** *46,* 1271–1278.

Newman, D. J.; Cragg, G. M.; Snader, K. M. Natural Products as Sources of New Drugs Over the Period 1981–2002. *J. Nat. Prod.* **2003,** *66,* 1022–1037.

Nowak, D.; Kasielski, M.; Antczak, A.; Pietras, T.; Bialasiewicz, P. Increased Content of Thiobarbituric Acid-Reactive Substances and Hydrogen Peroxide in the Expired Breath Condensate of Patients with Stable Chronic Obstructive Pulmonary Disease: No Significant Effect of Cigarette Smoking. *Respir. Med.* **1999,** *93,* 389–396.

Ozer, J.; Ratner, M.; Shaw, M.; Bailey, W.; Schomaker, S. The Current State of Serum Biomarkers of Hepatotoxicity. *Toxicology.* **2008,** *245,* 194–205.

Persson, T.; Popescu, B. O.; Cedazo-Minguez, A. Oxidative Stress in Alzheimer's Disease: Why Did Antioxidant Therapy Fail. *Oxid. Med. Cell. Longev.* **2014,** Article ID 427318, 1-11. http://dx.doi.org/10.1155/2014/427318

Ravanat, J. L.; Turesky, R. J.; Gremaud, E.; Trudel, L. J.; Stadler, R. H. Determination of 8-Oxoguanine in DNA by Gas Chromatography-Mass Spectrometry and HPLC-Electrochemical Detection: Overestimation of the Background Level of the Oxidized Base by the Gas Chromatography-Mass Spectrometry Assay. *Chem. Res. Toxicol.* **1995,** *18,* 1833–1836.

Rej, R. Aminotransferases in Disease. *Clin. Lab Med.* **1989,** *9,* 667–687.

Sahin, K.; Orhan, C.; Tuzcu, M.; Tuzcu, M.; Ali, S.; Sahin, N.; Hayirli, A. Epigallocatechin-3-Gallate Prevents Lipid Peroxidation and Enhances Antioxidant Defense System Via Modulating Hepatic Nuclear Transcription Factors in Heat-Stressed Quails. *Poult. Sci.* **2010,** *89,* 2251–2258.

Sahin, K.; Orhan, C.; Smith, M. O.; Sahin, N. Molecular Targets of Dietary Phytochemicals for the Alleviation of Heat Stress in Poultry. *Worlds Poult. Sci. J.* **2013,** *69,* 113–124.

Sakagishi, Y. Alanine aminotransferase (ALT). Nippon rinsho. *Jpn. J. Clin. Med.* **1995,** *53,* 1146–1150.

Saw, C. L.; Yang, A. Y.; Cheng, D. C.; Boyanapalli, S.S.;Su, Z.Y.;Khor, T.O.;Gao, S.;Wang, J.;Jiang, Z.H.;Kong, A.N.; Pharmacodynamics of Ginsenosides: Antioxidant Activities, Activation of Nrf2, and Potential Synergistic Effects of Combinations. *Chem. Res. Toxicol.* **2012,** *25,* 1574–1580.

Shacter, E.; Williams, J. A.; Stadtman, E. R.; Levine, R. L. Determination of Carbonyl Groups In Oxidized Proteins. In *Free Radicals: a Practical Approach;* Punchard, N. A., Kelly, F. J., Eds.; Oxford University Press, Oxford, 1996; pp 159–179.

Sharma, U. K.; Sharma, A. K.; Gupta, A.; Kumar, R.; Pandey, A.; Pandey, A. K. Pharmacological Activities of Cinnamaldehyde and Eugenol: Antioxidant, Cytotoxic and Anti-Leishmanial Studies. *Cell Mol. Biol.* **2017,** *63,* 73–78.

Shen, G. G.; Kong, A. N. Nrf2 Plays an Important Role in Coordinated Regulation of Phase II Drug Metabolism Enzymes and Phase III Drug Transporters. *Biopharm. Drug Dispos.* **2009,** *30,* 345–355.

Shigenaca, M. K.; Ames, B. N. Assay for 8-Hydroxy-2'-Deoxyguanisine: a Biomarker of in Vivo Oxidative DNA Damage. *Free Radical Biol. Med.* **1991,** *10*:211–216.

Shigenaga, M. K.; Aboujaode, E. N.; Chen, Q.; Ames, B. N. Assay of Oxidative DNA Damage Biomarkers 8-Oxo-2' Deoxyguanosine and 8-Oxoguanine in Nuclear DNA and Biological Fluids by High-Performance Liquid Chromatography with 1998, Electrochemical Detection. *Methods Enzymol.* **1994,** *234,* 16–33.

Sies, H. Oxidative Stress: from Basic Research to Clinical Application. *Am J. Med.* **1991,** *91,* 31–38.

Singh, A.; Bhat, T. K.; Sharma, O. P. Clinical Biochemistry of Hepatotoxicity. *J. Clinic. Toxicol.* **2011,** 1–19.

Siu, G. M.; Draper, H. H. Metabolism of Malonaldehyde in Vivo and in Vitro. *Lipids* **1982,** *17,* 349–355.

Stadtman, E. R. Metal Ion-Catalyzed Oxidation of Proteins: Biochemical Mechanism and Biological Consequences. *Free Radic. Biol. Med.* **1990,** *9,* 315–325.

Stadtman, E. R.; Berlett, B. S. Reactive Oxygen Mediated Protein Oxidation in Aging and Disease. *Drug Metab. Rev.* **1998,** *30,* 357–363.

Stadtman, E. R.; Levine, R. L. Free Radical-Mediated Oxidation of Free Amino Acids and Amino Acid Residues in Proteins. *Amino Acids* **2003,** *25,* 207–218.

Stadtman, E. R.; Moskovitz, J.; Levine, R. L. Oxidation of Methionine Residues of Proteins: Biological Consequences. *Antioxid Redox Signal* **2003,** *5,* 577–582.

Suc, I.; Meilhac, O.; Lajoie-Mazenc, I.; Vandaele, J.; Jurgens, G.; Salvayre, R.; Negre-Salvayre, A. Activation of EGF Receptor by Oxidized LDL. *FASEB J* **1998,** *12*, 665–671.

Takaya, K.; Suzuki, T.; Motohashi, H.; Onodera, K.; Satomi, S.; Kensler, T. W.; Yamamoto, M. Validation of the Multiple Sensor Mechanism of the Keap1-Nrf2 System. *Free Radical Bio. Med.* **2012,** *53*, 817–827.

Tate, S. S.; Meister, A. Gamma-Glutamyl-Transferase Transpeptidase from Kidney. *Meth. Enzymol.* **1985,** *113*, 400–419.

Tsukagoshi, H.; Kawata, T.; Shimizu, Y.; Ishizuka, T.; Dobashi, K. Mori M. 4-Hydroxy-2-Nonenal Enhances Fibronectin Production by Imr-90 Human Lung Fibroblasts Partly Via Activation of Epidermal Growth Factor Receptor-Linked Extracellular Signal-Regulated Kinase P44/42 Pathway. *Toxicol Appl. Pharmacol.* **2002,** *184*, 127–135.

Watanabe, M.; Yumi, O.; Itoh, Y.; Yasuda, K.; Kamachi, K.; Ratcliffe, R.G. Deamination Role of Inducible Glutamate Dehydrogenase Isoenzyme 7 in Brassica Napus Leaf Protoplasts. *Phytochemistry* **2011,** *72*, 587–593.

WHO. *Biomarkers in Risk Assessment: Validity and Validation*; WHO: Geneva,, 2001.

Wisean, H.; kaur, H.; Halliwell, B. DNA Damage and Cancer: Measurement and Mechanism. *Cancer Lett.* **1995,** *93*, 113–119.

Yang, R. Z.; Blaileanu, G.; Hansen, B. C.; Shuldiner, A. R.; Gong, D. W. CDNA Cloning, Genomic Structure, Chromosomal Mapping, and Functional Expression of a Novel Human Alanine Aminotransferase. *Genomics* **2002,** *79*, 445–450.

Yang, R.; Park, S.; Reagan, W. J.; Goldstein, R.; Zhong, S.; Lawton, M.; Rajamohan, F.; Qian, K.; Liu, L.; Gong, D. W. Alanine Aminotransferase Isoenzymes: Molecular Cloning and Quantitative Analysis of Tissue Expression in Rats and Serum Elevation in Liver Toxicity. *Hepatology* **2009,** *49*, 598–607.

# CHAPTER 9

# ANTIMICROBIAL MEDICINAL PLANTS AS EFFECTIVE NATURAL BIORESOURCES

JAYA VIKAS KURHEKAR

*Department of Microbiology, Dr. Patangrao Kadam Mahavidyalaya, Sangli, Maharashtra 416416, Maharashtra, India, E-mail: jaya_kurhekar@rediffmail.com; Mob.: +91 9423869169*

*ORCID ID: https://orcid.org/0000-0001-7705-9041*

## ABSTRACT

Our planet has been bestowed with abundant natural resources, some yet unexplored, unfortunately depleting fast! Medicinal plants were found highly efficient in their antimicrobial properties, which could be related to their inherent components as flavonoids, tannins, and alkaloids present in varying amounts. Antimicrobial efficiencies of medicinally significant plants were compared with the help of inhibitory zones and their inherent components extracted and identified with the help of analytical techniques. Present deliberation includes five medicinal plants along with their parts, known to have antimicrobial activities checked for their antimicrobial effects against common pathogens causing human infections. Efficient extracts showed encouraging inhibitory activities against common and burns wound pathogens. Wide biodiversity existing on earth has always been helping us in combating various existing and upcoming microbial infections, developing due to the human interference with nature. Protecting, preserving, and propagating helpful species, so as to manage our bioresources for the future generations should form our moral responsibility.

## 9.1  INTRODUCTION

Plants have undoubtedly proven to be the fountains of life. Their association with human beings is known since time immemorial. Humans have always tailored plants to make use of them for personal, basic, essential amenities such as food, shelter, clothing, and medicines. In addition, the plants purify air for all living beings, reducing pollution. Their therapeutic value is due to their inherent phytochemicals, a fact which has been exploited by man, and plants have been traditionally used as home remedies for curing microbial diseases. Extracts of plant parts are oldest medicines, known to mankind, more so because of their easy accessibility and availability, comparatively low costs and minor side effects. The importance of medicinal plants is more pronounced in the present scenario of population explosion, increasing standards of living and synthetic chemicals posing potential health hazards. Natural therapy is progressively being accepted as an effective therapeutic tool which is establishing an important link between humans and environment.

Pharmacognosy is an interdisciplinary subject, encompassing microbiology, medicine, Ayurveda, medical botany, and agriculture and horticulture. Documentation, identification, and various types of investigations of valuable plants are necessary to apply them in therapeutic capacity in diseased conditions. Plants help in improving and maintaining health related to psychological, social, and mental well-being as well as aiding in recreation as a hobby. They are responsible for an ecological dynamic stability (Baskar and Chezhiyan, 2002c). Plants, for ages, have been natural resources for the manufacture of medicines and drugs. About six thousand medicinal plants are employed, in various capacities, in the production of medicines (Redkar and Jolly, 2003). Ayurveda, Unani, and Siddha are Indian systems of medicines, gaining global popularity (Farooqi et al., 2003).

More than 20,000 medicinal plant species have been enlisted by World Health Organization (WHO), encouraging, recommending, and promoting the inclusion of herbal drugs in programs related to healthcare. Medicinal and aromatic plants (MAPs) possess antimicrobial properties as a result of which, a large number of phytochemical agents are being used in modern medicine, solely, as pure compounds or in combinations. Plant originating drugs are easily available, reasonably priced, time-tested, and comparatively much safer (Khan and Khanum, 2002).

Plants have always been used for medicinal purposes, elevating the quality of human life as well as other plants, recently for biocontrol in an eco-friendly manner, as biopesticides and biofertilizers. Since ancient times, plants were good sources of antimicrobial agents like emetine, quinine, and berberine. Phytomedicines are proving to be promising agents in the curing of intractable infectious diseases like opportunistic AIDS infections (Khan and Khanum, 2002). Our mother Earth is heading towards a major crisis, with reference to population explosion and resulting pollution, elevating the incidences of a variety of microbial infections in humans, affecting all human systems like the respiratory, digestive, genitourinary, and skin. Microbial infections are caused by very commonly occurring Gram-positive pathogens like *Enterococcus faecalis, Staphylococcus aureus*, and Gram-negative species of *Escherichia coli, Salmonella, Shigella, Pseudomonas*, and so forth. Antibiotics are employed for treatment of infections, but pose a number of disadvantages like leading to hypersensitivity in some, causing undesirable side effects, and besides being generally expensive. Wonder drugs, antibiotics such as ciprofloxacin, tetracycline, erythromycin, chloramphenicol, septran are presently impotent due to their excessive and irrational use, leading to the development of bacterial and hazardous side effects (Kaul, 2006). In this respect, medicinal plant products with broad-spectrum antimicrobial properties and minimum ill effects are being identified (Farnsworth et al., 1998; Banginwar and Tambekar, 2003). A popular Chinese folk wisdom song depicts psychological benefits obtained from interaction with plants;

*If you wish to be happy for a day, kill a pig;*
*If you wish to be happy for a week, take a wife;*
*If you wish to be happy for a lifetime, plant a garden.*

A growing plant is very powerful because it is alive and has various potent properties. The man has been using different types of herbs which are valued for their medicinal and magical properties, from time immemorial. Each of them has a distinct flavor, medicinal, and health-giving properties. Many varieties of medicinal plants are known to be natural sources of ointments, inhalations, and sneezing powders for curing various diseases (Dhawan and Rastogi, 1991; Varshney et al., 2001).

## 9.2 HISTORY OF MEDICINAL PLANTS

Charaka Samhita (900 BC), the ancient classic, is the oldest text which available on the complete treatment of diseases. It describes 341 medicinal plants and specifies the use of hundreds of herbs in the complete treatment of bacterial diseases like leprosy and tuberculosis. The next landmark in

Ayurveda is Sushruta Samhita, written in 600 BC, which enlists 395 medicinal plants. The first scientific literature on medicinal plants can be linked to the record on Egyptian Papyri dating from 2000 BC. A Chinese herbal, TZU I Pen Tshao Ching, appeared later in 500 BC (Handa and Kapoor, 1989). Various quotes depicting the significance of plants as medicines, have been recorded in Bible; "There are herbs that are harmless, the use of which will tide over many apparently serious difficulties.

Researchers in the field have continuously worked on and interpreted the therapeutic values cited in the ancient texts and compiled medicinal herbs used as antibacterial drugs. Ayurveda has given due to importance to the employment of medicinal plants in treating of various infectious diseases. Between fifth and sixth century, Dioscorides Anazarbus, in *Materia Medica*, compiled an assembly of 600 plants, their names, botanical and habitat descriptions, preparations, and medicinal and aromatic uses. Gaius Plinius in his "Natural history" gave an encyclopedic account of knowledge about the vegetable world. During 372–286 BC, systematic separation and characterization of herbs and other plants, according to their morphological traits first appeared in the publication of Historia Plantarum and De Causis Plantarum, written by Theophrastus of Eresus, a colleague of Aristotle in Plato's academy. In 1470, a monk Bartholomew de Glanville got the credit of compiling "The encyclopedia of Bartholomaeus Anglicus (1470)." In 1475, Hanns Bamler printed "The book of nature" at Augsburg. In the fifteenth century, "The Herbarium of Apuleius Platonicus" was compiled, based on the classical writings of Dioscorides and Pliny. Latin "Herbarius" (1484), German "Herbarius" (1485), and "Hortus Sanitatis" (1491) published in Mainz, Germany, were three books of great significance. One of the founders of the science of plant anatomy was Nehemiah Grew (1641–1712) (Agnes, 2003). An ancient India's Atharvavedic hymn appropriately depicts the importance of medicinal plants (Chomchalow and Henle, 1995);

*Well doth the wild boar know a plant; the mongoose knows the healing herbs,*
*I call to aid this man, the plant which serpents and gandharvas know.*
*Plants of Angirasas which hawks celebrate, plants which eagles are in an*
*aspiration*
*Plants known to swan and lesser fowls, plants known to all the birds that fly,*
*Plants that are known to sylval beasts, I call them all to aid this man*
*The multitudes of herbs all that are food and medicines for goat and sheep*
*So many plants, brought hither will shelter and defend this man and thee.*

(Griffiths: Atharvaveda, English translation)

Plants were used as medicines in China, Egypt, and Greece much before the beginning of Christian era. Between 15th and 17th century, in Europe, herbals recorded the medicinal value of plants. In the early 19th century, it was realized that medicinal properties of healing in plants were due to minute active ingredients. With the development of organic chemistry, in the twentieth century, extraction and fractionation techniques improved significantly. It became possible to isolate and identify many of the active constituents from plants.

## 9.3 IMPORTANCE OF MEDICINAL PLANTS

Medicine originates from natural plant products. The secondary metabolite serves the plant as a biochemical link between the producing plant and its surrounding environment. Plant cells are chemical factories which are highly efficient in producing secondary metabolites such as alkaloids, steroids, terpenoids, flavonoids, xanthones, and coumarins exhibiting significant biological properties (Baskar and Chezhiyan, 2002a). The human immune system is closely dependent on plant antioxidants, derived from medicinal plants (Devasagayam and Sainis, 2002). A 4-coumarins, caffeic, ferulic, and sinapic acids are natural cinnamic acids found in free forms and in esterified forms in plants. Methyl derivatives of xanthine; purine alkaloids – caffeine, theobromine, and theophylline, co-occur in a plant. Lipophilic alkylamides, polar caffeic acid derivatives, and high molecular weight polysaccharide material are immunostimulant (due to water-soluble alkaloids), anti-inflammatory, antibacterial, and antiviral. Caffeine is a CNS stimulant, used in combination, with other therapeutic agents, like chlorogenic acid which is decomposed to caffeic acid and quinic acid (Kokate et al., 2001).

Basic plant constituents like flavonoids are concluded as a major class of phenolic compounds present in many fruits and vegetables. They have a major role in the prevention of several forms of cancers, cardiovascular diseases, and cytotoxicity of low-density lipoproteins (LDL). They are hydrogen donating free radicals with free hydroxyl groups and catechol moiety in the ring B of flavonoid nucleus (Shetgiri and D'Mello, 2003). Flavon-3-ols like catechin (Shetgiri and D'Mello, 2003); flavonols like quercetin and rutin (Jadhav and Kharya, 2005) show very good free radical scavenging activity. Many traditional medicinal plants have been reported to have strong antiviral activity, some being used against viral infections (Jassim and Naji, 2003).

Most of the natural attacks on humans have a natural cure, as medicinal plants. These fantastic bioresources need to be protected, preserved for our future generations. The healing power of plants is due to their inherent magical phytochemicals like flavonoids which can be used for control and management of burns wounds (Kurhekar, 2016). In the ancient times, the knowledge of plants seems to be percolated by word of mouth through generations and later from written records. The herbalists' knowledge of plants today, may be a result of that transmission. Various plants are being used as healing agents from very early times and it is necessary to study them in detail to know the different types and recognize them accurately. Indiscriminate use of presently used antibiotics and other agents has resulted in the development of resistance in human beings, leading to a critical situation in related therapy (Bonjar, 2004). On this background, a study of a few medicinal plants, with reference to their antimicrobial characteristics have been presented Fig. 9.1), which may contribute towards strengthening the belief that nature provides the best cure against most natural invasions. This is an attempt towards creating awareness to the employment of medicinal plants, in times of common casualties and in the treatment of common infections, in a simple way.

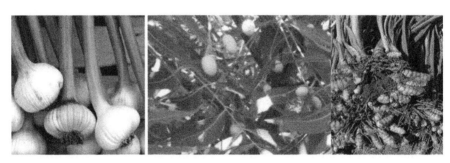

Plate 1: *Allium sativum*          Plate 2: *Azadirachta indica*          Plate 3: *Curcuma longa*

Plate 4: *Eucalyptus globulus*                    Plate 5: *Pongamia pinnata*

**FIGURE 9.1   (See color insert.)** Antimicrobial medicinal plants.

## 9.4 SOME PLANTS EFFECTIVE AGAINST MICROORGANISMS: A REVIEW

### 9.4.1 ALLIUM SATIVUM

A very popular spice, *Allium sativum* (garlic) (Plate 1), enhancing the taste of food and adding pep to it, has significant medicinal value. It belonging to Liliaceae, an herb with bulblets called cloves with whitish skin. The bulbs are reported to contain such as essential oil, sulfur, allyl disulfides, Allicin, and Allisatin I and II. It is a traditional component of Indian diet with a broad spectrum antibiotic action, inhibiting a wide range of organisms. It is known to be a wonder cure for many human ailments like excreting intestinal worms, curing spasms, stimulating the flow of saliva, aiding digestion, relieving gases and pains from bowels, promoting the flow of urine, removing mucous secretions from bronchial tubes, reducing fever, dissipating oedematous swelling, reducing cholesterol, giving relief from aching ear, skin infections, and delaying graying of hair. It is effective in curing upper respiratory tract infections. A natural antibiotic and antiseptic agent, it is observed to prevent the cancerous growth and reduce hypertension. Raw garlic juice is used in coconut oil, as garlic tea, as a dressing, and wound healing (Kurian, 1999). Leaves and bulbs of *A. sativum* are commonly used as a spice in kitchens and have a very strong aromatic flavor, which imparts a typical taste to food. Since olden times, due to its medicinal properties, *A. sativum* is applied as water, acetone, oil extracts, and bulblets. During the ancient days, in critical disease conditions such as typhoid, cholera, asthma, gastritis, chronic bronchitis, pneumonia, and tuberculosis, garlic was used for treatment (Kurian, 1999; Nadkarni, 2004). Wills studied inhibition of enzymes by its active principle Allicin (1956).

### 9.4.2 AZADIRACHTA INDICA

*Azadirachta indica* (neem or Indian lilac) is a popular remedy against many ailments, with extraordinary medicinal properties. A member of family Meliaceae, *A. indica* (Plate 2) is a drupe, with glabrous, fleshy, light yellow, smooth, and oval-elongated fruits. Every part of the tree; the bark, the leaves, fruits, lactic secretion, and seed oil are useful. Its bark decoction is used to expel worms from the body to reduce the sensation of burning in the chest, fever, cough, ulcers, inflammation, and leprosy. Particularly in pyorrhea, fresh tender twigs are used to clean teeth (Sonawane et al., 2006).

Infusion of leaves is used as an insecticide, for treating eye problems, swelling, inflammation of liver, and skin diseases. In ancient times, leaves of *A. indica* were crushed and used as poultices, ointments, and liniments against eczema, skin ulcers, leprosy lesions, burns wounds, gangrenous lesions, herpes lesions and scabies as anti-inflammatory, and antibacterial agent (Martindale, 1989). In many parts of India, it is used in bathing water regularly. Leaf paste of neem is applied on ringworm and wounds. The lactic secretion is useful against tuberculosis and leprosy, cures for diseases of the uterus in postnatal period, and as an anti-snake poison. Its seed oil is used as a local stimulant, antihelminthic and antiseptic. It is used for treating skin infections, headache, kills lice, for dressing in leprosy, tetanus, suppurating scrofulous glands, urticaria, chronic skin diseases like ringworms, eczema, erysipelas, scabies, and sloughing ulcers. Its young fruit is used as an astringent, tonic, laxative, emetic, and antihelminthic (Das et al., 1999). Consumption of ripe and unripe fruit functions as a purgative, to treat urinary problems, tumors, piles, and toothache. Its seeds are very rich in oils and fatty acids like stearic, oleic, lauric, butyric, valeric acids, volatile fatty principles (Das and Agarwal, 1985), with Aziridine, benzoic acid, resin, meliotannic acid, tannin, sterols, bakayanin, margosine, and sulfur (Sawant, 1974). Leaves, seeds, and bark are reported to carry allelochemicals to be used as pesticides (Kumar, 2006). Its active constituents are saturated and unsaturated fatty acid glycerols (Kokate et al., 2001). Azadirachtin (insect repellant), meliatriol and salanin (antifeedants), and nimbin and nimbidin (antivirals) present in neem are very useful to human life. Globally, flowers, tender twigs, barks, and seed oil are used as antifungal, antiseptic, blood purifying agents, in treating skin diseases, cosmetics, bio-pesticides (Baskar et al., 2002a), for skin allergy, as an antidote for insect bites, and asthma (Baskar and Chezhiyan, 2002b).

### 9.4.3   CURCUMA LONGA

*Curcuma longa* (turmeric) (Plate 3) is used every day in kitchens to impart color, flavor, and a typical taste to foods and vegetable preparations. Belonging to family Scitamineae, it has fibrous roots, modified stem, rhizome, and main stem, each of which has potential medicinal value. It is rich in vitamin A. It has fantastic medicinal applications, and has always been a significant addition to grandmother's medicinal purse, for ages! The plant is medicinally used to relieve gases, pain in the stomach, favorably modifies the process of nutrition and excretion, and aiding in normal functions of human body. It is used in reducing symptoms of bronchitis, cold, cough, as a gargling agent, diarrhea, fever, edema, chest pain, diabetes, scabies and pustules of smallpox, chicken-pox, measles, jaundice, and liver problems. It is used against formation of mucus, as an anti-snake venom and anti-scorpion venom. It is useful for treatment of urinary diseases, healing wounds, sprains, rheumatism, and pain of the facial nerves.

Fresh turmeric contains zingiberene, curcumin, alkalies, essential oil (antiseptic), ketone, alcohol, carmine (antacid), and p-tolylmethyl carbinol (Sawant, 1974). Since ancient times, turmeric is known to exhibit potential antioxidant and anti-inflammatory activities (Salimath et al., 1986; Chawla et al., 1987). Its extract is effective in conjunctivitis (Shrinivas and Prabhakaran, 1987), as a dressing material for wound healing, contraceptive, stimulation of lactation activity, and acts as counter-irritant to suppress milk secretion in mammary abscesses (Santhanum and Snagarajan, 1990). The rhizomes of *C. longa* show natural antibacterial, anti-inflammatory, antineoplastic and analgesic activities because of monoterpenoids, sesquiterpenoids, and curcuminoids (Tang and Eisenbrand, 1992). Chemically, Curcumin is diferuloylmethane, an effective natural anti-inflammatory agent (Reddy and Lokesh, 1992; Ciddi and Kaleab, 2005).

## 9.4.4  EUCALYPTUS GLOBULUS

*Eucalyptus globulus* (southern blue gum), (Plate 4), a member of family Myrtaceae, has about 400 species with same medicinal properties. Its colorless oil is medicinal but evaporates fast, initially with a strong camphor-like odor, later giving a sense of coldness, on consumption. The leaf extract is consumed internally and applied externally for cough, mucous, bronchitis, whooping cough, flu, fever, cold, to treat symptoms of asthma, respiratory problems, tuberculosis, indigestion, diarrhea and worms, typhoid, inflammation of mucous membrane with free discharge, urinary bladder, diabetes, malaria, inflammation of kidneys, nose-lining, joints, rheumatism, and infections of sciatic nerve. Its extract has applications as a disinfectant in washing hands, wounds, burns, sores, and ulcers. It provides a tone and vitality to the body. It is used to prevent pyorrhea. Its leaf-infusion, decoction, oil extract, and vapors are applied externally for various problems (Sawant, 1974). Its medicinal significance is because of its components; volatile oil, tannins, resins, eucalyptol, cajeputol, and cineol (Sathe, 1998).

### 9.4.5   PONGAMIA PINNATA

*Pongamia pinnata* (Poongam oil tree), (Plate 5), medicinally very significant, is classified under family Leguminosae. Each tree part has been observed to be medicinally useful. Freshly prepared root and bark decoction help in throwing out worms from the body, treating infections of the eyes, vagina, skin, piles, wounds, ulcers, itching, ascites and in enlargement of spleen, abdomen, tumor dissipation, and liver congestion. Very young sprouts and shoots extract strengthen the stomach, giving it a tone, enhances appetite, and reduces swellings and effects of poisons. A leaf infusion is a mild laxative, useful in digestion. Flower infusion is helpful in treating diabetes, liver congestion and biliary juices depletion, infections of hair scalp, whooping cough, and against snake venom (Das and Agarwal, 1991). Seeds decoction is known to purify the blood, cure an earache, complaints of the chest, fevers, hydrocele, and remedy in eyes, nose, skin infections, and wounds. Powder made from seeds helps as a tonic and febrifuge. Seed crush paste is externally applied on leprosy sore, skin diseases, and painful and rheumatic joints (Clare and Wood, 1992). Seed oil is used to relieve rheumatic pains, itches, leprosy sores, ulcers, urinary disorders, headaches, fevers, liver pain, wounds, and skin diseases (Sawant, 1974). It is antiparasitic and antiseptic, applied in scabies, herpes, eczema, ulcers, and maggot-infested sores. It has essential oil, fatty oil, bitter oil, pongamol, pongamin, glabrin, quasitine, and karangin as its components (Kurian, 1999). Crushed leaf paste can be soaked in cloth dressing for wounds and hot leaf extract relieves pains due to rheumatism, cleans sores, and ulcers.

### 9.5   EFFECTS OF PLANT EXTRACTS ON HUMAN PATHOGENS: A RESEARCH

### 9.5.1   MICROORGANISMS USED FOR THE STUDY

The antimicrobial potency of medicinal plant extracts, prepared in distilled water and acetone was observed against few representative human pathogens, commonly occurring in the environment around us.

**Pathogenic isolates:** Common pathogens isolated from pathological laboratory included Gram-positive and Gram-negative bacteria, yeast and fungal cultures. They were identified as *Micrococcus luteus, Bacillus subtilis, S. aureus,* and *E. faecalis* as Gram-positive; *Proteus vulgaris, Escherichia coli, Salmonella typhi, Salmonella paratyphi B, Klebsiella pneumoniae, Shigella* *flexneri, Serratia marcescens,* and *Pseudomonas aeruginosa,* as Gram-negative, *Candida albicans* as yeast and *Aspergillus niger* as a fungal pathogen.

**Burns wound isolates:** Six isolates from burns patients' wound swabs were identified as *S. aureus, Streptococcus pyogenes, Bacillus cereus, B. subtilis, E. faecalis*, and *P. aeruginosa.*

**Standard cultures:** Four standard cultures – *S. aureus* ATCC 29,213, *E. faecalis* ATCC 29,212, *E. coli* ATCC 25,928, and *P. aeruginosa* ATCC 27,853 were procured from Himedia laboratories, maintained and preserved on antibiotic assay medium slants and used for comparative studies on properties of medicinal plants, with reference to their antimicrobial potency.

## 9.5.2   STANDARD ANTIBIOTICS

A few antibiotics, which are generally prescribed for common infections, were chosen for this study, for comparative evaluation of antimicrobial efficiencies of medicinal plants. For sensitivity testing, antibiotic octadiscs number OD – 42 from Himedia and Gram-negative master multidiscs number MD – 2, Micro Master Laboratories were used to check the response of the study isolates under study and they were Amikacin, Augmentin, Cefuroxime, Cefoperazone, Ceftazidime, Cefotaxime, Chloramphenicol, Ciprofloxacin, Doxycycline Gentamicin, Lomefloxacin, Nalidixic acid (NA), Netilmicin, Nitrofurantoin, Norfloxacin, Ofloxacin, and Pefloxacin.

## 9.5.3   PREPARATION OF PLANT EXTRACTS

Known weights of plant material were used to prepare aqueous and acetone extracts from whole plant as well as different plant parts of the plants and used for checking the antimicrobial activity of each medicinal plant, depicted in the form of inhibitory zones, taken in triplicates, noted using standard test of variance for statistical analysis.

## 9.6   DISCUSSION OF RESULTS

### 9.6.1   A. SATIVUM

Garlic is suggested in Ayurveda as a plant with curing capacity. It is known to lower blood cholesterol, cure respiratory diseases, skin diseases and related disorders, and protects the human body from bacterial toxins responsible for symptoms of infections. Garlic is a spice with medicinal qualities, which adds a typical flavor to foods. Its aqueous extract inhibited the growth of pathological laboratory isolates of Gram-positive *S. aureus, B. subtilis*, and *E. faecalis;* Gram-negative *K. pneumoniae, E. coli, S. typhi, S. flexneri*, and *P. aeruginosa.* The acetone extract was found effective in inhibiting the growth of Gram-positive *S. aureus* and *E. faecalis*, Gram-negative, *E. coli, K. pneumoniae, S. typhi*, and *S. flexneri*, while yeast *C. albicans*, and fungal isolate *A. niger* were found to be resistant. *S. flexneri* was most sensitive to the aqueous extract (zone of inhibition - 22 mm) while *E. faecalis* to acetone extract (19 mm). *P. aeruginosa* was least sensitive to the aqueous extract (8.83 mm) and *S. typhi* to acetone extract (10.5 mm). Efficiencies of aqueous as well as acetone extracts inhibiting *E. faecalis* and *K. pneumoniae* were found statistically comparable, *S. aureus* significant, *E. coli* highly significant, and *S. typhi, S. flexneri* very highly significant. Comparatively, aqueous extract of *A. sativum* inhibiting eight of the laboratory isolates was equipotent to Norfloxacin in its antimicrobial action. The aqueous extract inhibited the growth of standard controls *S. aureus* ATCC 29213, *E. faecalis* ATCC 29212, and *E. coli* ATCC 25928 and the acetone extract inhibited *E. faecalis* ATCC 29212 and *E. coli* ATCC 25928. Preliminary chemical analytical investigation of the plant extract showed the presence of alkaloids, fats, glucose, proteins, reducing sugars, and triterpenoids.

A number of authors have reported numerous properties of garlic extract including antimicrobial (Jain, 1993), like inhibition of five resistant strains of *P. aeruginosa, K. aerogenes, Edwardsiella tarda, S. pneumoniae, Citrobacter freundii*, and *St. pyogenes*. Tumane et al. noted that aqueous extract of *A. sativum* exhibited high antimicrobial potency, inhibiting antibiotic resistant 8 isolates out of fourteen, bulb extract inhibiting 10 strains of *Vibrio*, 5 of *S. flexneri*, 25 of *S. aureus*, 20 of *E. coli*, 10 of *K. aerogenes, K. pneumoniae, P. mirabilis*, and *P. vulgaris*, and 5 of *S. typhimurium* and *Sal. paratyphi* (2002). Garlic extract inhibits pathogenic *Mycobacteria* (Delaha and Garagusi, 1985). Garlic has exhibited antimicrobial activity against drug resistant of *S. pyogenes, Streptococcus pneumoniae, C. freundii, E. tarda, Pseudomonas aeruginosa*, and *Klebsiella aerogenes* (Jain, 1993). The activity of garlic against the causative agent of chronic gastritis, gastric and duodenal ulcers – *Helicobacter pylori*, was reported. Garlic is more significant because it being chemically complex, with broad-spectrum activity, least chances of acquired antibiotic resistance are expected (O'Gara et al., 2000). Inhibition of fungal growth was reported by garlic extract (Tansey and Appleton, 1975). Garlic has been inhibitory to Methicillin-Resistant

*Staphylococcus aureus* (MRSA) from wounds (Almawlah, 2017). Garlic shows the presence of Allicin, volatile Diallylthiosulfinate inhibiting MDR strains, and lung pathogenic bacteria (Reiter et al., 2017).

### 9.6.2 A. INDICA

Neem is a multipurpose tree with numerous health benefits, known since ages. Every part of it is antimicrobial in nature, against microorganisms belonging to a wide range. Its screening for active ingredients acting against pathogens may help us in developing efficient, cost-effective and comparatively safe antimicrobial agents. Few medicinal plants like *A. indica* are as variedly used to keep human life healthy and comfortable. Aqueous *A. indica* extract was found to inhibit Gram-positive *B. subtilis, S. aureus,* and *E. faecalis;* Gram-negative, *S. typhi, E. coli, P. Vulgaris*, and *S. flexneri*. Acetone extract inhibited Gram-positive *B. subtilis, S. aureus*, and *E. faecalis;* Gram-negative, *E. coli,* yeast, and *C. albicans*; and fungus *A. niger. S. flexneri* was the most sensitive to the aqueous extract (zone of inhibition – 16 mm) while *A. niger* to acetone extract (19.16 mm). *S. typhi* was least sensitive to the aqueous (7.5 mm) and *B. subtilis* to acetone extract (7.83 mm), as concluded from the zone sizes. Comparative potencies of both the extracts inhibiting *S. aureus, E. faecalis,* and *E. coli* were statistically very highly significant while those for *B. subtilis* comparable. Aqueous *A. indica* extract showed 50% while acetone showed 43% efficiency.

Aqueous extract of *A. indica* inhibited seven of the pathological laboratory isolates and hence, was equipotent to NA in its antimicrobial action. Primary screening of isolates from swab samples of burns wounds showed that out of ten, five were sensitive to the aqueous extracts of *A. indica*. Of the seven burns wounds isolates, three were inhibited by the aqueous extract. Hence, the antimicrobial activity of *A. indica* can be concluded to be more than that of Ciprofloxacin, as depicted from zone sizes. Both extracts were found to be inhibitory for *E. faecalis* ATCC 29,212. Qualitative analysis of the plant extract showed alkaloids, glycosides, flavonoids, proteins, and reducing sugars to be present. Aqueous extract analysis by HPTLC probably showed peaks corresponding to terpenes, saponins at 200 nm, while Echinacoside, Rutin/Sinensetin, Chlorogenic acid/Eriocitrin, Eriodictyl, Cichoric acid at 366 nm, Cascarosides A, B, C, D, Sennosides, Glucofrangulins, Aloinosides at 254 nm flavonoids, xantho-eriodictyl, sily-christin, and taxifolinandiso-silybinat at 366 nm.

The extract of neem plant and its secretions have components which have spermicidal anti-nematodes, antifungal, antibacterial, and antimicrobial activity (Ahmed et al., 1995) *A. indica* was effective out of 27 plant extracts, out of 64 studied, against *S. flexneri, Sal. paratyphi, S. typhimurium, Vibrio, K. pneumoniae, and K. aerogenes* and against standard cultures of *B. stearothermophilus* NCIM-2328, *B. subtilis* NCIM-2063, *S. aureus* NCTC-3750, *P. mirabilis* NCIM-208, and *E. Coli* ATCC-1948 (Tumane et al., 2000), Anti-acne action of ethanolic leaf extract of *A. indica* was noted by Kumar and Khanum (2004) through inhibition of *Propionibacterium acne*. Leaves of the plant contain bioactive components, which inhibited dominating fungal growth of *Fusarium moniliforme, A. flavus,* and *A. niger* (Muley and Pawar, 2005) affecting green grams, grams and groundnuts seeds. Green leaf extract is more antimicrobial effective than dry leaves, young twigs or seeds (Bipte and Musaddiq, 2005). *Plasmodium falciparum* isolates showing resistance to chloroquine were inhibited by various neem extracts Mohite et al. (2005). Extract of neem pulp against different strains of human pathogens showed good efficiency and on screening showed saponins, reducing sugars and alkaloids. In Siddha therapy, it is prescribed for the treatment of enteric fever (Uma, 2017).

*A. indica* is a multipurpose tree. The leaf extract of had exhibited a potent antibacterial activity against various strains of bacterial pathogens (Hala, 2017).

### 9.6.3   C. LONGA

*C. longa* aqueous extract was found to inhibit Gram-positive, *S. aureus, B. subtilis,* and *E. faecalis*; and Gram-negative, *S. typhi, S. flexneri, P. aeruginosa, P. vulgaris, E. coli,* and *C. Albicans* Figs. 9.2 and 9.3). *A. niger* showed resistance. The acetone extract inhibited Gram-positive *M. luteus,* and *E. faecalis*; Gram-negative *Ser. marcescens. E. faecalis* was most sensitive to the aqueous extract (zone of inhibition – 19.66 mm) while *M. luteus* to acetone extract (19 mm). *P. aeruginosa* was least sensitive to the aqueous extract (15.33 mm) and *E. faecalis* to acetone extract (15.66 mm). Efficiencies of aqueous and acetone extracts inhibiting *E. faecalis* were found to be statistically comparable and significant; th        e  aqueous  extract  was 64% efficiency while acetone showed 21%. Aqueous extract of *C. longa* inhibited nine of the pathology laboratory isolates and hence, was equipotent to Gentamicin (G), Cephotaxime (AX), Netillin (NT), Ofloxacin (OF), and Pefloxacin (PF) in its antimicrobial action. Primary screening of response of organisms in swab samples of burns wounds showed that three out of ten and three out of seven isolates from burns wounds were susceptible to *C. longa* aqueous extract. Thus, the antimicrobial activity of *C. longa* can be concluded as more than that of Ciprofloxacin. The aqueous extract was inhibitory to standard cultures of *S. aureus* ATCC 29213 and *E. coli* ATCC 25928 while acetone extract to *S. aureus* ATCC 29213. Preliminary chemical investigation of the aqueous plant extract showed alkaloids, fats, glucose, and triterpenoids. HPTLC analysis showed peaks relating to probably coumarins and triterpenoids at 200 nm, Eriodictyl, Echinacosides, Caffeic acid derivatives, and Cichoric acid at 366 nm.

A commonly available and popular spice, *C. longa* kills intestinal pathogens, reducing the number of intestinal bacteria, helping normal flora of intestine to stabilize. It is digestible and shows no harmful side effects, increasing its advantage as a prebiotic.

Aqueous extract of *C. longa* shows inhibitory efficiency against wound pathogens and *Actinomycetal* species. *C. longa* roots constitute of curcuminoids, sesquiterpenoids, and monoterpenoids which have natural analgesic, antineoplastic, anti-inflammatory, and antibacterial properties (Tang and Eisenbrand, 1992). Soni et al. (1992) reported that *C. longa* extracts inhibit 90% production of aflatoxin in *Aspergillus parasiticus*. Chandi et al. (1999) reported activity of *C longa* roots and stem extract against *E. coli, S. aureus,* and other pathogens. *C longa* has essential oils which inhibit *E. coli, S. typhi, K. pneumoniae, S. aureus*, and *B. subtilis* (Dubey et al., 2005). Resistant strains of *Candida*, *Vibrio, Shigella, and Salmonella* were inhibited by turmeric (Joshi and Shete, 2006). Petroleum ether and Methanol fractions of *C. longa* roots proved better antimicrobial agents than crude extract *(Gupta et al., 2015)*.

## 9.6.4  E. GLOBULUS

A fragrant plant, *E. globulus* was observed to be a potent antimicrobial agent Fig. 9.2 and 9.3). Aqueous extract inhibited the growth of Gram-positive *B. subtilis, M. luteus,* Gram-negative isolates, *K. pneumoniae, S. flexneri, P. aeruginosa*, and *P. vulgaris*. The acetone extract inhibited Gram-positive *M. luteus, B. subtilis, E. faecalis,* and *S. aureus* while Gram-negative *K. pneumoniae, S. typhi, Sal. paratyphi B, S. flexneri, P. aeruginosa, and P. vulgaris*, and *C. albicans. B. subtilis* was most sensitive to the aqueous extract (inhibition zone diameter − 21.83 mm) while *M. luteus* to acetone extract (29 mm). *K. pneumoniae* (16 mm) was least sensitive to the aqueous extract and *B. subtilis* (14.33 mm) to acetone extract, as concluded from the zone sizes. Comparative efficiencies of extracts, aqueous and acetone, inhibiting

*S. flexneri* were statistically significant, those for *B. subtilis, M. luteus,* and *P. vulgaris* very highly significant while for *K. pneumoniae, P. aeruginosa* comparable. *E. globulus* aqueous extract showed 43% efficiency while that of acetone showed 79% efficiency. Aqueous extract of *E. globulus* inhibited six of the pathology laboratory isolates and hence, was equipotent to Ceftazidime (CT), Chloramphenicol (CHLO), and *Nitrofurantoin* (NF) in its antimicrobial action. Primary screening of response of organisms from swab samples of burns wounds showed that out of ten, three were sensitive to the aqueous extracts of *E. globulus*. Of the seven burns wounds isolates, four were inhibited by the aqueous extract. The aqueous extract was inhibitory to the standard control cultures of *S. aureus* ATCC 29213, *E. faecalis* ATCC 29212, and the acetone extract inhibited *E. faecalis* ATCC 29212. Chemical investigation of the plant extract depicted flavonoids, alkaloids, reducing sugars, and glycosides. Its aqueous extract on HPTLC analysis probably showed peaks corresponding to anthroquinones at 200 nm, Echinacosides, Eriodictyl, Cichoric acid, Cynarin, and Caffeic acid derivatives at 366 nm, flavonoids – Cascarosides A, B, C, D, sennosides, Glucofrangulins, and aloinosides at 254 nm and flavonoid – taxifolin at 366 nm. Akki et al. (2004) reported healing activity of *E. globulus* leaf extract and its aqueous extract found effective against wound pathogens – *Actinomycetal, Enterococcus,* and *Pseudomonas* species. Very encouraging results were noted by Timande and Nafde after checking essential oil of *Eucalyptus* against fungi *P. chrysogenum, A. terreus, A. niger,* and bacteria, *Corynebacterium* species NCIM 2640, *B. subtilis, B. cereus, E. coli, S. aureus, P. aeruginosa, P. vulgaris, Salmonella* species, and *Shigella* species (2004). Plant extracts were concluded as superior to commercial antifungal products which inhibit germination of spores; extracts found to be better, acetone extract showing 30% efficiency (Karade et al., 2001). Geraniol, Citronellal, and Citronellol are the active ingredients in *E. globulus* leaves, used in cosmetics, soaps, and perfumes (Baskar et al., 2002b).

### 9.6.5  P. PINNATA

Very encouraging results were observed in case of *P. pinnata*, with reference to its antimicrobial activity in pathological samples Figs. 9.2 and 9.3), as well as against burns wound samples. The aqueous extract was effective against Gram-positive bacteria, *S. aureus, B. subtilis, E. faecalis,* and *M. luteus,* Gram-negative *K. pneumoniae, E. coli, S. typhi, S. Flexneri,* and *C. albicans* and fungal isolate *A. niger*. The acetone extract was observed potent against bacterium *S. aureus* and fungus *A. niger* while Gram-negative isolates were resistant. *M. luteus* was observed to be very sensitive to the aqueous extract (zone diameter – 19 mm) while *A. niger* to the acetone extract (21.5 mm). Amongst the test isolates, *S. aureus* was least sensitive to the aqueous (8.33 mm) and acetone (19 mm) extracts, may be attributed to its property of showing resistance to antimicrobial agents. The *P. pinnata* aqueous extract showed 71% efficiency while that of acetone 14% efficiency. Thus, aqueous extract was concluded to be more potent than acetone extract against *S. aureus* and *A. niger*. On comparison, the *P. pinnata* aqueous extract can be said to be equivalent in efficiency to Lomefloxacin in its antimicrobial action. In primary screening, response of organisms from swab samples of burns wounds showed that out of ten, six were sensitive to the aqueous extracts of *P. pinnata*. Of the seven burns wounds isolates, five were inhibited by the aqueous extract. Both, aqueous and acetone extracts were checked to be effective against the standard control cultures, *S. aureus* ATCC 29,213 while *E. faecalis* ATCC 29,212, *E. coli ATCC 25,928*, and *Ps. aeruginosa* ATCC 27,853 were resistant. Primary qualitative analysis of aqueous extract depicted flavonoids, alkaloids, glucose, and proteins to be present. Analysis by HPTLC at 200 nm indicated the probable presence of anthraquinones and alkaloids, at 254 nm caffeic acid, at 366 nm Echinacosides, Cynarin, and Caffeic acid derivatives, at 254 nm flavonoids, Cascarosides A, B, C, D, sennosides, Glucofrangulins, and aloinosides and A – monoglycosides and frangulin A, B, and at 366 nm silychristin, taxifolin, and isosilybin.

Akki et al. reported the action of aqueous *P. pinnata* extract wound pathogens, killing *Actinomycetes, Bacillus, Enterococcus*, and *Pseudomonas* (2004). Very potent antifungal and antibacterial activities were observed by Wagh et al. (2005) against *A. Niger, Asp. fumigatus, P. aeruginosa*, and *S. aureus,* attributed to the presence of concentrated methyl ester, 9-octadecenoic acid. Seeds of *P. pinnata* showed an effect against *Staphylococcus epidermidis* and *Propionibacterium acnes* responsible for acne (Kumar et al., 2007). Few hospital isolates of pathogens were inhibited more by *P. pinnata* methanol and ethanol extracts, as compared to Ceftazidime (Mary et al., 2013).

Latest reports show that *P. pinnata* brings about the contraction of wounds and stimulates tensile strength, enhances the content of hexosamine and hydroxyproline, is antioxidative, induces production of cytokine and exhibits antimicrobial potency, and hence, can be used for healing of burns wounds (Dwivedi et al., 2017). The bark of stems has alkaloids, phenylpro-panoids, and flavonoids as contents and inhibited *K. pneumoniae, E. coli, E. faecalis, P. aeruginosa, S. aureus, and B. subtilis* (Krishna and Grampurohit, 2006). Aqueous extract of seeds of *P. pinnata* kill herpes simplex virus types 1 and 2 (Dweck, 1994).

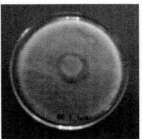
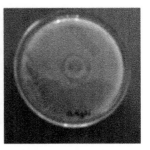

Effect of extract of        Effect of extract of          Effect of extract of *P.*
*P. pinnata* on *E. fecalis.*   *C. longa* on *E. fecalis.*   *pinnata* on *S. typhi.*

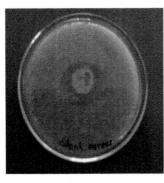

Effect of extract of *P. pinnata* on          Effect of extract of *P. pinnata*
*P. aeruginosa* from burns.                   on *S. aureus* from burns

**FIGURE 9.2**   Antimicrobial activity of plant extracts on pathogenic isolates (representative samples depicting the inhibitory zones).

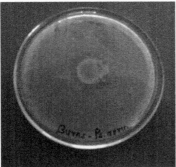
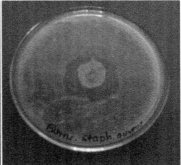

Effect of extract of *P. pinnata* on          Effect of extract of *P. pinnata*
*P. aeruginosa* from burns.                   on *S. aureus* from burns.

**FIGURE 9.3**   **(See color insert.)** Antimicrobial activity of plant extracts on pathogenic isolates from burn wounds.

## 9.7 CONCLUSION AND WAY FORWARD

Wide biodiversity exists on the earth and has always been helping us, in overcoming existing and upcoming microbial infections, resulting from human beings interfering with nature. Parasites, fungi, bacteria, and viruses are posing a menacing challenge to human beings, because of the resistance developing in them, against antimicrobial agents. Very few things are known about the positive, negative, and symbiotic relationships between the components of medicinal plants like essential oils, components of these oils and with antibiotics. If this knowledge is obtained, nanocapsules of essential oils could be prepared and used. This understanding may be employed for generating more potent and new agents against pathogenic organisms, which may help in solving multiple drug resistance of pathogens (Chouhan et al., 2017).One of their component ingredients, alkaloids is a bioactive molecule, which can be utilized as an effective therapeutic tool in the management of diseases (Kurhekar, 2017). The fact that extracts of aqueous origin are more efficient against microorganisms than acetone extracts, has inspired the future development and engineering of phytochemically active classes of agents inhibiting microorganisms (Kurhekar, 2016).

Protecting, preserving, and propagating such indigenous and potent species, so as to manage our bioresources for the future generations, in the present critical scenario, is our moral responsibility.

### KEYWORDS

- **antimicrobial properties**
- **inhibitory zones**
- **medicinal plants**
- **pathogens**
- **bioresources**

### REFERENCES

Akki, K. S.; Hukkeri, V. I.; Karadi, R. V.; Savadi, R. V.; Manohara, K. P. Wound Healing Activity of Eucalyptus Globulus Leaf Extract. *Indian J. Nat. Prod.* **2004,** *20*, 19–22.
Almawlah, I. Y. H. *Int. J. Chem. Tech. Res. CODEN (USA)* IJCRGG. **2017,** *10*(2), 604–611.

Banginwar, Y. S., Tambekar, D. H. Effect of Various Plant Extracts on the Growth of Vibrio Cholerae, *J. Microb. World* **2003**, *5*, 1–3.

Baskar, R. G.; Chezhiyan, N.*Plant compounds and Chemotherapy – Role of Biotechnology in Medicinal and Aromatic Plants;* Special Volume on Diseases, Ukaaz Publication: Hyderabad, 2002a; Vol. 6, p 124.

Baskar, R. G.; Chezhiyan, N. *Therapeutic Drugs from Fruit Plants – Role of Biotechnology in Medicinal and Aromatic Plants;* Special Volume on Diseases, Ukaaz Publication: Hyderabad, 2002b; Vol. 6, pp 77–88.

Baskar, R. G.; Chezhiyan, N. *Horticultural therapy – Role of Biotechnology in Medicinal and Aromatic Plants*; Special Volume on Diseases, Ukaaz Publication: Hyderabad, 2002c; Vol. 6, pp 207–208.

Baskar, R. G.; Ragavathathum, D. V.; Chezhiyan, N.; Khan I. A. *Medicinal Value of Vegetables – Role of Biotechnology in Medicinal and Aromatic Plants;* Special Volume on Diseases, Ukaaz Publication: Hyderabad, 2002b; Vol. 6, pp 250–252.

Baskar, R. G.; Ramesh, P. T.; Sagaya, A. R.; Chezhiyan, N. *Anti-disease properties of medicinal and aromatic plants – Role of Biotechnology in Medicinal and Aromatic Plants;* Special Volume on Diseases, Ukaaz Publication:. Hyderabad, 2002a;Vol. 6, pp 69–75.

Bipte, S.; Musaddiq, M. Studies on Anti-Microbial Activity of *Azadirachta indica* L. on Certain Foliar Pathogens, *J. Microb. World* **2005**, *7*, 28–31.

Bonjar, S. G. H. Anti-Yeast Activity of Some Plants Used in Traditional Herbal Medicine of Iran. *J. Biol. Sci.* **2004**, *4*, 212–215.

Chandi, C. R.; Dash, S. K.; Mishra, R.; Azeemoddin, G. Ibid.**1999**, *36*, 133.

Chomchalow, N.; Henle, H. V. *Medicinal and Aromatic Plants in Asia, Breeding and Improvement;* RAPA publication. Lebanon, NH: Science Publishers in Arrangement with Food and Agriculture Organization of the United Nations, Regional Office for Asia and the Pacific, Bangkok, 1995, p 83.

Chouhan, S.; Sharma, K.; Guleria, S. Antimicrobial Activity of Some Essential Oils – Present Status and Future Perspectives *Medicines* **2017**, *4*, 58.17.

Clare, V.; Wood, B. Trees in Society in Rural Karnataka, India, Overseas Development Administrative National Resources Institute in Association with Karnataka Forest Development. India **1992,** *47*:3.

Das, D.; Agarwal, V. S. *The Study, Exploitation and Identification*; Drug Plants of India, Kalyani Publishers: Ludhiana, 1985.

Das, D.; Agarwal, V. S. *Fruit Drug Plants of India*; Kalyani publishers: Ludhiana, 1991.

Das, B. K.; Mukherjee, S. C.; Sahu, B.B.; Murjani, G. Neem (*Azadirachta indica*) Extract as an Anti-Microbial Agent Against Pathogenic Bacteria, *Ind. J. Exp. Biol.* **1999,** *37*, 1097–1100.

Delaha, E. C.; Garagusi, V. F. Inhibition of Mycobacteria by Garlic Extract (Allium Sativum). *Antimicrob. Agents Chemother.* **1985**, *27*, 485–480.

Devasagayam, T. P. A.; Sainis, K. B. Immune System and Anti-Oxidants, Especially Those Derived from Indian Medicinal Plants, *Ind. J. Exp. Biol.* **2002**, *40*, 643.

Dhawan, B. N.; Rastogi, R. P. Recent Development of Indian Medicinal Plants. In *The Medicinal Plant Industry;* Wijesekara, R. O. B. Ed.; CRC Press: Boca Raton, 1991; pp 185–208.

Dubey, R. C.; Rana, A.; Shukla, R. K. Antibacterial Activity of Essential Oils of Some Medicinal Plants Against Certain Human Pathogens, *Indian Drugs* **2005**, *42*, 443–446.

Dweck, A. C. The Green Pharmacy Herbal Handbook, www.mothernature.com. 1994. Accessed Oct 8, 2017.

Dwivedi, D.; Dwivedi, M.; Malviya, S.; Singh, V. Evaluation of Wound Healing, Anti-Microbial and Antioxidant Potential of Pongamia Pinnatain Wistar Rats. *J. Tradit. Complementary Med.* **2017**, *7*, 79–85.

Farnsworth, N. R. Screening Plants for New Medicines. In *Biodiversity*; Wilson, E. O., Ed.; National Academy Press: Washington D. C, 1998; pp 83–97.

Farooqi, A. S.; Sreeramu, B. S.; Srinivasappa, K. N. Cultivation of Medicinal Plants in South India. *Agrobios Newsletter* **2003**, *1*, 21–22.

Gupta, A.; Mahajan, S.; Sharma, R. Evaluation of Antimicrobial Activity of Curcuma Longarhizome Extract Against *Staphylococcus aureus*. *Biotechnol. Rep.* **2015**, *6*, 51–55.

Hala, A. M. Antibacterial Activity of *Azadirachta Indica* (Neem) Leaf Extract Against Bacterial Pathogens in Sudan. *Afr. J. Med. Sci.* **2017**, *2*, 1.

Handa, S. S.; Kapoor, V. K. Pharmacognosy, Vallabh Prakashan, New Delhi., 1st Ed., 1989, 3–9.

Jadhav, R. B.; Kharya, M. D. Plant Flavonoids: A Versatile Class of Phyto-Constituents with Potential Anti-Inflammatory Activity, *Indian Drugs* **2005**, *42*, 485–495.

Jain, R. C. Antibacterial Activity of Garlic Extract. *Indian J. Med. Microbio.* **1993**, II, 26–31.

Jassim, S. A.; Naji, M. A. Novel Antiviral Agents: A Medicinal Plant Perspective, *J. App. Microbiol.* **2003**, *95*, 412–427.

Kaul, V. Many Antibiotics Have Gone Dud; *TNN, The Times of India*: Pune, 2006; p 10.

Khan, I. A.; Khanum, A. *Role of Biotechnology in Medicinal and Aromatic Plants*; Special Volume on Diseases, Ukaaz Publications: Hyderabad, 2002; Vol. 6, pp 4–6.

Kokate, C. K.; Purohit, A. P.; Gokhale, S. B. Pharmacognosy, 17th ed.; Nirali Prakashan: Pune, 2001, 58,*197*(445), 586–588.

Krishna, V. N.; Grampurohit, N. D. Studies on Alkaloids of Stem Bark of *Pongamia Pinnata* Linn. *Ind. Drugs* **2006**, *43*, 383–387.

Kumar, G. S.; Khanam, S. Anti-Acne Activity of Natural Products. *Indian J. Nat. Prod.* **2004**, *20*, 7.

Kumar, G. S.; Jayaveera, K. N.; Kumar, C. K.; Umachigi, P. S.; Swamy, B. M.; Kumar, D. V. Antimicrobial Effects of Indian Medicinal Plants Against Acne-Inducing Bacteria. *Tropical. J. Pharma. Res.* **2007**, *6*, 717–723.

Kurhekar, J. V. Flavonoids – the magical components of medicinal plants, ChemXpress. **2016**, *9*(2), 139–144.

Kurhekar, J. V. Alkaloids - the Healers in Medicinal Plants. *Int. J. Res. Biosciences.* **2017**, *6*(4), 1–7.

Kurian, J. C. Plants That Heal. Oriental Watchman Publishing House: Pune. xvi, 1st Ed. India, 1999.

Kurian, J. C. Plants That Heal. Oriental Watchman Publishing House: Pune. 5th edition. 2004.

Martindale. *The Extra Pharmacopoeia;* 29th ed.; The Pharmaceutical Press: 1989; pp 210–216.

Mary, S. R.; Dayanand, C. D.; Shetty, J.; Vegi, P. K.; Kutty, M. Evaluation of Antibacterial Activity of Pongamia Pinnata Linn on Pathogens of Clinical Isolates, *Am. J. Phytomed. Clin. Ther.* AJPCT **2013**, *1*(8), 645–651.

Mohite, J. B.; Bodhankar, M. G.; Sharma, V.; Urhekar, A. D. Effect of Herbal Extracts on Chloroquine Resistant Plasmodium Falciparum Isolates. *J. Microb. World* **2005**, *7*, 227–232.

Muley, S. M.; Pawar, P. V. *Utilization of Azadirachta indica A. Juss and Trigonella foenum graceum L. to Control Seed-borne Pathogens of Soyabean*, Proceedings, National Conference on Bioactive Compounds: New Frontiers and Therapeutic Usage, Nanded. 2005, p 31.

Nadkarni, K. M. *Indian Plants and Drugs and their Medicinal Properties and Uses*; Shrishti Book Distributors: New Delhi, 2004, pp 26–27.

O'Gara, E. A.; Hill, D. J.; Maslin, D. J. Activities of Garlic Oil, Garlic Powder and Their Diallya Constituents Against Helicobacter Pylori. *Appl. Environ. Microbiol.* **2000,** *66,* 2269–2273.

Reddy, A. C. H.; Lokesh, B. R. Studies on Spice Principles as Antioxidants in the Inhibition of Lipid Peroxidation of Rat Liver Microsomes. *Mol. Cell. Biochem.***1992,** *111,* 117–124.

Redkar, R. G.; Jolly, C. I. Natural Products as Anticancer Agents, *Indian Drugs* **2003,** *40,* 619–626.

Reiter, J.; Natalja L.; Linden, M. V. D.; Gruhlke, M.; Martin, C.; Slusarenko, A. *J. Molecules.* **1711,** *2017*(22), 1–14.

Santhanum, G.; Snagarajan, S. Wound Healing Activity of *Curcuma aromatica* and Piper Betel. *Fitoterapia* **1990,** *61,* 458–459.

Sathe, A. V. *Gharguti Aushadhe*; 15th ed.; Ganesh Printers: Pune, 1998; pp 336–337.

Sawant, S. Y. Maharashtratil Divya Vanaushadhi. Continental Publications: Pune, 1974.

Shetgiri, P. P.; D'Mello, P. M. Antioxidant Properties of Flavonoids, a Comparative Study, *Indian. Drugs* **2003,** *40,* 567–569.

Shrinivas, C.; Prabhakaran, K. V.S. Clinical Bacteriological Study of *C. longa* on Conjunctivitis. *Antiseptic* **1987,** *84,* 166–168.

Sonawane, Y. D.; Ansari, Z.; Mamude, Y. B. Utilization of Some Medicinal Plants of Baglan Taluka of Nasik District (Maharashtra State).*J. Swamy Bot – Cl.* **2006,** *23,* 173–174.

Soni, K. B.; Rajan, A.; Kuttan, R. Reversal of Aflatoxin Induced Liver Damage by Turmeric and Curcumin. *Cancer Lett.***1992,** *66,* 115–121.

Tang, W.; Eisenbrand, G. Chinese Drugs of Plant Origin, Springer-Verlag: Berlin and Heidelberg, Germany, 1992; pp 401–415.

Tansey, M. R.; Appleton, J. A. Inhibition of Fungal Growth by Garlic Extract. *Mycologia* **1975,** *67,* 409–413.

Timande, S. P.; Nafde, S. K. Antimicrobial Activity of Some Essential Oils, *J. Microb. World* **2004,** *6,* 162–167.

Uma, A. P. Antimicrobial Activity in Pulp Extract of Neem (*Azadirachta Indica* Linn.), *Int. J. Curr. Res. Biol. Med.* **2017,** *2*(5), 43–45.

Varshney, A.; Dhawan, V.; Shrivastava, P. S. Ginseng: Wonder Drug of the World, In *Role of Biotechnology in Medicinal and Aromatic Plants*; Special Volume on Diseases, Ukaaz Publication: Hyderabad, 2001; Vol. 6, pp 26–41.

Wagh, P.; Rai, M.; Marta, C.; Teixeira, D. *Chemical Composition and Anti-microbial Activity of Oils extracted from Trigonella foenum-graecum L. and Pongamia pinnata L. Pierre,* Proceedings, National Conference on Bioactive Compounds: New Frontiers and Therapeutic Usage, Nanded. 2005.

Wills, B. G. Enzyme Inhibition by Allicin, the Active Principle of Garlic. *Biochem. J.* **1956,** *63,* 514–520.

**CHAPTER 10**

# MEDICINAL PLANTS WITH ANTIVENOM ACTIVITIES

HABIBU TIJJANI[1,*] and CHUKWUEBUKA EGBUNA[2]

[1]Department of Biochemistry, Natural Product Research Laboratory, Bauchi State University, Gadau, Bauchi State, Nigeria, Tel.: +2348037327138

[2]Department of Biochemistry, Chukwuemeka Odumegwu Ojukwu University, Anambra State, Nigeria

*Corresponding author. E-mail: tijjanihabibu@basug.edu.ng
*ORCID: https://orcid.org/0000-0001-5466-322X

## ABSTRACT

Animals producing venoms pose challenges to public health across the globe. However, the degree of severity, caused by their envenomation varies from one another. Their major pathophysiology is characterized by tissue damages, edema, hemorrhage, hyperglycemia, hypertension or hypotension, hypothermia, myonecrosis, nephrotoxicity, cardiotoxicity as well as other systemic response. This chapter documents the various plants used in the treatment of venoms from snakes and scorpions. A literature search was conducted using the keywords "phytochemicals," "venom," "antivenom activities," and "envenomation," in different combinations. Articles were selected based on the appearance of relevant keywords in their titles and/ or abstracts. The active principles identified in these literature searches include 4-nerolidylcatechol, arjunolic acid, neolignans, piperine, quercetin-3-O-rhamnoside, β-sitosterol among others. Their mechanisms of action include the inhibition of enzymes such as phospholipase, ability to neutralize some metallo- and serine proteinases in venoms, their antihemorrhagic, anticoagulant, anti-inflammatory, and anti-edematogenic effects.

## 10.1  INTRODUCTION

Animals producing venoms pose challenges to public health across the globe. However, the degree of severity, caused by their envenomation varies from one another. Their venomous productions and its effect have fascinated the human race for millennia (Mackessy, 2010). Snake and scorpions are among animals producing venoms, contributing to morbidity and mortality in humans and animals most especially in regions where they are prominent.

Snakebites occur in several tropic and subtropic countries where they are considered as neglected public health issues. There are about 5.4 million snakebite cases every year, with 137,880 deaths and several other complications such as amputations and other permanent disabilities each year (WHO, 2017). Snake venoms are a complex mixture of enzymes and toxins, which are injected into their prey in order to incapacitate, immobilize, and digest them or as a means to provide defense for themselves against predator (Murari et al., 2005). Their composition varies and may include acetylcholinesterase, ATPase, hyaluronidases, metalloproteases, nucleosidases, nucleotidase, acid or alkaline phosphatases, phosphodiesterase, phospholipases, serine proteases, and transaminase (Aird, 2002).

While snakes are known for their bites, scorpions are known for their sting. There are about 1.2 million scorpion stings and estimated 3250 deaths every year, representing 0.27% (Chippaux and Goyffon, 2008). They are also a public health concern, especially in many underdeveloped tropical and subtropical countries, including Africa, Middle East, South India, and South Latin America (Khatony et al., 2015; Queiroz et al., 2015). Scorpion venoms are heterogeneous mixtures of cardiotoxins, nephrotoxins, neurotoxins, and hemolytic substances which, when injected into humans exert deleterious effects to humans (Bawaskar and Bawaskar, 2012).

Scorpions and snakes have specialized mechanism for delivery of their venoms into their respective host. Once ingested, they exert their pathophysiological effects, which are characterized by severe pains (Uawonggul et al., 2006), inflammatory reactions (Uawonggul et al., 2006), tissue damages, edema, hemorrhage, hyperglycemia, hypertension or hypotension, hypothermia, myonecrosis, nephrotoxicity, cardiotoxicity as well as other systemic response (Escalante et al., 2011; Markland and Swenson, 2013). Treatments with specific antivenoms are the recognized standard treatment methods. The therapeutic options available for alleviation of scorpion stings and its symptoms include the use of angiotensin-converting enzyme inhibitors, sodium channel blockers, antivenoms, insulin, and prazosin (Fatani et al., 2000; Bawaskar and Bawaskar, 2007; Krishnan et al., 2007; Deshpande et al., 2008). However, they have side effects which may be absent in plant-based treatments. Therefore, it is important to make plant-based treatments a widely sort treatment option in areas where antivenoms are not readily available. This chapter documents the various medicinal plants used in the treatment of venoms from snakes and scorpions with the aim to identify the various active principles that have been isolated from them.

## 10.2   METHOD

A literature search was conducted in Google Scholar, PubMed, and Science Direct using the keywords "phytochemicals," "venom," "anti-venom activities," "medicinal plant," and "envenomation" in different combinations. Articles were selected based on the appearance of relevant keywords in their titles and/or abstracts. No restriction was made as to the period or language of publication. The listed plant was selected with at least one or more scientific evidence of their antivenom activities, the plant names were then cross-checked in http://www.theplantlist.org.

## 10.3   RESULT AND DISCUSSION

Literature search identified 217 plants for treatment of snakebites (92.31%) and scorpion stings (7.69%) distributed over 77 families (Table 10.1). The Leguminosae 24 (11.06%) has the highest listing followed by the Apocynaceae 12 (5.53%), Compositae 11 (5.07%), Euphorbiaceae 9 (4.15%), Lamiaceae 8 (3.69%), and Rubiaceae 8 (3.69%) (Fig. 10.1). Leaves, roots, bark, and stem were among the frequently used part of plants (Fig. 10.2). The most used solvents of extraction include water, ethanol, methanol, and ethyl acetate (Fig. 10.3).

Venomous animals such as snake and scorpions have adapted to their various environments and have developed the means to survive in these environments. Snakes, for example, effectively capture and immobilize their prays by the rapid dissemination of their venoms which are target specific into blood and tissue system, this process is a prerequisite for action of this toxin (Kini, 1997; Gutiérrez and Rucavado, 2000) and thereafter, the debilitating and life-threatening events which eventually result in morbidity and mortality. Morbidity symptoms of snake envenomations, which vary in extents and presentations, may include coagulopathy, gangrene, swelling, to severe mouth bleeding, edema, and to widespread loss of skeletal muscle

**TABLE 10.1** Medicinal Plants with Antivenom Activities.

| S/ No. | Names of plants | Family | Extraction solvent | Source of envenomation | Parts used | References |
|---|---|---|---|---|---|---|
| 1. | *Andrographis paniculata* (Burm.f.) Nees | Acanthaceae | Methanol, ethanol | Snake, scorpion | Roots, leaves | Uawonggul et al. (2006); Kuma-rapppan et al. (2011); Alam (2014) |
| 2. | *Clinacanthus nutans* (Burm.f.) Lindau | Acanthaceae | Aqueous | Scorpion | Leaves | Uawonggul et al. (2006) |
| 3. | *Acorus calamus* | Acoraceae | Aqueous | Snake | Root | Meenatcisundaram and Sindhu (2011) |
| 4. | *Blutaparon portulacoides* (A. St.-Hil.) Mears | Amaranthaceae | Ethanol | Snake | Aerial parts | Pereira et al. (2009) |
| 5. | *Pupalia lappacea* Juss | Amaranthaceae | Aqueous | Snake | Herba | Molander et al. (2014) |
| 6. | *Allium sativum* L. | Amaryllidaceae | Aqueous | Snake | Bulbs | Kuriakose et al. (2012) |
| 7. | *Crinum jagus* (J. Thomps.) Dandy | Amaryllidaceae | Methanol | Snake | Bulb | Ode and Asuzu (2006) |
| 8. | *Anacardium occidentale* L. | Anacardiaceae | Aqueous | Snake | Bark | Ushanandini et al. (2009) |
| 9. | *Mangifera indica* L. | Anacardiaceae | Ethanol aqueous | Snake | Fruits, stem bark | Pithayanukul et al., 2009; Dhananjaya et al. (2011) |
| 10. | *Lannea acida* A. Rich. | Anacardiaceae | Ethanol | Snake | Cortex | Molander et al. (2014) |
| 11. | *Annona senegalensis* Pers | Annonaceae | Methanol | Snake | Root bark | Adzu et al. (2005) |
| 12. | *Hydrocotyle javanica* Thunb. | Apiaceae | Ethanol | Snake | Leaves | Kumarapppan et al. (2011) |
| 13. | *Thapsia garganica* L. | Apiaceae | Methanol | Scorpion | Leaves | Bouimeja et al. (2018) |
| 14. | *Eryngium creticum* Lam. | Apiaceae | Aqueous, ethanol | Snake, scorpion | Leaves | Alkofahi et al. (1997) |
| 15. | *Tylophora indica* (Burm.f.) Merr. | Apocynaceae | Aqueous | Snake | Leaves, roots | Sakthivel et al. (2013) |
| 16. | *Tabernaemontana alternifolia* (Roxb) | Apocynaceae | Methanol | Snake | Root | Shrikanth et al. (2014a) |

**TABLE 10.1** *(Continued)*

| S/ No. | Names of plants | Family | Extraction solvent | Source of envenomation | Parts used | References |
|---|---|---|---|---|---|---|
| 17. | *Carissa spinarum* L. | Apocynaceae | Methanol | Snake | Leaves | Janardhan et al. (2015) |
| 18. | *Rauvolfia serpentine* (L.) Benth. ex Kurz | Apocynaceae | Aqueous | Snake | Whole plant | James et al. (2013) |
| 19. | *Tabernaemontana catharinensis* | Apocynaceae | Aqueous | Snake | Root bark | Batina et al. (2000) |
| 20. | *Hemidesmus indicus* (L.) R. Br. ex Schult. | Apocynaceae | Methanol | Snake | Roots | Alam et al. (1996); Chatterjee et al. (2006) |
| 21. | *Peschiera fuchsiaefolia* (A. DC.) Miers | Apocynaceae | Aqueous | Snake | Root bark | Batina et al. (1997) |
| 22. | *Mandevilla illustris* (Vell.) Woodson | Apocynaceae | Aqueous | Snake | Roots, leaves | Biondo et al. (2004) |
| 23. | *Mandevilla velutina* | Apocynaceae | Aqueous | Snake | Rhizomes | Rizzini et al. (1988); Biondo et al. (2004) |
| 24. | *Strophanthus gratus* Baill. | Apocynaceae | Aqueous | Snake | Leaves | Houghton and Skari (1994) |
| 25. | *Strophanthus hispidus* DC. | Apocynaceae | Aqueous | Snake | Leaves | Houghton and Skari (1994) |
| 26. | *Allamanda cathartica* L. | Apocynaceae | Ethanol | Snake | Whole plant | Otero et al. (2000b) |
| 27. | *Philodendron megalophyllum* Schott | Araceae | Aqueous | Snake | Liana | Moura et al. (2013) |
| 28. | *Dracontium dubium* Kunth | Araceae | Aqueous, alcoholic | Snake | Tubers | Caro et al. (2017) |
| 29. | *Dracontium croatii* G. H. Zhu | Araceae | Aqueous | Snake | Rhizomes | Rizzini et al. (1988); Otero et al. (2000a); Núñez et al. (2004b) |
| 30. | *Pinellia ternata* (Thunb.) Breitenbach | Araceae | | Snake | Rhizome | Mors et al. (2000) |
| 31. | *Philodendron tripartitum* (Jacq.) Schott | Araceae | Ethanol | Snake | Leaves | Otero et al. (2000b) |

**TABLE 10.1** (Continued)

| S/No. | Names of plants | Family | Extraction solvent | Source of envenomation | Parts used | References |
|---|---|---|---|---|---|---|
| 32. | Astrocaryum vulgare Mart. | Arecaceae | | Snake | Fruits | Bernard et al. (2001) |
| 33. | Bactris gasipaes Kunth | Arecaceae | | Snake | Fruits | Bernard et al. (2001) |
| 34. | Aristolochia bracteolata Lam. | Aristolochiaceae | Aqueous | Scorpion, Snake | Leaves, roots | Sakthivel et al. (2013) |
| 35. | Aristolochia indica L. | Aristolochiaceae | Methanol | Snake | Whole plant | Meenatchisundaram et al. (2008) |
| 36. | Aristolochia elegans Mast. | Aristolochiaceae | Hexane, methanol | Scorpion | Root | Jiménez-Ferrer et al. (2005) |
| 37. | Pergularia daemia (Forsk.) chiov. | Asclepiadaceae | Aqueous | Snake | Leaves | Raghavamma et al. (2016) |
| 38. | Caltopis gigantea | Asclepiadaceae | Methanol | Snake | Whole Plant | Chacko et al. (2012) |
| 39. | Gymnema sylvestre (Retz.) R. Br. ex Sm. | Asclepiadaceae | Aqueous | Snake | Plant part | Kini and Gowda, (1982a); Kini and Gowda (1982b) |
| 40. | Baccharis trimera (Less) DC | Asteraceae | Aqueous | Snake | Aerial | Bernard et al. (2001); Núñez et al. (2004a) |
| 41. | Lychnophora pinaster | Asteraceae | | Snake | Leaves | Melo et al. (2003) |
| 42. | Solidago chilensis | Asteraceae | | Snake | Aerial | Melo et al. (2003) |
| 43. | Artemisia absinthium L. | Asteraceae | Methanol | Snake | Aerial | Nalbantsoy et al. (2013) |
| 44. | Chaptalia nutans | Asteraceae | Aqueous | Snake | Leaves | Badilla et al. (2006) |
| 45. | Balanites aegyptiaca | Balanitaceae | Acetone, methanolic | Snake | Stem bark | Wufem et al. (2007) |
| 46. | Betula alba L. | Betulaceae | Extract | Snake | Compound | Bernard et al. (2001); Núñez et al. (2005) |
| 47. | Crescentia cujete L. | Bignoniaceae | Ethanol | Snake | Fruit | Otero et al. (2000b); Shastry et al. (2012) |

**TABLE 10.1** *(Continued)*

| S/ No. | Names of plants | Family | Extraction solvent | Source of envenomation | Parts used | References |
|---|---|---|---|---|---|---|
| 48. | *Tabebuia rosea* (Bertol.) Bertero ex A. DC. | Bignoniaceae | Aqueous | Snake | Stems | Castro et al. (1999); Otero et al. (2000a); Núñez et al. (2004b) |
| 49. | *Bixa orellana* L. | Bixaceae | Ethanol | Snake | Leaves | Otero et al. (2000a); Núñez et al. (2004b) |
| 50. | *Cordia verbenacea* A. DC. | Boraginaceae | Methanol | Snake | Rosmarinic acid | Ticli et al. (2005) |
| 51. | *Ehretia buxifolia* Roxb. | Boraginaceae | Aqueous | Snake | Roots | Mors et al. (2000) |
| 52. | *Cordia macleodii* Hook.f. & Thomson | Boraginaceae | Ethanol | Snake | Bark | Soni and Bodakhe (2014) |
| 53. | *Argusia argentea* (L.f.) Heine | Boraginaceae | Methanol | Snake | Leaves | Aung et al. (2010) |
| 54. | *Bursera simaruba* (L.) Sarg. | Burseraceae | Ethanol, ethyl acetate, aqueous | Snake | Aerial | Castro et al. (1999) |
| 55. | *Caraipa minor* Huber | Calophyllaceae | | Snake | Bark | Bernard et al. (2001) |
| 56. | *Mesua ferrea* L. | Calophyllaceae | Aqueous | Scorpion | Leaves | Uawonggul et al. (2006) |
| 57. | *Crateva magna* (Lour.) DC. | Capparidaceae | Ethanol | Snake | Stem bark | Kumarapppan et al. (2011) |
| 58. | *Maytenus ilicifolia* Mart. ex Reissek | Celestraceae | | Snake | Leaves | Bernard et al. (2001) |
| 59. | *Clusia torresii* Standl. | Clusiaceae | Ethanol, ethyl acetate, aqueous | Snake | Aerial | Castro et al. (1999) |
| 60. | *Combretum leprosum* Mart. | Combretaceae | Ethanol | Snake | Root | Fernandes et al. (2014) |
| 61. | *Guiera senegalensis* J. F. Gmel. | Combretaceae | Methanol | Snake | Leaves | Abubakar et al. (2000) |
| 62. | *Combretum molle R. Br. ex G. Don* | Combretaceae | Aqueous | Snake | Folium | Molander et al. (2014) |
| 63. | *Eclipta prostrata* (L.) L. | Compositae | Hydroalcoholic | Scorpion, snake | Leaves, whole plant | Melo et al. (1994); Jalalia et al. (2006) |

**TABLE 10.1** *(Continued)*

| S/No. | Names of plants | Family | Extraction solvent | Source of envenomation | Parts used | References |
|---|---|---|---|---|---|---|
| 64. | *Pluchea indica* Less. | Compositae | Methanol | Snake | Roots | Melo et al. (1994); Alam et al. (1996); Gomes et al. (2007) |
| 65. | *Cynara scolymus* L. | Compositae | Aqueous | Snake | Leaves | Pereira et al. (1991) |
| 66. | *Elephantopus scaber* L. | Compositae | Aqueous | Snake | Leaves | Pereira et al. (1991) |
| 67. | *Vernonia condensata* Baker | Compositae | Aqueous | Snake | Leaves | Pereira et al. (1991) |
| 68. | *Calendula officinalis* L. | Compositae | | Snake | Flower | Mors et al. (2000) |
| 69. | *Echinacea purpurea* (L.) Moench | Compositae | Aqueous | Snake | Root | Chaves et al. (2007) |
| 70. | *Eclipta alba* (L.) Hassk. | Compositae | Methanol | Snake | Aerial parts, roots | Diogo et al. (2009) |
| 71. | *Mikania glomerata* | Compositae | Aqueous | Snake | Roots, stems, leaves | Pereira et al. (1991) |
| 72. | *Pseudelephantopus spicatus* | Compositae | Ethanol | Snake | Whole plant | Otero et al. (2000b) |
| 73. | *Neurolaena lobata* (L.) R. Br. ex Cass. | Compositae | Ethanol | Snake | Whole plant | Otero et al. (2000b) |
| 74. | *Connarus favosus* Planch. | Connaraceae | Aqueous | Snake | Bark | Silva et al. (2016) |
| 75. | *Ipomoea asarifolia* (Desr.) Roem. & Schult. | Convolvulaceae | Aqueous | Scorpion | Leaves | Lima et al. (2014) |
| 76. | *Ipomoea aquatica* Forssk | Convolvulaceae | Aqueous | Scorpion | Leaves | Uawonggul et al. (2006) |
| 77. | *Ipomoea cairica* (L.) Sweet | Convolvulaceae | Ethanol | Snake | Leaves, branches, stems | Otero et al. (2000b) |
| 78. | *Costus spicatus* (Jacq.) Sw. | Costaceae | | Snake | Leaves | Bernard et al. (2001) |
| 79. | *Costus lasius* Loes. | Costaceae | Ethanol | Snake | Whole plant | Otero et al. (2000a) |
| 80. | *Kalanchoe brasiliensis* | Crassulaceae | Hydroethanol | Snake | Leaves | Fernandes et al. (2016) |

**TABLE 10.1** *(Continued)*

| Sl No. | Names of plants | Family | Extraction solvent | Source of envenomation | Parts used | References |
|---|---|---|---|---|---|---|
| 81. | *Kalanchoe pinnata* | Crassulaceae | Hydroethanol | Snake | Leaves | Fernandes et al. (2016) |
| 82. | *Bryophyllum adelae* (Hamet) A. Berger | Crassulaceae | | Snake | Leaves | Bernard et al. (2001) |
| 83. | *Wilbrandia ebracteata* Cogn. | Cucurbitaceae | Aqueous | Snake | Root | Pereira et al. (1991) |
| 84. | *Scleria pterota* C. Presl ex C. B. Clarke | Cyperaceae | | Snake | Leaves | Soares et al. (2004) |
| 85. | *Davilla elliptica* A. St.-Hil. | Dilleniaceae | Methanol | Snake | Leaves | Nishijima et al. (2009) |
| 86. | *Diospyros kaki* L.f. | Ebenaceae | Aqueous | Snake | Fruits, seed | Okonogi et al. (1979); Mors et al. (2000) |
| 87. | *Acalypha indica* L. | Euphorbiaceae | Ethanol | Scorpion Snake | Leaf | Shirwaikar et al. (2004) |
| 88. | *Jatropha gossypiifolia* L. | Euphorbiaceae | Aqueous | Snake | Leaf | Félix-Silva et al. (2018) |
| 89. | *Jatropha mollissima* (Pohl) Baill. | Euphorbiaceae | Aqueous | Snake | Leaf | Félix-Silva et al. (2018) |
| 90. | *Emblica officinalis* Gaertn. | Euphorbiaceae | Methanol | Snake | Roots | Alam and Gomes (2003); Sarkhel et al. (2011) |
| 91. | *Euphorbia hirta* L. | Euphorbiaceae | Methanol | Snake | Whole plant | Gopi et al. (2015); Gopi et al. (2016) |
| 92. | *Croton draco* Schltdl. | Euphorbiaceae | Aqueous | Snake | Stems | Castro et al. (1999) |
| 93. | *Croton urucurana* Baill. | Euphorbiaceae | Aqueous | Snake | | Esmeraldino et al. (2005) |
| 94. | *Euphorbia neriifolia* L. | Euphorbiaceae | Aqueous | Scorpion | Leaves | Uawonggul et al. (2006) |
| 95. | *Ricinus communis* L. | Euphorbiaceae | Aqueous | Scorpion | Seed | Uawonggul et al. (2006) |
| 96. | *Columnea kalbreyeriana* Mast. | Gesneriaceae | Ethanol | Snake | Whole plant | Otero et al. (2000b) |
| 97. | *Heliconia curtispatha* | Heliconiaceae | Ethanol | Snake | Rhizome | Otero et al. (2000a); Núñez et al. (2004b) |

**TABLE 10.1** *(Continued)*

| Sl No. | Names of plants | Family | Extraction solvent | Source of envenomation | Parts used | References |
|---|---|---|---|---|---|---|
| 98. | *Trichomanes elegans* Rich. | Hymenophyllaceae | Ethanol | Snake | Whole plant | Otero et al. (2000a); Otero et al. (2000b) |
| 99. | *Hypericum brasiliense* Choisy | Hypericaceae | Ethanol | Snake | Whole plant | Assafim et al. (2011) |
| 100. | *Leucas aspera* (Willd.) Link | Lamiaceae | Methanol aqueous | Snake | Leaves, roots | Sakthivel et al. (2013); Gopi et al. (2014) |
| 101. | *Marsypianthes chamaedrys* (Vahl) Kuntze | Lamiaceae | Aqueous | Snake | Whole plant | Castro et al. (1999); Mors et al. (2000) |
| 102. | *Ocimum sanctum* L. | Lamiaceae | Aqueous | Snake | Leaves | Kuriakose et al. (2012) |
| 103. | *Vitex negundo* L. | Lamiaceae | Methanol | Snake | Roots | Alam and Gomes (2003) |
| 104. | *Ocimum basilicum L.* | Lamiaceae | | Snake | Leaves | Bernard et al. (2001) |
| 105. | *Plectranthus amboinicus (Lour.) Spreng.* | Lamiaceae | Aqueous | Scorpion | Leaves | Uawonggul et al. (2006) |
| 106. | *Hyptis capitata* Jacq. | font of | Ethanol | Snake | Whole plant | Otero et al. (2000b) |
| 107. | *Ocimum micranthum* Willd. | Lamiaceae | Ethanol | Snake | Whole plant | Otero et al. (2000b) |
| 108. | *Aniba fragrans* Ducke | Lauraceae | Aqueous | Snake | Leaves, bark | Moura et al. (2017) |
| 109. | *Persea americana* Mill. | Lauraceae | Ethanol, ethyl-acetat, aqueous | Snake | Seeds | Castro et al. (1999) |
| 110. | *Phoebe brenesii* Standl. | Lauraceae | Aqueous | Snake | Stems | Castro et al. (1999) |
| 111. | *Barringtonia acutangula* (L.) Gaertn. | Lecythidaceae | Aqueous | Scorpion | Stem bark | Uawonggul et al. (2006) |
| 112. | *Plathymenia reticulata* Benth. | Leguminosae | Condensed-Tannin | Snake | Barks | Moura et al. (2016) |
| 113. | *Cassia hirsute* L. | Leguminosae | | Snake | Leaves, root | Bernard et al. (2001) |
| 114. | *Platymiscium pleiostachyum* | Leguminosae | Ethanol, ethyl acetate, aqueous | Snake | Leaves | Castro et al. (1999) |

**TABLE 10.1** (Continued)

| S/ No. | Names of plants | Family | Extraction solvent | Source of envenomation | Parts used | References |
|---|---|---|---|---|---|---|
| 115. | *Bauhinia thonningii* Schumach. | Leguminosae | Ethanol | Snake | Radix | Molander et al. (2014) |
| 116. | *Tamarindus indica* L. | Leguminosae | Aqueous | Scorpion Snake | Stem, leaves, whole plant, seed | Ushanandini et al. (2006) |
| 117. | *Mucuna pruriens* (L.) DC | Leguminosae | Aqueous | Snake | Seed | Aguiyi et al. (1999); Tan et al. (2009) |
| 118. | *Schizolobium parahyba* (Vell.) S. F. Blake | Leguminosae | Aqueous | Snake | Leaves | Vale et al. (2011) |
| 119. | *Bauhinia forficata* Link | Leguminosae | Aqueous | Snake | Aerial parts | Oliveira et al. (2005) |
| 120. | *Harpalyce brasiliana* Benth. | Leguminosae | Aqueous | Snake | Roots | da Silrva et al. (1997) |
| 121. | *Apuleia leiocarpa* (Vogel) J. F. Macbr. | Leguminosae | Aqueous | Snake | Bark | Pereira et al. (1991); Ruppelt et al. (1991) |
| 122. | *Periandra mediterranea* (Vell.) Taub. | Leguminosae | Aqueous | Snake | Root | Pereira et al. (1991) |
| 123. | *Dipteryx alata* Vogel | Leguminosae | Methanol | Snake | Bark | Nazato et al. (2010) |
| 124. | *Periandra pujalu* Emmerich & L. M.de Senna | Leguminosae | Aqueous extracts | Snake | Root | Pereira et al. (1991) |
| 125. | *Mimosa pudica* L. | Leguminosae | Aqueous, alcoholic | Snake | Root | Mahanta and Mukherjee (2001); Meenatchisundaram and Michael (2009) |
| 126. | *Bowdichia major* (Mart.) Mart. ex Benth. | Leguminosae | | Snake | Seeds | Bernard et al. (2001) |
| 127. | *Brongniartia podalyrioides* Kunth | Leguminosae | Methylene | Snake | Roots | Reyes-Chilpa et al. (1994); Mors et al. (2000) |
| 128. | *Brongniartia intermedia* | Leguminosae | Methylene | Snake | Roots | Reyes-Chilpa et al. (1994) |

**TABLE 10.1**   (Continued)

| S/No. | Names of plants | Family | Extraction solvent | Source of envenomation | Parts used | References |
|---|---|---|---|---|---|---|
| 129. | *Brownea angustiflora* Little | Leguminosae | Ethanol | Snake | Stem bark | Otero et al. (2000a) |
| 130. | *Caesalpinia bonduc* (L.) Roxb. | Leguminosae | | Snake | Seed, leaves | Mors et al. (2000); Datte et al. (2004) |
| 131. | *Parkia biglobosa* (Jacq.) G. Don | Leguminosae | Water-methanol | Snake | Stem bark | Asuzu and Harvey (2003) |
| 132. | *Pentaclethra macroloba* (Willd.) Kuntze | Leguminosae | Aqueous | Snake | Barks | da Silva et al. (2007) |
| 133. | *Galactia glaucescens* Kunth | Leguminosae | Ethanol | Snake | Leaves | Dal Belo et al. (2008) |
| 134. | *Entada Africana* Guill. & Perr. | Leguminosae | | Snake | Bark | Bernard et al. (2001) |
| 135. | *Senna dariensis* | Leguminosae | Ethanol | Snake | Whole plant | Otero et al. (2000b) |
| 136. | *Gloriosa superba* L. | Liliaceae | Ethanol | Snake | Tubers | Kumarapppan et al. (2011) |
| 137. | *Strychnos nux-vomica* L. | Loganiaceae | Ethanol | Snake | Seed | Chatterjee et al. (2004) |
| 138. | *Strychnos innocua* Delile | Loganiaceae | Aqueous | Snake | Folium | Molander et al. (2014) |
| 139. | *Strychnos xinguensis* Krukoff | Loganiaceae | Ethanol | Snake | Stem | Otero et al. (2000b) |
| 140. | *Struthanthus flexicaulis* (Mart. ex Schult. f.) Mart. | Loranthaceae | | Snake | Stem | Bernard et al. (2001) |
| 141. | *Struthanthus orbicularis* (Kunth) Eichler | Loranthaceae | Ethanol | Snake | Leaves | Otero et al. (2000a) |
| 142. | *Byrsonima crassa* Nied. | Malpighiaceae | Methanol | Snake | Leaves | Nishijima et al. (2009) |
| 143. | *Abutilon indicum* (L.) Sweet | Malvaceae | Methanol | Snake | Leaves | Shrikanth et al. (2014b) |
| 144. | *Hibiscus aethiopicus* L. | Malvaceae | Aqueous | Snake | Whole plant | Hasson et al. (2010) |
| 145. | *Althaea officinalis* L. | Malvaceae | | Snake | Leaves | Bernard et al. (2001) |
| 146. | *Grewia mollis* Jass. | Malvaceae | Aqueous | Snake | Cortex | Molander et al. (2014) |
| 147. | *Sida acuta* Burm.f. | Malvaceae | Ethanol | Snake | Whole plant | Otero et al. (2000a); Otero et al. (2000b) |

**TABLE 10.1** *(Continued)*

| S/ No. | Names of plants | Family | Extraction solvent | Source of envenomation | Parts used | References |
|---|---|---|---|---|---|---|
| 148. | *Bellucia dichotoma* Cogn. | Melastomataceae | Aqueous | Snake | Bark | Moura et al. (2014); Moura et al. (2017) |
| 149. | *Mouriri pusa* Gardner ex Gardner | Melastomataceae | Methanol | Snake | Leaves | Nishijima et al. (2009) |
| 150. | *Azadirachta indica* A. Juss. | Meliaceae | Aqueous | Snale | Bark | Kuriakose et al. (2012) |
| 151. | *Abuta grandifolia* (Mart.) Sandwith | Menispermaceae | Aqueous | Snake | Stem | Rizzini et al. (1988); Bernard et al. (2001) |
| 152. | *Cissampelos pareira* L. | Menispermaceae | Aqueous, Ethanol | Snake | Whole plant | Verrastro et al. (2017) |
| 153. | *Dorstenia brasiliensis* Lam. | Moraceae | Aqueous | Snake | Root | Pereira et al. (1991) |
| 154. | *Ficus nymphaeifolia* Mill. | Moraceae | Aqueous | Snake | Leaves, branches, stems | Otero et al. (2000b); Núñez et al. (2004b) |
| 155. | *Morus Alba* L. | Moraceae | Aqueous | Snake | Stems and leaves | Mors et al. (2000); Bernard et al. (2001) |
| 156. | *Dorstenia reniformis* Pohl ex Miq. | Moraceae | | Snake | Whole plant | Bernard et al. (2001) |
| 157. | *Castilla elastica* | Moraceae | Ethanol | Snake | Whole plant | Otero et al. (2000b) |
| 158. | *Musa paradisiaca* L. | Musaceae | Juice | Snake | Stems sap | Borges et al. (2005) |
| 159. | *Virola koschnyi* Warb. | Myristicaceae | Ethanol, ethyl acetate, aqueous | Snake | Aerial | Castro et al. (1999) |
| 160. | *Pimenta dioica* (L.) Merr. | Myrtaceae | Ethanol, ethyl acetate, aqueous | Snake | Aerial | Castro et al. (1999) |
| 161. | *Nymphaea rudgeana* G. Mey. | Nymphaeaceae | | Snake | Leaves | Bernard et al. (2001) |
| 162. | *Calamus* sp. | Palmae | Aqueous | Scorpion | Whole plant | Uawonggul et al. (2006) |
| 163. | *Passiflora laurifolia* L. | Passifloraceae | Aqueous | Scorpion | Leaves | Uawonggul et al. (2006) |
| 164. | *Passiflora quadrangularis* | Passifloraceae | Ethanol | Snake | Leaves | Otero et al. (2000b) |

**TABLE 10.1** *(Continued)*

| S/ No. | Names of plants | Family | Extraction solvent | Source of envenomation | Parts used | References |
|---|---|---|---|---|---|---|
| 165. | *Phyllanthus klotzschianus* Müll. Arg. | Phyllanthaceae | Aqueous | Snake | Leaves | Pereira et al. (1991) |
| 166. | *Bridelia ndellensis* Beille | Phyllanthaceae | Aqueous ethanol | Snake | Bark | Mostafa et al. (2006) |
| 167. | *Petiveria alliacea* L. | Phytolaccaceae | Aqueous | Snake | Leaves | Bernard et al. (2001) |
| 168. | *Piper longum* L. | Piperaceae | Ethanol | Snake | Fruits | Shenoy et al. (2013) |
| 169. | *Piper umbellatum* L. | Piperaceae | Methyl-terbutyl ether | Snake | Whole plant | Núñez et al. (2005) |
| 170. | *Piper peltatum* L. | Piperaceae | Methyl-terbutyl ether | Snake | Whole plant | Núñez et al. (2005) |
| 171. | *Piper pulchrum* C. DC. | Piperaceae | Ethanol | Snake | Whole plant | Otero et al. (2000b) |
| 172. | *Piper arboreum* | Piperaceae | Ethanol | Snake | Leaves | Otero et al. (2000b) |
| 173. | *Scoparia dulcis* L. | Plantaginaceae | | Snake | Whole plant | Bernard et al. (2001) |
| 174. | *Bredemeyera floribunda* Willd. | Polygalaceae | Aqueous | Snake | Roots | Pereira et al. (1991) |
| 175. | *Rumex* sp. | Polygonaceae | Aqueous | Scorpion | Whole plant | Uawonggul et al. (2006) |
| 176. | *Pleopeltis percussa* (Cav.) Hook. & Grev. | Polypodiaceae | Ethanol, ethyl acetate, aqueous | Snake | Whole plant | Otero et al. (2000a); Núñez et al. (2004b) |
| 177. | *Ziziphus joazeiro* Mart. | Rhamnaceae | | Snake | Bark | Bernard et al. (2001) |
| 178. | *Canthium parviflorum* Lam. | Rubiaceae | Ethyl acetate, methanol | Snake | Root | Shrikanth et al. (2017) |
| 179. | *Bouvardia ternifolia* Schltdl | Rubiaceae | Hexane, methanol | Scorpion | Root | Jiménez-Ferrer et al. (2005) |
| 180. | *Gonzalagunia panamensis* (Cav.) K. Schum. | Rubiaceae | | Snake | Leveas | Otero et al. (2000b); Núñez et al. (2004b); Gomes et al. (2010) |
| 181. | *Mitragyna stipulosa* (DC.) Kuntze | Rubiaceae | Ethanol | Snake | Bark | Fatima et al. (2002) |

**TABLE 10.1** *(Continued)*

| S/ No. | Names of plants | Family | Extraction solvent | Source of envenomation | Parts used | References |
|---|---|---|---|---|---|---|
| 182. | *Schumanniophyton magnificum* Harms | Rubiaceae | Methanol, aqueous | Snake | Stem, root bark | Houghton et al. (1992); Houghton and Skari (1994) |
| 183. | *Ophiorrhiza mungos* L. | Rubiaceae | Aqueous | Snake | Root | Krishnan et al. (2014) |
| 184. | *Genipa americana* L. | Rubiaceae | Aqueous | Snake | Bark | Bernard et al. (2001) |
| 185. | *Uncaria tomentosa* (Willd. ex Schult.) DC. | Rubiaceae | Aqueous | Snake | Root | Badilla et al. (2006) |
| 186. | *Citrus limon* (L.) Osbeck | Rutaceae | Ethanol | Snake | Ripe fruits | Otero et al. (2000a); Núñez et al. (2004b) |
| 187. | *Casearia sylvestris* | Salicaceae | Aqueous | Snake | Leaves | Mors et al. (2000); Soares et al. (2004) |
| 188. | *Casearia mariquitensis* Kunth | Salicaceae | Aqueous | Snake | Leaves | Izidoro et al. (2003) |
| 189. | *Azima tetracantha* Lam. | Salvodoraceae | Ethylacetate | Snake | Leaves | Janardhan et al. (2014) |
| 190. | *Cardiospermum halicacabum* L. | Sapindaceae | Aqueous | Snake | Tender shoots | Chandra et al. (2011) |
| 191. | *Sapindus saponaria* L. | Sapindaceae | Ethanol, ethyl acetate, aqueous | Snake | Aerial | Castro et al. (1999) |
| 192. | *Serjania erecta* Radlk. | Sapindaceae | Methanol | Snake | Stem, leaf | Fernandes et al. (2011) |
| 193. | *Sapindus rarak* DC. | Sapindaceae | Aqueous | Scorpion | Leaves | Uawonggul et al. (2006) |
| 194. | *Glycoxylon huberi* Ducke | Sapotaceae | | Snake | Bark | Bernard et al. (2001) |
| 195. | *Selaginella articulate* (Kunze) Spring | Selaginellaceae | Ethanol | Snake | Whole plant | Otero et al. (2000b) |
| 196. | *Picrasma quassioides* (D. Don) Benn. | Simaroubaceae | | Snake | Plant | Liang (1987) |
| 197. | *Siparuna thecaphora* (Poepp. & Endl.) A. DC. | Siparunaceae | Ethanol | Snake | Whole plant | Otero et al. (2000b) |

**TABLE 10.1** *(Continued)*

| S/No. | Names of plants | Family | Extraction solvent | Source of envenomation | Parts used | References |
|---|---|---|---|---|---|---|
| 198. | *Smilax cuculmeca* | Smilacaceae | Ethanol, ethyl acetate, aqueous | Snake | Aerial | Castro et al. (1999) |
| 199. | *Withania somnifera* (L.) Dunal | Solanaceae | Aqueous | Scorpion | Leaf, root | Machiah and Gowda (2002) |
| 200. | *Brunfelsia uniflora* (Pohl) D. Don | Solanaceae | Aqueous | Snake | Leaves | Pereira et al. (1991) |
| 201. | *Capsicum frutescens* | Solanaceae | Ethanol | Snake | Ripe fruits | Otero et al. (2000b) |
| 202. | *Symplocos cochinchinensis* (Lour.) S. Moore ssp. laurina | Symplocaceae | Methanol | Snake | Leaves | Lakshmi and Vadivu (2010) |
| 203. | *Symplocos racemosa* Roxb. | Symplocaceae | Methanol | Snake | Whole plant | Ahmad et al. (2003) |
| 204. | *Thea sinensis* L. | Theaceae | Extract | Snake | Whole plant | Hung et al. (2004) |
| 205. | *Cecropia palmate* Willd. | Urticaceae | | Snake | Leaves, latex | Bernard et al. (2001) |
| 206. | *Loasa speciosa* | Urticaceae | Aqueous | Snake | Leaves | Badilla et al. (2006) |
| 207. | *Stachytarpheta dichotoma* (Ruiz & Pav.) Vahl | Verbenaceae | Aqueous | Snake | Whole plant | Pereira et al. (1991) |
| 208. | *Stachytarpheta jamaicensis* (L.) Vahl | Verbenaceae | Aqueous | Snake | Leaves | Castro et al. (1999) |
| 209. | *Clerodendrun viscosum* | Verbenaceae | Alcoholic | Snake | Root | Lobo et al. (2006) |
| 210. | *Cissus sicyoides* L. | Vitaceae | | Snake | Creeper | Bernard et al. (2001) |
| 211. | *Vitis vinifera* L. | Vitaceae | Methanol | Snake | Seed | Mahadeswaraswamy et al. (2008) |
| 212. | *Cissus assamica* | Vitaceae | Ethanol | Snake | Root | Yang et al. (1998) |
| 213. | *Renealmia alpinia* Rottb. | Zingiberaceae | Ethanol | Snake | Leaves Rhizome | Otero et al. (2000a); Patiño et al. (2012); Patiño et al. (2013) |
| 214. | *Curcuma longa* L. | Zingiberaceae | Aqueous | Snake | Rhizome | Ferreira et al. (1992); Chethankumar and Srinivas (2008) |

**TABLE 10.1** *(Continued)*

| S/ No. | Names of plants | Family | Extraction solvent | Source of envenomation | Parts used | References |
|---|---|---|---|---|---|---|
| 215. | *Curcuma aromatic* Salisb | Zingiberaceae | Methanol | Snake | Root | Alam (2014) |
| 216. | *Curcuma zedoaria* (Christm.) Roscoe | Zingiberaceae | Methanol | Snake | Roots | Alam (2014) |
| 217. | *Curcuma zedoaroides* A. Chav. & Tanee | Zingiberaceae | Acetone | Snake | Rhizome | Lattmann et al. (2010) |

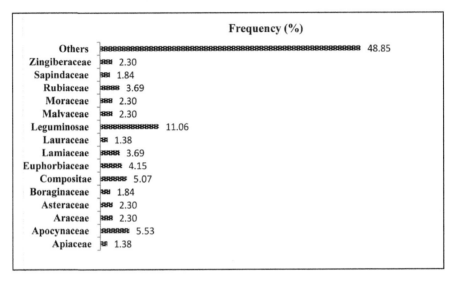

**FIGURE 10.1** Medicinal plant family for the treatment of envenomation.

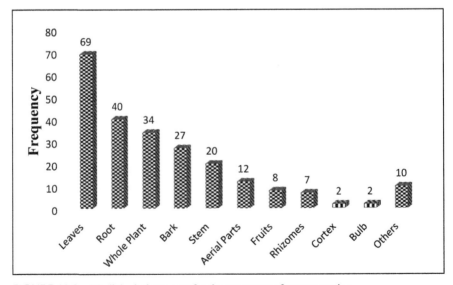

**FIGURE 10.2** Medicinal plant parts for the treatment of envenomation.

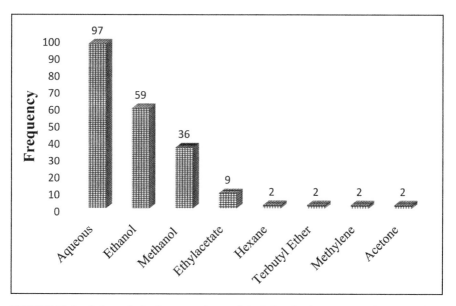

**FIGURE 10.3**   Solvent choice for processing medicinal plant for the treatment of envenomation.

integrity (Nanjaraj et al., 2013). Studies using plant-based preparations have shown the extent to which they can be used to treat complications arising from snake envenomations. In contrast, their purified forms express direct interaction of venom toxins with identified compounds in the crude extract. This helps to reduce toxicity and increase the survival time of envenom experimental animals. Therefore, making pure compounds isolated from plants have higher popularity over crude extracts in venom research (Nanjaraj et al., 2013). Several of such compounds isolated from medicinal plants have shown good inhibitory properties against snake or scorpion venom (Table 10.2, Fig. 10.4). These isolated compounds and their plant sources are discussed further.

Tannins extracted from persimmon the fruit of *Diospyros kaki* (Ebenaceae) has been reported to improve the survival time of mice treated with *Laticauda semifasciata* and *Trimeresurus flavoviridis* venoms. They also inhibited the swelling produced in the feet of mice treated with *L. semifasciata* (Okonogi et al., 1979). The mechanism by which tannins act as inhibitors of the activity of phospholipases and/or neutralization of the proteolysis is probably through the formation of hydrogen bonds in the active site of the enzyme site, by binding at the site of cation cofactor, or probably by the chelation of enzyme cofactors (Gutiérrez and Ownby, 2003;

**TABLE 10.2**  Antivenom Active Principles Isolated from Medicinal Plants.

| S/No. | Medicinal Plant Name | Family | Active Principles | References |
|---|---|---|---|---|
| 1. | *M. indica* L. | Anacardiaceae | Pentagalloylglucopyranose | Pithayanukul et al. (2009)1 |
| 2. | *M. indica* L. | Anacardiaceae | Methyl gallate | Pithayanukul et al. (2009) |
| 3. | *M. indica* L. | Anacardiaceae | Gallic acid | Pithayanukul et al. (2009) |
| 4. | *Angelica sinensis* | Apiaceae | Vanillic acid | Dhananjaya et al. (2006) |
| 5. | *Pimpinell anisum* L. | Apiaceae | Hydroxy anisic acid | Alam and Gomes (1998c) |
| 6. | *H. indicus* (L.) R. Br. ex Schult. | Apocynaceae | Lupeol | Chatterjee et al. (2006) |
| 7. | *H. indicus* (L.) R. Br. ex Schult. | Apocynaceae | 2-hydroxy-4-methoxy-benzoic acid | Alam and Gomes (1998a) |
| 8. | *T. catharinensis* | Apocynaceae | 12-methoxy-4-methylvoachalotine | Batina et al. (2000) |
| 9. | *A. elegans* Mast. | Aristolochiaceae | Neolignans | Zamilpa et al. (2014) |
| 10. | *Aristolochia radix* | Aristolochiaceae | Aristolochic acid | Chandra et al. (2002) |
| 11. | *G. sylvestre* | Asclepiadaceae | Gymnenic acid | Kini and Gowda (1982a); Kini and Gowda (1982b) |
| 12. | *P. daemia* (Forsk.) chiov. | Asclepiadaceae | β-sitosterol | Raghavamma et al. (2016) |
| 13. | *B. trimera* (Less) DC | Asteraceae | Clerodane | Januario et al. (2004) |
| 14. | *B. alba* L. | Betulaceae | Betulinic acid | Bernard et al. (2001); Soares et al. (2005) |
| 15. | *A. argentea* | Boraginaceae | Rosmarinic acid | Aung et al. (2010) |
| 16. | *C. verbenacea* A. DC. | Boraginaceae | Rosmarinic acid | Ticli et al. (2005) |
| 17. | *E. buxifolia* | Boraginaceae | Ehretianone | Selvanayagam et al. (1996) |
| 18. | *C. leprosum* Mart. | Combretaceae | Arjunolic acid | Fernandes et al. (2014) |
| 19. | *E. alba* (L.) Hassk. | Compositae | Wedelolactone | Diogo et al. (2009) |
| 20. | *E. alba* (L.) Hassk. | Compositae | Demethylwedelolactone | Diogo et al. (2009) |

**TABLE 10.2** *(Continued)*

| S/No. | Medicinal Plant Name | Family | Active Principles | References |
|---|---|---|---|---|
| 21. | *E. prostrata* (L.) L. | Compositae | β-sitosterol | Gomes et al. (2007) |
| 22. | *E. prostrata* (L.) L. | Compositae | Wedelolactone | Melo and Ownby (1999) |
| 23. | *P. indica* Less. | Compositae | Stigmasterol | Gomes et al. (2007) |
| 24. | *P. indica* Less. | Compositae | β-sitosterol | Gomes et al. (2007) |
| 25. | *I. asarifolia* (Desr.) | Convolvulaceae | Rutin | Lindahl and Tagesson (1997); Lima et al. (2014) |
| 26. | *E. officinalis* Gaertn. | Euphorbiaceae | Di-iso-butyl phthalate | Sarkhel et al. (2011) |
| 27. | *E. hirta* L. | Euphorbiaceae | Quercetin-3-O-rhamnoside | Gopi et al. (2016) |
| 28. | *Tragia involucrate* | Euphorbiaceae | 2,4-dimethylhexane, 2-methylnonane, 2,6-dimethylheptane | Samy et al. (2012) |
| 29. | *Pueraria lobata* (Kudzu) | Fabaceae | Stigmasterol, curcumin, tectoridin | Nirmal et al. (2008) |
| 30. | *Glycine max* | Fabaceae | Genistein | Dharmappa et al. (2010) |
| 31 | *Glycyrrhiza glabra* | Fabaceae | Glycyrrhizin | Assafim et al. (2006) |
| 32. | *M. pudica* | Fabaceae | Mimosine | Mahadeswaraswamy et al. (2011) |
| 33. | *Hydrangea macrophylla* | Hydrangeaceae | Hydrangenol | Kakegawa et al. (1988) |
| 34. | *L. aspera (Willd.) Link* | Lamiaceae | Leucasin | Sakthivel et al. (2013) |
| 35. | *Sideritis mugronensis* | Lamiaceae | Hypolaetin-8-glucoside | Alcaraz and Hoult (1985) |
| 36. | *V. negundo* | Lamiaceae | Tris (2,4-di-tert-butylphenyl) phosphate | Vinuchakkaravarthy et al. (2011) |
| 37. | *S. parahyba* (Vell.) S. F. Blake | Leguminosae | Myricetin-3-O-glucoside | Vale et al. (2011) |
| 38. | *S. parahyba* (Vell.) S. F. Blake | Leguminosae | Gallocatechin | Vale et al. (2011) |
| 39. | *B. intermedia* | Leguminosae | (-)-Edunol | Reyes-Chilpa et al. (1994) |

**TABLE 10.2** *(Continued)*

| S/No. | Medicinal Plant Name | Family | Active Principles | References |
|---|---|---|---|---|
| 40. | *B. podalyrioides* Kunth | Leguminosae | (-)-Edunol | Reyes-Chilpa et al. (1994) |
| 41. | *H. brasiliana* Benth. | Leguminosae | Cabenegrin A-I | Da Silrva et al. (1997) |
| 42. | *H. brasiliana* Benth. | Leguminosae | Cabenegrin A-II | Da Silrva et al. (1997) |
| 43. | *H. brasiliana* Benth. | Leguminosae | Edunol | Da Silrva et al. (1997) |
| 44. | *P. macroloba* (Willd.) Kuntze | Leguminosae | Macrolobin-A | da Silva et al. (2007) |
| 45. | *P. macroloba* (Willd.) Kuntze | Leguminosae | Macrolobin-B | da Silva et al. (2007) |
| 46. | *B. crassa* Nied. | Malpighiaceae | Myricetin | Nishijima et al. (2009) |
| 47. | *B. crassa* Nied. | Malpighiaceae | Amenthoflavone | Nishijima et al. (2009) |
| 48. | *Marchantia berteroana* | Marchantiaceae | Isoscutellarein | Alcaraz and Hoult (1985) |
| 49. | *B. ndellensis* Beille | Phyllanthaceae | Quinovic acid | Mostafa et al. (2006) |
| 50. | *P. longum* L. | Piperaceae | Piperine | Shenoy et al. (2013) |
| 51. | *P. peltatum* L. | Piperaceae | 4-nerolidylcatechol | Núñez et al. (2005) |
| 52. | *P. umbellatum* L. | Piperaceae | 4-nerolidylcatechol | Núñez et al. (2005) |
| 53. | *B. floribunda* | Polygalaceae | Bredemeyerosides B, bredemeyerosides D | Daros et al. (1996); Pereira et al. (1996) |
| 54. | *M. stipulosa* (DC.) Kuntze | Rubiaceae | Quinovic acid | Fatima et al. (2002) |
| 55. | *S. magnificum* Harms | Rubiaceae | Schumanniofoside | Akunyili and Akubue (1986) |
| 56. | *C. sylvestris* | Salicaceae | Ellagic acid | da Silva et al. (2008) |
| 57. | *C. halicacabum* L. | Sapindaceae | Berberine | Chandra et al. (2011) |
| 58. | *W. somnifera* (L.) Dunal | Solanaceae | Glycoprotein | Machiah and Gowda (2002); Machiah et al. (2006) |
| 59. | *Anisodus tanguticus* | Solanaceae | Anisodamine | Li et al. (1999) |

**TABLE 10.2** *(Continued)*

| S/No. | Medicinal Plant Name | Family | Active Principles | References |
|---|---|---|---|---|
| 60. | *S. racemosa* Roxb. | Symplocaceae | Benzoylsalireposide | Ahmad et al. (2003) |
| 61. | *S. racemosa* Roxb. | Symplocaceae | Salireposide | Ahmad et al. (2003) |
| 62. | *C. assamica* | Vitaceae | Resveratrol | Yang et al. (1998) |
| 63. | *C. longa* L. | Zingiberaceae | Ar-turmerone | Ferreira et al. (1992) |
| 64. | *C. longa* L. | Zingiberaceae | Turmerin | Chethankumar and Srinivas (2008) |
| 65. | *C. zedoaroides A. Chav. & Tanee* | Zingiberaceae | Dialdehyde | Lattmann et al. (2010) |
| 66. | *R. alpinia* Rottb. | Zingiberaceae | Pinostrobin | Gómez-Betancur et al. (2014) |

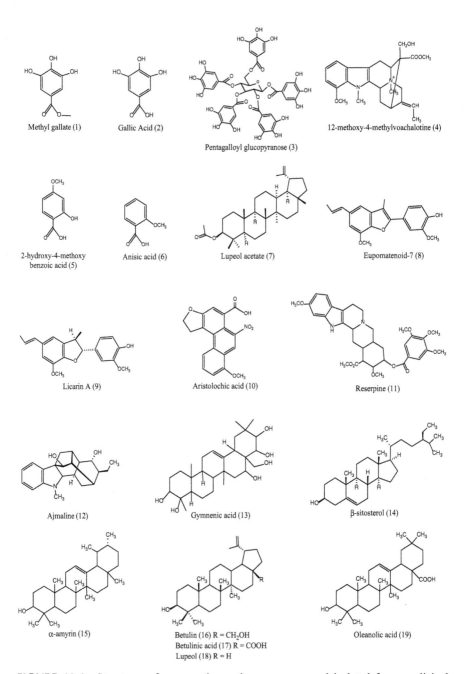

**FIGURE 10.4** Structures of some active antivenom compound isolated from medicinal plants.

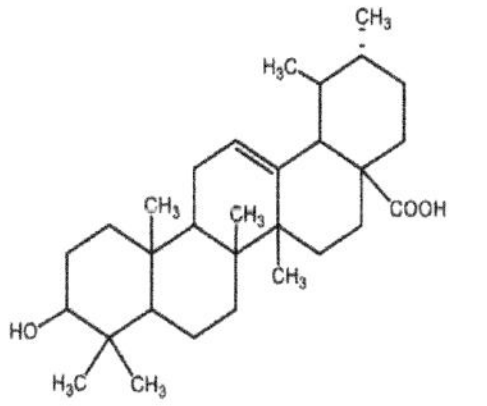

Ursolic acid (20)

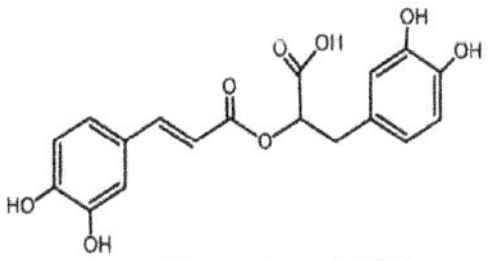

Rosmarinic acid (21)

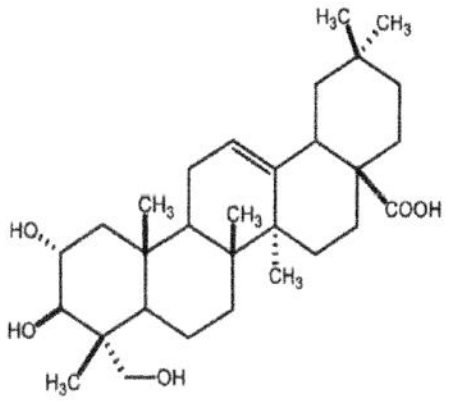

Arjunolic acid (22)

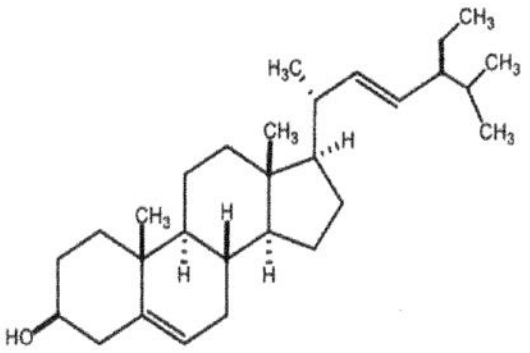

Stigmasterol (23)

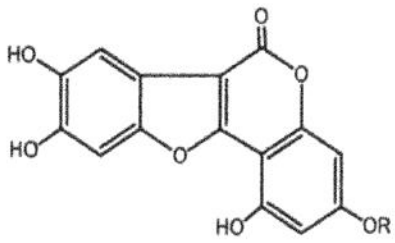

Wedelolactone (24) R = CH$_3$
Demethylwedelolactone (25) R = H

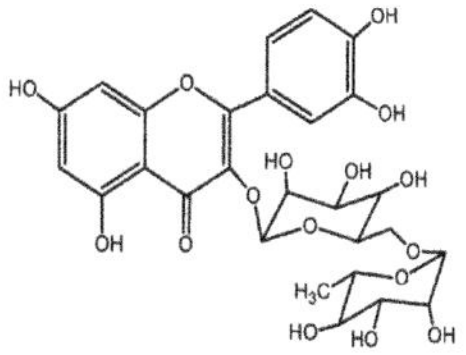

Rutin (26)

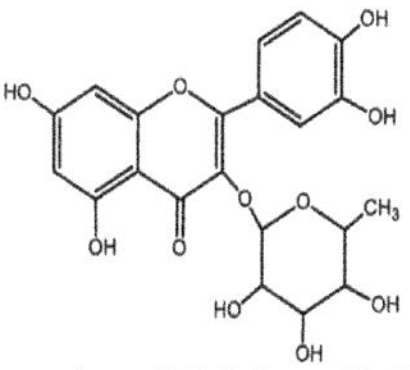

Quercetin-3-O-rhamnoside (27)

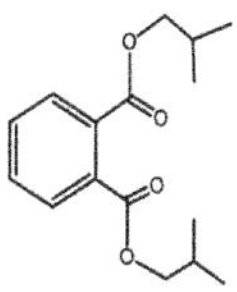

Iso-butyl phthalate (28)

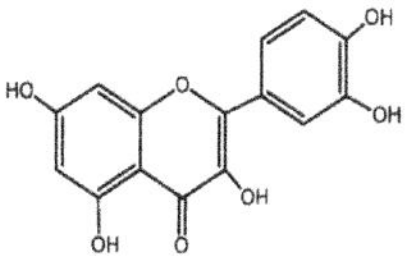

Quercetin (29)

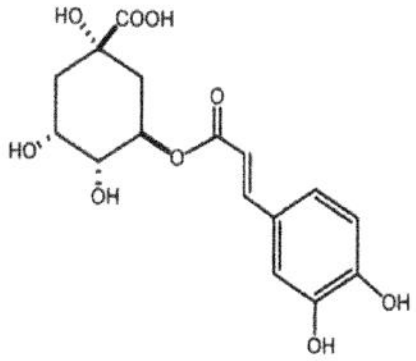

Chlorogenic acid (30)

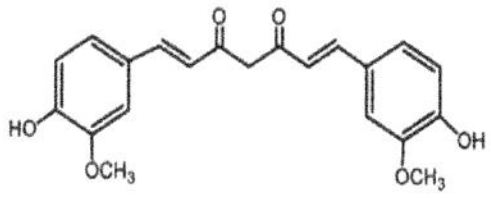

Curcumin (31)

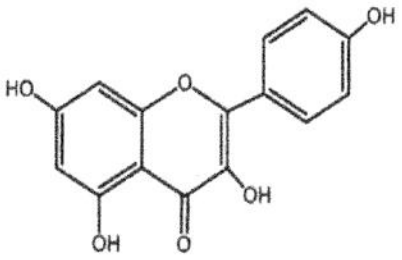

Kaemferol (32)

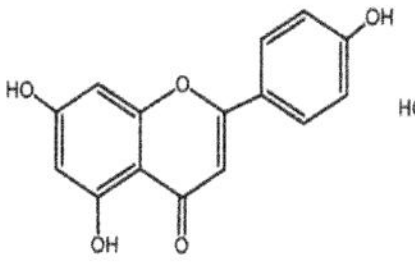

Apigenin (33)

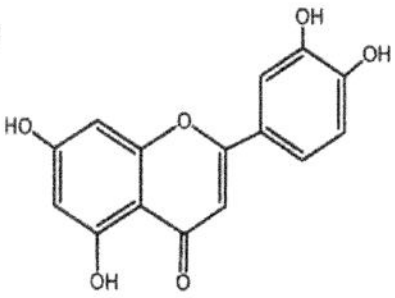

Luteolin (34)

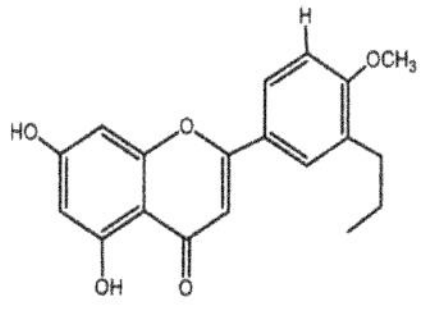

Leucasin (35)

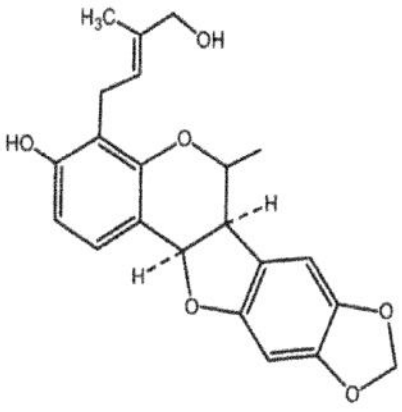

Cabenegrin A-I (36)

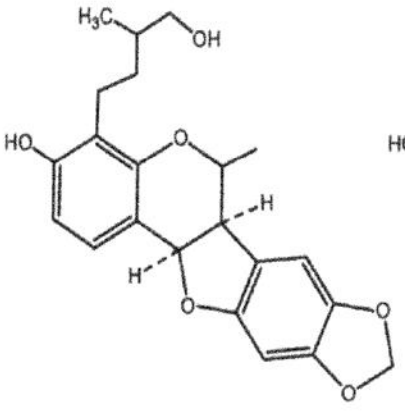

Cabenegrin A-II (37)

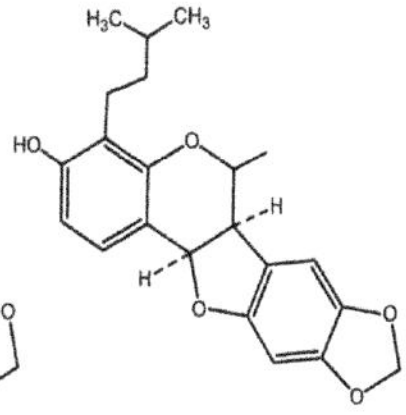

Edunol (38)

**FIGURE 10.4** *(Continued)*

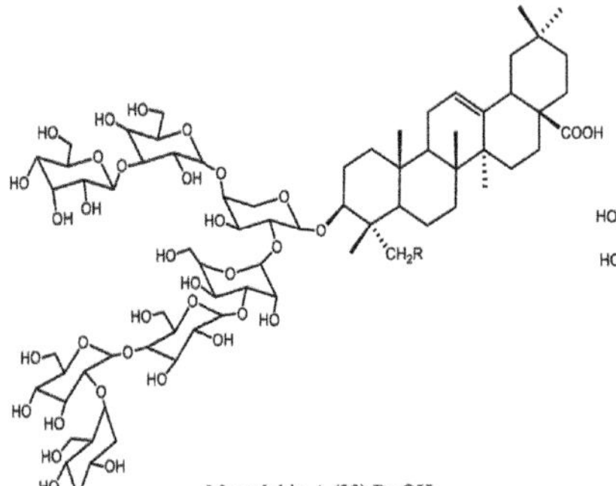

Macrolobin-A (39) R=OH
Macrolobin-B (40) R=H

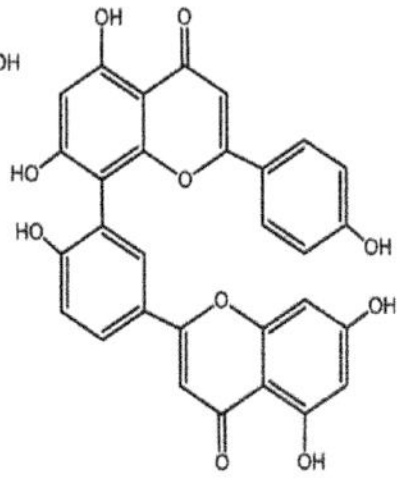

Amenthoflavone (41)

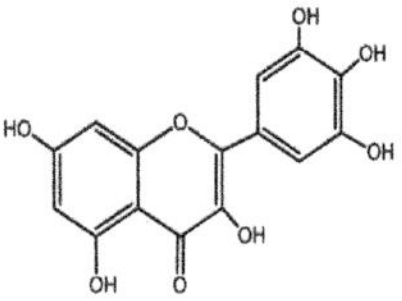

Myricetin (42)

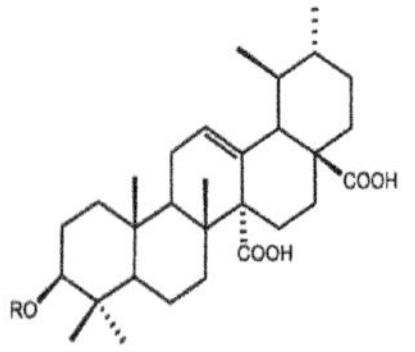

Quinovic acid-L-rhamnose (43)
Quinovic acid-D-fucose (44)
Quinovic acid-D-glucopyranosyl-β-D-fucose (45)

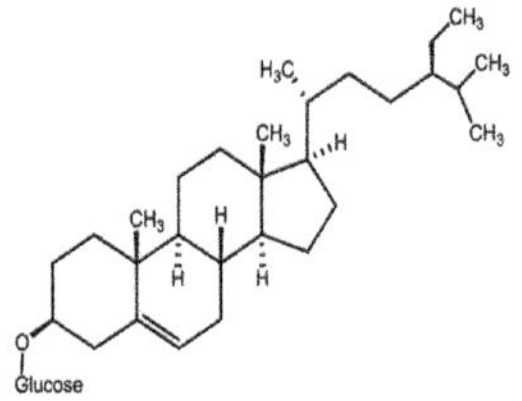

β-sitosterol glucopyranoside (46)

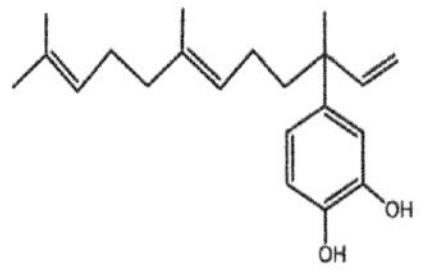

4-nerolidylcatechol (47)

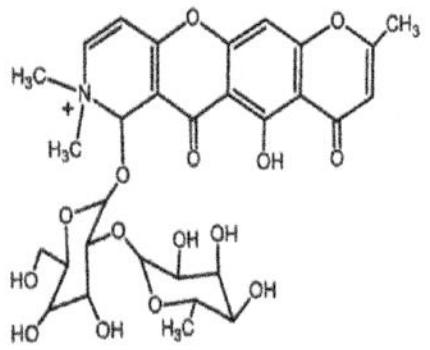

Schumanniofoside (48)

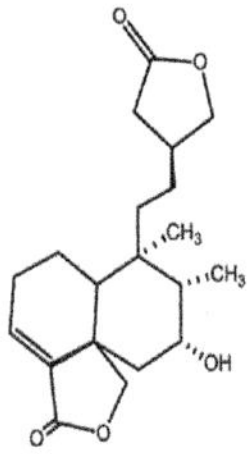

Clerodane diterpenoid (49)

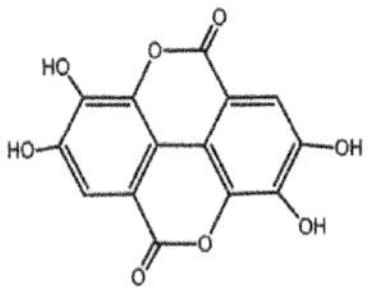

Ellagic acid (50)

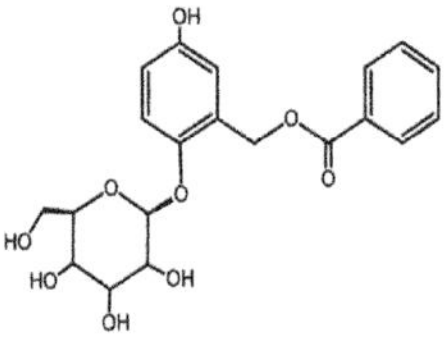

Salireposide (51)

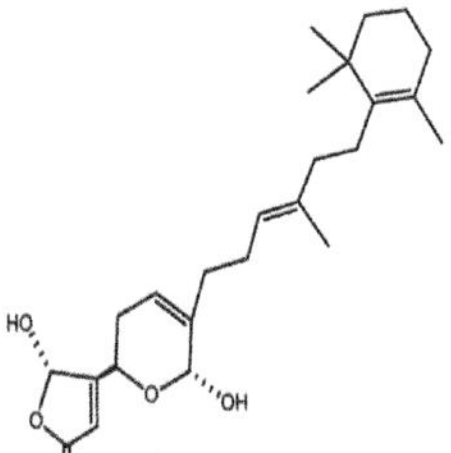

Manoalide (52)

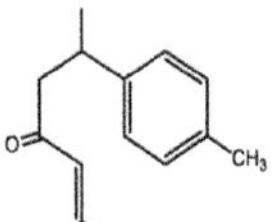

ar-Turmerone (53)

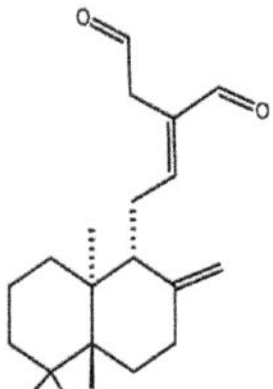

trans-Curcuma C20 dialdehyde (54)

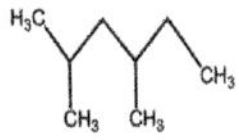

2,4-dimethyl hexane (55)

**FIGURE 10.4**   *(Continued)*

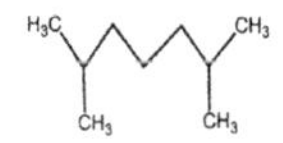

2,6-dimethyl heptane (56)     2-methyl nonane (57) 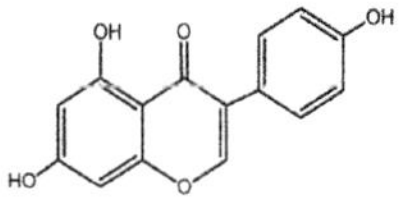    Genistein (58) 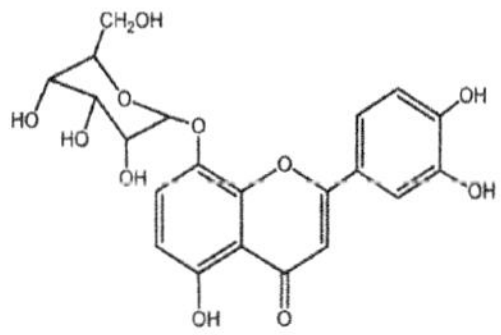    Hypolaetin-8-glucoside (59)

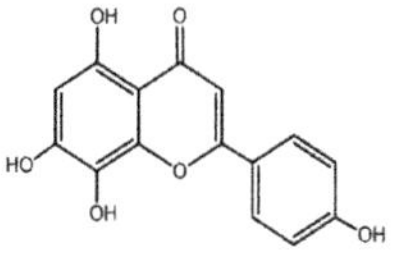

Isoscutellarein (60) 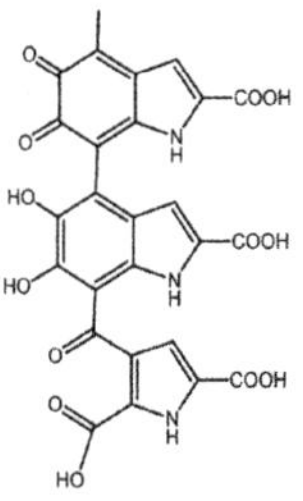    Melanin (61) 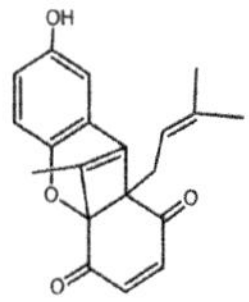    Ehretianone (62) 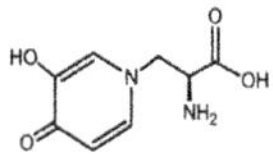    Mimosine (63)

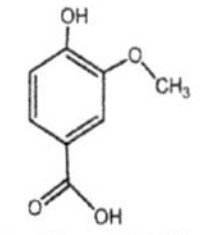

Vanillic acid (64) 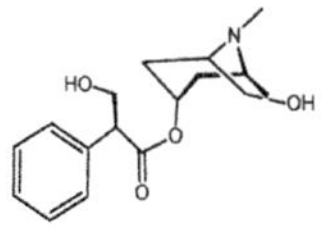    Anisodamine (65) 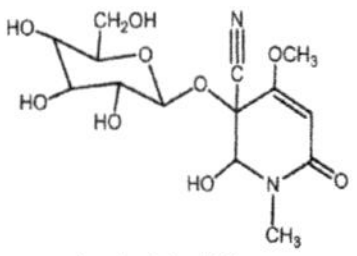    Acalyphin (66) 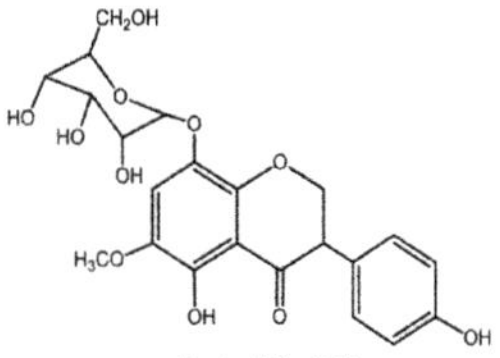    Tectoridin (67)

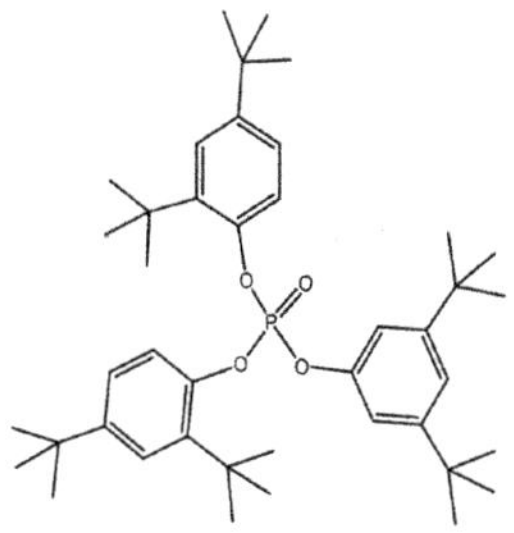

Tris (2,4-di-tert-butylphenyl) phosphate (68) 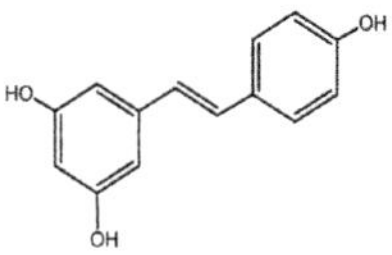    Resveratrol (69) 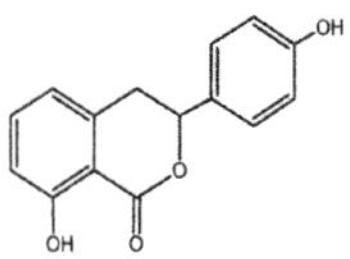    Hydrangenol (70)

**FIGURE 10.4**    *(Continued)*

Leanpolchareanchai et al., 2009). Phenolics, on the other hand, are known for their ability to bind with proteins, especially their complex polyphenolic forms. Methyl gallate (1), gallic acid (2), and pentagalloylglucopyranose (3) are phenolics isolated from the ethanolic seed kernel extract of Thai mango (*Mangifera indica* L.). Their inhibitory effects were studied against the proteolytic effects of *Calloselasma rhodostoma* and *Naja naja kaouthia* venoms (Pithayanukul et al., 2009). The pentagalloylglucopyranose (tannic acid) was found to exalt potent activity in a dose-dependent manner against caseinolytic and fibrinogenolytic activities of the Malayan pit viper and Thai cobra venoms in an in vitro test.

The Apocynaceae families are fairly represented in plant families identified with antivenom properties. Lupeol, 2-hydroxy-4-methoxy-benzoic acid, and 12-methoxy-4-methylvoachalotine are among the phytochemicals that have been isolated and characterized from the Apocynaceae. 12-methoxy-4-methylvoachalotine (4) isolated from *Tabernaemontana catharinensis* have been reported to inhibit myotoxic phospholipases $A_2$ (PLA$_2$) activities (Batina et al., 2000). PLA$_2$s are multitoxic class of enzymes that act locally and systematically to induce a wide range of pathophysiological alterations in the envenomed victim (Gowda and Middlebrook, 1994; Gutiérrez and Ownby, 2003). They are mytotoxic when they exert their actions on skeletal muscles. The indole alkaloid isolates of *T. catharinensis* reduce the lethality demonstrated by *Crotalus durissus terrificus* venom in vivo (Batina et al., 2000). Similarly, from the Apocynaceae was isolated a hydroxy anisic acid (2-hydroxy-4-methoxy-benzoic acid) (5) from the methanolic root extract of *Hemidesmus indicus* (Alam and Gomes, 1998a). When tested in vivo against the venomic effects of viper venom-induced inflammation, it effectively neutralizes the inflammation, hemorrhage, fibrinolytic, and lethality of the venom (Alam and Gomes, 1998b). The compound similarly reduces the free radicals generated by the snake venom, expressing its antioxidant activities. Hydroxy anisic acid was observed to be corrosive in nature, even though it is effective in the management of bite from Russell's viper. However, anisic acid (6) which lacks the phenolic hydroxyl group is noncorrosive and nontoxic which also maintained the potency to neutralize lung hemorrhage induced by PLA$_2$ VRV-PL-VIIIa of Russell's viper venom. Co-crystallization of anisic acid with PLA$_2$ shows that anisic acid binds at the active site of PLA$_2$ (Sekar et al., 2003).

Lupeol acetate (7) has been isolated from the methanolic extract of the roots of *H. indicus* (L.) (Chatterjee et al., 2006). The isolated compound was able to significantly neutralize the lethality of *Daboia russelli* venom. It reduces the observed defibrinogenation, edema, and hemorrhage induced by

*D. russelli* venom. In the studies, *D. russelli* venom-induced changes in lipid peroxidation and superoxide dismutase activities, which were neutralized by the administration of lupeol acetate. Lupeol acetate also inhibited the lethality, hemorrhage, defibrinogen in addition to their antivenom activities. Furthermore, lupoel derivatives have similarly been reported as a better anti-inflammatory drug compared with indomethacin (Reddy et al., 2009). In in vivo, in vitro, and in silico studies by Katkar et al. (2015), synthetic lupeol derivative efficiently neutralized the *Echis carinatus* venom-induced hemorrhage and dermo/myonecrosis, neutralizing their venom toxins and protecting collagen from venomous damages.

*Aristolochia elegans*, *Bouvardia ternifolia*, and *Vitex mollis* are among the medicinal plants used by Mexicans, traditionally for the treatment of scorpion stings. In their in vivo antitoxin activities, only *A. elegans* and *B. ternifolia* extracts were able to modify the lethality induced by scorpion venom (Jiménez-Ferrer et al., 2005). Further, in the search for the active antivenom compounds, neolignans of several forms were isolated from *A. elegans*, which were the active principles (s) responsible for *A. elegans* ability to neutralize the effects of *Centruroides limpidus limpidus* scorpion venom. In the n-hexane root fraction, two neolignans; eupomatenoid-7 (8) and licarin A (9) were isolated, characterized, and identified (Zamilpa et al., 2014). Other isolated compounds could not produce the antivenom activities of these two compounds.

Alkaloids are an important class of compounds with analgesic properties. Aristolochic acid (10), an alkaloid extracted from the *Aristolochia radix* demonstrated antivenom activities against *Vipera russelli* venom. It inhibited the enzymatic and pharmacological activities of $PLA_2$ enzymes, its edematonic effects, hyaluronidase spreading property, hemorrhagic, and extended the survival time of experimental animals (Girish and Kemparaju, 2005; Girish and Kemparaju, 2006). Alongside aristolochic acid, other alkaloids with *Naja naja* venom hyaluronidase activities include reserpine (11) and ajmaline (12) (Vishwanath et al., 1988; Chandra et al., 2002). The inhibitory activities of ajmaline and reserpine were partial while aristolochic acid completely inhibited the activity in a noncompetitive manner.

Gymnenic acid (13) and β-sitosterol (14) are two antivenom compounds isolated from the Asclepiadaceae. Gymnenic acid is a triterpenoid glycoside extracted from *Gymnema sylvestre*, a plant known in Indian ethnobotany for the treatment of snake envenomation. Its potassium salt inhibits ATPase activities from *Naja naja* venom (Kini, and Gowda, 1982a; Kini, and Gowda, 1982b). Raghavamma et al. (2016) reported the presence of β-sitosterol, lupeol, β-amyrin, and α-amyrin (15) from the leaf extract of

*Pergularia daemia* (Forsk.) chiov. through gas chromatography–mass spectrometry analysis. Among the phytochemicals, lupeol acetate has already established antivenom activities (Chatterjee et al., 2006), while their in silico studies with β-sitosterol showed that it can bind to the active site of PLA$_2$ and L-amino acid oxidase (Raghavamma et al., 2016).

Betulin (16) and betulinic acid (17) are among pentacyclic triterpenes that have demonstrated good binding energy in silico inhibitory effect on PLA$_2$ enzyme with betulinic acid having a better inhibitory activity, interacting directly with the catalytic site of the enzyme (Bernard et al., 2001). Both betulinic acid and betulin have been isolated from *Betula alba*. Other pentacyclic triterpenes from plant sources with activities against snake venom include lupeol (18), friedelin, oleanolic acid (19), taraxasterol, taraxerol, ursolic acid (20), and α-amyrin (Mors et al., 2000; Nataraju et al., 2007).

*Cordia verbenacea* (Cv) (Boraginaceae) is medicinally active against snake venom. Its methanolic extract effectively inhibited induced edema and myotoxic activities by *Bothrops jararacussu* snake venom (Ticli et al., 2005). An ester of caffeic acid and 3, 4-dihydroxyphenyllactic acid (2-*O*-cafeoil-3-(3, 4-di-hydroxy-phenyl)-*R*-lactic acid) better known as rosmarinic acid was isolated as the active principle. Rosmarinic acid (21) has also been isolated from leaves of *Argusia argentea* (Aung et al., 2010). Its activities against snake venom was demonstrated in vivo through its ability to inhibit hemorrhage induced by crude venom of *T. flavoviridis*, *Crotalus atrox*, *Gloydius blomhoffii,* and *Bitis arietans* (Aung et al., 2010).

Fernandes et al (2014) investigated the effect of two methods of treatment with *Combretum leprosum* extract and one of its isolated compounds; arjunolic acid (22) against the venomic activities of *B. jararacussu* and *Bothrops jararaca* venoms. Investigations were carried out on their effects on collagenase, edematogenic, hemorrhagic, hyaluronidase, phospholipase, myotoxic, procoagulant, proteolytic, and lethal activities induced by these venoms in Swiss mice. In the pretreatment methods, arjunolic acid presented active antivenom activities, reducing the lethality of *B. jararacussu* up to 70% and preincubation prevented mortality of all experimental animals (Fernandes et al., 2014). The extract when administered to the animals decreased the edema formation in preincubation and pretreatment. Both crude extract and isolated compound presented significant effects of some of the enzymatic activities from snake venom in a concentration-dependent manner (Fernandes et al., 2014).

Gomes et al. (2007) isolated two active antivenom compounds, namely β-sitosterol and stigmasterol (23) from the methanolic root extract of *Pluchea indica* (less). The methanolic extract was found to neutralize the hemorrhagic, defibrinogenation, edema, and lethality posed by exposure to viper venom. Other activities observed include the active fraction ability to ameliorate the induced change in lipid peroxidation and superoxide dismutase activities. From the methanolic extract of *Eclipta alba*, two coumestans were similarly isolated. Wedelolactone (24) and demethylwedelolactone (25) are active against the mytoxic activities from *durissus terrificus* and *B. jararacussu* basic PLA$_2$ enzyme (Diogo et al., 2009). Melo and Ownby (1999) have also demonstrated the inhibitory effects of wedelolactone isolated from the *Eclipta prostrata* plant. Wedelolactone was active against South American crotalid venoms among which are *B. jararaca, B. jararacussu, C. d. terrificus, Lachesis muta,* as well as against the North American *Agkistrodon contortrix* and crotalids *Crotalus viridis viridis* (Martz, 1992; Melo et al., 1994). Similarly, wedelolactone analog was reported to inhibit myotoxicity induced by *B. jararacussu* venom, where it antagonizes the creatine kinase (CK) release from the skeletal muscle (da Silva et al., 2001). In other studies by Melo et al. (1994), the antihemorrhagic and anti-myotoxic effects of sitosterol, stigmasterol, and wedelolactone isolated form *E. prostrata* were investigated. Wedelolactone effectively neutralized the in vitro myotoxic effects of *B. jararaca, Bothrops jararacuçu,* and *L. muta* venoms when compared to stigmasterol and sitosterol. In the in vivo myotoxic studies, crude extract of *E. prostrata* and wedelolactone also effectively inhibited the crotalid venoms by their preincubation. Similarly, the intravenous administration with wedelolactone to mice attenuated the increase in activities of plasma CK. Aside *E. prostrata* extract, pure wedelolactone also inhibited the hemorrhagic and proteolytic activities caused by *B. jararaca* venom. *E. prostrata* extract was also tested using the venoms from *C. rhodostoma* Kuhl, the Malayan pit viper venom (Pithayanukul et al., 2004). The partially purified ethyl acetate extract was effective in a dose-dependent manner against the proteolytic and hemorrhagic activities of Malayan pit viper venom. Pithayanukul et al. (2004) reported wedelolactone (47%) as the major constituent in *E. prostrata* ethyl acetate extract.

The presence of flavonol glycoside, rutin (26) has been established in ethyl acetate fraction of the leaf extract of *Ipomoea asarifolia* (Lima et al., 2014). When evaluated for its possible anti-scorpion and anti-snake venom activities, rutin was effective against inflammations posed by scorpion venom, reduce leukocytes migration and cytokines expression caused by *Tityus serrulatus* venom in male and female BALB/c mice (Lima et al., 2014). Rutin also expresses anti-snake venom activities by efficiently inhibiting PLA$_2$ enzymes from *Vipera russelli* and *C. atrox*, but to a lesser extent when compared to PLA$_2$ enzymes from *Naja naja* (Lindahl and Tagesson, 1997).

Among the identified compound from the Euphorbiaceae family are quercetin-3-O-rhamnoside (27) and Di-iso-butyl phthalate (28). Quercetin-3-O-rhamnoside has both quercetin plus a rhamnose ring, identified and extracted from the whole plant methanolic extract of the *Euphorbia hirta* (Gopi et al., 2016). Its activities show a significant inhibition of 93% against hyaluronidase from *Naja naja* venom in a dose-dependent manner. Snake venom hyaluronidases are the major enzymes responsible for the degradation of hyaluronic acid (glycosaminoglycans) of extracellular matrix. They also degrade connective tissues surrounding the blood vessels, capillaries, and smooth muscles, thereby resulting in the loss of structural integrity of their target organs. Hyaluronidases are, therefore, referred to as the "spreading factors" because of their ability to facilitate diffusion of venom toxins from bitten site to target organ regions (Girish et al., 2004). Quercetin-3-O-rhamnoside significantly inhibited hyaluronidase from *Naja naja* venom and similarly inhibited completely the hemolytic, hemorrhage, phospholipase-A2 and protease activity induced by the venom at a ratio of 1:20 w/w of venom: quercetin-3-O-rhamnoside. Flavanoids possess good anti-allergic, anti-inflammatory, antioxidant properties and have been reported to inhibit the activities of snake venom PLA$_2$ and hyaluronidase activities (Kuppusamy et al., 1990; Kuppusamy and Das, 1991; Mors et al., 2000; Girish and Kemparaju, 2005). Quercetin (29) alone, has been reported to completely inhibit snake hyaluronidase enzyme activities. Another example of flavonoids with antivenom activities includes flavone, chlorogenic acid (30), curcumin (31), kaempferol (32), apigenin (33), and luteolin (34) (Kuppusamy and Das, 1991; Yingprasertchai et al., 2003; Girish and Kemparaju, 2005).

Studies have isolated Di-iso-butyl phthalate from the root extract of *Emblica officinalis* (Sarkhel et al., 2011). *E. officinalis* belongs to the Euphorbiaceae family. Di-iso-butyl phthalate isolated from it has activities against viper and cobra venoms, it neutralized the venom lethality, coagulant activity, defibrinogenation, fibrinolysis, free radical, hemorrhage, and myotoxic activity generated by induced venoms. Sarkhel et al. (2011) further reported the beneficial effect of Di-iso-butyl phthalate, in its ability to decrease the levels of creatinine kinase and lactate dehydrogenase enzymes.

Chatterjee et al. (2004) isolated a small straight chain compound from the whole seed extract of *Strychnos nux vomica* (Loganiaceae). The compound was described to have methyl and amide radicals. It was effective in neutralizing the lethality, hemorrhage, and defibrinogenating activities induced by

*D. russelli* and *Naja kaouthia* venom. In addition, the extract effectively antagonized *D. russelii* venom induced defibrinogenating, hemorrhagic, lethal, edema, $PLA_2$ enzyme activity and *N. kaouthia* induced cardiotoxic, lethal, neurotoxic, $PLA_2$ enzyme activity. Methanolic extract of Leucas aspera (Lamiaceae) leaves was active against venom protease and hyaluronidase activities. The active component from in silico studies identified leucasin (35) in *L. aspera* extract as the active component (Sakthivel et al., 2013). Leucasin neutralized $PLA_2$ type I enzyme from snake venom and possessed a dose-dependent antioxidant activity.

The plant known as cabeça-de-negro (negro's head) from Northeast Brazilian folk medicine (*Específico Pessoa*) contains two phenolic pterocarpans isolated from it. They are cabenegrin A-I (36) and cabenegrin A-II (37), with prenylated A-rings (Nakagawa et al., 1982). Another pterocarpan named edunol (38) isolated from the root of *Harpalyce brasiliana*, was active with properties to inhibit myotoxic, proteolytic, and $PLA_2$ activities (da Silrva et al., 1997). Edunol was also isolated from the roots of Brongniartia podalyrioides and Brongniartia intermedia (Reyes-Chilpa et al., 1994). Edunol reduced the mortality rate of mice injected by Bothrops atrox. Isolated edunol, cabenegrins A-I and A-II have similar molecular structures.

Macrolobin-A (39) and macrolobin-B (40) are both antiproteolytic and antihemorrhagic compounds isolated from *Pentaclethra macroloba* (da Silva et al., 2007). They are potent triterpenoid saponin snake venom inhibitors and are active in neutralizing the fibrinolytic activities of class P-I and P-III metalloproteases isolated from *Bothrops neuwiedi* and *B. jararacussu* venoms. Vale et al. (2011) investigated the protective effect of flavonoids from the leaf extract of *Schizolobium parahyba* (Leguminosae) against snake venoms. They isolated catechin, gallocatechin, isoquercetin, and myricetin-3-O-glucoside from its aqueous extract. Their results also indicated the antihemorrhagic and anti-fibrinogenolytic activities of gallocatechin and myricetin-3-O-glucoside against *Bothrops* crude venoms and isolated snake venom metalloproteinase.

Gallocatechin inhibited the myotoxic effects of *Bothrops alternatus* and *B. neuwiedi* venoms. Nishijima et al. (2009) similarly identified quercetin, amenthoflavone (41) alongside myricetin (42) as active principles in *Byrsonima crassa* (Malpighiaceae). Myricetin is potent against hemorrhagic effects of *B. jararaca*.

Isolated from the ethyl acetate extract of *Bridelia ndellensis* barks are quinovic acid glycoside with different sugar moiety. Methyl and ethyl gallate were similarly isolated from the extract (Mostafa et al., 2006). The L-rhamnose (43), D-fucose (44), and D-glucopyranosyl (1→4) β-D-fucose

(45) glycosidic derivatives of quinovic acid were all significantly active in inhibition of snake venom phosphodiesterase-I. Fatima et al. (2002) in their report also isolated series of triterpenoids from bark extract of *Mitragyna stipulosa*, among them are quinovic acid, quinovic acid derivatives, and β-sitosterol glucopyranoside (46). The significant inhibitory activity of quinovic acid against snake venom phosphodiesterase I was also reported in *M. stipulosa*.

Piperine is an alkaloid and a constituent of the ethanolic fruit extract of *Piper longum* L. (Shenoy et al., 2013). Piperine interacted with phospholipase $A_2$ from Russell's viper venom. Its possible means of action may be from its ability to release platelet-activating factor and histamine from mast cells (Shenoy et al., 2013). Similarly, 4-nerolidylcatechol (47), is a member of the *Piper* spp. with inhibitory effects towards $PLA_2$ and myotoxic effects from purified myotoxins from *Bothrops* venoms (Núñez et al., 2005). 4-nerolidylcatechol inhibited serine proteinase, $PLA_2$ activity of enzymes classified under group II *Bothrops* toxins. Its mechanism could be via the binding to the active site or its proximity, it could also be via the modification of conserved residues that are critical for catalysis (Núñez et al., 2005).

Schumanniofoside alkaloid (48) from the bark extract of *Schumanniophyton magnificum* (Rubiaceae) was isolated (Akunyili and Akubue, 1986). Schumanniofoside was reported to be responsible for in vivo protective effect observed by the extract against *Naja melanoleuca* toxin. However, Houghton et al. (1992) reported a peptide with a molecular weight of 6000 Da, with anti-cardiotoxin activities from *S. magnificum*. Machiah and Gowda (2002) isolated a glycoprotein inhibitor from *Withania somnifera* with molecular mass of 27 kDa. *W. somnifera* inhibited toxic effects of $PLA_2$ enzymes from cobra venom. The glycoprotein is an acidic glycoprotein similar to the α-chain of the snake plasma phospholipase inhibitors but in contrast, it has a single subunit (Lizano et al., 2003). From the leaves of *Azadirachta indica*, a phospholipase $A_2$ inhibitor was isolated. It was termed *A. indica* phospholipase A2 inhibitor, and demonstrated good inhibition against $PLA_2$ from the cobra and Russell's viper venom in a dose-dependent manner (Mukherjee et al., 2008).

Clerodane diterpenoid (49) from *Baccharis trimera* (Less) possesses antiproteolytic and antihemorrhagic properties. Clerodane partially inhibits the edema induced by $PLA_2$ enzymes, crude venom, and metalloproteases of *Bothrops* snake venoms (Januario et al., 2004). Similarly, it partially inhibited metalloproteases-induced edema from *B. neuwiedi* and *B. jararacussu* venoms, neutralized their caseinolytic and fibrinogenolytic activities (Januario et al., 2004).

Da Silva et al. (2008) identified ellagic acid derivatives from the aqueous extract of *Casearia sylvestris* and determined its antivenom activities and inhibitory concentration against *B. jararacussu* snake venom. Ellagic acid (50) was effective in inhibiting the induced edema, myotoxicity, and enzymatic activities from the tested toxins. Isoquinoline alkaloid named berberine was isolated from *Cardiospermum halicacabum* (Chandra et al., 2011). Chandra et al. (2011) analyzed the inhibitory effect of $PLA_2$ from Russell's viper venom using X-ray crystallographic studies. Berberine was found to form complex with $PLA_2$, positioning itself in the active site of the $PLA_2$ enzyme.

Phenolic glycoside named benzoylsalireposide and salireposide were isolated from Symplocos racemosa (Ahmad et al., 2003). Other compounds isolated include oleonolic acid, β-amyrin, β-sitosterol, and β-sitosterol glycoside. Both benzoylsalireposide and salireposide (51) were found to be active against snake venom phosphodiesterase I with $IC_{50}$ of $171 \pm 0.002$ µM and $544 \pm 0.0021$ µM, respectively.

Chethankumar and Srinivas (2008) isolated active turmerin, a protein with relative molecular mass of 14 kDa from *Curcuma longa* L. The protein is rich in proline and inhibited the activities of $PLA_2$ enzymes of *Naja naja* venom in a 1:2.5 molar ratio of tested snake venom ($NV\text{-}PLA_2$) and turmerin, respectively. Similarly, Lombardo and Dennis (1985) have earlier identified manoalide (52) from turmeric as anti-cobra $PLA_2$ compound. Furthermore, Ferreira et al. (1992) isolated ar-turmerone (53) from *C. longa* L. through a silica gel fraction of its hexane extract. Ar-turmerone was active against the lethality caused by *Crotalus durissimus terrificus* venom in mice and reduced the hemorrhagic effect observed in mice after treatment with *Bothrops jururaca* venom. Likewise, from *Curcuma zedoaroides* was isolated a curcuma dialdehyde (Lattmann et al., 2010). The $C_{20}$ dialdehyde (54) was found to be active in vitro and in vivo for antivenom activity against the King cobra venom. The compound significantly antagonizes the effect of snake venom on the inhibition of neuromuscular transmission using isolated rat phrenic nerve-hemidiaphragm preparations.

*Renealmia alpinia* is used as a medicinal plant traditionally for the treatment of snakebites and as analgesic in the treatment of pain. Gómez-Betancur et al. (2014) isolated pinostrobin from the active crude extract from *R. alpinia* leaves. Pinostrobin may mitigate the antivenom activities of *Bothrops asper* venom. It prevented some venom-induced local tissue damage due to hemorrhagic effects, it's also responsible for the analgesic and anti-inflammatory activities of *R. alpinia* extract.

Mukherjee et al. (1997) investigated the membrane stabilizing effect of plant-based oil from nuts, seeds, and whole grains against viper venom $PLA_2$ enzyme. α-Tocopherol stabilized membrane following envenomation by the snake venom. Short-chain hydrocarbon of 8 carbon (2, 4-dimethylhexane) (55), 9 carbon (2, 6-dimethylheptane) (56) and 10 carbon (2-methylnonane) (57) isolated from *Tragia involucrate* investigated for antivenom activities in mice, expressed $PLA_2$ neutralization activities (Samy et al., 2012). Genistein (58) extracted from *Glycine max*, expressed good activities against *Naja naja* enzymes. Genistein showed inhibitory properties against phospholipase-I and RV-PL-V induced in mouse and against paw edema formations (Dharmappa et al., 2010). The flavonoids hypolaetin-8-glucoside (59), hypolaetin, isoscutellarein (60), kaempferol have been reported with phospholipase $A_2$ activities. Alcaraz and Hoult (1985) as well tested these flavonoid effects on soybean lipoxygenase. Hypolaetin-8-glucoside was isolated from *Sideritis mugronensis* (Lamiaceae).

Extract from the whole plant of *Thea sinensis* L. contains phospholipase $A_2$ inhibitory and antivenin activity. Melanin (61) extracted from the plant was identified as that component with the observed activities (Hung et al., 2004). Glycyrrhizin from *Glycyrrhiza glabra* reduces venom-induced bleeding as well as thrombus formation and thrombus weight of *B. jararaca* (Assafim et al., 2006). The aqueous root extract of Fukien tea plant (*Ehretia buxifolia*) has been reported as an anti-snake venom plant (Mors et al. 2000). From the plant was isolated ehretianone (62), a quinonoid xanthene with anti-snake venom activity against *E. carinatus* venom in mice (Selvanayagam et al., 1996). Mimosine (63), a β-3-(3-hydroxy-4-oxopyridyl) α-amino propionic acid, isolated and purified from *Mimosa pudica* inhibits in vitro hyaluronidase activity of *V. russelii* venom (Mahadeswaraswamy et al., 2011). Dhananjaya et al. (2006) reported the 5'nucleotidase and anticoagulant effects against *Naja naja* venom by vanillic acid (64). Similarly, anisodamine (65) from Zang Qie (*Anisodus tanguticus*) relieves venom-induced microcirculatory and associated renal failure (Li et al., 1999).

Other studies have employed the use of molecular docking to evaluate the antivemon activities of isolated phytochemical components from plants. For example, acalyphin (66) was evaluated for its potency in inhibiting $PA_2$ activities (Nirmal et al., 2008). Acalyphin is present in the extract of Indian acalypha plant (*Acalypha indica*). Similarly, Tectoridin (67) from *Pueraria lobata* (Kudzu) and Tris (2, 4-di-tert-butylphenyl) phosphate (68) from *Vitex negundo* were also evaluated for $PA_2$ inhibitory activities through docking studies (Nirmal et al., 2008; Vinuchakkaravarthy et al., 2011). Study on the endothelin antagonist effect of Chinese medicinal herbs have identified resveratrol (3, 4'5-trihydroxytransstilbene) (69) from *Cissus assamica* as an active plant component for treating snakebites (Yang et al., 1998). Attempts have been made in the modifications of plant-derived compound in order to obtain improved activities from them when compared to their parent compounds. Hydrangenol (70) from the *Hydrangea macrophylla* is a typical example where two of its derivatives (BT-1 and BT-2) have shown promising hyaluronidase activity (Kakegawa et al., 1988)

## 10.4   CONCLUSION

No doubt that many reviews have documented the various medicinal plants used in the treatment of snakebite and scorpion stings. Most of these reviews are based on ethnobotanical surveys. We presented a compilation of both medicinally active plants with reported antivenom activities either by in vitro or in vivo studies, or both, alongside the structures of the various active principles that have been isolated. Our report could be of help in identifying medicinal plants requiring further studies.

## KEYWORDS

- **active principles**
- **antivenoms**
- **ethnomedicine**
- **phytochemicals**
- **envenomation**

## REFERENCES

Abubakar, M. S.; Sule, M. I.; Pateh, U. U.; Abdurahman, E. M.; Haruna, A. K.; Jahun, B. M. In Vitro Snake Venom Detoxifying Action of the Leaf Extract of *Guiera senegalensis*. *J. Ethnopharmacology* **2000,** *69,* 253.

Adzu, B.; Abubakar, M. S.; Izebe, K. S.; Akumka, D. D.; Gamaniel, K. S.; Effect of *Annona senegalensis* Root Bark Extracts on *Naja nigricotlis nigricotlis* Venom In Rats. *J. Ethnopharmacology* **2005,** *96,* 507–513.

Aguiyi, J. C.; Igweh, A. C.; Egesie, U. G.; Leoncini, R. Studies on Possible Protection Against Snake Venom using *Mucuna pruriens* Protein Immunization. *Fitoterapia* **1999,** *70,* 21–24.

Ahmad, V. U.; Abbasi, M. A.; Hussain, H.; Akhtar, M. N.; Farooq, U.; Fatima, N.; Choudhary, M. I. Phenolic Glycosides from *Symplocos racemosa*: Natural Inhibitors of Phosphodiesterase I. *Phytochemistry* **2003**, *63*, 217–220.

Aird, S. D. Ophidian Envenomation Strategies and the Role of Purines. *Toxicon* **2002**, *40*, 335–393.

Akunyili, D. N.; Akubue, P. I. Schumanniofoside, the AntiSnake Venom Principle from the Stem Bark of Schumanniophyton magnificum Harms. J Ethnopharmacology **1986**, 18, 167–172.

Alam, M. I.; Auddy, B.; Gomes, A. Viper Venom Neutralizing by Indian Medicinal Plant (*Hemidesmus indicus* and *Pluchea indica*) Root Extract. *Phytother. Res.* **1996**, *10*, 58–61.

Alam, M. I.; Gomes, A. Adjuvant Effects and Antiserum Action Potentiation by a (herbal) Compound 2-hydroxy-4-methoxy Benzoic Acid Isolated from the Root Extract of the Indian Medicinal Plant 'Sarsaparilla' (*Hemidesmus indicus* R.Br.). *Toxicon* **1998a**, *36*(10), 1423–1431.

Alam, M. I.; Gomes, A. Viper Venom-Induced Inflammation and Inhibition of Free Radical Formation by Pure Compound (2-hydroxy-4-methoxy Benzoic Acid) Isolated and Purified from Anantamul (*Hemidesmus indicus* R. Br) Root Extract. *Toxicon* **1998b**, *36*, 207–215.

Alam, M. I.; Gomes, A. An Experimental Study on Evaluation of Chemical Antagonists Induced Snake Venom Neutralization. *Indian J. Med. Res.* **1998c**, *107*, 142–146.

Alam, M. I.; Gomes, A. J. Snake Venom Neutralization by Indian Medicinal Plants (*Vitex negundo* and *Emblica officinalis*) Fruit Extract. *Ethnopharmacology* **2003**, *86*, 75.

Alam, M. Inhibition of Toxic Effects of Viper and Cobra Venom by Indian Medicinal Plants. *Pharmacol. Pharm.* **2014**, *5*, 828–837.

Alcaraz, M. J.; Hoult, J. R. Effects of Hypolaetin-8-Glucoside and Related Flavonoids on Soybean Lipoxygenase and Snake Venom Phospholipase A2. *Arch. Int. Pharmacodyn. Ther.* **1985**, *278*(1), 4–12.

Alkofahi, A.; Sallal, A. J.; Disi, A. M. Effect of *Eryngium creticum* on the Haemolytic Activities of Snake and Scorpion Venoms. *Phytother. Res.* **1997**, *12*, 540.

Assafim, M.; Ferreira, M. S.; Frattani, F. S.; et al. Counteracting Effect of Glycyrrhizin on the Hemostatic Abnormalities Induced by *Bothrops jararaca* Snake Venom. *Br. J. Pharmacol.* **2006**. *148*(6), 807–813.

Assafim, M.; Coriolano, E. C.; Benedito, S. E.; et al. *Hypericum brasiliense* Plant Extract Neutralizes Some Biological Effects of *Bothrops jararaca* Snake Venom. *J. Venom Res.* **2011**, *2*, 11–16.

Asuzu, I. U.; Harvey, A. L. The Antisnake Venom Activities of *Parkia biglobosa* (*Mimosaceae*) Stem Bark Extract. *Toxicon* **2003**, *42*, 763.

Aung, H. T.; Nikai, T.; Niwa, M.; Takaya, Y. Rosmarinic Acid in *Argusia argentea* Inhibits Snake Venom-Induced Hemorrhage. *J. Nat. Med.* **2010**, *64*(4), 482–486.

Badilla, B.; Chaves, F.; Mora, G.; Poveda, J. L. Edema Induced by *Bothrops asper* (Squamata: Viperidae) Snake Venom and Its Inhibition by *Costa Rican* Plant Extracts. *Rev. Biol. Trop.* **2006**, *54*(2), 245–252.

Batina, M. F.; Giglio, J. R.; Sampaio, S. V. Methodological Care In The Evaluation of The $LD_{50}$ and of The Neutralization of The Lethal Effect of Crotalus Durissus Terrificus Venom by the Plant *Peschiera fuchsiaefolia (Apocynaceae)*. *J. Venomous Anim. Toxins* **1997**, *3*, 22.

Batina M, F.; Cintra, A. C.; Veronese, E. L.; Lavrador, M. A.; Giglio, J. R.; Pereira, P. S.; Dias, D. A.; França, S. C.; Sampaio, S. V. Inhibition of the lethal and Myotoxic Activities of *Crotalus durissus terrificus* Venom by *Tabernaemontana catharinensis*: Identification of One of the Active Components. *Planta Med.* **2000**, *66*(5), 424–428.

Bawaskar, H. S.; Bawaskar, P. H. Utility of Scorpion Antivenin vs Prazosin in the Management of Severe *Mesobuthus tamulus* (Indian Red Scorpion) Envenoming at Rural Setting. *J. Assoc. Physicians India* **2007**, *55,* 14–21.

Bawaskar, H. S.; Bawaskar, P. H. Scorpion Sting: Update. *J. Assoc. Physicians India* **2012**, *60,* 46–55.

Bernard, P.; Scior, T.; Didier, B.; Hibert, M.; Berthon, J. Y. Ethnopharmacology and Bioinformatic Combination for Leads Discovery: Application to Phospholipase A(2) Inhibitors. *Phytochemistry* **2001**, *58*(6), 865–874.

Biondo, R.; Soares, A. M.; Bertoni, B. W.; França, S. C.; Pereira, A. M. S. Direct Organogenesis of *Mandevilla illustris* (Vell) Woodson and Effects of its Aqueous Extract on the Enzymatic and Toxic Activities of *Crotalus durissus terrificus* snake venom. *Plant Cell Rep.* **2004**, *22,* 549.

Borges, M. H.; Alves, D. L. F.; Raslan, D. S.; Piló-Veloso, D.; Rodrigues, V. M.; Homsi-Brandeburgo, M. I.; de Lima, M. E. Neutralizing Properties of *Musa paradisiaca* L. (*Musaceae*) Juice on Phospholipase A2, Myotoxic, Hemorrhagic and Lethal Activities of Crotalidae Venoms. *J. Ethnopharmacology* **2005**, *98,* 21.

Bouimeja, B.; El Hidan, M. A.; Touloun, O.; Ait Laaradia, M.; Ait Dra, L.; El Khoudri, N.; Chait, A.; Boumezzough, A. Anti-Scorpion Venom Activity of *Thapsia garganica* Methanolic Extract: Histopathological and Biochemical Evidences. *J. Ethnopharmacology* **2018**, *30*(211), 340–347.

Caro, D.; Ocampo, Y.; Castro, J.; Barrios, L.; Salas, R.; Franco, L. A. Protective Effect of *Dracontium dubium* Against *Bothrops asper* venom. *Biomed. Pharmacother.* **2017**, *89,* 1105–1114.

Castro, O.; Gutiérrez, J. M.; Barrios, M.; Castro, I.; Romero, M.; Umaña, E. Neutralization of the Hemorrhagic Effect Induced by *Bothrops asper* (Serpentes: *Viperidae*) Venom with Tropical Plant Extracts. *Rev. Biol. Trop.* **1999**, *47,* 605.

Chacko, N.; Ibrahim, M.; Shetty, P.; Shastry, C. S. Evaluation of Antivenom Activity of *Calotropis Gigantea* Plant Extract Against *Vipera Russelli* Snake Venom. *Int. J. Pharm. Sci. Res.* **2012**, *3*(7), 2272–2279.

Chandra, N. D.; Prasanth, G. K.; Singh, N.; Kumar, S.; Jithesh, O.; Sadasivan, C.; Sharma, S.; Tej Singh, P.; Haridas, M. Identification of a Novel and Potent Inhibitor of Phospholipase A$_2$ in a Medicinal Plant: Crystal Structure at 1.93 Å and Surface Plasmon Resonance Analysis of Phospholipase A$_2$ Complexed with Berberine. *Biochim. Biophys. Acta* **2011**, *1814,* 657–663.

Chandra, V.; Jasti, J.; Kaur, P.; Srinivasan, A.; Betzel, C. H.; Singh, T. P. Structural Basis of Phospholipase A2 Inhibition for the Synthesis of Prostaglandins by the Plant Alkaloid Aristolochic Acid from A 1.7 A Crystal Structure. *Biochemistry* **2002**, *41*(36), 10914–10919.

Chatterjee, I.; Chakravarty, A. K.; Gomes, A. Antisnake Venom Activity of Ethanolic Seed Extract of *Strychnos nux vomica* Linn. *Indian J. Exp. Biol.* **2004**, *42*(5), 468–475.

Chatterjee, I.; Chakravarty, A. K.; Gomes, A. *Daboia russellii* and *Naja kaouthia* Venom Neutralization by Lupeol Acetate Isolated from the Root Extract of Indian Sarsaparilla *Hemidesmus indicus* R.Br. *J. Ethnopharmacology* **2006**, *106,* 38–43.

Chaves, F.; Chacón, M.; Badilla, B.; Arévalo, C. Effect of *Echinacea purpurea* (*Asteraceae*) Aqueous Extract on Antibody Response to *Bothrops asper* Venom and Immune Cell Response. *Rev. Biol. Trop.* **2007**, *55,* 113–119.

Chethankumar, M.; Srinivas, L. New Biological Activity Against Phospholipase A2 by Turmerin, a Protein from *Curcuma longa* L. *Biol. Chem.* **2008**, *389*(3), 299–303.

Chippaux, J. P.; Goyffon, M. Epidemiology of Scorpionism: A Global Appraisal. Acta Trop. **2008,** *107*(2), 71–79.

da Silva, A. J.; Coelho, A. L.; Simas, A. B.; Moraes, R. A.; Pinheiro, D. A.; Fernandes, F. F.; Arruda, E. Z.; Costa, P. R.; Melo, P. A. Synthesis and Pharmacological Evaluation of Prenylated and Benzylated Pterocarpans Against Snake Venom. *Bioorg. Med. Chem. Lett.* **2004,** *14*(2), 431–435.

da Silva, A. J. M.; Melo, P. A.; Silva, N. M.; Brito, F. V.; Buarque, C. D.; De Souza, D. V.; Rodrigues, V. P.; Pocas, E. S.; Noel, F.; Albuquerque, E. X.; Costa, P. R. Synthesis and Preliminary Pharmacological Evaluation of Coumestans with Different Patterns of Oxygenation. *Bioorg. Med. Chem. Lett.* **2001,** *11*(3), 283–286.

da Silva, G. L.; Matos, F. J. A.; Silveira, E. R. 4'-Dehydroxycabenegrin A-I from Roots of *Harpalyce brasiliana*. Phytochemistry *1997, 46, 1059*–1062.

da Silva, J. O.; Fernandes, R. S.; Ticli, F. K.; Oliveira, C. Z.; Mazzi, M. V.; Franco, J. J.; Giuliatti, S.; Pereira, P. S.; Soares, A. M.; Sampaio, S. V. Triterpenoid Saponins, New Metalloprotease Snake Venom Inhibitors Isolated from *Pentaclethra macroloba*. *Toxicon* **2007,** *50*(2), 283–291.

da Silva, S. L.; Calgarotto, A. K.; Chaar, J. S.; Marangoni, S. Isolation and Characterization of Ellagic Acid Derivatives Isolated from *Casearia sylvestris* SW Aqueous Extract with Anti-PLA$_2$ Activity. *Toxicon* **2008,** *52*(6), 655–666.

Dal Belo, C. A.; Colares, A. V.; Leite, G. B.; Ticli, F. K.; Sampaio, S. V.; Cintra, A. C.; Rodrigues-Simioni, L.; Dos Santos, M. G. Antineurotoxic Activity of *Galactia glaucescens* Against *Crotalus durissus terrificus* Venom. *Fitoterapia* **2008,** *79,* 378–380.

Daros, M. R.; Matos, F. J. A.; Parente, J. P. *Planta Med.* **1996,** *62,* 523.

Datte, J. Y.; Yapo, P. A.; Kouame-Koffi, G. G.; Kati-Coulibaly, S.; Amoikon, K. E.; Offoumou, A. M. Leaf Extract of *Caesalpinia bonduc* Roxb. (*Caesalpiniaceae*) Induces an Increase of Contractile Force in Rat Skeletal Muscle in Situ. *Phytomedicine* **2004,** *11,* 235.

Deshpande, S. B.; Pandey, R.; Tiwari, A. K. Pathophysiological Approach to the Management of Scorpion Envenomation. *Indian J. Physiol. Pharmacol.* **2008,** *52,* 311–314.

Dhananjaya, B. L.; Nataraju, A.; Rajesh, R.; et al. Anticoagulant Effect of *Naja naja* Venom 5'Nucleotidase: Demonstration Through the use of Novel Specific Inhibitor Vanillic Acid. *Toxicon* **2006,** *48,* 411–421.

Dhananjaya, B. L.; Zameer, F.; Girish, K. S.; D'Souza, C. J. M. Anti-Venom Potential of Aqueous Extract of Stem Bark of *Mangifera indica* L. Against *Daboia russellii* (Russell's viper) Venom. *Indian J. Biochem. Biophys.* **2011,** *48,* 175–183.

Dharmappa, K. K.; Mohamed, R.; Shivaprasad, H. V.; Vishwanath, B. S. Genistein, a Potent Inhibitor of Secretory Phospholipase A2: A New Insight in Down Regulation of Inflammation. *Inflammopharmacology* **2010,** *18,* 25–31.

Diogo, L. C.; Fernandes, R. S.; Marcussi, S.; Menaldo, D. L.; Roberto, P. G.; Matrangulo, P. V.; Pereira, P. S.; França, S. C.; Giuliatti, S.; Soares, A. M.; Lourenço, M. V. Inhibition of Snake Venoms and Phospholipases A(2) by Extracts from Native and Genetically Modified *Eclipta alba*: Isolation of Active Coumestans. *Basic Clin. Pharmacol. Toxicol.* **2009,** *104,* 293–299.

Escalante, T.; Ortiz, N.; Rucavado, A.; Sanchez, E. F.; Richardson, M.; et al. Role of Collagens and Perlecan in Microvascular Stability: Exploring the Mechanism of Capillary Vessel Damage by Snake Venom Metalloproteinases. *PLoS One.* **2011,** *6*(12), e28017.

Esmeraldino, L. E.; Souza, A. M.; Sampaio, S. V. Evaluation of the Effect of Aqueous Extract of *Croton urucurana* Baillon (*Euphorbiaceae*) on the Hemorrhagic Activity Induced by the

Venom of *Bothrops jararaca*, using New Techniques to Quantify Hemorrhagic Activity in Rat Skin. *Phytomedicine* **2005**, *12*, 570–576.

Fatani, A. J.; Harvey, A. L.; Furman, B. L.; Rowan, E. G. The effects of lignocaine on actions of the venom from the yellow scorpion *Leiurus quinquestriatus* in Vivo and in Vitro. *Toxicon* **2000**, *38*, 1787–1801.

Fatima, N.; Tapondjou, L. A.; Lontsi, D.; Sondengam, B. L.; Atta-Ur-Rahman; Choudhary, M. I. Quinovic Acid Glycosides from *Mitragyna stipulosa* – First Examples of Natural Inhibitors of Snake Venom Phosphodiesterase I. *Nat. Prod. Lett.* **2002**, *16*, 389.

Félix-Silva, J.; Gomes, J. A. S.; Fernandes, J. M.; Moura, A. K. C.; Menezes, Y. A. S.; Santos, E. C. G.; Tambourgi, D. V.; Silva-Junior, A. A.; Zucolotto, S. M.; Fernandes-Pedrosa, M. F.; Comparison of Two *Jatropha species* (*Euphorbiaceae*) used Popularly to Treat Snakebites in Northeastern Brazil: Chemical Profile, Inhibitory Activity Against *Bothrops* Erythromelas Venom and Antibacterial Activity. *J Ethnopharmacol* **2018**, *1*(213), 12–20.

Fernandes, F. A. F.; Tomaz, A. M.; El-Kik, Z. C.; Monteiro-Machado, M.; Strauch, A. M.; Cons, L. B.; Tavares-Henriques, S.; Cintra, C. O. A.; Matheus, Facundo, A. V.; Paulo, A. M. Counteraction of *Bothrops* Snake Venoms by *Combretum leprosum* Root Extract and Arjunolic Acid. *J. Ethnopharmacology* **2014**, *155*, 552–562.

Fernandes, J. M.; Félix-Silva, J.; Da Cunha, L. M.; Gomes, J. A. dS.; Siqueira, E. M. dS.; Gimenes, L. P.; et al. Inhibitory Effects of Hydroethanolic Leaf Extracts of *Kalanchoe brasiliensis* and *Kalanchoe pinnata* (*Crassulaceae*) Against Local Effects Induced by *Bothrops jararaca* Snake Venom. *PLoS One* **2016**, *11*, 12.

Fernandes, R. S.; Costa, T. R.; Marcussi, S.; Bernardes, C. P.; Menaldo, D. L.; Rodriguéz Gonzaléz, I. I.; Pereira, P. S.; Soares, A. M. Neutralization of Pharmacological and Toxic Activities of *Bothrops jararacussu* Snake Venom and Isolated Myotoxins by *Serjania erecta* Methanolic Extract and its Fractions. *J. Venomous Anim. Toxins Incl. Trop. Dis.* **2011**, *17*(1), 85–93.

Ferreira, L. A. F.; Henriques, O. B.; Andreoni, A. A. S.; Vital, G. R. F.; Campos, M. M. C.; Habermehl, G. G.; De Moraes, V. L. G. Antivenom and Biological Effects of Ar-Turmerone Isolated from *Curcuma longa* (*Zingiberaceae*). *Toxicon* **1992**, *30*, 1211–1218.

Girish, K. S.; Shashidharamurthy, R.; Nagaraju, S.; Gowda, T. V.; Kemparaju, K. Isolation and Characterization of Hyaluronidase a "Spreading Factor" from Indian Cobra (*Naja naja*) Venom. *Biochimie* **2004**, *86*, 193–202.

Girish, K. S.; Kemparaju, K. Inhibition of *Naja naja* venom Hyaluronidase by Plant-Derived Bioactive Components and Polysaccharides. *Biochem.* (*Mosc.*) **2005**, *70*, 948–952.

Girish, K. S.; Kemparaju, K. Inhibition of *Naja naja* Venom Hyaluronidase: Role in the Management of Poisonous Bite. *Life Sci.* **2006**, *78*, 1433–1440.

Gómez-Betancur, I.; Benjumea, D.; Patiño, A.; Jiménez, N.; Osorio, E. Inhibition of the Toxic Effects of *Bothrops asper* venom by Pinostrobin, a Flavanone Isolated from *Renealmia alpinia* (Rottb.) MAAS. *J. Ethnopharmacology* **2014**, *155*, 1609–1615.

Gomes, A.; Saha, A.; Chatterjee, I.; Chakravarty, A. K. Viper and Cobra Venom Neutralization by Beta-Sitosterol and Stigmasterol Isolated from the Root Extract of *Pluchea indica* Less. (*Asteraceae*). *Phytomedicine* **2007**, *14*, 637–643.

Gomes, A.; Das, R.; Sarkhel, S.; Mishra, R.; Mukherjee, S.; Bhattacharya, S.; Gomes, A. Herbs and Herbal Constituents Active Against Snake Bite. *Indian J. Exp. Biol.* **2010**, *48*, 865–878.

Gopi, K.; Renu, K.; Jayaraman, G. Inhibition of *Naja naja* Venom Enzymes by the Methanolic Extract of *Leucas aspera* and its Chemical Profile by GC-MS. *Toxicol. Rep.* **2014**, *1*, 667–673.

Gopi, K.; Renu, K.; Vishwanath B. S.; Jayaraman G. Protective Effect of *Euphorbia hirta* and its Components Against Snake Venom Induced Lethality. *J. Ethnopharmacology.* **2015**, *165*, 180–90.

Gopi, K.; Anbarasu, K.; Renu, K.; Jayanthi, S.; Vishwanath, B. S.; Jayaraman, G. Quercetin-3-O-rhamnoside from *Euphorbia hirta* Protects Against Snake Venom Induced Toxicity. *BBA - General Subjects.* **2016**, *1860*(7), 1528–40.

Gowda, T. V.; Middlebrook, J. L. Monoclonal Antibodies to VRV-PLVIIIa, a Basic Multitoxic Phospholipase A2 from *Vipera russelii* Venom. *Toxicon* **1994**, *32*(8), 955–964.

Gutiérrez, J. M.; Rucavado, A. Snake Venom Metalloproteinases: Their Role in the Pathogenesis of Local Tissue Damage. *Biochimie* **2000**, *82*, 841–845.

Gutiérrez, J. M.; Ownby, C. L. Skeletal Muscle Degeneration Induced by Venom Phospholipases A2: Insights into the Mechanisms of Local and Systemic Myotoxicity. *Toxicon* **2003**, *42*(8), 915–931.

Hasson, S. S.; Al-Jabri, A. A.; Sallam, T. A.; Al-Balushi, M. S.; Mothana R. A. A. Antisnake Venom Activity of *Hibiscus Aethiopicus* L. Against *Echis Ocellatus* and *Naja n. Nigricollis*. *J. Toxicol* **2010**, Article ID 837864.

Houghton, P. J.; Osibogun, I. M.; Bansal, S. A Peptide from *Schumanniophyton magnificum* with Anti-Cobra Venom Activity. *Planta Med*. **1992**, *58*, 263–265.

Houghton, P. J.; Skari, K. P. The Effect on Blood Clotting of Some West African Plants Used Against Snakebite. *J. Ethnopharmacology* **1994**, *44*, 99–108.

Hung, Y. C.; Sava, V.; Hong, M. Y.; Huang, G. S. Inhibitory Effects on Phospholipase A2 and Antivenin Activity of Melanin Extracted from *Thea sinensis* Linn. *Life Sci*. **2004**, *74*, 2037–2047.

Izidoro, L. F.; Rodrigues, V. M.; Rodrigues, R. S.; Ferro, E. V.; Hamaguchi, A.; Giglio, J. R.; Homsi-Brandeburgo, M. I. Neutralization of some Hematological and Hemostatic Alterations Induced by Neuwiedase, a Metalloproteinase Isolated from *Bothrops neuwiedi pauloensis* Snake Venom, by the Aqueous Extract from *Casearia mariquitensis* (*Flacourtiaceae*). *Biochimie* **2003**, *85*, 669–675.

Jalalia, A.; Vatanpour, H.; Bagheri khalili, M.; et al. The Antitoxicity Effects of *Parkinsonia aculeate* Against Scorpion Venom (*Bothotus saulcyi*): in vivo and in vitro Studies. *J. Med. Plant*. **2006**, *5*, 59–69.

James, T.; Dinesh, M. D.; Uma, M. S.; Vadivelan, R.; Shrestha, A.; Meenatchisundaram, S.; Shanmugam, V. In vivo and in vitro Neutralizing Potential of *Rauvolfia serpentina* Plant Extract Against *Daboia russelli* Venom. *Adv. Biol. Res*. **2013**, *7*(6), 276–281.

Janardhan, B.; Shrikanth, V. M.; Mirajkar, K. K.; More, S. S. In Vitro Screening and Evaluation of Antivenom Phytochemicals from *Azima tetracantha* Lam. Leaves Against *Bungarus caeruleus* and *Vipera russelli*. *J. Venomous Anim. Toxins Incl. Trop. Dis*. **2014**, *20*, 12.

Janardhan, B.; Shrikanth, V. M.; Mirajkar, K. K.; More, S. S. In Vitro Anti-Snake Venom Properties of *Carisssa spinarum* Linn Leaf Extracts. *J. Herbs, Spices Med. Plants* **2015**, *21*(3), 283–293.

Januario, A. H.; Santos, S. L.; Marcussi, S.; Mazzi, M. V.; Pietro, R. C.; Sato, D. N.; Ellena, J.; Sampaio, S. V.; França, S. C, Soares, A. M. Neo-Clerodane Diterpenoid, a New Metalloprotease Snake Venom Inhibitor from *Baccharis trimera* (*Asteraceae*): Anti-Proteolytic and Anti-Hemorrhagic Properties. *Chem. Biol. Interact* **2004**, *150*(3), 243–251.

Jiménez-Ferrer, J. E.; Pérez-Terán, Y. Y.; Román-Ramos, R.; Tortoriello, J. Antitoxin Activity of Plants used in Mexican Traditional Medicine Against Scorpion Poisoning. *Phytomedicine* **2005**, *12*, 116–122.

Kakegawa, H.; Matsumoto, H.; Satoh, T. Inhibitory Effects of Hydrangenol Derivatives on the Activation of Hyaluronidase and their Antiallergic Activities. *Planta Med.* **1988,** *54*(5), 385–389.

Katkar, G. D.; Sharma, R. D.; Vishalakshi, G. J.; Naveenkumar, S. K.; Madhur, G.; Narender, T.; Girish, K. S.; Kemparaju, K.; Thushara, R. M. Lupeol Derivative Mitigates *Echis carinatus* Venom-Induced Tissue Destruction by Neutralizing Venom Toxins and Protecting Collagen and Angiogenic Receptors on Inflammatory Cells. *Biochim. Biophys. Acta.* **2015,** *1850,* 2393–2409.

Khatony, A.; Abdi, A.; Fatahpour, T.; Towhidi, F. The Epidemiology of Scorpion Stings in Tropical Areas of Kermanshah Province, Iran, During 2008 and 2009. *J. Venomous Anim. Toxins Incl. Trop. Dis.* **2015,** *21,* 45.

Kini, R. M.; Gowda, T. V. Studies on Snake Venom Enzymes: Part I. Purification of ATPases, a Toxic Component of *Naja naja* Venom and Its Inhibition by Potassium Gymnemate. *Indian J. Biochem. Biophys.* **1982a,** *19,* 152–154.

Kini, R. M.; Gowda, T. V. Studies on Snake Venom Enzymes: Part II – Partial Characterization of ATPases from Russell's Viper (*Vipera russelli*) Venom and their Interaction with Potassium Gymnemate. *Indian J. Biochem. Biophys.* **1982b,** *19,* 342–346.

Kini, R. M. *Venom Phospholipase A2 Enzymes: Structure, Function and Mechanism;* John Wiley: New York, 1997.

Krishnan, A.; Sonawane, R. V.; Karnad, D. R. Captopril in the Treatment of Cardiovascular Manifestations of Indian Red Scorpion (*Mesobuthus tamulus* Concanesis Pocock) Envenomation. *J. Assoc. Phys. India* **2007,** *55,* 22–26.

Krishnan, S. A.; Dileepkumar, R.; Nair, A. S.; Oommen, O. V. Studies on Neutralizing Effect of *Ophiorrhiza mungos* Root Extract Against *Daboia russelii* Venom. *J. Ethnopharmacology* **2014,** *151,* 543–547.

Kumarapppan, C.; Jaswanth, A.; Kumarasunderi, K.; Antihaemolytic and Snake Venom Neutralizing Effect of Some Indian Medicinal Plants. *Asian Pac. J. Trop. Med.* **2011,** *4,* 743–747.

Kuppusamy, U. R.; Khoo, H. E.; Das, N. P. Structure-Activity Studies of Flavonoids as Inhibitors of Hyaluronidase. *Biochem. Pharmacol.* **1990,** *40,* 397–401.

Kuppusamy, U. R.; Das, N. P. Inhibitory Effects of Flavonoids on Several Venom Hyaluronidases. *Experientia* **1991,** *47,* 1196–2000.

Kuriakose, B. B.; Aleykutty, N. A.; Nitha, B. Evaluation of Venom Neutralising Capacity of Indian Medicinal Plants by in Vitro Methods. *Asian J. Pharm. Health Sci.* **2012,** *2*(4), 552–554.

Lakshmi, K. S.; Vadivu, R. The Anti-Snake Venom Activity of the Leaves of *Symplocos cochinchinensis* (Lour.) S. Moore ssp. Laurina (*Symplocaceae*). *Der Pharmacia Lett.* **2010,** *2*(4), 77–81.

Lattmann, E.; Sattayasai, J.; Sattayasai, N.; Staaf, A.; Phimmasone, S.; Schwalbe, C. H.; Chaveerach, A. In-vitro and in-vivo Antivenin Activity of 2-[2-(5,5,8a-trimethyl-2-methylene-decahydronaphthalen-1-yl)-ethylidene]-succinaldehyde Against *Ophiophagus hannah* Venom. *J. Pharm. Pharmacol.* **2010,** *62,* 257–262.

Leanpolchareanchai, J.; Pithayanukul, P.; Bavovada, R.; Saparpakorn, P. Molecular Docking Studies and Anti-Enzymatic Activities of Thai Mango Seed Kernel Extract Against Snake Venoms. *Molecules* **2009,** *14*(4), 1404–1422.

Li, Q. B.; Pan, R.; Wang, G. F.; Tang, S. X. Anisodamine as an Effective Drug to Treat Snakebites. *J. Nat. Toxins* **1999,** *8*(3), 327–330.

Liang, W. F. Anti-Snake Bite Action of *Picrasma quassioides*. *Bull. Chin. Mater. Med.* **1987,** *12,* 54.

Lima, M. C.; Bitencourt, M.; Furtado, A. A.; Oliveira Rocha, H. A.; Oliveira, R. M.; da Silva-Júnior, A. A.; et al. *Ipomoea asarifolia* Neutralizes Inflammation Induced by *Tityus serrulatus* Scorpion Venom. *J. Ethnopharmacol.* **2014,** *153,* 890–895.

Lindahl, M.; Tagesson, C. Flavonoids as Phospholipase A2 Inhibitors: Importance of their Structure for Selective Inhibition of Group II Phospholipase A2. *Inflammation* **1997,** *21*(3), 347–356.

Lizano, S.; Domont, G. B.; Perales, J. Natural Phospholipase A(2) Myotoxin Inhibitor Proteins from Snakes, Mammals and Plants. *Toxicon* **2003,** *42,* 963–977.

Lobo, R.; Punitha, I. S. R.; Rajendran, K.; Shirwaikar, A. Preliminary Study on the Antisnake Venom Activity of Alcoholic Root Extract of *Clerodendrum Viscosum* (Vent.) in *Naja naja* Venom. *Nat. Prod. Sci.* **2006,** *12*(3), 153–156.

Lombardo, D.; Dennis, E. A. Cobra Venom Phospholipase A2 Inhibition by Manoalide. A Novel Type of Phospholipase Inhibitor. *J. Biol. Chem.* **1985,** *260*(12), 7234–7240.

Machiah, D.; Gowda, T. V. Purification and Characterization of a Glycoprotein Inhibitor of Toxic Phospholipase from *Withania somnifera. Arch. Biochem. Biophys.* **2002,** *408*(1), 42–50.

Machiah, D. K.; Girish, K. S.; Gowda, T. V. A Glycoprotein from a Folk Medicinal Plant, *Withania somnifera,* Inhibits Hyaluronidase Activity of Snake Venoms. *Comp. Biochem. Physiol. (Part C)* **2006,** *143,* 158–161.

Mackessy, P. S. Reptile Toxinology, Systematics, and Venom Gland Structure. In *Handbook of Venoms and Toxins of Reptiles;* Stephen, P. M., Ed.; CRC Press, Taylor & Francis: Boca Raton, 2010; pp 3–23.

Mahadeswaraswamy, Y. H.; Nagaraju, S.; Girish, K. S.; Kemparaju, K. Local Tissue Destruction and Procoagulation Properties of *Echis carinatus* Venom: Inhibition by *Vitis vinifera* Seed Methanol Extract. *Phytother. Res.* **2008,** *22,* 963–969.

Mahadeswaraswamy, Y. H, Manjula, B.; Devaraja, S. et al. *Daboia russelii* Venom Hyaluronidase: Purification, Characterization and Inhibition by β-3-(3-hydroxy-4-oxopyridyl) α-amino-propionic Acid. *Curr. Top Med Chem.* **2011,** *11*(20), 2556–2565.

Mahanta, M.; Mukherjee, A. K. Neutralisation of Lethality, Myotoxicity and Toxic Enzymes of *Naja kaouthia* Venom by *Mimosa pudica* Root Extracts. *J. Ethnopharmacology* **2001,** *75,* 55–60.

Markland, F. S., Jr.; Swenson, S. Snake Venom Metalloproteinases. *Toxicon* **2013,** *62,* 3–18.

Martz, W. Plants with a Reputation Against Snakebite. *Toxicon* **1992,** *30,* 1131–1142.

Meenatchisundaram, S.; Prajish, P. G.; Subbraj, T.; Michael, A. Studies on Antivenom Activity of *Andrographis paniculata* and *Aristolochia indica* Plant Extracts Against *Echis carinatus* Venom. *Internet J. Toxicol.* **2008,** *6,* 1.

Meenatchisundaram, S.; Michael, A. Preliminary Studies on Antivenom Activity of *Mimosa pudica* Root Extracts Against Russell's Viper and Saw Scaled Viper Venom by in vivo and in vitro Methods. *Pharmacologyonline* **2009,** *2,* 372–378.

Meenatcisundaram, S.; Sindhu, M. In vivo and in vitro Studies on Neutralizing Effects of *Acorus calamus* and *Withania somnifera* Root Extracts Against *Echis carinatus* Venom. *Iran. J. Pharmacol. Ther.* **2011,** *10*(1), 26–30.

Melo, M. M.; Merfort, I.; Habermehl, G. G.; Ferreira, K. M.; Uso De Extratos De Plantas No Tratamento Local De Pele De Coelho Após Envenenamento Botrópico Experimental. *J. Bras Fitomed.* **2003,** 1(1), 100–106.

Melo, P. A.; do Nascimento, M. C.; Mors, W. B.; Suarez-kurtz, G. Inhibition of the Myotoxic and Hemorrhagic Activities of Crotalid Venoms by *Eclipta Prostrata* (*Asteraceae*) Extracts and Constituents. *Toxicon* **1994**, *32*(5), 595–603.

Melo, P. A.; Ownby, C. L. Ability of Wedelolactone, Heparin, and Para-Bromophenacyl Bromide to Antagonize the Myotoxic Effects of Two Crotaline Venoms and their PLA$_2$ Myotoxins. *Toxicon* **1999**, *37*(1), 199–215.

Molander, M.; Nielsen, L.; Søgaard, S.; Staerk, D.; Rønsted, N.; Diallo, D.; Chifundera, K. Z.; Van Staden, J.; Jäger, A. K. Hyaluronidase, Phospholipase A2 and Protease Inhibitory Activity of Plants Used in Traditional Treatment of Snakebite-Induced Tissue Necrosis In Mali, DR Congo and South Africa. *J. Ethnopharmacol* **2014**, *157*, 171–180.

Mors, W. B.; Nascimento, M. C.; Pereira, B. M.; Pereira, N. A. Plant Natural Products Active Against Snake Bite – the Molecular Approach. *Phytochem.* **2000**, 55(6), 627–642.

Mostafa, M.; Nahar, N.; Mosihuzzaman, M.; Sokeng, S. D.; Fatima, N.; Atta-Ur-Rahman, Choudhary, M. I. Phosphodiesterase-I Inhibitor Quinovic Acid Glycosides from *Bridelia ndellensis*. *Nat. Prod. Res.* **2006**, *20*, 686–692.

Moura, M. V.; Guimarães, da C. N.; Batista, T. L.; Freitas-de-Sousa, A. L.; Martins, de S. J.; de Souza, C. S.; de Almeida, D. O. P.; Monteiro, M. W.; de Oliveira, B. R.; DosSantos, C. M.; Mourão, H. V. R., Assessment of the Anti-Snakebite Properties of Extracts of *Aniba fragrans* Ducke (*Lauraceae*) Used in Folkmedicine as Complementary Treatment in Cases of Envenomation by *Bothrops* Atrox. *J. Ethnopharmacology.* **2017**, *213*, 350-358.

Moura, V. M.; Sousa, L. A. F.; Oliveira, R. B.; Moura-da-Silva, A. M.; Chalkidis, H. M.; Silva, M. N.; Pacheco, S.; Mourão, R. H. V. Inhibition of the Principal Enzymatic and Biological Effects of the Crude Venom of *Bothrops atrox* by Plant Extracts. *J. Med. Plants Res.* **2013**, *7*, 2330–2337.

Moura, V. M.; Bezerra, A. N. S.; Mourão, R. H. V.; Lameiras, J. L. V.; Raposo, J. D. A.; Sousa, R. L.; Boechat, A. L.; Oliveira, R. B.; Chalkidis, H.de M.; Dos-Santosa, M. C. A Comparison of the Ability of *Bellucia dichotoma* Cogn. (*Melastomataceae*) Extract to Inhibit the Local Effects of *Bothrops atrox* Venom When Pre-Incubated and When Used According to Traditional Methods. *Toxicon* **2014**, *85*, 59–68.

Moura, V. M.; Silva, W. C. R.; Raposo, D. A. J.; Freitas-de-Sousa, A. L.; Dos-Santos, C. M.; Oliveira, B. R.; Mourão, V. R. H. The inhibitory Potential of the Condensed-Tannin-Rich Fraction of *Plathymenia reticulata* Benth. (*Fabaceae*) Against *Bothrops atrox* Envenomation. *J. Ethnopharmacology* **2016**, *183*, 136–142.

Mukherjee, A. K.; Ghosal, S. K.; Maity, C. R. Lysosomal Membrane Stabilization by Alpha-Tocopherol Against the Damaging Action of *Vipera russelli* Venom Phospholipase A2. *Cell. Mol. Life Sci.* **1997**, *53*(2), 152–155.

Mukherjee, A. K.; Doley, R.; Saikia, D. Isolation of a Snake Venom Phospholipase A2 (PLA2) Inhibitor (AIPLAI) from Leaves of *Azadirachta indica* (Neem): Mechanism of PLA2 Inhibition by AIPLAI in Vitro Condition. *Toxicon* **2008**, *51*(8), 1548–1553.

Murari, S. K.; Frey, F. J.; Frey, B. M.; Gowda, T. V.; Vishwanath, B. S. Use of *Pavo cristatus* Feather Extract for the Better Management of Snakebites: Neutralization of Inflammatory Reactions. *J. Ethnopharmacology* **2005**, *99*, 229–237.

Nakagawa M.; Nakanishi, K.; Darko, L. L.; Vick, J. A. Structures of Cabenegrins A-I and A-II, Potent Anti-Snake Venoms. *Tetrahedron Lett.* **1982**, *23*, 3855–3858.

Nalbantsoy, A.; Erel, S. B.; Köksal, Ç.; Göçmen, B.; Yıldız, M. Z.; Yavasoglu, N. Ü. K. Viper Venom Induced Inflammation with *MontiVipera xanthina* (Gray, 1849) and the Anti-Snake Venom Activities of *Artemisia absinthium* L. in Rat. *Toxicon* **2013**, *65*, 34–40.

Nanjaraj, A. N. Urs; Yariswamy, M.; Joshi, V.; Nataraju, A.; Gowda, T. V.; Vishwanath, B. S. Implications of Phytochemicals in Snakebite Management: Present Status and Future Prospective. *Toxin Rev. Early Online* **2013**, *33*, 1–24.

Nataraju, A.; Raghavendra Gowda, C. D.; Rajesh, R.; Vishwanath, B. S. Group IIA Secretory PLA2 Inhibition by Ursolic Acid: A Potent Anti-Inflammatory Molecule. *Curr. Top. Med. Chem.* **2007**, *7*, 801–809.

Nazato, V. S.; Rubem-Mauro, L.; Vieira, N. A. G.; et al. In Vitro Antiophidian Properties of *Dipteryx alata* Vogel Bark Extracts. *Molecules* **2010**, *15*(9), 5956–5970.

Nirmal, N.; Praba, G. O.; Velmurugan, D. Modeling Studies on Phospholipase A$_2$-Inhibitor Complexes. *Indian J. Biochem. Biophys.* **2008**, *45*(4), 256–262.

Nishijima, C. M.; Rodrigues, C. M.; Silva, M. A.; Lopes-Ferreira, M.; Vilegas, W.; Hiruma-Lima, C. A. Anti-Hemorrhagic Activity of Four Brazilian Vegetable Species Against *Bothrops jararaca* Venom. *Molecules* **2009**, *14*, 1072–1080.

Núñez, V.; Otero, R.; Barona, J.; Fonnegra, R.; Jimenez, S.; Osorio, R. G.; Quintana, J. C.; Diaz, A. Inhibition of the Toxic Effects of *Lachesis muta*, *Crotalus Durissus Cumanensis* and *Micrurus Mipartitus* Snake Venoms by Plant Extracts. *Pharm. Biol.* **2004a**, *42*, 49.

Núñez, V.; Otero, R.; Barona, J.; Saldarriaga, M.; Osorio, R. G.; Fonnegra, R.; Jiménez, S. L.; Diaz, A.; Quintana, J. C. Neutralization of the Edema-Forming, Defibrinating and Coagulant Effects of *Bothrops Asper* Venom by Extracts of Plants Used by Healers in Colombia. *Braz. J. Med. Biol. Res.* **2004b**, *37*, 969.

Núñez, V.; Castro, V.; Murillo, R.; Ponce-Soto, L. A.; Merfort, I.; Lomonte, B. Inhibitory Effects of *Piper umbellatum* and *Piper peltatum* Extracts Towards Myotoxic Phospholipases A2 from *Bothrops* Snake Venoms: Isolation of 4-Nerolidylcatechol as Active Principle. *Phytochemistry.* **2005**, *66*, 1017–1025.

Ode, O. J.; Asuzu, I. U. The Anti-Snake Venom Activities of the Methanolic Extract of the Bulb of *Crinum jagus* (*Amaryllidaceae*). *Toxicon* **2006**, *48*, 331–342.

Okonogi, T.; Hattori, Z.; Ogiso, A.; Mitsui, S. Detoxification by Persimmon Tannin of Snake Venoms and Bacterial Toxins. *Toxicon* **1979**, *17*, 524–527.

Oliveira, C. Z.; Maiorano, V. A.; Marcussi, S.; et al. Anticoagulant and Antifibrinogenolytic Properties of the Aqueous Extract from *Bauhinia Forficata* Against Snake Venoms. *J. Ethnopharmacology* **2005**, *98*(1–2), 213–216.

Otero, R.; Núñez, V.; Jiménez, S. L.; Fonnegra, R.; Osorio, R. G.; García, M. E.; et al. Snakebites and Ethnobotany in the Northwest Region of Colombia: Part II: Neutralization of Lethal and Enzymatic Effects of *Bothrops atrox* Venom. *J. Ethnopharmacology* **2000a**, *71*, 505–511.

Otero, R.; Nuñez, V.; Barona, J.; Fonnegra, R.; Jimenez, S. L.; Osorio, R. G.; et al. Snakebites and Ethnobotany in the Northwest Region of Colombia. Part III: Neutralization of the Haemorrhagic Effect of *Bothrops atrox* Venom. *J Ethnopharmacology* **2000b**, *73*, 233–241.

Patiño, A. C.; López, J.; Aristizabal, M.; Quintana, J. C.; Benjumea, D. Evaluation of the Inhibitory Effect of Extracts from Leaves of *Renealmia alpinia* Rottb. Maas (*Zingiberaceae*) on the Venom of *Bothrops asper* (mapana). *Biomédica* **2012**, *32*, 365–374.

Patiño, A. C.; Benjumea, D. M.; Pereañez, J. A. Inhibition of Venom Serine Proteinase and Metalloproteinase Activities by *Renealmia alpinia* (*Zingiberaceae*) Extracts: Comparison of Wild and in Vitro Propagated Plants. *J. Ethnopharmacology* **2013**, *149*, 590–596.

Pereira, B. M. R.; Daros, M. R.; Parente, J. P.; Matos, F. J. A. *Phytother. Res.* **1996**, *10*, 666.

Pereira, I. C.; Barbosa, A. M.; Salvador, M. J.; et al. Anti-Inflammatory Activity of *Blutaparon portulacoides* Ethanolic Extract Against the Inflammatory Reaction Induced by *Bothrops jararacussu* Venom and Isolated Myotoxins BthTX-I and II. *J. Venomous Anim. Toxins Incl. Trop. Dis.* **2009**, *15*(3), 527–545.

Pereira, N. A.; Ruppelt, B. M.; do Nascimento, M. C.; Parente, J. P.; Mors, W. B. An Update on Plants Used against Snakebite, Brasilianisch-Deutsches Symposium fur Naturstoffchemie. *Hanover* **1991**, pp 48–51.

Pithayanukul, P.; Laovachirasuwan, S.; Bavovada, R.; Pakmanee, N.; Suttisri, R. Anti-Venom Potential of Butanolic Extract of *Eclipta Prostrata* Against Malayan Pit Viper Venom. *J. Ethnopharmacology* **2004**, *90*(2–3), 347–352.

Pithayanukul, P.; Leanpolchareanchai, J.; Saparpakorn, P. Molecular Docking Studies and Anti-Snake Venom Metalloproteinase Activity of Thai Mango Seed Kernel Extract. *Molecules* **2009**, *14*, 3198–3213.

Queiroz, A. M.; Sampaio, V. S.; Mendonça, I.; Fé, N. F.; Sachett, J.; Ferreira, L. C.; et al. Severity of Scorpion Stings in the Western Brazilian Amazon: A Case-Control Study. PLoS One **2015**, *10*(6), e0128819.

Raghavamma, S. T. V.; Rao, N. R.; Rao, G. D. Inhibitory Potential of Important Phytochemicals from *Pergularia daemia* (Forsk.) Chiov., on Snake Venom (*Naja naja*). *J. Genet. Eng. Biotechnol.* **2016**, *14*, 211–217.

Reddy, K. P.; Singh, A. B.; Puri, A.; Srivastava, A. K.; Narender, T. Synthesis of Novel Triterpenoid (lupeol) Derivatives and their in Vivo Antihyperglycemic and Antidyslipidemic Activity. *Bioorg. Med. Chem. Lett.* **2009**, *19*, 4463–4466.

Reyes-Chilpa, R.; Gomez-Garibay, F.; Quijano, L.; Magos-Guerrero, G. A.; Rios, T. Preliminary Results on the Protective Effect of (-)-edunol, a Pterocarpan from *Brongniartia podalyrioides* (*Leguminosae*), Against *Bothrops atrox* Venom in Mice. *J. Ethnopharmacology* **1994**, *42*, 199.

Rizzini, C. T.; Mors, W. B.; Pereira, N. A. Brazilian Plants So-Believed Active Against Animal-Venons, Especially Anti-Snake Venoms. *Rev. Bras. Farm.* **1988**, *69*, 82–86.

Ruppelt, B. M.; Pereira, E. F.; Gonçalves, L. C.; Pereira, N. A. Pharmacological Screening of Plants Recommended by Folk Medicine as Anti-Snake Venom – I. Analgesic and Anti-Inflammatory Activities. *Mem. Inst. Oswaldo Cruz.* **1991**, *86*, 203.

Sakthivel, G.; Dey, A.; Nongalleima, Kh.; Chavali, M.; Isaac, R. S. R.; Singh, N. S.; Deb, L. In vitro and in vivo Evaluation of Polyherbal Formulation Against Russell's Viper and Cobra Venom and Screening of Bioactive Components by Docking Studies. *Evidence-Based Complementary Altern. Med.* **2013**, 12, Article ID 781216.

Samy, R. P.; Gopalakrishnakone, P.; Chow, V. T. Therapeutic Application of Natural Inhibitors Against Snake Venom Phospholipase $A_2$. *Bioinformation* **2012**, *8*(1), 48–57.

Sarkhel, S.; Chakravarty, A. K.; Das, R.; Aparna Gomes; Gomes, A. Snake Venom Neutralizing Factor from the Root Extract of *Emblica officinalis* Linn. *Orient. Pharm. Exp. Med.* **2011**, *11*, 25–33.

Sekar, K.; Vaijayanthi, M. S.; Yogavel, M.; et al. Crystal Structures of the Free and Anisic Acid Bound Triple Mutant of Phospholipase A2. *J. Mol. Biol.* **2003**, *333*(2), 367–376.

Selvanayagam, Z. E.; Gnanavendhan, S. G.; Balakrishna, K.; Rao, R. B.; Sivaraman, J.; Subramanian, K.; Puri, R.; Puri, R. K. Ehretianone, A Novel Quinonoid Xanthene from *Ehretia buxifolia* with Antisnake Venom Activity. *J. Nat. Prod.* **1996**, *59*(7), 664–667.

Shastry, C. S.; Bhalodia M. M.; Aswathanarayana, B. J. Antivenom Activity of Ethanolic Extract of *Crescentia cujete* Fruit. *Int. J. Phytomed.* **2012**, *4*, 108–114.

Shenoy, P. A.; Nipate, S. S.; Sonpetkar, J. M.; Salvi, N. C.; Waghmare, A. B.; Chaudhari, P. D.; Anti-Snake Venom Activities of Ethanolic Extract of Fruits of *Piper longum* L.

(*Piperaceae*) Against Russell's Viper Venom: Characterization of Piperine as Active Principle. *J. Ethnopharmacology* **2013**, *147*, 373–382.

Shirwaikar, A.; Rajendran, K.; Bodla, R.; Kumar, C. D. Neutralization Potential of Viper *Russelli Russelli* (Russell's viper) Venom by Ethanol Leaf Extract of *Acalypha indica*. *J. Ethnopharmacology* **2004**, *94*, 267–273.

Shrikanth, V. M.; Bhavya, J.; Mirjakar, K. M.; More, S. S. In Vitro Evaluation of Active Phytochemicals from *Tabernaemontana alternifolia* (Roxb) Root Against the *Naja naja* and *Echis carinatus* Indian Snake Venom. *J. Biol. Act. Prod. Nat.* **2014a**, *4*(4), 286–294.

Shrikanth, V. M.; Janardhan, B.; More, S. S.; Muddapur, U. M.; Mirajkar, K. K. In Vitro Anti Snake Venom Potential of *Abutilon indicum* Linn Leaf Extracts Against *Echis carinatus* (Indian saw scaled viper). *J. Pharmacogn. Phytochem.* **2014b**, *3*(1), 111–117.

Shrikanth, V. M.; Janardhan, B.; More, S. S. Biochemical and Pharmacological Neutralization of Indian Saw-Scaled Viper Snake Venom by *Canthium parviflorum* Extracts. *Indian J. Biochem. Biophys.* **2017**, *54*, 173–185.

Silva, T. P. D.; Moura, V. M.; Souza, M. C. S.; Santos, V. N. S.; Silva, K. A. M. M. D.; Mendes, M. G. G.; Nunez, C. V.; Almeida, P. D. O.; Lima, E. S.; Mourão, R. H. V.; Dos-Santos, M. C. *Connarus favosus* Planch.: An Inhibitor of the Hemorrhagic Activity of Bothrops Atrox Venom and a Potential Antioxidant and Antibacterial Agent. *J. Ethnopharmacology* **2016**, *183*, 166–175.

Soares, A. M.; Januario, A. H.; Lourenco, M. V.; Pereira, A. M. S.; Pereira, P. S. Neutralizing Effects of Brazilian Plants Against Snake Venoms. *Drugs Future* **2004**, *29*, 1105.

Soares, A. M.; Ticli, F. K.; Marcussi, S.; Lourenço, M. V.; Januario, A. H.; Sampaio, S. V.; Giglio, J. R.; Lomonte, B.; Pereira, P. S. Medicinal Plants with Inhibitory Properties Against Snake Venoms. *Curr. Med. Chem.* **2005**, *12*, 2625–2641.

Soni, P.; Bodakhe, S. H. Antivenom Potential of Ethanolic Extract of *Cordia macleodii* Bark Against *Naja* Venom. *Asian Pac. J. Trop. Biomed.* **2014**, *4*(Suppl 1), S449-S454.

Tan, N. H.; Fung, S. Y.; Sim, S. M.; Marinello, E.; Guerranti, R.; Aguiyi, J. C. The Protective Effect of *Mucuna pruriens* Seeds Against Snake Venom Poisoning. *J. Ethnopharmacology* **2009**, *123*, 356–358.

Ticli, F. K.; Hage, L. I.; Cambraia, R. S.; Pereira, P. S.; Magro, A. J.; Fontes, M. R.; Stabeli, R. G.; Giglio, J. R.; Franca, S. C.; Soares, A. M.; Sampaio, S. V. Rosmarinic Acid, a New Snake Venom Phospholipase A2 Inhibitor from *Cordia verbenacea* (*Boraginaceae*): Antiserum Action Potentiation and Molecular Interaction. *Toxicon* **2005**, *46*(3), 318–327.

Uawonggul, N.; Chaveerach, A.; Thammasirirak, S.; Arkaravichien, T.; Chuachan, C.; Daduang, S, Screening of Plants Acting Against *Heterometrus laoticus* Scorpion Venom Activity on fibroblast Cell Lysis. *J. Ethnopharmacology* **2006**, *103*, 201–207.

Ushanandini, S.; Nagaraju, S.; Harish Kumar, K.; Vedavathi, M.; Machiah, D. K.; Kemparaju, K.; Vishwanath, B. S.; Gowda, T. V.; Girish, K. S. The Anti-Snake Venom Properties of *Tamarindus indica* (*Leguminosae*) Seed Extract. *Phytother. Res.* **2006**, *20*, 851–858.

Ushanandini, S.; Nagaraju, S.; Nayaka, S. C.; Kumar, K. H.; Kemparaju, K.; Girish, K. S. The Anti-Ophidian Properties of Anacardium occidentale Bark Extract. *Immunopharmacol. Immunotoxicol.* **2009**, *31*(4), 607–615.

Vale, F. L. H.; Mendes, M. M.; Fernandes, R. S.; Costa, T. R.; S. Hage-Melim, L. I.; A. Sousa, M.; Hamaguchi, A.; Homsi-Brandeburgo, M. I.; Franca, S. C.; Silva, C. H.; Pereira, P. S.; Soares, A. M.; Rodrigues, V. M. Protective Effect of *Schizolobium parahyba* Flavonoids Against Snake Venoms and Isolated Toxins. *Curr. Top. Med. Chem.* **2011**, *11*(20), 2566–2577.

Ricciardi, V. B.; Torres, A. M.; Ricciardi, G.; Teibler, P.; Maruñak, S.; Barnaba, C.; Larcher, R.; Nicolini, G.; Dellacassa, E. The effects of *Cissampelos pareira* Extract on Envenomation Induced by *Bothrops diporus* Snake Venom. *J. Ethnopharmacology.* **2017**, *212*, 36-42.

Vinuchakkaravarthy, T.; Kumaravel, K. P.; Ravichandran, S.; Velmurugan, D. Active Compound from the Leaves of *Vitex negundo* L. Shows Anti-Inflammatory Activity with Evidence of Inhibition for Secretory Phospholipase $A_2$ through Molecular Docking. *Bioinformation* **2011**, *7*(4), 199–206.

Vishwanath, B. S.; Fawzy, A. A.; Franson, R. C. Edema-Inducing Activity of Phospholipase $A_2$ Purified from Human Synovial Fluid and Inhibition by Aristolochic Acid. *Inflammation* **1988**, *12*(6), 549–561.

World Health Organisation (WHO) (2017) Snakebite Envenoming. http://www.who.int/mediacentre/factsheets/fs337/en/ (Accessed Sept 01, 2017).

Wufem, B. M.; Adamu, H. M.; Cham, Y. A.; Kela, S. L. Preliminary Studies on the Antivenin Potential and Phytochemical Analysis of the Crude Extracts of *Balanites aegyptiaca* (Linn.) Delile on Albino Rats. *Nat. Prod. Radiance* **2007**, *6*(1), 18–21.

Yang, L. C.; Wang, F.; Liu, M. A Study of an Endothelin Antagonist from a Chinese Anti-Snake Venom Medicinal Herb. *J. Cardiovasc. Pharmacol.* **1998**. *31*(Suppl 1), S249–S250.

Yingprasertchai, S.; Bunyasrisawat, S.; Ratanabanangkoon, K. Hyaluronidase Inhibitors (sodium cromoglycate and sodium auro-thiomalate) Reduce the Local Tissue Damage and Prolong the Survival Time of Mice Injected with *Naja kaouthia* and *Calloselasma rhodostoma* Venoms. *Toxicon* **2003**, *42*, 635–646.

Zamilpa, A.; Abarca-Vargas, R.; Ventura-Zapata, E.; Osuna-Torres, L.; Zavala, M. A.; Herrera-Ruiz, M.; Jiménez-Ferrer, E.; González-Cortazar, M. Neolignans from *Aristolochia elegans* as Antagonists of the Neurotropic Effect of Scorpion Venom. *J. Ethnopharmacology* **2014**, *157*, 156–160.

# CHAPTER 11

# MEDICINAL POTENTIALS OF GREEN TEA

FREDERICK O. UJAH

*Department of Chemical Sciences, College of Natural and Applied Sciences, University of Mkar, Gboko, Benue State, Nigeria, Tel.: +2348039234148, E-mail: oyiujah2004@yahoo.com*

*ORCID: https://orcid.org/0000-0002-5150-3703*

## ABSTRACT

Tea, the most popular beverage known to man, has phytonutritional and medicinal benefits whose secret lies in its rich catechins (polyphenols). Catechins possess powerful antioxidant properties with the least caffeine content and are known for their anticarcinogenic, antimutagenic, antibacterial and antiviral properties. Green tea is also rich in amino acids (such as L-threonine and theogallin), succinate and gallic acid which give the tea its unique sweetness and umami flavor. This chapter details the chemistry of green tea, factors that influence the chemo-nutritional composition during processing, the pharmacokinetic properties of grccn tea, its detoxification activity and possible side effects and risks among others.

## 11.1  INTRODUCTION

God commanded wealth into nature a part of which is green tea. Green tea is produced from the leaves of *Camellia sinensis* (L.), an evergreen shrub typical to Southeast Asia. Although its ancestry began in China, it is native to Southern China, North India, Myanmar and Cambodia (Hicks et al., 2001). The idea of tea first came from China (Wang and Tang, 2016). Tea is manufactured in four basic forms; green, white, oolong and black tea, from the leaves of the same plant under different conditions during processing.

White tea contains the highest amount of catechins among the four and is the least processed. This is because the tender leaves or buds are steamed immediately after harvesting to inactivate the polyphenol oxidase that destroys catechins. The extent of how rich a tea is to health can be related to its catechin content. Most Americans and Europeans prefer taking black tea while the Japanese and Northern Chinese will go for green tea. Southern China and Taiwan are used to Oolong tea.

Green tea is the most consumed beverage in the world (Gomikawa et al., 2008). An estimated three billion kg of tea is produced and consumed annually in the world. The approximate percentage consumption of black, green and oolong tea in western countries, Asia, and Southern China is 78%, 20% and 2%, respectively (Parley et al., 2012). Of all the tea forms, the secret of the phytonutritional benefits of the green tea lies in its rich catechins (polyphenols), which possess powerful antioxidant properties with the least caffeine content. Catechins have been reported to possess anticarcinogenic, antimutagenic, antibacterial, and antiviral properties (Archana and Jayanthi, 2011).

## 11.2   BRIEF DESCRIPTION OF GREEN TEA

The plant is a green shrub of the family Theaceae and genus *Camelia*. Its leaves are dark green and shiny, opposite and round, flowers are large, pink, red or white and fruits are small and brown. Two basic botanical varieties exist: Chinese tea's shrub (*Camelia sinensis*) as well as Indian tea tree (*Camellia assamica*) (Chu et al., 1997). Harvesting can be done throughout the year but the best teas are from the spring collection as they are delicate and the most aromatic then. However, the teas prepared from the young and tender leaflets of top twigs show suitable gustatory features and have a special aroma (Chu et al., 1997). Initially, the locals took tea leaf infusion for its medicinal impor- tance, later on, it became the most popular beverage known to man.

## 11.3   THE CHEMISTRY OF GREEN TEA

The level of chemical constituents in green tea can be influenced by geographical conditions (such as climate, season), horticultural practices, and the maturity of the leaf (position of the leaf on the harvested shoot) (Pastore, 2005). The chemicals in green tea that contribute to full-bodied sweetness and umami flavor are amino acids of which L-threonine constitutes more than 50%. This gives tea its uniqueness (Chaturvedula et al., 2011). Besides, its chemical structure is similar to that of glutamine which is known for its distinct sweetness and umami flavor. Tea leaves also contain other amino acids such as alanine, arginine, serine and asparagine. Kaneko et al. (2006) claimed that about 70% of this savory taste in green tea is credited to threonine, and also postulated that threonine, theogallin, succinate and gallic acid are the main constituents responsible for the taste. Chaturvedula et al. (2011) stated that amino acids such as arginine and alanine, in particular, besides caffeine and catechins contribute to its unique bitterness. Polyphenols are the main components of tea leaves. The fresh tea leaf contains theophylline (0.02–0.04%), theobromine (0.15–0.20%), caffeine (about 3.50% of dry weight), organic acids (1.50%), lignin (6.5%), threonine (4%) and free amino acids (1–5.5%), chlorophyll (0.5%) and other pigments and a lot of flavored compounds (Graham, 1992). Furthermore, varieties of other constituents exist, such as phenolic acids, alkaloids, flavones, minerals, carbohydrates, enzymes and vitamins (Chaturvedula et al., 2011). Tea also contains kaempferol, flavonols, mainly quercetin, myricetin and their glycosides. The most favorable health implications associated to green tea are polyphenols, notably the catechins, which represent 25–35% of the green tea dry weight (Abdel- Rahman et al., 2011).

The proximate nutritional analysis of dry leaves of green tea revealed 92.20% dry matter, 7.80% moisture, 82.40% organic matter, 8.72% ether extract, 18.15% crude protein, 19.32% crude fiber, 36.21% nitrogen-free extract, 9.80% total minerals and 3002 kcal kg-1 calculated metabolizable energy and amino acids (Abdo et al., 2010). From the study of Karori et al. (2007), green tea leaf contains volatile oils, lipids, ascorbic acid, vitamins E, K, A, low levels of B vitamins and vitamin C, amino acids, alkaloids (caffeine, theophylline and theobromine), polysaccharides, polyphenols (catechins and flavonoids), chlorophyll, carotenoids, minerals, and other uncharacterized compounds. Polyphenolic compounds (Figs. 11.1 and 11.2) including epigallocatechin (EGC) (19.28%), epicatechin gallate (ECG) (4.69%), epicatechin (EC) (6.99%) and epigallocatechin gallate (EGCG) (69.04%) are collectively referred to as catechins and are enriched with versatile health-promoting properties (Wu and Wei, 2002).

The amount of catechins present in green tea depends on how the leaves are processed before drying. The degree of fermenting and heating of tea leaves during processing can result in polymerization of catechins. This results in conformational changes and hence changes in its characteristic properties. Other factors influencing the catechin content are the tea type (either blended or decaffeinated), method of preparation of the infusion, the geographical location, soil condition and other agricultural practices (Wu and Wei, 2002).

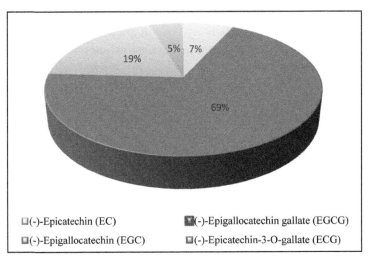

FIGURE 11.1 **(See color insert.)** Relative percentage of polyphenol in green tea.

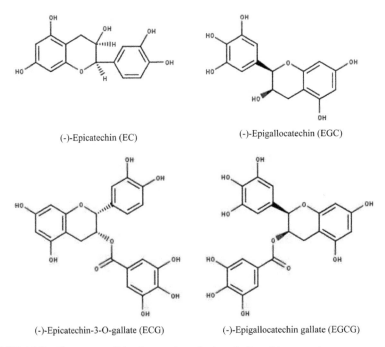

FIGURE 11.2 Structures of the four main polyphenols found in green tea.

It has been estimated that 90 mg of EGCG (which is about 25–100 times more potent antioxidant than vitamin C and E) can be obtained from 2.5 g of green tea leaves prepared in 200 mL of water which is the single most studied catechin in relation to the health contributing potentials (Cabrera et al., 2006). Green tea contains more than four times the amount of catechins than black tea, that is, about 70 mg/100 mL as compared to 15 mg/100 mL of black tea (Wu and Wei, 2002). The mechanism of action of EGCG is consequence upon its nonspecific ability to denature proteins. ECG undergoes high methylation compared to EGC and EGCG, whereas EGCG is less conjugated than EGC and EC (Chow et al., 2001).

Structure-activity studies explain that the presence of the galloyl ring in the 3-position and the trihydroxyphenyl B ring of catechins is responsible for its antioxidant properties (Guo et al., 1996). Tea leaf contains flavones and water-soluble flavanols which contribute to the yellow color in green tea infusion. The flavanols are about 1.3–1.5% dry weight of the leaf; these include kaempferol, quercetin, isoquercetin, myricetin, myricitrin, rutin, kaempferitin, and so forth and their glycosides such as coumarylquinic acid and chlorogenic acid. The flavone compounds (0.02% in dry weight of the leaf) are apigenin, isovitexin, vitexin, saponarin, vicenin-2, and so forth, besides the water-soluble anthocyanins (Chaturvedula et al., 2011).

Green tea is rich in mineral elements such as magnesium, silver, copper, zinc, manganese, iron, titanium, sodium, potassium, aluminum, bromium, nickel, chromium and phosphorus (Gramza et al., 2005). These metals promote the antioxidant property of green tea which is essential for health. The concentration of toxic metals such Ag, Ne, Cr and Pb ($1.477 > 0.100 > 0.0096 > 0.00$ mg/mL, respectively) is within the acceptable daily intake. Among these, Ag is relatively higher than the other heavy metals (Tahir and Moeen, 2011).

## 11.4  GREEN TEA PROCESSING

Commercially, teas are classified into three major categories: unfermented (white and green teas), semi-fermented (oolong tea) and fully fermented (black tea) form, obtained from the same tea plant. They differ in their appearance, taste, flavor, biological properties, and chemical compositions according to the extent of the fermentation process. After harvesting, the tea leaf immediately begins to lose turgor and enzymatic oxidation starts, if not dried quickly. During oxidation process, the leaves turn dark as chlorophylls are broken down and tannins are released. In the tea industry, this enzymatic reaction is stopped by either heating or exposing the leaves to sunlight or applying warm air to the leaves and then panfrying to stop all further processes. This thermal processing inhibits the polyphenol oxidase responsible for this phenomenon. The enzyme (polyphenol oxidase) converts catechins in green tea to theaflavins, thearubigins and other complex polyphenols which are characteristic of fermented teas. Carotenoids and unsaturated acids are responsible for the tea's aroma. A simple indicator of the fermentation degree can be changes in color of the tea leaves. According to Werkhoven (1978), processing involves the following steps.

**Withering:** This is the first step in tea processing whereby the leaves lose moisture after plucking. Oolong and black teas are withered to about 55% of the original leaf weight to concentrate the catechins content. White tea is produced from the buds and young tea leaves, immediately steamed to prevent enzymatic oxidation while green tea is prepared from the maturated leaves.

**Rolling:** This has to do with rupturing of cell walls to allow the polyphenols to become oxidized. The degree of oxidation/polymerization of catechins is lower in oolong tea.

**Fermenting**: This is the oxidation of the polyphenols by polyphenol oxidase. The extent of this process determines the forms of tea products.

**Firing:** This is to roast the leaves until a dark brown or black color with suitable aroma appears. It also stops the fermentation process.

**Drying:** This is to reduce the moisture content. Once harvested, the leaves are dried to prevent oxidation either by pan-firing or steam-drying. The dried leaves are rolled to control the release of natural substances and flavor and to produce a uniformly rounded leaf. The moisture content is reduced to 5% before packaging.

**Sifting/Grading:** Immediately after drying, the leaves are sifted and graded. This is done according to shape and size through a series of progressively finer meshes. The main grades are leaf grade, consisting of larger and longer particles which yield a light infusion, and the broken grade are smaller particles that produce darker and stronger infusions.

## 11.4.1  ENZYMATIC ACTIVITY ON TEA POLYPHENOLS DURING PROCESSING

During the manufacturing process, the monomeric catechins in green tea are subjected to the enzymatic activity of polyphenol oxidase (Halder et al., 1998). This leads to oxidation of catechins to quinones which undergo polymerization to bisflavans and to more complex structures of theaflavins, thearubigins and higher molecular mass compounds (Lin et al., 1998).

Theaflavins and their gallates appear in black tea due to oxidative condensation between (-)-EC and (-)-EGC (Tanaka et al., 2002). Although the process of fermentation greatly affects the level of catechins and the levels of the aflavins and thearubigens in tea leaves, the alkaloid content is not reasonably altered (Schulz et al., 1999). The level of caffeine increases in the process of 85% fermentation from 8.69 to 16.03 mg/100 mg of leaf dry weight (Lin et al., 1998a). Generally, the degree of tea fermentation is 0–85% (0% – green, 85% – black). Physical and biochemical factors such as temperature, oxygenation, and enzymatic activity contribute to the transformations of polyphenols (Halder et al., 1998).

## 11.4.2 FACTORS INFLUENCING CATECHIN CONTENTS IN TEA LEAVES EXTRACTS

Increase in temperature with time increases catechin content in tea processing. A study revealed that the highest quantity of catechins was extracted at a temperature of 77–80°C (Khokhar and Magnusdottir, 2002) and epimerization of C, EGCG, ECG, EGC and EC gives rise to suitable epimers: (-)-gallocatechin gallate (GCG), (-)- catechin gallate (CG), (-)-gallocatechin (GC) as well as (-)-C (Chen et al., 2001). pH is another physical factor that influences catechin content. An increase in pH of tea leaves infusion results in a considerable change in its chemical composition: For instance, 30.7% aflavins content at pH 4.9 decreases to an undetectable percent at pH 9.45 while, EGCG, ECG and EGC were reduced to about 97, 45.5 and 3.0%, respectively. However, the percent content of gallic acid, caffeine, and EC was observed to increase. Catechins are unstable in alkaline medium and almost completely disintegrated in several minutes but show large stability at pH < 4 (Friedman and Jurgens, 2000).

## 11.5 BENEFICIAL EFFECTS AND ITS MECHANISMS IN BIOLOGICAL SYSTEM

Green tea is known to contain flavanols or proanthocyanidin derivatives (Yamamoto, 1997). The most abundant catechin is EGCG although green tea contains as many as twelve catechins out of which eight are prominent; (-)-EC, (+)-catechin (C), (-)-EGC, (-)-Gallocatechin (GC), (-)-ECG, (-)-catechin gallate (CG), (-)-epigallocatechin gallate (EGCG), and (-)-gallocatechin gallate (GCG). These polyphenols can have varying effects, depending on the compounds they react with. They are found to reduce free radicals, such as 2, 2-diphenyl-1-picrylhydrazyl, to become oxidized. However, when EGCG reacts with cancer cells, it oxidizes itself and attacks cysteine residues of proteins in cancerous cells, thus inhibiting their growth, induces apoptosis (programmed cell death), or inhibits angiogenesis (Ishii, 2008). Their antioxidant activity is ranked as ECG>EGCG>EGC>GA>EC>C (Guo et al., 1996). Also, green tea reduces dietary nonheme iron absorption by chelating prooxidant metals like iron (Fe). The ratio of EGC, EGCG, ECG, or EC to iron is 3:2, 2:1, 2:1, and 3:1, respectively (Guo et al., 1996). It is however not well known how catechins differentiate between normal and cancerous cells and cause apoptosis only in the latter.

## 11.5.1   BIOACTIVITY OF CATECHINS

The bioactivity of green tea extract (GTE) is determined by its bioavailability in humans following ingestion (Holst et al., 2008). Variations in the bioavailability of green tea catechins among individuals can be substantial and the reason may be due to differences in colonic microflora and genetic polymorphisms in the enzymes involved in metabolism of polyphenol (Scalbert et al., 2000). The majority of ingested GTE-catechins (98%) in the liver and gut microbiota undergo extensive conjugation and appear transiently (60–120 min after ingestion) in the plasma (Kroon et al., 2004). Conjugated forms are chemically different from the free forms found in beverages (Kroon et al., 2004). Therefore, they are likely to have different physiological and biological effects in the human system. Insufficient information exists on the bioavailability of free catechins in plasma following GTE intake. During digestion, in the colon, they are catabolized to ring fission products and phenolic acids known as valerolactones (Aura, 2008). The biological effects of these catabolites in vivo are not known but GTE is known to take longer peak (8–15 h) in the plasma (Aura, 2008).

## 11.5.2   PHARMACOKINETICS

According to Auger et al. (2008), the bioactivity of green tea constituents following oral administration is dosed dependent. They are metabolized and cleared from the body mainly by the kidneys. Demeule et al. (2000) reported that excretion of EGCG is mainly through the bile; EGC and EC are through the bile and urine. It is also reported that the plasma half-life of EGCG is 5.5 h. The absorption of EGCG may be reduced by multidrug resistance (MDR1) gene product in the apical membranes of intestinal cells by transporting it back into lumen of the intestine (Jodoin et al., 2002). In physiological conditions, it is very likely that EGCG is oxidatively decomposed, but not (+)-catechin. The decomposition of EGC and EGCG was found to be short even at pH 7.4 (Zhu et al., 1997). Although EGCG has more intense activity, it appears to be inferior in bioavailability due to its possible decomposition during intestinal absorption and in blood. The major fractions of orally administered catechins are present as sulfate and glucuronide conjugates in human blood (Lee et al., 1995). Less active but more stable (+)-catechin is preferable as a hepatoprotective agent than EGCG. Although green tea catechins have high probability to reduce membrane fluidity, it is presumed that there is no possibility for catechins to show unfavorable actions in vivo because of metabolic conjugation in addition to low bioavailability (Tsuchiya et al. 1999). Catechins metabolites such as 5-(3/,4/,5/-Trihydroxyphenyl)-γ-valerolactone and 5-(3/,4/-Dihydroxyphenyl)-γ valerolactone were detected in plasma, feces, and urine of human volunteers after ingestion of green tea (Li et al., 2000). They were produced by intestinal microorganisms from EGC and EC respectively (Li et al., 2000). Other metabolites such as 3-Methoxy-4-hydroxy-hippuric acid, vanillic acid, 4-Hydroxybenzoic acid and 3, 4-Dihydroxybenzoic acid have also been detected by Pietta et al., (1998) in urine. Catechins undergo glucuronidation, sulfation and O-methylation in the liver of animals and humans (Hollman et al., 1997). Lu et al. (2003) studied the cytosolic catechol-O-methyltransferase (COMT) concerned with the addition of methyl group on EGCG and EGC in humans, mice, and rats. EGCG is readily methylated to 4/4/-Di-O-methyl-EGCG, and EGC to 4/-O-methyl-EGC. The Km and Vmax values for EGC methylation are higher than EGCG. Rat liver cytosol had higher COMT activity than that of humans or mice. The liver has higher specific methylation activity compared to the small intestine.

The pharmacokinetic (concentration-time curves) properties of catechins in plasma were conducted upon intravenous administered of green tea (25 mg/kg) and pure EGCG fractions (10 mg/kg) in rats and it revealed that beta-elimination half-lives (T1/2β) were found to be 212, 45, and 41 min; clearance was 2.0, 7.0, and 13.9 mL/min/kg, and apparent distribution volumes (Vd) were 1.5, 2.1, and 3.6 dl/kg for EGCG, EGC, and EC, respectively. In comparison, EGCG had a shorter T1/2β (135 min), a larger clearance (72.5 mL/min/kg), and a larger volume (22.5 dl/kg) than the other two. With intragastrical administration (200 mg/kg), approximately 0.1, 13.7, and 31.2% of EGCG, EGC, and EC were detected in the plasma compartment. The EGCG level was found to be the highest in the intestine samples and declined with a T1/2 of 173 min. EGC and EC levels were found to be the highest in the kidneys and declined rapidly with T1/2 of 29 and 28 min, respectively. The liver and lung levels of EGCG, EGC, and EC are lower to that of intestine and the kidney (Chen et al., 1997). This simply suggests why EGCG is mainly excreted through bile, while EGC and EC are through urine and bile.

## 11.5.3   DETOXIFICATION ACTIVITY

Green tea administration has been demonstrated to enhance glucuronidation reaction of phase II liver detoxification in humans (Bu-Abbas et al., 1995). Glucuronic acid is conjugated with toxins to facilitate their elimination from the body via bile (Luper, 1999). Exclusive administration of green tea in rats increased glucuronidation by 100%. Bu-Abbas et al. (1995) suggested that the increase in glucuronidation might contribute to the green tea anticarcino-genic effect by facilitating the metabolism of the chemical carcinogens into inactive, readily excreteable products.

## 11.6   GREEN TEA SIDE EFFECTS AND RISKS

Green tea has several health benefits notwithstanding, its physiological effects in high dosage may cause some unknown adverse effects which may not be the same for all individuals. EGCG is cytotoxic, hence high consumption can exert acute cytotoxicity in liver cells (Schmidt et al., 2005). Studies suggest higher intake might cause oxidative damage to the DNA of hamster in pancreas and liver (Takabayashi et al., 2004). This is because EGCG acts as a prooxidant rather than an antioxidant in pancreatic β cells in vivo which could be detrimental for diabetic animals when used to control hypergly-cemia. It has also been noted to induce a thyroid enlargement (goiter) in normal rats (Satoh et al., 2002). Furthermore, due to its caffeine content, the presence of aluminum and the effects of polyphenols on iron bioavailability, green tea should not be taken by patients suffering from major cardiovascular problems, patients on certain drugs, pregnant and breastfeeding women or should drink no more than one or two cups per day, because caffeine can cause an increase in heart rhythm and interferes with some drugs due to its diuretic effects. Also, people on blood thinners such as Coumadin/Warfarin should take green tea with caution because of its vitamin K content. It is also advisable to avoid taking green tea and aspirin because they interact to reduce the clotting effect of platelets. Green tea supplements are not under the regulation of Food and Drug Administration and may also contain other substances detrimental to health. Decaffeinated teas have high fluoride content and fluoride has been listed as a cancer promoter. High fluoride is reported to cause renal and neurological damage especially in the presence of aluminum and several bone disorders (osteoporosis, arthritis) (PFPC, 1999).

## 11.6.1 PRECAUTIONS AND POSSIBLE INTERACTIONS

Green tea may inhibit the actions of adenosine, interfere with the sedative effects of medications commonly used to treat anxiety, such as Diazepam (Valium) and Lorazepam (Ativan), and may cause agitation, tremors, insomnia, and weight loss when taken with Ephedrine. Caffeinated green tea may also interfere with a number of other medications including: birth control pills, Acetaminophen, Carbamazepine, Dipyridamole, Estrogen, Fluvoxamine, Methotrexate, Mexiletine, Phenobarbital, Theophylline, Verapamil, and so forth.

## KEYWORDS

- *Camelia sinensis*
- polyphenol oxidase
- green tea
- catechins
- anticarcinogenic

## REFERENCES

Abdel-Rahman, A.; Anyangwe, N.; Carlacci, L.; Casper, S.; Danam, R. P. The Safety and Regulation of Natural Products Used as Foods and Food Ingredients. *Toxicol. Sci.* **2011,** *123*, 333–348.

Abdo, Z. M. A.; Hassan, R. A.; El- Salam, A. A.; Helmy, S. A. Effect of Adding Green Tea and its Aqueous Extract as Natural Antioxidants to Laying Hen Diet on Productive, Reproductive Performance and Egg Quality During Storage and its Content of Cholesterol. *Egypt. Poult. Sci. J.* **2010,** *30*, 1121–1149.

Archana, S.; Jayanthi, A. Comparative Analysis of Antimicrobial Activity of Leaf Extract from Freshgreen Tea, Commercial Green Tea and Black Tea on Pathogens. *J. App. Pharm. Sci.* **2011,** *1*(8)*,* 149–152.

Auger, C.; Mullen, W.; Hara, Y.; Crozier, A. *J Nutr.* **2008,** *138*(8), 1535S–1542S.

Aura, A. M. Microbial Metabolism of Dietary Phenolic Compounds in the Colon. *Phytochem. Rev.* **2008,** *7,* 22.

Bu-Abbas, A.; Clifford M. N.; Ioannides, C.; Walker, R. *Food Chem. Toxicol,* **1995,** *33,* 27–30.

Cabrera, C.; Reyes, A.; Rafael G. Beneficial Effects of Green Tea- A review. *J. Am. Coll. Nutr.* **2006,** *25*(2), 79–99.

Chaturvedula, V. S. P.; Prakash, I. The Aroma, Taste, Color and Bioactive Constituents of Tea. *J. Med. Plants Res.* **2011,** *5*(11), 2110–2124.

Chen, L.; Lee, M. J.; Li, H.; Yang C. S. Absorption, Distribution, Elimination of Tea Polyphenols in Rats. *Drug Metab. Dispos.* **1997,** *25*(9), 1045–1050.

Chen, Z. Y.; Zhu, Q. Y.; Tang, D.; Huang, Y. Degradation of Green Tea Catechins in Tea Drinks. *J. Agric. Food Chem.* **2001,** *49,* 477–482.

Chow, H. H.; Cai, Y.; Alberts, D. S. Phase I Pharmacokinetic Study of Tea Polyphenols Following Single-Dose Administration of Epigallocatechin gallate and Polyphenon E. *Cancer Epidemiol. Biomarkers Prev.* **2001,** *10,* 53–58.

Chu, D. C.; Juneja, L. R.; Kim, M.; Yamamotro, T. *Chemistry and Applications of Green Tea;* CRC. Press: New York, USA, 1997; pp 13–16.

Demeule, M.; Brossard, M.; Page, M.; Gingras, D.; Beliveau, R. *Biochem. Biophys. Acta,* **2000,** *1478,* 51–60.

Friedman, M.; Jurgens, H. S. Effect of pH on the Stability of Plant Phenolic Compounds. *J. Agric. Food Chem.* **2000,** *48,* 2101–2110.

Gomikawa, S.; Ishikawa, Y.; Hayase, W.; Haratake, Y.; Hirano, N.; Matuura, H.; Mizowaki, A.; Murakami, A.; Yamamoto, M. Effect of Ground Green Tea Drinking for 2 Weeks on the Susceptibility of Plasma and LDL to the Oxidation Ex vivo in Healthy Volunteers. *Kobe J. Med. Sci,* **2008,** *54*(1), E62–E72

Graham, H. N. Green Tea Composition, Consumption and Polyphenol Chemistry. *Prev. Med.* **1992,** *21,* 334–350.

Gramza, A.; Korczak, J.; Amarowicz, R. Tea Polyphenols- Their Antioxidant Properties and Biological Activity- A Review. *Pol. J. Food nutr. Sci.* **2005,** *14/55*(3), 219–235.

Guo, Q.; Zhao, B.; Li, M.; Shen, S.; Xin, W. Studies on Protective Mechanisms of Four Components of Green Tea Polyphenols Against Lipid Peroxidation in Synaptosomes. *Biochimica. Biophys. Acta,* **1996,** *1304*(3), 210–222.

Halder, J.; Tamuli, P.; Bhaduri, A. N. Isolation and Characterization of Polyphenoloxidase from Indian Tea Leaf (*Camellia sinensis*). *J. Nutr. Biochem.* **1998,** *9,* 75–80.

Hicks, A. Review of Gobal Tea Production and the Impact on Industry of the Asian Economic Situation. *Assumption Univ. J. Technol.* **2001,** *5*(2), 1–8.

Hollman, P. C.; Tijburg, L. B.; Yang, C. S. *Crit. Rev. Food Sci. Nutr.* **1997,** *37,* 719–738.

Holst, B.; Williamson, G. Nutrients and Phytochemicals: from Bioavailability to Bioefficacy beyond Antioxidants. *Curr. Opin. Biotechnol.* **2008,** *19,* 73–82.

Ishii, T.; Mori,T.; Tanaka,T.; Mizuno,D.; Yamaji, R.; kumazawa, S.; Nakayama, T.; Akagawa, W. Covalent modification of proteins by green tea polyphenol (-) – epigallo-catechin-3-gallate through autoxidation. Free Radical Biology and Medicine, **2008,** *45,* 1384–1394.

Jodoin, J.; Demeule, M.; Beliveau, R. *Biochem. Biophys. Acta,* **2002,** *1542,* 149–159.

Kaneko, S.; Kumazawa, K.; Masuda, H.; Henze, A.; Hofmann, T. Molecular and Sensory Studies on the Umami Taste of Japanese Green Tea. *J. Agric. Food Chem.* **2006,** *54,* 2688–2694.

Karori, S. M.; Wachira, F. N.; Wanyoko, J. K.; Ngure, R. M. Antioxidant Capacity of Different Types of Tea Products. *Afr. J. Biotechnol.* **2007,** *6,* 2287–2296.

Khokhar, S.; Magnusdottir, S. G. Total Phenol, Catechin, and Caffeine Contents of Teas Commonly Consumed in the United Kingdom. *J. Agric. Food Chem.* **2002,** *50,* 565–570.

Kroon, P. A.; Clifford, M. N.; Crozier, A.; Hels, O.; Kovacs, E. M.; Rycroft, J. A.; Frandsen, E.; Mela, D. J.; Astrup, A. How should we Assess the Effects of Exposure to Dietary Polyphenols in vitro? *Am. J. Clin. Nutr.* **2004,** *80,* 15–21.

Lee, M. J.; Wang, Z. Y.; Li, H.; Chen, L.; Sun, Y.; Gobbo, S.; et al. *Cancer Epidemiol. Biomarkers Prev.* **1995,** *4,* 393–399.

Li, C.; Lee, M-J.; Sheng, S.; Meng, X.; Prabhu, S.; Winnik, B.; et al. *Chem. Res. Toxicol.* **2000,** *13,* 177–184.

Lin, J. K.; Lin, C-L.; Liang, Y-C.; Lin-Shiau, S-Y.; Juan, I-M. Survey of Catechins, Gallic Acid and Methylxanthines in Green, Oolong, Pu-erh and Black Teas. *J. Agric. Food Chem.* **1998,** *46,* 3635–3642.

Lu, H., Meng, X., Li, C., Sang, S., Patten, C., Sheng, S., Glucuronides of Tea Catechins: Enzymology of Biosynthesis and Biological Activities. Drug Metabolism and Disposition, **2003,** *31* (4), 452-61.

Luper, S. *Altern. Med. Rev.* **1999,** *4,* 178–189.

Parley, M.; Bansal, N.; Bansal, S. Is life-Span Under our Control? *Int. Res. J. Pharm.* **2012,** *2,* 40–48.

Pastore, R. Green and White Tea Max: A Closer Look at the Benefits of Green and White Tea. *Pastore Formulations,* **2005,** 1-19.

PFPC (Parents of Fluoride Poisoned Children). Green Tea, Fluoride & the Thyroid. Available: http://poisonfluoride.com/pfpc/html/green_tea___.html (accessed Jan 18, 2018).

Pietta, P. G.; Simonetti, P.; Gardana, C.; Brusamolino, A.; Morazzoni, P.; Bombardelli, E. *BioFactors,* **1998,** *8,* 111–118.

Satoh, K.; Sakamoto, Y.; Ogata, A.; Nagai, F.; Mikuriya, H.; Numazawa, M.; Yamada, K.; Aoki, N. Inhibition of Aromatase Activity by Green Tea Extract Catechins and their Endocrinological Effects of Oral Administration in Rats. *Food Chem. Toxicol.* **2002,** *40,* 925–933.

Scalbert, A.; Williamson, G. Dietary Intake and Bioavailability of Polyphenols. *J. Nutr.* **2000,** *130*(8S Suppl), 2073S–2085S.

Schmidt, M.; Schmitz, H. J.; Baumgart, A.; Guedon, D.; Netsch, M. I.; Kreuter, M. H.; Schmidlin, C. B.; Schrenk, D. Toxicity of Green Tea Extracts and their Constituents in Rat Hepatocytes in Primary Culture. *Food Chem. Toxicol.* **2005,** *43,* 307–314.

Schulz, H.; Engelhardt, U. H.; Wegent, A.; Drews, H.; Lapczynski, S. Application of Near-Infrared Reflectance Spectroscopy to the Simultaneous Prediction of Alkaloids and Phenolic Substances in Green Tea Leaves. *J. Agric. Food Chem.* **1999,** *47,* 5064–5067.

Tahir, A.; Moeen, R. Comparison of Antibacterial Activity of Water and Ethanol Extracts of *Camellia Sinensis (L.)* Kuntze Against Dental Caries and Detection of Antibacterial Components. *J. Med. Plants Res.* **2011,** *5*(18), 4504–4510.

Takabayashi, F.; Tahara, S.; Kaneko, T.; Harada, N. Effect of Green Tea Catechins on Oxidative DNA Damage of Hamster Pancreas and Liver Induced by N-Nitrosobis (2-Oxopropyl) Amine and/or Oxidized Soybean Oil. *BioFactors.* **2004,** *21,* 335–337.

Tanaka, T.; Mine, C.; Inoue, K.; Matsuda, M.; Kouno, I. Synthesis of Theaflavin from Epicatechin and Epigallocatechin by Plant Homogenates and Role of Epicatechin Quinone in the Synthesis and Degradation of Theaflavin. *J. Agric. Food Chem.* **2002,** *50,* 2142–2148.

Tsuchiya, H. *Pharmacology,* **1999,** *59,* 34–44.

Wang, M.; Tang, Y. Why tea is Chinese to a tee. *China Daily,* People's Republic of china. 2016.

Werkhoven, J. *Tea Processing;* 1st ed.; FAO: Rome, Italy, 1978.

Wu, C. D.; Wei, G. X. Tea as a Functional Food for Oral Health. *Nutrition,* **2002,** *18,* 443–444.

Yamamoto, T.; Juneja, L. R.; Chu, D.-C.; Kim, M. Chemistry and Applications of Green Tea. CRC Press: Boca Raton, 1997; pp 109–112.

Zhu, Q. Y.; Zhang, A.; Tsang, D.; Huang, Y.; Chen, Z-Y. Stability of Green Tea Catechins. *J. Agric. Food Chem.* **1997,** *45,* 4624–4628.

# CHAPTER 12

# ANTIOXIDANT POTENTIALS OF CINNAMON

S. ZAFAR HAIDER*, HEMA LOHANI, DOLLI CHAUHAN, and NIRPENDRA K. CHAUHAN

*Centre for Aromatic Plants (CAP), Industrial Estate Selaqui, Dehradun Uttarakhand 248011, India, Mob.: +91 9450743795*

*\*Corresponding author. E-mail: zafarhrdi@gmail.com; zafarhaider.1@rediffmail.com*
*\*ORCID: https://orcid.org/0000-0002-3061-9264*

## ABSTRACT

The antioxidant potential of cinnamon has received a great deal of attention in preventing numerous severe diseases due to the adverse effects of oxidative stress. Owing to its high antioxidant potential and medicinal value, cinnamon has become an important ingredient of spice and the demand is increasing day-by-day in the global market. There are four species, namely *Cinnamomum zeylanicum* (syn. *C. verum*), *C. cassia* (syn. *C. aromaticum*), *C. burmannii*, and *C. loureiroi* with commercial importance, especially for cinnamon bark. *C. tamala*, growing mainly in India and neighboring countries is also used as a common spice in trade for aromatic, fragrant, and spicy leaves. Numerous works have been carried out to explore antioxidant properties in cinnamon bark, leaf, and essential oil using different assays such as DPPH, FRAP, ABTS, and so forth. This chapter discusses the antioxidant potentials of cinnamon by consulting current literature. It also presents the various assay methods and prepares systematically with sources a comprehensive list of significant research in this domain.

## 12.1   INTRODUCTION

The antioxidant potentials of many medicinal, aromatic, and spice plants have received a great deal of attention in preventing several diseases due to undesirable effects of oxidative stress. A large number of these plants have long been a source of exogenous (i.e., dietary) antioxidants. Antioxidants are those substances which significantly delays or prevents the oxidation of oxidizable substrates when present at lower concentrations than the substrate (Halliwell, 2007). The physiological role of antioxidants is to protect tissues and prevent damages to cellular components which occur due to the chemical reactions involving free radicals. Generally, there are two antioxidant sources, natural and synthetic. Plants are the source of natural or dietary antioxidants which accelerate metabolism in body and having many health benefits, while the synthetic antioxidants, used in certain food products require more efforts in metabolism and may cause harmful effects in humans. The presence of secondary metabolites in many medicinal and aromatic plants such as phenolics (phenolic acids, flavonoids, tannins, and coumarins), terpenoids, carotenoids, folic acid, ascorbic acid, tocopherols, and so forth, provide protection against a number of diseases (Gulcin, 2012; Kasote et al., 2015).

Owing to high antioxidant potential and medicinal value, cinnamon has become an important ingredient of spice and the demand is increasing day by day in global market. The presence of secondary metabolites in cinnamon, such as phenols is essential since they have shown potential antioxidant activity, both in vitro and in vivo investigations (Myburgh, 2014). Hence, the demand for natural antioxidants and food preservatives is increasing rapidly (Peschel et al., 2006). Cinnamon is reported to have antidiabetic (Lu et al., 2011), lipid-lowering, cardiovascular-disease-lowering properties (Vangala-pati et al., 2012) and also useful to lower blood glucose, serum cholesterol, and blood pressure (Ranasinghe et al., 2013). According to the Food and Agriculture Organization of the United Nations, the total world production of cinnamon has reached 2.24 lakh metric tonnes and the cultivated area has also increased up to 2.84 lakh hectare in cinnamon producing countries, up to the year 2016 (FAOSTAT, 2016).

## 12.2   DESCRIPTION OF COMMERCIAL CINNAMON

The Genus, *Cinnamomum* belongs to the family Lauraceae which comprises approximately 250–350 species found naturally in the subtropical Asia,

Australia, and South America (Pravin et al., 2013; Lohani et al., 2015). Out of these reported species, only four species viz. *Cinnamomum zeylanicum* syn. *C. verum*, *C. cassia* syn., *C. aromaticum*, *C. burmannii,* and *C. loureiroi* have commercial importance, especially for cinnamon bark. Among these, Ceylon cinnamon (*C. zeylanicum*) and Cassia cinnamon (*C. cassia*) are the most popular (Fig. 12.1). The bark and leaves of *C. tamala* are aromatic, fragrant, and used as spices, therefore, it is common spice in trade (Dhar et al., 2002). Owing to its high medicinal value and as an important ingredient of spice, the demand for *C. tamala* is increasing rapidly. The commercial products of cinnamon are bark quills, bark chips, leaf, fruits, and essential oils. The details of various commercial cinnamon species are described in Table 12.1.

**FIGURE 12.1 (See color insert.)** Bark of cinnamon (A) Ceylon cinnamon; (B) Chinese cassia; (C) Indonesian cinnamon; (D) Vietnamese cassia; and (E) Indian cassia.

**TABLE 12.1** Description of Various Commercial Species of Genus *Cinnamomum*.

| Cinnamo-mum spp. | Other names | Description | | | Coumarin content (g/kg) | References |
|---|---|---|---|---|---|---|
| | | Origin | Taste (Bark) | Color (Bark) | | |
| *C. zeylanicum* syn., *C. verum* | True cinnamon, Ceylon cinnamon, Mexican cinnamon, and Dalchini | Sri Lanka | Mild sweet | Light to medium reddish brown | 0.017 | http://www. cinnamonvogue.com |
| *C. cassia* syn., *C. aromaticum* | Cassia cinnamon, Chinese cinnamon | China | Spicy bitter | Dark reddish brown | 0.31 | http://www. cinnamonvogue.com |
| *C. burmannii* | Indonesian cinnamon, Korintje cinnamon, and Padang cassia | Indonesia | Spicy | Dark reddish brown | 2.15 | http://www. cinnamonvogue.com |
| *C. loureiroi* | Vietnamese cassia, Vietnamese cinnamon, and Saigon cinnamon | Vietnam | Spicy sweet | Dark reddish brown | 6.97 | http://www. cinnamonvogue.com |
| *C. tamala* | Indian bay leaf, Indian cassia, Tejpat, and Dalchini | India | Sharp bitter | Light Brown | - | Anonymous, 1992 |

## 12.3  ASSAY METHODS FOR ANTIOXIDANTS

Various analytical methods for the evaluation of antioxidant activity have been published in a number of research articles. For the evaluation of in vitro antioxidant activity, 1,1, di-phenyl-2-picrylhydrazyl (DPPH) assay method is the most frequently used followed by ferric reducing antioxidant power (FRAP) and 2,2-azino-bis (3-ethylbenzothiazoline-6-sulfonic acid (ABTS). Some other in vitro assay methods used by researchers to explore antioxidant potential are reducing power assay, oxygen radical absorption capacity (ORAC), potassium ferricyanide reducing power (PFRAP), hydroxyl radical averting capacity (HORAC), total peroxyl radical trapping antioxidant parameter (TRAP), cupric ion reducing antioxidant capacity (CUPRAC), lipid peroxidation inhibition capacity (LPIC), hydrogen peroxide scavenging assay ($H_2O_2$), β-carotene linoleic acid, xanthine oxidase, metal chelating activity, hydroxyl radical scavenging assay, peroxy nitrite radical scavenging assay, nitric oxide scavenging assay, and so forth (Boga et al., 2011; Jayanthi and Lalitha, 2011; Pisoschi and Negulescu, 2011; Rakshit and Ramalingam, 2011; Abeysekera et al., 2013; Parul et al., 2013; Radi, 2013; Shintani, 2013; Ueno et al., 2014; Kostic et al., 2015; Sivaraj et al., 2015; Mustaffa et al., 2016; Sharma et al., 2016). In addition, lipid peroxidation (LPO), glutathione peroxidase (GPx) estimation, γ-glutamyltranspeptidase (GGT), glutathione reductase (GR), and ferric reducing ability assay methods are reported to evaluate in vivo antioxidant potential (ASantos et al., 2016).

## 12.4  ANTIOXIDANT PROPERTIES OF COMMERCIAL CINNAMON

A number of investigations on the antioxidant potential of cinnamon were carried out by workers and their findings pointed out that cinnamon has a very high potential in countering life-threatening problems such as peroxidation of polyunsaturated fatty acids in the phospholipid membranes, formation of cytotoxic peroxides, oxidation of proteins, denaturation of DNA, and so forth (Vijayan and Thampuran, 2012).

A review of scientific literature was carried out to explore the previous works on the antioxidant potential of commercial cinnamon by various assays, such as DPPH free radical scavenging activity, FRAP, ABTS, reducing power assay, and so forth, and described in Table 12.2. Studies revealed that many antioxidant activity assays have been performed on methanol extracts, a most used solvent (Mathew and Abraham, 2006; Sultana et al., 2010; Sudan et al., 2013). Antioxidant potential of cinnamon

**TABLE 12.2** Antioxidant Potentials of *Cinnamomum* Species by Different Assays

| S/ No. | Species | Plant parts | Extract | Assay methods used and brief of antioxidant potential | References |
|---|---|---|---|---|---|
| | C. verum | Bark | Ethanol | **DPPH:** $93.4 \pm 0.21\%$ inhibition. <br> **Total antioxidant:** $0.93 \pm 0.15$ Abs. | Anal et al. (2014) |
| | C. verum | Bark | Essential oil | **DPPH:** $91.4 \pm 0.002\%$ inhibition at the concentration of 5 mg/mL | Lin et al. (2009) |
| | C. verum | Leaf | Methanol | **DPPH:** The $EC_{50}$ value found to be 4.21 µg/mL. <br> **ABTS:** Increased with increasing concentration. <br> **Reducing power assay:** 2.727 at a dose of 1 mg. | Mathew and Abraham (2006) |
| | C. verum | Leaf, bark | Methanol, ethyl acetate | **DPPH:** Methanol extract of bark exhibited good antioxidant activity with $IC_{50}$ value 76.5 µg/mL. <br> **Metal chelating assay:** Methanol extract of bark exhibited good antioxidant activity with $IC_{50}$ value 72.3 µg/mL | Mazimba et al. (2015) |
| | C. verum | Leaf | Aqueous, Ethanol | **DPPH:** Ethanol and aqueous extract exhibited almost equal antioxidant activity with $IC_{50}$ value 13.53 µg/mL and 13.3 µg/mL, respectively. <br> **Super radical scavenging assay:** Ethanol extract exhibited good antioxidant activity with $IC_{50}$ value 119.7µg/mL | Pandey and Chandra (2015) |
| | C. verum | Bark | Aqueous | **DPPH:** $EC_{50}$ value 0.321 mg/mL. <br> **Metal Chelating Assay:** $39.05 \pm 0.17$ at 1 mg concentration. <br> **Reducing Power Assay:** $EC_{50}$ value 10.05 mg/mL | Rakshit and Ramalingam (2011) |
| | C. verum | Bark | Essential oil, Methanol, Ethanol | **DPPH:** Essential oil exhibited good antioxidant activity with $89.6 \pm 5.23\%$ inhibition at the concentration ranging from 50 to 200 µg/mL <br> **ABTS:** Essential oil exhibited good antioxidant activity with $1023 \pm 51.3$ µm Trolox/g | Ramadan et al. (2014) |

**TABLE 12.2** *(Continued)*

| S/ No. | Species | Plant parts | Extract | Assay methods used and brief of antioxidant potential | References |
|---|---|---|---|---|---|
| | *C. verum* | Bark | Aqueous, Methanol, Chloroform | **DPPH:** Methanol extract exhibited good antioxidant activity with $IC_{50}$ value $111.5 \pm 0.62$ μg/mL <br><br> **Chelation Power on Ferrous Ion:** Methanol extract exhibited good antioxidant activity with $IC_{50}$ value $108.7 \pm 0.53$ μg/mL | Sudan et al. (2013) |
| | *C. verum* | Bark | Methanol | **DPPH:** $IC_{50}$ value 159.14 μg/mL at the concentration ranging from 500 to 0.98 μg/mL (decreasing order). | Sultana et al. (2010) |
| | *C. zeylanicum* | Leaf, Bark | Ethanol, Dichloromethane: Methanol (DCM: M) | **DPPH:** Ethanol extract of bark exhibited strong antioxidant activity, that is, $107.7 \pm 2.01$ mg Trolox. <br><br> **ABTS:** Ethanol extract of leaf exhibited strong antioxidant activity that is, $121.8 \pm 3.20$ mg Trolox. <br><br> **FRAP:** Ethanol extract of leaf exhibited strong antioxidant activity that is, $125.7 \pm 3.21$ mg Trolox. <br><br> **ORAC:** Ethanol extract of leaf exhibited strong antioxidant activity that is, $44.74 \pm 0.36$ mg Trolox. | Abeysekera et al. (2013) |
| | *C. zeylanicum* | Bark | Essential oil | **DPPH:** 13.02% inhibition. | Aliakbarlu et al. (2013) |
| | *C. zeylanicum* | Bark | Ethyl acetate | **DPPH:** $86.408 \pm 0.23\%$ inhibition at the concentration 25 mg/mL <br><br> **FRAP:** $122.833 \pm 6.59\%$ inhibition at the concentration 25 mg/mL | Asimi et al. (2013) |
| | *C. zeylanicum* | Bark | Essential oil | **DPPH:** $IC_{50}$ value 13.1 g/mL at the concentration ranging from 5 to 200 μg. | El-Baroty et al. (2010) |
| | *C. zeylanicum* | Bark | Water, ether, methanol | **β-Carotene linoleic acid system:** Methanolic extract showed 95.5% inhibition at 44 mg/mL | Mancini-Filho et al. (1998) |

**TABLE 12.2**  (Continued)

| S/ No. | Species | Plant parts | Extract | Assay methods used and brief of antioxidant potential | References |
|---|---|---|---|---|---|
| | C. zeylanicum | Bark | Water, ethanol, methanol | **DPPH:** Strong antioxidant activity exhibited by methanol extract that is, 18.16±1.34 mM TE/g DW. **ABTS:** Methanol extract exhibited strong antioxidant activity1.92±0.02 mm TE/g DW. **Reducing power assay:** The reducing power of methanol extract was observed 0.668. | Georgieva and Mihaylova (2014) |
| | C. zeylanicum | Bark | Water | **DPPH:** Strong antioxidant activity shown at the concentration 0.29–5.66 mg/mL **FRAP:** 78.28–84.30% inhibition. **Reducing Power Assay:** 0.22–2.19 mM/mg reducing capacity. | Nanasombat et al. (2011) |
| | C. zeylanicum | Bark | Essential oil | **DPPH:** 14% inhibition at the concentration of 50 g/mL **FRAP:** <1 mmol/L at the concentration ranging from 50 to 5 g/mL (decreasing order). | Politeo et al. (2006) |
| | C. zeylanicum | Leaf, bark | Essential oil | **DPPH:** Maximum 73.9% and 82.1% inhibition of leaf and bark oil at 25 µl concentration. **Hydroxyl radical scavenging:** Highest % inhibition of leaf and bark oil are 72.2% and 79.6%, respectively. | Singh et al. (2007) |
| | C. zeylanicum | Bark | Aqueous | **DPPH:** 84.39±5.90 at 120 µg/mL concentration. **ABTS:** 54.84±3.83 at 12 µg/mL concentration. **FRAP:** 0.219±0.010 at 60 µg/mL concentration. **Nitric oxide radical scavenging activity:** 55.61±3.89 at 120 µg/mL concentration. **Hydroxyl radical scavenging Activity:** 67.06±4.69 at 60 µg/mL **Reducing power assay:** 0.219±0.010 at 60 µg/mL concentration. | Sivaraj et al. (2015) |

**TABLE 12.2** (Continued)

| Sl No. | Species | Plant parts | Extract | Assay methods used and brief of antioxidant potential | References |
|---|---|---|---|---|---|
| | C. zeylanicum | Bark | Methanol | **DPPH:** $92.5 \pm 0.303\%$ at 40µL concentration. | Slowianek and Leszczynska (2016) |
| | C. zeylanicum | Bark | Essential oil | **DPPH:** $IC_{50}$ value −79.54 µL | Valizadeh et al. (2015) |
| | C. zeylanicum | Bark | Water, methanol, chloroform | **DPPH:** Methanolic extract showed strong antioxidant activity with 80% inhibition.<br><br>**ABTS:** Methanolic extract showed strong antioxidant activity with 68% inhibition.<br><br>**H$_2$O$_2$ Scavenging Assay:** Methanolic extract exhibited strong antioxidant activity with 79% inhibition. | Varalakshmi et al. (2012) |
| | C. zeylanicum | Leaf | Essential oil | **ABTS:** $96.45 \pm 0.01\%$ inhibition at 1 mg/mL concentration.<br><br>**Reducing power assay:** $119.42 \pm 0.68\%$ at 10 mg/mL concentration inhibition. | Wang et al. (2010) |
| | C. zeylanicum | Bark | Essential oil | **Linoleic acid peroxidation:** 94.42% inhibition in Essential oil. | Womeni et al.(2013) |
| | C. aromaticum | Leaf | Acetone | **DPPH:** 76.37% at the concentration of 25µL.<br><br>**Reducing power assay:** 87.2% and 83.5% (conjugated diene method) at the concentration of 25µL. | Singh et al. (2007) |
| | | | Essential oil | **DPPH:** 67.22% at the concentration of 25µL.<br><br>**Reducing power assay:** 68.5% and 70.22% (conjugated diene method) at the concentration of 25µL. | |
| | C. cassia | Bark | Methanol | **DPPH:** Methanol extract possess high increase in antioxidant activity with $IC_{50}$ value $42.03 \pm 0.06$ (µg/L) at the concentration ranging from 5.0 to 50.0 µg/L.<br><br>**ABTS:** $IC_{50}$ value $5.13 \pm 0.07$ (µg/L) at the concentration ranging from 5.0 to 50.0 (µg/L). | Brodowska et al. (2016) |

**TABLE 12.2** (Continued)

| Sl No. | Species | Plant parts | Extract | Assay methods used and brief of antioxidant potential | References |
|---|---|---|---|---|---|
| | | | Essential oil | **DPPH:** IC$_{50}$ value 147.23±0.04 (µg/L) at the concentration ranging from 5.0 to 50.0 µg/L <br> **ABTS:** IC$_{50}$ value 64.51±0.09 at the concentration ranging from 5.0 to 50.0 µg/L | Karadagli et al. (2014) |
| | C. cassia | Bark | Water | **DPPH:** IC$_{50}$ value 76.68 µg/mL at the concentration ranging from 0.01 to 100 µg/mL | Kumar and Prakash (2012) |
| | C. cassia | Bark | Ethanol | **DPPH:** The % inhibition was found to be 86.6±4.33 mg/100g DW at the concentration ranging from 0.02 to 2 mg/mL | |
| | C. cassia | Bark | Water, ethanol | **Xanthine oxidase inhibition test:** Ethanol extract revealed the strongest antioxidant activity with IC$_{50}$ value 0.09 mg/mL | Lin et al. (2003) |
| | C. cassia | Leaf, bark, and fruits | Ethanol, SCFE | **DPPH:** Ethanol extract exhibited strongest antioxidant activity with IC$_{50}$ value 0.072–0.208 mg/mL <br> **Trolox equivalent antioxidant capacity:** SCFE extract exhibited strongest antioxidant activity with IC$_{50}$ value 6.789–58.335 mmol Trolox/g | Yang et al. (2012) |
| | C. burmannii | Bark | Aqueous, ethanol, ethyl acetate | **DPPH:** Aqueous infusion exhibited higher antioxidant activity with IC$_{50}$ value 3.03±0.22 µg/mL | Ervina et al. (2016) |
| | C. tamala | Leaf | Ethanol | **DPPH:** IC$_{50}$ value 13.55 µg/mL | Akter et al. (2015) |
| | C. tamala | Leaf | Methanol, hexane, ethyl acetate | **ABTS:** Hexane extract exhibited highest antioxidant activity with IC$_{50}$ value 38.86±1.64 µg/mL <br> **Super oxide radical scavenging assay:** Hexane extract exhibited highest antioxidant activity with IC$_{50}$ value 187.48±2.0 µg/mL <br> **Hydroxyl radical scavenging assay:** Hexane extract exhibited highest antioxidant activity with IC$_{50}$ value 493±3.2 µg/mL | Chaurasia and Tripathi (2011) |

**TABLE 12.2** (Continued)

| S/ No. | Species | Plant parts | Extract | Assay methods used and brief of antioxidant potential | References |
|---|---|---|---|---|---|
| | C. tamala | Leaf | Methanol | **DPPH:** 80% inhibition capacity at the concentration of 220 µg GAE<br><br>**Hydroxyl radical scavenging assay:** 67.2±4.3% inhibition capacity at the concentration of 220 µg GAE | Devi et al. (2007) |
| | C. tamala | Leaf | Essential oil | **DPPH:** 250±1.2 µg/mL at the concentration ranging from 1.0 to 50.0 g/L<br><br>**$H_2O_2$ scavenging power assay:** 180±1.4 µg/mL at the concentration ranging from 20 to 60 µg/mL | Kumar et al. (2012) |
| | C. tamala | Leaf | Essential oil | **DPPH:** Cinnamaldehyde and eugenol chemotypes exhibited maximum scavenging activity at 0.8% concentration i.e. 32.2 and 37.5%, respectively. | Padalia et al. (2012) |
| | C. tamala | Leaf | Ethanol | **DPPH:** 90.89±2.07% inhibition at the concentration ranging from 200 to 800 µg/mL | Palanisamy et al. (2011) |
| | C. tamala | Leaf | Aqueous, methanol, chloroform | **DPPH:** Methanol extract possess higher antioxidant activity with $IC_{50}$ value 175±0.31 µg/mL<br><br>**Chelation power ferrous assay:** Methanol extract exhibited high antioxidant activity with $IC_{50}$ value 114.2±0.46 µg/mL | Sudan et al. (2013) |
| | C. tamala | Leaf | Methanol | **DPPH:** $IC_{50}$ value 157.58 µg/mL at the concentration ranging from 0.98 to 500 µg/mL (decreasing order). | Sultana et al. (2010) |
| | C. tamala | Leaf | Ethanol | **FRAP:** 0.40 m M/100 g at the concentration of 50 mg/mL | Thakur et al. (2013) |

was observed not only in bark but also in leaves and essential oil. Mazimba et al. (2015) evaluated antioxidant potential of *C. verum* bark and leaves and found that the methanolic extract is more effective than ethyl acetate solvent as it extracted higher concentration of phenolics that exhibited better antioxidant activities. *C. verum* bark, extracted by the ultrasonic method has shown higher DPPH radical scavenging activity and total antioxidant activity than vacuum microwave method (Anal et al., 2014). Methanolic extract of *C. zeylanicum* bark with DPPH assay detected good antioxidant activity (Slowianek and Leszcznska, 2016). Varalakshmi et al. (2012) and Georgieva and Mihaylova (2014) also assessed the methanolic and aqueous extract of *C. zeylanicum* bark and reported potent antioxidant activity in methanolic extracts using different assays, while Ghosh et al. (2015) evaluated the antioxidant activity of pectic polysaccharides extracted from bark, using DPPH and FRAP assays and revealed that polysaccharides could be used as natural antioxidants by the food industry. Abeysekera et al. (2013) used dichloromethane: methanol (DCM: M) and ethanolic extracts of leaf and bark of *C. zeylanicum* and observed that the ethanolic extract of both leaf and bark had high antioxidant activity by using DPPH, ABTS, FRAP, and ORAC assay methods. Significant antioxidant activity in methanolic extracts of *C. cassia* was observed using DPPH and reducing power assay (Kumar et al., 2012). Karadagli et al. (2014) determined the antioxidant activity in a water extract of *C. cassia* against $H_2O_2$ induced oxidative DNA damage using DPPH assay. Yang et al. (2012) investigated the antioxidant activities of barks and leaves of *C. cassia* extracted with ethanol and super-critical fluid extraction (SCFE), by using the DPPH scavenging assay.

In another study, the dichloromethane, ethanol, and water extracts of *C. cassia* were tested to determine the antioxidant activity using CUPRAC assay in which the ethanolic extracts were found to be more effective (Boga et al., 2011). Ervina et al. (2016) compared the in vitro antioxidant activity of infusion, extracts, and fractions of Indonesian cinnamon (*C. burmannii*) bark by DPPH assay and calculated the $IC_{50}$ value (inhibitory concentration). The lowest values showed higher antioxidant activity. They also found that the antioxidant activity of infusion and ethanolic extract were higher than others. The antioxidant potential of ethnolic extract of *C. tamala* was assessed by FRAP assay and found to possess strong anti-oxidant activity (Thakur et al., 2013). The ethanolic extract showed the antioxidant FRAP value of 0.40 mm/100 g. In another study, the ethanolic extract of *C. tamala* leaves was not only found to have an increased anti-oxidant activity but also significantly decreased the blood glucose level in streptozotocin induced diabetes (Palanisamy et al., 2011). Devi et al.

(2007) reported that methanolic extract of *C. tamala* leaves effectively scavenge ROS and suppresse $Fe^{+2}$ ascorbate-induced lipid peroxidation in brain synaptosomes. Similarly, Reddy and Lokesh (1992) obsereved that cinnamon at high doses (600 μg/mL) inhibited lipid peroxidation of rat microsomes. Singh et al. (2007) observed the scavenging effects of acetone extracts of *C. tamala* and *C. aromaticum* by using DPPH, reducing power and chelating effect assays.

Many researchers have also explored the antioxidant properties of essential oils extracted from cinnamon bark and leaves and found strong activity by using various assay methods (Singh et al., 2007; Lin et al., 2009; Kumar et al., 2012; Aliakbarlu et al., 2013; Ramadan et al., 2014). Brodowska et al. (2016) examined the antioxidant activity of essential oils and methanolic extract of *C. cassia* bark using DPPH and ABTS assay. The inhibitory effects of essential oil ($147.23 \pm 0.04$ and $64.51 \pm 0.09$ mg/L) and extract ($5.13 \pm 0.07$ and $64.51 \pm 0.09$ mg/L) were observed by using DPPH and ABTS methods, respectively. Padalia et al. (2012) studied the antioxidant potential in the essential oil of two *C. tamala* chemotypes (Cinnamaldehyde and eugenol). Both oil chemotypes have shown maximum DPPH scavenging activity, that is, 32.2 and 37.5% at 0.8% concentration, respectively. The antioxidant activity of oils suggests that it could serve as a source of compounds with preservative phenomenon.

## 12.5  CONCLUSION

This review provides information on commercial species of the genus *Cinnamomum*, which shows potential antioxidant activity. Cinnamon barks, leaves, and essential oils are having plenty of medicinal properties, especially their antioxidant potential and therefore, the global demand of cinnamon is increasing rapidly. The antioxidant potential of cinnamon has been assessed by the researchers using various assay methods. The methanol extract of cinnamon bark and leaves was found with the highest frequency for antioxidant study. The cinnamon of commerce is a good source of natural antioxidant. Due to the presence of phenols, flavonoids, and antioxidant compounds, cinnamon has many health benefits and considered as the most valuable spice. Antidiabetic and anti-inflammatory properties are also reported in literature. A daily consumption of cinnamon can have a positive effect on blood sugar level and digestion. The essential oil of cinnamon can be used as a natural food preservative in the food industry.

## KEYWORDS

- antioxidant
- cinnamon
- essential oil
- DPPH
- FRAP

## REFERENCES

Abeysekera, W. P. K. M.; Premakumara, G. A. S.; Ratnasooriya, W. D. In Vitro Antioxidant Properties of Leaf and Bark Extracts of Ceylon Cinnamon (*Cinnamomum zeylanicum* Blume). *Trop. Agric. Res.* **2013,** *24*(2), 128–138.

Akter, S.; Ali, M. A.; Barman, R. K.; Rahman, B. M.; Wahed, M. I. I. In Vitro, Antioxidant and Cytotoxic Activity of Ethanolic Extract of *Cinnamomum tamala* (Tejpat) Leaves. *Int. J. Pharm. Sci. Res.* **2015,** *6*(3), 531–536.

Aliakbarlu, J.; Sadaghiani, S. K.; Mohammadi, S. Comparative Evaluation of Antioxidant and Anti Food-Borne Bacterial Activities of Essential Oils from Some Spices Commonly Consumed in Iran. *Food Sci. Biotechnol.* **2013,** *22*(6), 1487–1493.

Anal, A. K.; Jaisanti, S.; Noomhorn, A. Enhanced Yield of Phenolic Extracts from Banana Peels and Cinnamon (*Cinnamomum verum*) and Their Antioxidative Potential in Fish Oil. *J. Food Sci. Technol.* **2014,** *51*(10), 2632–2639.

Anonymous, Cinnamomum Schaeff. (Lauraceae). In *The Wealth of India - Raw Materials;* Publications and Information Directorate, Council of Scientific and Industrial Research: New Delhi, 1992; Vol. 3, pp 572–589.

Asimi, O. A.; Sahu, N. P.; Pal. A. K. Antioxidant Activity and Antimicrobial Property of Some Indian Spices. *Int. J. Sci. Res. Publ.* **2013,** *3*(3), 1–8.

Boga, M.; Hacibekiroglu I.; Kolak U. Antioxidant and ant cholinesterase activities of eleven edible Plants. *Pharm. Biol.* **2011,** *49*(3), 290–295.

Brodowska, K. M.; Brodowska, A. J.; Smigielski, K.; Chruscinska, E. L. Antioxidant Profile of Essential Oil and Extracts of Cinnamon Bark (*Cinnamomum cassia*). *Eur. J. Biol. Res.* **2016,** *6*(4), 310–316.

Chaurasia, J. K.; Tripathi, Y. B. Chemical Characterization of Various Fractions of Leaves of *Cinnamomum tamala* Linn. Toward Their Antioxidant, Hypoglycemic and Anti-Inflammatory Property. *Immunopharmacol. Immunotoxicol.* **2011,** *33*(3), 466–472.

Devi, S. L.; Kannappan, S.; Anuradha, C. V. Evaluation of in Vitro Antioxidant Activity of Indian Bay Leaf *Cinnamomum tamala* (Buch.-Ham.) T. Nees and Eberm Using Rat Brain Synaptosomes as Model System. *Indian J. Exp. Biol.* **2007,** *45*(9), 778–784.

Dhar, U.; Manjkhola, S.; Joshi, M.; Bhat, A.; Bisht, A. K.; Joshi, M. Current Status and Future Strategy for the Development of Medicinal Plant Sector in Uttaranchal, India. *Curr. Sci.* **2002,** *83*(8), 956–963.

El-Baroty, G. S.; El-Baky, H. H. A.; Farag, R. S.; Saleh, M. A. Characterization of Antioxidant and Antimicrobial Compounds of Cinnamon and Ginger Essential Oils. *Afr. J. Biochem. Res.* **2010,** *4*(6), 167–174.

Ervina, M.; Nawu,Y. E.; Esar, S. Y. Comparison of in Vitro Antioxidant Activity of Infusion, Extract and Fractions of Indonesian Cinnamon (*Cinnamomum burmannii*) Bark. *Int. Food Res. J.* **2016,** *23*(3), 1346–1350.

Georgieva, L.; Mihaylova, D. Evaluation of the *in vitro* antioxidant potential of extracts obtained from *Cinnamomum zeylanicum* barks. *Sci. Work Russ. Univ.* **2014,** *53*(10.2), 41–45.

Ghosh, T.; Basu, A.; Adhikari, D.; Roy, D.; Pal, A.K. Antioxidant Activity and Structural Features of *Cinnamomum zeylanicum*. *3 Biotech.* **2015,** *5*(6), 939–947.

Gulcin, I. Antioxidant activity of food constituents: An overview. *Arch. Toxicol.* **2012,** *86*(3), 345–391.

Halliwell, B. Biochemistry of Oxidative Stress. *Biochem. Soc. Trans.* **2007,** *35*(5), 1147–1150.

Jayanthi, P.; Lalitha, P. Reducing Power of the Solvent Extracts of Eichhornia Crassipes (Mart.) Solms. *Int. J. Pharm. Pharm. Sci.* **2011,** *3*(3), 126–128.

Karadagli, S. S.; Agrap, B.; Erciyas, F. L. Investigation of the Protective Effect of *Cinnamomum cassia* Bark Extract Against H2o2-Induced Oxidant DNA Damage in Human Peripheral Blood Lymphocytes and Antioxidant Activity. *Marmara Pharm. J.* **2014,** *18*, 43–48.

Kasote, D. M.; Katyare, S. S.; Hegde, M. V.; Bae, H. Significance of Antioxidant Potential of Plants and Its Relevance to Therapeutic Applications. *Int. J. Biol. Sci.* **2015,** *11*(8), 982–991.

Kostic, D. A.; Dimitrijevic, D. S.; Gordana, S.; Stojanovic, G. S.; Palic, I. R.; Dordevic, A. S.; Ickovski, J. D. Xanthine oxidase: isolation, assays of activity, and inhibition. *J. Chem.* **2015,** Article ID 294858. 1–8. http://dx.doi.org/10.1155/2015/294858

Kumar, U.; Prakash V. Comparative Analysis of Antioxidant Activity and Phytochemical Screening of Some Indian Medicinal Plants. *Int. J. Pharm. Pharm. Sci.* **2012,** *4*(3), 291–295.

Kumar, S.; Vasudeva, N.; Sharma, S. GC-MS Analysis and Screening of Antidiabetic, Antioxidant and Hypolipidemic Potential of *Cinnamomum tamala* Oil in Streptozotocin Induced Diabetes Mellitus in Rats. *Cardiovasc. Diabetol.* **2012,** *11*(95), 1–11.

Lin, C. C.; Wu, S. J.; Chanq, C. H.; Ng, L. T. Antioxidant Activity of *Cinnamomum cassia*. *Phytother. Res.* **2003,** *17*(7), 726–730.

Lin, C.; Yu, C.; Wu, S.; Yih, K. DPPH Free Radical Scavenging Activity; Total Phenolic Contents and Chemical Composition Analysis of Forty Two Kinds of Essential Oils. *J. Food Drug Anal.* **2009,** *17*(5), 386–395.

Lohani, H.; Singh, S. K.; Bhandari, U.; Haider, S. Z.; Gwari, G.; Chauhan, N. Chemical Polymorphism in *Cinnamomum tamala* (Buch.-Ham.) Nees. and Eberm. Growing in Uttarakhand Himalaya (India). *J. Chem. Pharm. Res.* **2015,** *7*(8), 67–71.

Lu, Z.; Jia, Q.; Wang, R.; Wu, X.; Wu, Y.; Huang, C.; Li, Y. Hypoglycemic Activities of A- and B-Type Procyanidin Oligomer-Rich Extracts from Different Cinnamon Barks. *Phytomed.* **2011,** *18*(4), 298–302.

Mancini-Filho, J.; Van-Koiij, A.; Mancini, D. A.; Cozzolino, F. F.; Torres, R. P. Antioxidant Activity of Cinnamon (*Cinnamomum zeylanicum* Breyne) Extracts. *Boll. Chim. Farm.* **1998,** *137*(11), 443–447.

Mathew, S.; Abraham, T. E. In Vitro Antioxidant Activity and Scavenging Effects of Cinnamomum Verum Leaf Extract Assayed by Different Methodologies. *Food. Chem. Toxicol.* **2006,** *44*(2), 198–206.

Mazimba, O.; Wale, K.; Kwape, T. E.; Mihigo, S. O.; Kokengo, B. M. Cinnamomum Verum Ethylacetate and Methanol Extracts Antioxidants and Antimicrobial Activity. *J. Med. Plants Stud.* **2015,** *3*(3), 28–32.

Mustaffa, F.; Hassan, Z.; Abdullah, M. Antioxidant Activity of *Cinnamomum iners* Leaves Standardized Extract. *Asian J. Biochem.* **2016**, *11*(2), 90–96.

Myburgh, K. H. Polyphenol Supplementation: Benefits for Exercise Performance or Oxidative Stress? Sports Med. **2014**, 1, S57–70.

Nanasombat, S.; Wimuttigosol, P. Antimicrobial and Antioxidant Activity of Spice Essential Oils. *Food Sci. Biotechnol.* **2011**, *20*(1), 45–53.

Padalia, R. C.; Verma, R. S.; Sah, A.; Karki, N.; Chauhan, A.; Sakia, D.; Krishna, B. Study on Chemotypic Variations in Essential Oil of *Cinnamomum tamala* and Their Antibacterial and Antioxidant Potential. *J. Essent. Oil Bearing Plants.* **2012**, *15*(5), 800–808.

Palanisamy, P.; Srinath, K. R.; Kumar, D. Y.; Chowdhary, P. Evaluation of Antioxidant and Anti Diabetic Activities of *Cinnamomum tamala* Leaves in Streptozotocin-Induced Diabetic Rats. *Int. Res. J. Pharm.* **2011**, *2*(12), 157–162.

Pandey, M.; Chandra, D. R. Evaluation of Ethanol and Aqueous Extracts of *Cinnamomum tamala* Leaf Galls for Potential Antioxidant and Analgesic Activity. *Indian J. Pharm. Sci.* **2015**, *77*(2), 243–247.

Parul, R.; Kundu, S. K.; Saha, P. In Vitro Nitric Oxide Scavenging Activity of Methanol Extracts of Three Bangladeshi Medicinal Plants. *The Pharma Innovation.* **2013**, *1*(12), 83–88.

Peschel, F.; Sanchez; Rabaneda,W.; Dieckmann, A.; Plescher, I.; Gartzia. An Industrial Approach in the Search for Natural Antioxidants from Vegetable and Fruits Wastes. *Food Chem.* **2006**, *97*(1), 137–150.

Pisoschi, A. M.; Negulescu, G. P. Methods for Total Antioxidant Determination: a Review. *Biochem. Anal. Biochem.* **2011**, *1*(1), 1–10.

Politeo, O.; Jukic, M.; Milos, M. Chemical Composition and Antioxidant Activity of Essential Olis of Twelve Spice Plants. *Croat. Chem. Acta.* **2006**, *79*(4), 545–552.

Pravin, B.; Krishnkant, L.; Shreyas, J.; Ajay, K.; Priyanka, G. Recent Pharmacological Review on *Cinnamomum tamala. Res. J. Pharm. Biol. and Chem. Sci.* **2013**, *4*(4), 916–921.

Radi, R. Peroxynitrite, A. Stealthy Biological Oxidant. *J. Biol. Chem.* **2013**, *288*(37), 26464–26472.

Rakshit, M.; Ramalingam, C. In Vitro Antibacterial and Antioxidant Activity of Cinnamomum Verum Aqueous Bark Extract Inreference to Its Total Phenol Content as Natural Preservative to Food. *Int. J. Biol. Biotech.* **2011**, *8*(4), 529–537.

Ramadan, M. M.; Yehia, H. A.; Shaheen, M. S.; El-Fattah, M. A. Aroma Volatiles, Antibacterial, Antifungal and Antioxidant Properties of Essential Oils Obtained from Some Spices Widely Consumed in Egypt. *American-Eurasian J. Agric. Environ. Sci.*, **2014**, *14*(6), 486–494.

Ranasinghe, P.; Pigera, S.; Premakumara, G. A. S.; Galappaththy, P.; Constantine, G. R.; Katulanda, P. Medicinal Properties of 'True' Cinnamon (*Cinnamomum zeylanicum*): a Systematic Review. *BMC Complement. Altern. Med.* **2013**, *13*, 275.

Reddy, A. C.; Lokesh, R. R. Studies on Spice Principles as Antioxidants in the Inhibition of Lipid Peroxidation of Rat Liver Microsomes. *Mol. Cell. Biochem.* **1992**, *111*(1-2), 117–124.

Santos, C.; Pires, M. D. A.; Santos, D.; Payan-Carreira, R. Distribution of Superoxide Dismutase 1 and Glutathione Peroxidise 1 in the Cyclic Canine Endometrium. *Theriogenology.* **2016**, *86*(3), 738–748.

Sharma, U. K.; Sharama, A. K.; Pandey, A. K. Medicinal Attributes of Major Phenyl Propanoids Present in Cinnamon. *BMC Compliment. Altern. Med.* **2016**, *16*, 156.

Shintani, H. Determination of Xanthine Oxidase. *Pharm. Anal. Acta.* **2013**, *S7:004*, 1–2.

Singh, G.; Maurya, S.; De Lampasona, M. P.; Catalan, C. A. A Comparison of Chemical, Antioxidant and Antimicrobial Studies of Cinnamon Leaf and Bark Volatile Oils, Oleoresins and Their Constituents. *Food Chem. Toxicol.* **2007,** *45*(9), 1650–1661.

Sivaraj, C.; Abirami, K.; Nishanthika, T. K.; Purana, K. N.; Arumugam, P., Iqbal, S. Antioxidant Activities and Thin Layer Chromatographic Analysis of Aqueous Extract of Barks of *Cinnamomum zeylanicum* Blume. *Am. J. Phytomedicine Clin. Ther.* **2015,** *3*(10), 654–665.

Slowianek, M.; Leszczynska, J. Antioxidant Properties of Selected Culinary Spices. *Herba Pol.* **2016,** *62*(1), 29–41.

Sudan, R.; Bhagat, M.; Gupta, S.; Chitrarakha; Devi, T. Comparative Analysis of Cytotoxic and Antioxidant Potential of Edible *Cinnamomum verum* (Bark) and *Cinnamomum tamala* (Indian Bay Leaf). *Free Radic. Antioxid.* **2013,** *3*, S70–S73.

Sultana, S.; Ripa, F. A.; Hamid, K. Comparative Antioxidant Activity Study of Some Commonly Used Spices in Bangladesh. *Pak. J. Biol. Sci.* **2010,** *13*(7), 340–343.

Thakur, R.; Yadav, K.; Khadka, K. B. Study of Antioxidant, Antibacterial and Anti-Inflammatory Activity of Cinnamon (*Cinnamomum tamala*), Ginger (Zingiber of ficinale) and Turmeric (*Curcuma longa*). *Am. J. Life Sci.* **2013,** *1*(6), 273–277.

Ueno, H.; Yamakur, S.; Arastoo, R. S.; Oshima, T.; Kokubo, K. Systematic Evaluation and Mechanistic Investigation of Antioxidant Activity of Fullerenols Using β-Carotene Bleaching Assay. *J. Nanomater.* **2014,** *2014*, 1–7.

Valizadeh, S.; Katiraee, F.; Magmoud, R.; Fakheri, T.; Mardani, K. Biological Properties of *Cinnamomum zeylanicum* Essential Oil: Phytochemical Component; Antioxidant and Antimicrobial Activities. *Int. J. Food Nutr. Safety.* **2015,** *6*(3), 174–184.

Vangalapati, M.; Sree Satya, N.; Surya Prakash, D. V.; Avanigadda, S. A Review of Pharmacological Activities and Clinical Effects of Cinnamon Species. *Res. J. Pharm. Biol. Chem. Sci.* **2012,** *3*(1), 653–666.

Varalakshmi, B.; Anand, V.; Kumar, V. K.; Prasana, R. In Vitro Antioxidant Activity of *Cinnamomum zeylanicum* Linn Bark. *Int. J. Sci. Innov. and Discov.* **2012,** *2*(3), 154–164.

Vijayan, K. K.; Thampuran, R. V. A. Pharmacology and Toxicology of Cinnamon and Cassia. In *Cinnamon and Cassia: The Genus Cinnamomum;* Ravindran, P. N., Babu, K. N., Shylaja, M., Eds.; CRC Press: Boca Raton, Florida, 2012; pp 259–284.

Wang, H. F.; Yih, K. H.; And, K.F.; Huang, K. F. Comparative Study of Antioxidant Activity of Forty Five Commonly Used Essential Oils and Their Potential Active Components. *J. Food Drug Anal.* **2010,** *18*(1), 24–33.

Womeni, H. M.; Djikeng, F. T.; Tiencheu, B.; Linder, M. Antioxidant Potential of Methanolic Extracts and Powders of Some Cameroonian Spices During Accelerated Storage of Soyabean Oil. *Adv. Biol. Chem.* **2013,** *3*(3), 304–313.

www.cinnamonvogue. Types of Cinnamon. (Available https://www.cinnamonvogue.com/Types_of_Cinnamon_1.html) (Retrieved: January 18, 2018).

www.fao.org/statistics/en/ (FAOSTAT). Available http://www.fao.org/faostat/en/#data/QC) (Retrieved February 7, 2018).

Yang, C. H.; Li, R. X.; Chang, L. Y. Antioxidant Activity of Various Parts of *Cinnamomum cassia* Extracted with Different Extraction Methods. *Molecules.* **2012,** *17*(6), 7294–7304.

# CHAPTER 13

# PHYTOCHEMICAL AS HOPE FOR THE TREATMENT OF HEPATIC AND NEURONAL DISORDERS

SWAGATA DAS[1], PRAREETA MAHAPATRA[1], PRIYANKA KUMARI[1], PREM PRAKASH KUSHWAHA[1], PUSHPENDRA SINGH[2], and SHASHANK KUMAR[1,*]

[1]*School of Basic and Applied Sciences, Department of Biochemistry and Microbial Sciences, Central University of Punjab, Bathinda, Punjab, 151001, India, Mob.: +91 9335647413*

[2]*National Institute of Pathology, New Delhi, India*

*Corresponding author. E-mail: shashankbiochemau@gmail.com, shashank.kumar@cupb.edu.in*
*ORCID: https://orcid.org/0000-0002-9622-0512*

## ABSTRACT

Hepatic disorders and neuronal disorders are very common nowadays, and their prevalence is spread to different populations throughout the world. Various acute and chronic hepatic disorders like hepatitis, steatosis, cirrhosis, and cancer are a common cause for concern due to the elevation in their occurrences. Different drug therapies are available for certain diseases, showing excellent efficiency, but most of them are cost intensive and with side effects. Thus there is an urgent need for alternatives which are cost-effective and with no side effects. Plant-derived phytochemicals are gaining importance nowadays. They are usually cost-effective and have lesser toxicity or side effects. The present chapter discusses the various forms of hepatic and neurological disorders such as fatty liver, hepatitis C virus, Parkinson's and Alzheimer's diseases. Emphasis was laid on the phytochemicals involved in the mitigation of these diseases. Also, the chapter provides a mechanistic approach involved in targeting hepatic and neuronal diseases using phytochemicals.

## 13.1    INTRODUCTION

The liver is the largest organ of the human body comprising of multiple functions including the production of proteins and blood clotting factors, glycogen synthesis, manufacture of cholesterol and triglyceride, and bile production. It is also the largest gland of the body which plays an important role in the metabolism, storage, and detoxification of various substances in our body. Having a numerous function, along with its correlation, maintenance, and proper functioning of other associated organs, any dysfunction of the liver can cause a serious health hazard. Hepatic (liver) disorders may include infections such as hepatitis, cirrhosis, cancers, fascioliasis, and damage due to medication or toxins. Excessive alcohol consumption may also be a serious threat to the liver. Disease associated with the neuronal degeneration is known as neurodegenerative disease. Various endogenous and exogenous factors are known to be involved in neurodegenerative disease. Phytochemicals are from time immortal being used for their medicinal properties. Because of their non-toxicity/less toxicity and cost-effectiveness, natural products are nowadays very important therapeutic agents.

## 13.2    HEPATITIS

Hepatitis refers to a condition of viral infection of the liver which leads to inflammation. It may be of various types including A, B, C, D, and E. Hepatitis A and B spread through the fecal-oral route and blood, respectively. Both A and B types have vaccinations available for its prevention. Hepatitis C caused by hepatitis C virus (HCV) is mainly a blood-borne virus. It can cause acute as well as chronic infection. Chronic HCV infection may increase the risk of liver cirrhosis and hepatocellular carcinoma. Hepatitis D caused by hepatitis D virus can propagate in the body only with the help of hepatitis B virus (HBV). Either its superinfection or coinfection with HBV generates a serious threat than the infection of HBV alone. HBV vaccination is a way to prevent hepatitis D. Hepatitis E infection is food or waterborne and vaccine against hepatitis E virus has been licensed only in China. There are also other forms of hepatitis-like alcoholic hepatitis and autoimmune hepatitis. Alcoholic hepatitis is related to excessive alcohol consumption which leads to liver infection either through viral interference or liver cirrhosis. Autoimmune hepatitis is another type of chronic hepatitis in which the immune cells attack the host (hepatic) cells leading to liver damage or cirrhosis.

The new World Health Organization (WHO) data (2017) reveal that about an estimated 325 million people worldwide are living with chronic HBV or HCV infection and WHO global hepatitis report (2017) indicates that a large majority of these people lack access to the testing and treatment. Thus, millions of people are at the risk of a slow progression to chronic liver disease, cancer, and death. The mortality rate has been increasing for viral hepatitis causing about 1.34 million deaths in 2015. Thankfully, the vaccination for infection of hepatitis A and B are made accessible throughout the world which has made brilliant progress in the deterioration of the viral infection of hepatitis A virus and HBV. Although about 1.75 million people were newly infected with HCV in 2015, resulting in the total number of infected people living with hepatitis C to 71 million. Thus, researchers have been taking a deep interest in the drug discovery of HCV which is of urgent need based on the statistics. This led to the screening for phytochemicals and plant-derived compounds which can be potent competitors of antiviral drugs against hepatitis.

## 13.3   ALCOHOLIC AND NONALCOHOLIC LIVER DISEASE

Alcoholic liver disease is another common type of liver disease, and alcohol toxicity remains the third most common cause of morbidity rate around the world (Pal and Ray, 2016). It is also the leading cause of the high rate of mortality resulting from liver damage. Most common types of the alcoholic liver diseases include mainly fatty liver (steatosis), alcoholic hepatitis, and liver cirrhosis. Fatty liver or steatosis is the condition of accumulation of fat cells or triglycerides in the liver cells. Steatosis is diagnosed if the fat deposition in liver exceeds by 5–10% by weight. Fatty liver can be classified into two types: alcoholic fatty liver and nonalcoholic fatty liver. Alcoholic hepatitis is a condition of the fatty liver leading to inflammation and liver cirrhosis as a result of high alcoholic consumption. (O'Shea et al., 2010). The pathophysiological mechanisms found to be involved are:

- Decrease in the mitochondrial beta-oxidation;
- Increase in fatty acid synthesis along with the enhanced delivery of fatty acid to liver; and
- Minimal incorporation and export of triglycerides as very low-density lipoprotein (VLDL)

There are two important mechanisms for alcohol metabolism for alcohol-induced fatty acid synthesis within the liver, including alcohol dehydrogenase (ADH) and cytochrome P450 (CYP) 2E1. ADH which is a cytosolic enzyme present in the liver converts alcohol to acetaldehyde which subsequently gets metabolized to form acetate via a mitochondrial enzyme acetaldehyde dehydrogenase. Both steps lead to reactions involving the reduction of NAD to NADH and an increase in this NADH/NAD ratio shows significant effects on various metabolic reactions (Rasineni and Casey, 2012). Nonalcoholic fatty liver disease (NAFLD) is the condition of hepatic steatosis or fatty liver in the absence of excessive alcohol use. NAFLD is considered to be associated with diabetes, obesity, and hyperlipidemia. The prevalence of NAFLD has been increasing in parallel with the prevalence of obesity, diabetes, and metabolic syndrome and reached an estimate of about 15–20% in general population (Le et al., 2017).

## 13.4   LIVER CIRRHOSIS AND HEPATIC CARCINOMA

Liver cirrhosis refers to the condition of chronic liver failure which involves loss of liver cell resulting in scarring due to inflammation. It is associated with chronic infection of hepatitis, alcoholic or NAFLD, toxins, copper ion accumulation, and other autoimmune diseases of the liver. It is a serious condition of liver damage and can only be prevented from further damage or else liver transplantation is the only option for its treatment. Hepatocellular carcinoma or hepatic cancer is another serious condition of the chronic liver disease. It is the sixth common type of cancer, and its incidence is fifth highest in men and ninth highest in women. The only proven treatments are sorafenib, regorafenib, and lenvatinib along with the targeted therapy agents and chemotherapy.

## 13.5   PHYTOCHEMICALS AND HEPATIC DISORDERS

For thousands of centuries, plants have been used for healing. Plant-derived products were majorly used as foods or botanical potions and extracted powders which have been used successfully in the cure and prevention of diseases throughout history. Many written records about medicinal plants, almost 5000 years ago by the Sumerians and later on the evolution of the plant-based medicine system, primarily based on indigenous plants, led to the establishment of well-known traditional medicine systems, namely – the

Ayurvedic and Unani of the Indian subcontinent, Chinese and Tibetan of Asia, the Amazonian of South America, the Native American of North America, local systems within Africa, and several others came into light (Mamedov et al., 2012; see Chapters 1–4 of this volume). But the synthetic chemistry-dominated pharmaceutical industry took over completely in the 20th century and replaced natural extracts with synthetic molecules that had no connection to natural products. There was a spectacular rise of the pharmaceutical industry and had triumphed over with a tremendous impact on disease treatment and as well as prevention, saving countless lives. Thus modern pharmaceutical drugs became one of the outstanding achievements of the 20th century (Raskin et al., 2002).

However, these pharmaceutical drugs come with many pros and cons and most importantly its cons could be highly felt in comparison to the naturally derived products. The benefits of modern drugs could only be utilized primarily in developed countries, being highly cost intensive. Not being able to access the modern healthcare products, the developing countries continued to rely on ethnobotanical remedies as primary medicines. Moreover, some of the significant pharmaceutical drugs come with many side effects (Ratini 2017). Most of the externally used drugs triggers allergic reactions causing itching, rashes, and sometimes may lead to life-threatening anaphylactic reactions. For drugs taken internally, the most common set of side effects involves the gastrointestinal system causing nausea or an upset stomach. Some drugs trigger side effects due to their chemical structure like the common allergy drug diphenhydramine which sometimes suppresses the activity of acetylcholine leading to drowsiness along with other side effects including dry mouth. Some other drugs like warfarin for the prevention of blood clots may lead to serious internal bleeding. Some major side effects can be caused by drug interactions with other edibles. Thus all these pharmaceutical drugs should be strictly approved by Food and Drug Administration (FDA) for consumption. Even so, they were highly preferable over the traditional plant-derived products. But, recently the use of plants for the discovery of pharmacologically active compounds is gaining importance along with drugs being derived directly or indirectly from plants. The emphasis shifted gradually from extraction of medicinal plant compounds to making these compounds or their analogs synthetically in the 20th century. Natural products were screened as templates for structure optimization programs for designing novel drugs. Despite the current preoccupation with synthetic chemistry for drug designing and manufacture, the contribution of plants to disease treatment and prevention is still enormous. Phytochemicals are bioactive metabolites present naturally in plants showing biological significance in plants, playing an essential role to defense mechanism of plants by inhibiting or killing the interacting pathogen. The secretion of these compounds varies from plant to plant. These compounds are essential and currently many scientific researches and studies have been going on in the antiviral, antibacterial, anticancer, antifungal, antioxidant, and numerous other medicinal properties endowed by these compounds.

## 13.6 PHYTOCHEMICAL INHIBITORS OF HEPATITIS C VIRUS LIFE CYCLE

The HCV genome is a positive single-stranded RNA of size 9.6 kb. It has about six major genotypes with a series of subtypes which provides it a high degree of genetic variability. The HCV genome has a 5′ UTR region of about 324–341 nucleotides in length which involves internal ribosomal entry site, important for cap-dependent translation of viral RNA. The entry site leads to the translation of the open reading frame encoding 3010 amino acid polyprotein precursor which is cleaved by the host and viral proteases into 10 specific viral proteins of the order NH2-Core-E1-E2-p7-NS2-NS3-NS4A-NS4B-NS5A-NS5B-COOH (Ashfaq and Idrees, 2014). Researches have been mainly on the structural proteins (core, E1 and E2), NS3/4A complex, and nonstructural 5B protein (NS5B) RNA-dependent RNA polymerases (RdRps) as they are considered the best targets for developing novel molecular inhibitors. The core is the key component for assembly and packaging of HCV RNA genome through high-order oligomers that are linked with lipid droplets and the endoplasmic reticulum with a number of proteins forming essential particles for viral assembly. It is the least variable component of all other components of HCV which plays significant role in cell proliferation. It controls RNA-binding activity along with its capacity to form protein homomultimers ultimately inducing viral transformation. Thus, the inhibitors of the viral capsid assembly interfere with the uncoating of viral elements on infection which results in the destabilization of the assembled virion particle by the formation of new particles. Nonstructural protein NS3 which has separate protease and helicase domains functions in complex with cofactor NS4A and is important in the hepatitis C viral replication (McGivern, 2015). The NS3/4A serine protease domain is required to cleave the polyprotein at four sites generating viral proteins essential for replicating the genome of the viral RNA. It also cleaves the adaptor proteins mitochondrial antiviral signaling protein and TIR-domain-containing adapter-inducing interferon-β to block the activation of interferon gene expression through the retinoic acid-inducible gene I and toll-like receptor 3 pathways. Thus the NS3 protease is considered a promising target as its inhibition is involved with the interference of polyprotein processing and restoration of antiviral signaling. The helicase domain of NS3/4A has NTPase and 3'-5' RNA unwinding activity. The HCV RNA synthesis involves the ATP-dependent RNA unwinding activity of NS3A helicase enzyme. Further studies have also implicated its role in viral assembly (Kumar and Kao, 2006) The replication of viral RNA involves a membrane-associated replicase enzyme complex which consists of host and viral protein components. The catalytic subunit of this complex is the HCV encoded NS5B and it contains all the sequence motifs that are highly conserved among all the known RdRps. About 150 nucleotide sequences of the HCV RNA contain signals that are essential for RdRp binding and replication of viral RNA at the 3' terminal of the viral genome.

Many standard treatments have been proposed against HCV, of which, the combination therapy of pegylated interferon-α with a nucleoside analogue ribavirin which was later provided in addition with NS3 protease inhibitors boceprevir and telaprevir (Khan et al., 2013) and a US FDA approved drug named Sofosbuvir (brand name – Sovaldi) which targeted NS5B HCV enzyme for inhibition are well-known and gave effective results (Mirza et al., 2015). But these standards had a few side effects and did not show 100% efficiency in all the genotypes and subsequent subtypes due to its high genomic variability. Moreover, they had cost issues which made them accessible to only a small fraction of the population, especially in developed countries. Thus there is an urgent need for alternative therapies which are cost-effective, showing efficiency in all the genotypes and subtypes, and have fewer side effects. Recently, the focus has been shifted to screening for plant-derived phytochemicals which could be a potent competitor of drug therapy with maximum efficiency as they are highly cost-effective and have fewer side effects.

Some plants known for their hepatoprotective activity have been screened for its anti HCV activity along with a screening of compounds that renders this property (Table 13.1). Many important medicinal plants like *Boerhavia diffusa*, *Azadirachta indicia,* and *Eclipta alba* which are used in Ayurvedic treatment in India since ancient times showed potent anti-HCV activity along with other hepatoprotective activities on its screening. *B. diffusa* contained dehydrorotenoid compound namely Boeravinone H. which inhibited HCV entry and infection in the cell based on the in vitro cell culture studies. *Azadirachta indica* was screened for phytochemicals and antiviral compounds by a group of researchers and found that it contains a compound named 3-deacetyl-3-cinnamoyl-azadirachtin which effectively interacts with NS3/4A serine protease binding with various amino acids in the catalytic site including gun 526 and can be considered as a potent inhibitor of NS3/4A serine protease (Ashfaq et al., 2015). *E. alba* (Asteraceae) showed diverse medicinal properties along with properties like enhanced longevity, rejuvenation, and restoration of liver function. Plant extracts were isolated and screened for their ability to inhibit HCV replicase enzyme NS5B. Effective inhibition was rendered by the presence of phytocomponents like wedelolactone, luteolin, and apigenin. Other compounds like 3, 4-dihydroxybenzoic acid and 4-hydroxy benzoic acid showed inhibition at higher concentrations and are mostly involved in enhancing cell proliferation. Moreover, these compounds show synergistic effects in combination, although at higher concentrations they are cytotoxic. Researchers suggested that *E. alba* contained a cocktail of phytochemicals involved in anti-HCV activity of which some compounds being cytotoxic at higher concentration while others render cytoprotective activity (Manvar et al., 2012). Several flavanoids (a class of secondary metabolites) have been screened for their potential inhibition of NS5B polymerase enzyme by molecular interactions with the residues in its active site. Compounds showing strong-binding affinity include naringenin, tryphanthrine, dicoumarine, swertianin, luteolin, apigenin, honokiol, diosmetin, luteolin, thaliporphine, and oxymatrine. Out of these, the high-scoring ligands like dicoumarin and swertianin were found to be fit for human consumption (Mirza et al., 2015). Further studies showed that four phytochemical inhibitors extracted from various medicinal plants gave efficient results including epigallocatechin gallate (EGCG), ladanein, naringenin, and silybin which targeted to disrupt interactions between the HCV core and other proteins (Fig. 13.1). EGCG showed interactions with the capsid core while silybin is known to inhibit NS5B RdRp. Ladanein is one of the most active flavones which is known to inhibit receptor interaction after viral entry. Naringenin blocks the assembly of intracellular viral particles by disrupting the core dimerization which is necessary for the core oligomers to form and remain stable (Mathew et al., 2014). A number of medicinal plants used in traditional Sudanese medicine *like Boswellia carterii, Acacia nilotica, Quercus infectoria, Embelia schimperi, Trachyspermum ammi, Piper cubeba, and Syzygium aromaticum* as well as other plants like *Amelanchier alnifolia, Solanum nigrum, Viola yedoensis, Phyllanthus amarus* are found to be important for their antiviral properties having a number of phytochemical inhibitors with different structural and functional targets (Ravikumar et al., 2011).

**TABLE 13.1**    Phytochemicals and Target Site in Viral Life Cycle

| Phytochemicals | Class | Significant step | Function |
|---|---|---|---|
| Epigallocatechin gallate | Catechin | Viral entry | Disruption of glycoprotein and inhibit cell–cell transmission |
| Naringenin | Flavonoid | Viral assembly | Blocks assembly by suppression of core protein |
| Quercitin | | Viral replication | Reduce internal ribosomal entry site activity by inhibiting NS3 protease |
| Ladanein | | Viral attachment and entry | Inhibition of receptor-mediated viral entry, endocytosis, and membrane fusion |
| Diosgenin | Sapogenin | Viral replication | Inhibit TF3 and signal transducer |
| Silymarin/silybin | Flavonolignan | Viral entry and replication | Inhibit viral core and NS5B polymerase |
| Luteolin/apigenin | Flavonoids | Viral replication | Inhibit NS5B polymerase |
| Iridoids | Monoterpene | Viral entry | Block E2 protein with CD81 |

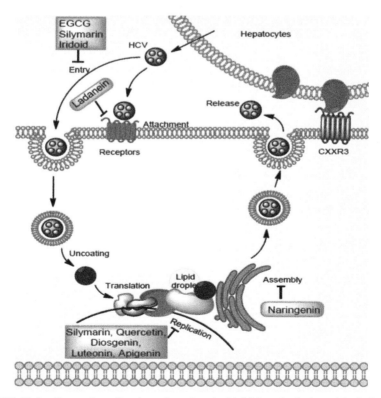

**FIGURE 13.1**    Target sites of various phytochemical inhibitors in the hepatitis C virus life cycle.

## 13.7  FATTY LIVER

Consumption of excess alcohol cause fatty liver by altering the ratio of $NADH/NAD^+$ redox potential in the liver, which results in inhibition of fatty acid oxidation and tricarboxylic acid cycle. Recent studies reveal that the substitution of ethanol impairs fatty acid oxidation and leads to lipogenesis additionally, interfering with the DNA-binding and transcription-activating properties of peroxisome proliferator-activated receptor α (PPAR-α). Ethanol also activates sterol regulatory element-binding protein 1 (SREBP1), which induces an array of lipogenic enzymes. Recent research has been going on to provide new therapeutic targets to reverse these effects for alcoholic fatty liver (You et al., 2004). Recently, some researchers show that aldose reductase (AR)/polyol pathway plays a significant role in fatty liver disease (FLD) development involving the production of hepatic fructose, gut microbial endotoxin-induced cytokine release, PPAR-α activity, and CYP2E1 expression which on the contrary are responsible for FLD.

NAFLD is another common cause of chronic liver disease which involves the accumulation of triglycerides in the hepatocytes resulting in inflammation, fibrosis, and cirrhosis. It is associated with other common diseases like obesity, diabetes mellitus, and so forth. (Pettinelli et al., 2011). The alterations in the metabolism of insulin are mainly responsible for the accumulation of fatty acid in the liver along with alterations in expression of certain transcriptional factors associated with lipogenesis and lipolysis of the liver. Triglycerides (storage lipid molecules) are utilized through β oxidation, esterification, and storage as lipid droplets or the packaging and export as VLDL. Hence, any changes in these pathways can cause accumulation of fatty acids in liver (Dowman et al., 2010).

With the rise in the incidence of fatty liver disease, it is essential to sort for all possible options for its treatment. As natural or plant-derived medicines, isoflavones have gained importance in the treatment of fatty liver (Fig. 13.2). Isoflavones are reported to be potent AR inhibitors such that it helps in the suppression of fructose production in the liver, enhance PPAR-α-mediated fatty acid breakdown, enhance the CYP2E1-mediated oxidative stress, and regulation of gut microbial endotoxin-mediated cytokine production which are the associated pathways resulting in FLD. Some studies show the significance of isoflavones as potent AR inhibitors and out of them genistein, daidzein, and puerarin are commonly recognized (Madeswaran et al., 2012). Tectoridin also exhibited AR inhibition to a good extent. Moreover, biochanin A showed a better binding affinity with AR than epalrestat (synthetic AR-specific inhibitor) (Huang et al., 2013). Out of these, genistein, daidzein, biochanin A, and formononetin are common PPAR-α agonist blocking the activity of AR (Qiu et al., 2012). Also, genistein is also known to inhibit CYP2E1 activity (Mueller et al., 2010). Another group of researchers studied the inhibitory effects of phytochemicals present in *Rhododendron oldhamii* which plays an important role in enhancing lipid oxidation by decreasing lipogenesis. The presence of flavonoids (2R, 3R)-astibilin, hyposide, guaijaverin, and quercetin was observed in the leaf extract. Out of these phytochemicals, (2R, 3R)-astibilin and hyposide showed potent inhibitory actions against the fat accumulation resulting in the reduced formation of foamy cells extending from hepatic portal vein to the central vein. These phytochemicals played a significant role in the downregulation of the SREBP1 gene expression which ultimately increased CPT1a, *PPAR-α,* and *PPAR-γ* activity in the liver.

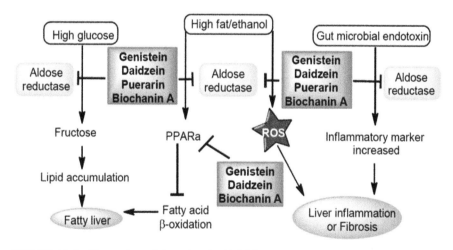

**FIGURE 13.2**    Target sites of phytochemical inhibitors in metabolic pathways of fatty liver.

## 13.8   NEURODEGENERATIVE DISEASES

Different neuronal disorders are described in the following section.

### 13.8.1   AGEING AND OXIDATIVE STRESS

Ageing is a complicated mechanism which implements both morphological and physiological changes in our body. Imbalance in cell oxidation/reduction status can be described as oxidative stress results in the production of moderately reduced oxygen intermediates. Reactive oxygen species (ROS) are derivatives of oxygen that are more reactive than molecular oxygen, such as superoxide, hydrogen peroxide, and hydroxyl radical. ROS production and oxidative damage lead to the development of age-related disorders (Barja, 2004). The brain consumes 20% oxygen through respiration. Due to its high oxygen consumption, the brain is a more affected organ to oxidative damage. Great extent of polyunsaturated fatty acids (PUFAs) in neuronal membranes forms cytotoxic aldehydes which initiates lipid peroxidation reactions. Oxidative stress and free radical hypothesis are widely accepted theories against aging. The natural neuronal membrane contains a high amount of PUFAs which enhance the brain tissues susceptibility for lipid peroxidation. This reaction forms various cytotoxic aldehydes like malondialdehyde and 4-hydroxynonenal. Numerous organisms have their mechanisms to reduce the ROS level. Various phytochemicals have been reported as a scavenging agent present in the dietary foods.

### 13.8.2  ACUTE CEREBROVASCULAR ATTACK (STROKE)

The sudden death of some brain cells, due to lack of oxygen, when the blood flow to the brain is impaired by blockage or rupture of an artery to the brain, is known as acute cerebrovascular attack, also referred to as a stroke in which poor blood flow to the brain results in cell death (Hachinski, 2007). There are three main types of stroke. Ischemic stroke is the most common type of stroke among all. In this stroke, blood clot prevents blood and oxygen from reaching the brain (Ginsberg, 2003). Hemorrhagic stroke occurs when a weakened blood vessel ruptures and normally occurs as a result of aneurysms or arteriovenous malformations. Transient ischemic attacks is referred to as a mini stroke, these occur after blood flow fails to reach part of the brain. Normal blood flow resumes after a short amount of time and symptoms cease.

### 13.8.3  NEURONAL CANCER

Brain tumor arises in the glial cell, present in the central nervous system (CNS). It belongs to the heterogeneous brain tumor family. Gliomas include both CNS tumor (30%) and malignant brain tumors (80%). Glioma was classified by microscopic examination of histological parts of tumors which differentiate cancer cell according to microscopic similarity, a subtype of gliomas like astrocytomas, oligodendroglioma, and ependymomas. The recent 2016 WHO classifications of tumors incorporate molecular biomarkers and histological appearance to define specific glioma in the CNS. Brain tumors are linked with poor prognosis and acute degradation in the patients which results in low survival rates. Glioma cells match with glial progenitor cells, which migrate and differentiate into the different types of glial cells. These glial cells are very proliferative in developing CNS. This mechanism plays a role in the migration of neuroepithelial cell during embryogenesis and is significant for glioma invasion. These data tell us about spatial and temporal variation in the signaling pathway and causes functional and phenotypic variation occur in glioma cells. This variation affects communication between neighboring malignant and non-malignant cells as well as different parts of brain tissue.

Wounds that do not heal for a long time can be seen as a tumor. The process of migration, proliferation inflammation, angiogenesis, invasion, activation of astrocyte occurs in both tumor and wound healing in response to any impairment. Cell migration and invasion are detected in gliomas but also follow active gliosis. On the other side, astrocytes play a role against any damage to CNS for healing progress. Numerous factors are produced by tumor-associated macrophage (TAM) that promotes survival and proliferation of cancer cells as well as also suppresses the anticancerous immunity. Glioma cells secrete colony stimulating factor which stimulates TAM to enhance invasion. TAM secretes chemokines, cytokines, and growth factors which stimulate several signaling pathways and which terminate glioma cells in the direction of more rapid action. Change in glucose metabolism also leads to a cancerous cell. Warburg effect/aerobic glycolysis are a crucial hallmark of tumor development. Aerobic glycolysis is less effective to gain energy as compared to mitochondrial oxidative metabolism. As a result, acidic environment generated by glycolysis, due to this cancer cells have an advantage concerning normal parenchyma. Acid-induced toxicity is a fundamental element mandatory for tumor invasion, hence hallmark of invasive cancer. The high amount of glycolysis and lactate extraction are features of glioma cells.

### 13.8.4 ALZHEIMER'S DISEASE

Alzheimer's is the most common neurodegenerative disease. It destroys memory and importantly mental functions in old age (after 65 years of age).

Memory damage can be detected in the early stage of disease, whereas sensory and motor functions are not damaged even in the following stages (Hebert et al., 2003).

### 13.8.5  PARKINSON'S DISEASE (PD)

Parkinson's disease (PD) was first described by Dr. James Parkinson as "shaking palsy" (Fig. 13.3). It is a slowly progressive degenerative neurologic disease, and it commonly develops in the old age of 50's and more. Its main symptoms involve:

- Tremor: Involuntary shaking movement of limbs, fingers, and other body parts. These are the initial symptoms of the PD patients.
- Rigidity: decrease in the flexibility of the muscles.
- Akinesia: loss of power of involuntary movement.
- Bradykinesia: slowness of movement and loss of ability to adjust body position.
- Postural instability: inability to walk or stand straight.

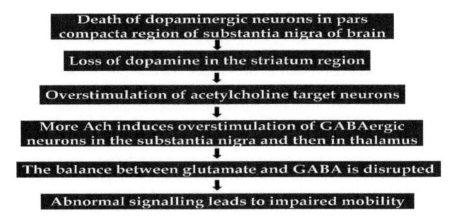

**FIGURE 13.3**  The biochemical sequence of Parkinson's disease (PD).

### 13.8.5.1  PARKINSON'S AT MOLECULAR LEVEL

The main cause of the disease was discovered to be the fall in the level of dopamine in the brain (Fig. 13.4). Most of the researchers were thinking that this disease is not associated with any of the genetic reasons but later on it was found that some genes like SNCA that codes for α-synuclein, LRRK 2 which codes for leucine-rich repeat kinase, PINK-1 which is PTEN-induced kinase 1, and so forth are responsible for causing the death of dopaminergic neurons when they are mutated.

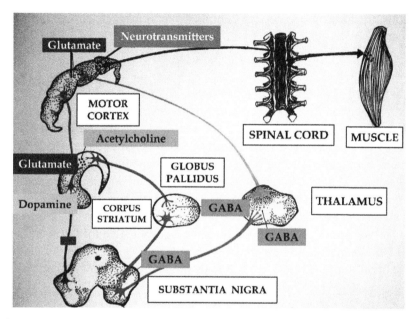

**FIGURE 13.4**   Pathophysiology of PD involving various part of the brain.

### i. SNCA gene

It was first described and associated with the familial Parkinson's in 1997. The gene is located on the q-arm of chromosome number 4. The gene codes for a protein called the α-synuclein which is a 140-residue cytosolic and lipid-binding protein. In its native condition or normal condition, it remains unfolded. It is abundant in the brain and the smaller amount found in the heart, muscles, and other tissues. In the brain, it is found in the presynaptic terminals of neurons. Here, they interact with fats and proteins and play an important role in maintaining an adequate and balanced supply of synaptic vesicles which carry the neurotransmitters and are responsible for relaying signals. It is also supposed to regulate the release of dopamine, a neurotransmitter that is important for various voluntary and involuntary movements of the body. Mutation in this gene can cause the death of dopaminergic neurons since the mutation can cause aggregation of alpha-synuclein protein leading to proteasomal and lysosomal system dysfunction and reduced mitochondrial activity by affecting its complexes (Hayashita, 2006). Subsequent duplication of this gene is responsible for its misfolding which leads to disease progression and development of different non-motor features. Mutation in and over-expression of this gene is supposed to be toxic to dopaminergic neurons, as dopamine-synuclein complex may inhibit chaperone-mediated autophagy.

## ii. LRRK2

It is expanded as leucine-rich repeat kinase 2 gene, identified by two groups in 2004. Mutations are often associated with late onset of the disease.

It is a large gene consisting of 51 exons encoded by 2527 amino acids protein with several functional domains. Its level regulates glutamate[1] transmission. It regulates plasticity caused due to dopamine and striatal signal transduction. Mutation in LRRK2 causes the death of dopaminergic neurons. It can transfer a phosphate group from ATP to a downstream substrate to activate it. It is the only known gene coding for a protein having both kinase and GTPase domain. Mutation in this gene is responsible for the decrease in the GTPase activity and impaired kinase activity. Most frequent mutation is on C.6055G and P. G2019S. It leads to dysfunctioning of presynaptic protein sorting and axonal trafficking.

## iii. PARKIN (PARK 2)

Parkin gene is a part of ubiquitin-ligase complex which is a part of ubiquitin proteasome system whose function is to control the targeting of proteins for degradation. A homozygous mutation in the parkin gene disrupts the activity of ubiquitin ligase, which then inhibits the activity of the proteasome complex due to which toxic substrates get accumulated, and they lead to the formation of Lewy bodies and finally lead to the death of dopaminergic neurons.

## iv. PINK 1

It is expanded as PTEN (phosphate and tensin homolog)-induced putative kinase 1, a serine/threonine-protein kinase encoded by the PINK1 gene. Mutation in this gene causes phosphorylation patterns and proteasomal stress, and this alters the cleavage pattern of PINK1, which leads to the

---

[1]Glutamate is an excitatory neurotransmitter which relays the signals from thalamus to motor cortex to corpus striatum.

accumulation of Lewy bodies. It interacts with the gene DJ-1 and gets reemitted to the pathway after oxidative damage due to PINK dysfunction.

### v. DJ-1

DJ-1 codes for a molecular chaperone and is an intracellular sensor of oxidative stress. It regulates D2 receptor signalling and has direct action as an antioxidant. Mutation in DJ-1 shows increased changes which occur due to the metabolism of energy. Overexpression of DJ-1 protects against mitochondria complex 1 inhibitors. Therefore, they can abrogate the effect, and as a result, the cells can abort apoptosis (can lead to cancer).

### 13.8.5.2   TREATMENTS AVAILABLE FOR PD

There is no complete cure for this disease till date, but treatments are available to reduce the effect of the disease or to reduce the symptoms. Types of treatments depend on the intensity of the disease (stage of the disease). It includes flexible exercises for the healthy movement of the joints, medications which are done in the early stages of the disease, and includes the temporary replenishment of dopamine or using mimicry molecules of dopamine such as the gel formulation, the levodopa, and surgical therapies like inserting of electrodes into targeted areas of the brain and then impulse is generated. Other surgical therapy like pallidotomy and thalamotomy exists.

### 13.8.5.3   CLINICAL STUDIES

Current clinical studies are focused on the application of dopamine mimicry compound. The researchers of national Parkinson's foundation studied a group of people who have had PD for more than 20 years to try to understand the factors responsible for their long-term survival such as the age of onset of disease, medication given, living situation and their symptoms, and so forth. On the molecular basis, there are no simple tests to diagnose PD, so a person has to go through many screenings to be diagnosed to know whether he/she has this disease. But it can be easier if doctors come to know the biomarkers for this disease so that they can test the person's blood or serum to test the presence of these biomarkers that can diagnose PD at an earlier stage. The following scientists are engaged in the clinical trial studies on PD patients:

- Professor Perdita Banan, University of Manchester, is searching for biomarkers on skin.
- Professor Simon Lovestone, Oxford University, is searching for biomarkers in proteins of the blood sample of the patients.

Recently it was found that mutations in genes like PINK-1, DJ-1, and LRRK2 are related to cancers (especially leukemia) and PD. It was found that a drug named Nitolib has the potential to treat PD. Some other drugs used to treat PD are Levodopa-carbidopa, Levodopa-carbidopa CR, Ropinirole, Pramipexole, Rotigotine, and so forth.

### 13.8.5.4   PHYTOCHEMICALS STUDIED IN PD

Several phytochemicals are known to treat or manage PD (Table 13.2). Some are discussed below:

### 1. Crocin

Crocin is a carotenoid found in flower *Crocus* and *Gardenia* which is primarily responsible for the color of saffron. It acts as an antioxidant by quenching free radicals, protecting cells and tissues against oxidation (Fig. 13.5). Pretreatment of cells with crocin inhibits ROS generation in the dopaminergic terminal and hence prevents the death of dopaminergic neurons (Zhang, 2015).

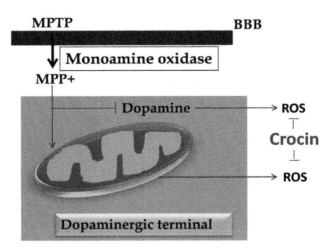

**FIGURE 13.5**   Mode of action of crocin.

## 2. Geraniol

Geraniol is a monoterpenoid and alcohol. It is the primary part of palmarosa oil, rose oil, citronella oil, lemon, and much other essential oils. The phyto-chemical geraniol acts as an antioxidant, and it can cross the blood–brain barrier. It has cytoprotective characteristics against oxidative stress produced due to various neurotoxins. It helps to restore the membrane potential of mitochondria (Fig. 13.6). Case 1: In the blood–brain barrier, monoamine oxidase-B (MAO-B) is present which is responsible for converting non-toxic mitochondrial permeability transition pore (MPTP) to toxic $MPP^+$ (Rekha, 2013). The administration of geraniol in the early stage of PD can destabilize the enzyme MAO-B, and hence toxin formation can be put to an end. Case 2: It can prevent the aggregation of α-synuclein and hence prevents the formation of α-synuclein inclusions, which is a major cause of PD and other neurological disorders.

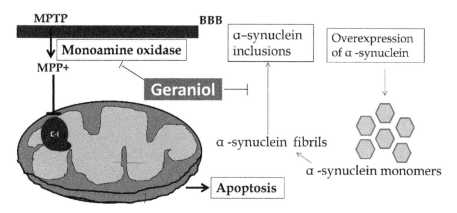

**FIGURE 13.6**    Mode of action of geraniol.

## 3. Safranal

Safranal is an organic compound isolated from saffron, a spice that contains the stigma of Crocus flower which is completely responsible for the aroma of saffron. Safranal inhibits rotenone-induced cell death. Rotenone significantly increases the release of lactate dehydrogenase into the surrounding. It decreases the membrane potential of mitochondria and increases the production of ROS, and an anaerobic state of respiration is established (Fig. 13.7) (Pan et al., 2016).

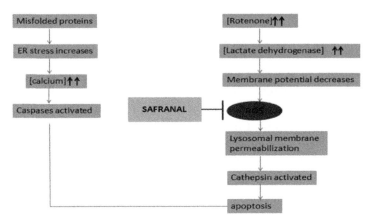

**FIGURE 13.7**    Mode of action of safranal.

## 4. Catalpol

Catalpol is a natural product which falls into the category of iridoid glycosides, a simple monoterpenes with an attached glucose moiety. It is found in the plants that belong to families like Scrophulariaceae, Lamiaceae, and Bignoniaceae, and so forth. MPP+ is found to be responsible for the opening of MPTP and release of cytochrome c in brain and mitochondria (Fig. 13.8). This is responsible for the mitochondrial destruction. Catalpol treatment reduces the activity of MPP+ regarding the opening of MPTP (Bi et al., 2008). On the other hand, Ca2+ concentration is also elevated in case of neuronal cells of PD patients, which is also responsible for the opening of MPTP. Catalpol also restrains the pressure of overloading Ca2+.

## 5. Tanshinone

Tanshinone is a diterpenoid and a major lipophilic bioactive compound found in *Salvia miltiorrhiza*. Tanshinone mediates its neuroprotective activity against 6-OHDA-induced oxidative stress via the Nrf2-ARE pathway (Fig. 13.9). The transcription factor Nrf2 plays an important role in the induction of cytoprotective genes like those that encode for endogenous antioxidants such as heme oxygenase-1, glutathione cysteine ligase regulatory subunit and glutathione cysteine ligase modulatory subunit. On the other hand, 6-OHDA is a neurotoxin that is initiated by extracellular autooxidation of oxidative products that are generated. The loss of Nrf2-mediated transcription increases the chances of dopaminergic neurons to undergo oxidative stress. Under normal conditions, Nrf2 is regulated posttranslationally and constitutively by its antagonist Keap1 (Jing et al., 2016). The actual function of Nrf2 begins when the cell is exposed to oxidative stress or electrophilic substances. As a result, the Keap1 is modified, detaches from Nrf2, and proceeds into the nucleus and transactivates many target genes. Under resting metabolic conditions, Nrf2 is negatively regulated through targeted ubiquitination mediated by Keap1.

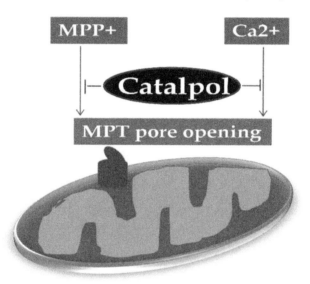

**FIGURE 13.8**    Mode of action of catalpol.

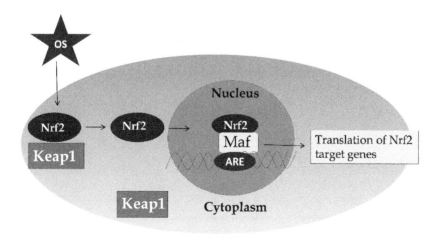

**FIGURE 13.9**    Mode of action of tanshinone.

**TABLE 13.2** Phytochemicals Involved in the Treatment of Neuronal Diseases

| Plant | Phytochemicals | Mode of action |
|---|---|---|
| Ginkgo biloba | Quercetin, kaempferol and isorhamnetin, ginkgolides (A, B, C, J, M) and Bilobalide | Ginkgolides are antagonist of platelet-activating factor, reduce platelet activation and aggregation, improve blood circulation |
| Panax ginseng | Ginsenosides -Rb1 group -Rg1 group | It attenuates Aβ and glutamate-induced toxicity, enhance clearance of Aβ by stimulating the phagocytic activity of microglia, promotes neuron survival |
| Scutellaria baicalensis | Baicalin, Baicalein, Wogonin | Protect neurons from oxidative damage in cerebral ischemia/reperfusion injury inhibit lipid peroxidation of neuronal membranes prevent excitotoxicity induced by glutamate |
| Ayurvedic medicine | | |
| Curcuma longa | Curcumin | Anti-inflammatory, antioxidant, antiviral, and antidiabetic activity |
| Mediterranean traditional diets | | |
| Grape | Phenylpropanoids, isoprenoids, and alkaloids | Anti-amyloidogenic, antioxidant, anti-inflammatory activity |
| Salvia officinalis | β-pinene, 1,8-cineole, camphor, borneol, caryophyllene, and Linalool | Warming and contractions for coughs, in labor pains and ulcers |
| International beverages: coffee and tea | | |
| Coffee | Caffeine | Depressant effects, decrease heart rate, decrease blood pressure |
| Tea | Catechins, flavonols, proanthocyanidins | Antioxidants, scavenging of free radicals, activation of signaling pathways |

Under conditions of stress, Keap1 is oxidized, and releases Nrf2, which is stabilized by DJ-1 and translocates to the nucleus via ARE enhancers do activates a range of antioxidant enzymes and hence decreases the oxidative stress (Pan, 2016).

## ACKNOWLEDGMENT

Shashank Kumar acknowledges Central University of Punjab, Bathinda and University Grants Commission, India for providing necessary infrastructure facility and financial support in the form of UGC-BSR Research Start-Up-Grant, GP: 87 [No. F.30–372/2017 (BSR)], respectively. Prem Prakash Kushwaha acknowledges financial support from University Grants Commission, India in the form of CSIR-UGC Junior Research fellowship. Swagata Das, Prareeta Mahapatra, and Priyanka Kumari1 acknowledge Central University of Punjab, Bathinda, India for providing necessary infrastructure facility.

## KEYWORDS

- cancer
- hepatic
- inhibitor
- neuronal
- phytochemical

## REFERENCES

Ashfaq, U. A.; Idrees, S. Medicinal Plants Against Hepatitis C Virus. *World J. Gastroenterol.* **2014,** *20*(11), 2941–2947.

Ashfaq, U. A.; Jalil, A.; Qamar, M. T. U. Antiviral Phytochemicals Identification from *Azadirachta indica* Leaves Against HCV NS3 Protease: An in Silico Approach. *Nat. Prod. Res.* **2015,** *30*(16), 1866–1869.

Barja, G. Free Radicals and Aging. *Trends Neurosci.* **2004,** *23,* 209–216.

Ben-Shlomo, Y.; Bhatia, K. 2004. Using Monoamine Oxidase Type B Inhibitors in Parkinson's Disease. *Br. Med. J.* **2004,** *329,* 581–582.

Bi, J.; Wang, X.; Chen, L.; Hao, S.; An, L.; Jiang, B.; Guo L. Catalpol Protects Mesencephalic Neurons Against MPTP Induced Neurotoxicity Via Attenuation of Mitochondrial Dysfunction and MAO-B Activity, *Toxicol. Vitro,* **2008,** *22,* 1883–1889.

Bovenberg, M. S. S.; et al. Cell-Based Immunotherapy Against Gliomas: From Bench to Bedside. *Mol. Ther.* **2013**, *21*, 1297–1305. DOI:10.1038/mt.2013.80.

DeFeudis, F. V.; Drieu, K. *Ginkgo biloba* Extract (EGb 761) and CNS Functions Basic Studies and Clinical Applications. *Curr. Drug Targets*, **2000**, *1*, 25–58.

Dowman, J. K.; Tomlinson, J. W.; Newsome P. N. Pathogenesis of Non-Alcoholic Fatty Liver Disease. *QJM*, **2010**, *103*(2), 71–83.

Elbaz, A. et al. A Risk Tables for Parkinsonism and Parkinson's Disease. *J. Clin. Epidemiol.* **2002**, *55*, 25–31.

Fitzpatrict, A. L.; et al. Recruitment of the Elderly into Pharmacologic Prevention trial: the Ginkgo Evaluation of Memory Study Experience. *Contemp. Clin. Trials*, **2006**, *27*, 541–553.

Ginsberg, M. D. Adventures in the Pathophysiology of Brain Ischemia: Penumbra, Gene Expression, Neuroprotection: the 2002 Thomas Willis Lecture. *Stroke*, **2003**, *34*, 214–223.

Hachinski, V. Stroke and Vascular Cognitive Impairment: A Transcisciplinary, Translational and Transactional Approach. *Stroke*, **2007**, *38*, 1396–1403.

Hayashita-Kinoh, H.; Yamada, M.; Yokota, T.; Mizuno, Y.; Mochizuki H. Down-Regulation of alpha-Synuclein Expression can Rescue Dopaminergic Cells from Cell Death in the Substantia Nigra of Parkinson's Disease Rat Model. *Biochem. Biophys. Res. Commun.* **2006**, *341*, 1088–1095.

Hebert, L. E.; et al. Alzheimer Disease in the US Population: Prevalence Estimates Using the 2000 Census. *Arch. Neurol.* **2003**, *60*, 1119–1122.

Huang, Q.; Huang, R.; Zhang, S.; Lin, J.; Wei, L.; He, M.; Zhuo, L.; Lin, X. Protective Effect of Genistein Isolated from Hydrocotyle Sibthorpioides on Hepatic Injury and Fibrosis Induced by Chronic Alcohol in Rats. Toxicol. Lett. **2013**, *217*, 102–110.

Jing, X.; Wei, X.; Ren, M.; Wang, L.; Zhang, X.; Lou, H. Neuroprotective Effects of Tanshinone I Against 6-OHDA Induced Oxidative Stress in Cellular and Mouse Model of Parkinson's Disease Through Upregulating Nrf2. *Neurochem. Res.* **2016**, *41*, 779–786.

Khan, M.; Mousoud, M. S.; Qasim, M.; Khan, M. A.; Zubair, M.; Idrees, S.; Ashraf, A.; Ashfaq, U. A. Molecular Screening of Phytochemicals from *Amelanchier Alnifolia* Against HCV NS3 Protease/Helicase using Computational Docking Techniques. *Bioinformation*, **2013**, *9*(19), 978–982.

Kumar, C. T. R.; Kao, C. C. *Hepatitis C Viruses: Genomes and Molecular Biology;* Tan, S. L., Ed.; Horizon Bioscience: Norfolk (UK), 2006.

Le, M. H.; Devaki, P.; Ha, N. B.; Jun, D. W.; Te, H. S.; Cheung, R. C.; Nguyen, M. H. Prevalence of Non-Alcoholic Fatty Liver Disease and Risk Factors for Advanced Fibrosis and Mortality in the United States. 2017, 12(3), 1–13.

Madeswaran, A.; Umamaheswari, M.; Asokkumar, K.; Sivashanmugam, T.; Subhadradevi, V.; Jagannath, P. In Silico Docking Studies of Aldose Reductase Inhibitory Activity of Commercially Available Flavonoids. Bangladesh J Pharmacol. **2012**, *7*, 266–271.

Mamedov, N. Medicinal Plants Studies: History, Challenges and Prospective. *Med. Aromat. Plants*, **2012**, *1*, 133.

Manvar, D.; Mishra, M.; Kumar, S.; Pandey, V. N. Identification and Evaluation of Anti Hepatitis C Virus Phytochemicals from *Eclipta Alba*. *J. Ethnopharmacol.* **2012**, *144*(3), 545–554.

Mathew, S.; Faheem, M.; Archunan, G.; Ilyas, M.; Begum, N.; Jahangir, S.; Qadri, I.; Qahtani, M. A.; Mathew, S. In Silico Studies of Medicinal Compounds Against Hepatitis C Capsid Protein from North India. *Bioinf. Biol. Insights*, **2014**, *8*, 159–168. DOI: 10.4137/BB.

McGivern, D. R.; Masaki, T.; Lovell, W.; Hamlett, C.; Saalau-Bethell, S.; Graham, B. Protease Inhibitors Block Multiple Functions of the NS3/4A Protease-Helicase During the Hepatitis C Virus Life Cycle. *J. Virol.* **2015**, *89*(10), 5362–5370.

Mirza, M. U.; Ghori, N. U. L.; Ikram, N.; Adil, A. R.; Manzoor, S. Pharmacoinformatics Approach for Investigation of Alternative Potential Hepatitis C Virus Nonstructural Protein 5B Inhibitors. *Drug Des. Dev. Ther.* **2015,** *9,* 1825–1841.

Mueller, M.; Hobiger, S.; Jungbauer, A. Red Clover Extract: A Source for Substances that Activate Peroxisome Proliferator-Activated Receptor Alpha and Ameliorate the Cytokine Secretion Profile of Lipopolysaccharide-Stimulated Macrophages. Menopause, **2010,** *17,* 379–387.

Muralikrishnan, D.; Mohanakumar, K. P. Neuroprotection by Bromocriptine Against 1-Methyl-4-Phenyl-1,2,3,6-Tetrahydropyridine-induced Neurotoxicity in Mice. *FASEB J.* **1998,** *12,* 905–912.

O'Shea, R. S.; Dasarathy, S.; McCullough, A. J. The Practice Guideline Committee of the AASLD and the Practice Parameters Committee of the American College of Gastroenterology. Alcoholic Liver Disease. *Hepatology,* **2010,** *51*(1), 307–328.

Pal, P.; Ray, S. Alcoholic Liver Disease: A Comprehensive Review. *Emerg. Med. J.* **2016,** *1*(2), 85–92.

Pan, P. K.; Qiao L. Y.; Wen, X. N. Safranal Prevents Rotenone-Induced Oxidative Stress and Apoptosis in an in Vitro Model of Parkinson's Disease Through Regulating Keap1/Nrf2 Signalling Pathway. *Cell Mol. Biol.* **2016,** *62,* 11–17.

Pettinelli, P.; Obregon, A. M.; Videla, L. A. Molecular Mechanisms of Steatosis in Nonalcoholic Fatty Liver Disease. *Nutr. Hosp.* **2011,** *3,* 441–450.

Qiu, L.; Lin, B.; Lin, Z.; Lin, Y.; Lin, M.; Yang, X. Biochanin A Ameliorates the Cytokine Secretion Profile of Lipopolysaccharide-Stimulated Macrophages by a PPARγ-Dependent Pathway. Mol. Med. Rep. **2012,** *5,* 217–222.

Rasineni, K.; Casey, C. A. Molecular Mechanism of Alcoholic Fatty Liver. *Indian J. Pharmacol.* **2012,** *44*(3), 299–303.

Raskin, I.; Ribnicky, D. M.; Komarnytsky, S.; Ilic, N.; Poulev, A.; Borisjuk, N. Brinker, A.; Moreno, D. A.; Ripoll, C.; Yakoby, N.; O'Neal, J. M.; Cornwell, T.; Pastor, I.; Fridlender, B. Plants and Human Health in the Twenty-First Century. *Trends Biotechnol.* **2002,** *20*(12), 522–531.

Ratini, M. Drug side effects explained. Web MD Medical Reference. 2017.

Ravikumar, Y. S.; Ray, U.; Nandhitha, M.; Perween, A.; Naika, H. R.; Khanna, N.; Das, S. Inhibition of Hepatitis C Virus Replication by Herbal Extract: *Phyllanthus amarus* as Potent Natural Source. *Virus Res.* **2011,** *158,* 89–97.

Rekha, K.; Selvakumar, G.; Satha, K.; Sivakamasundari, R. Geraniol Attenuates A-Synuclein Expression and Neuromuscular Impairment Through Increase Dopamine Content in MPTP Intoxicated Mice by Dose Dependent Manner, *Biochem. Biophys. Res. Commun.,* **2013,** *440*(4), 664–670.

You, M.; Crabb, D. W. Recent Advances in Alcoholic Liver Disease II. Minireview: Molecular Mechanisms of Alcoholic Fatty Liver. *Am. J. Physiol. Gastrointest. Liver Physiol.* **2004,** *287*(1), G1-G6.

Zhang, G. F.; Zhang, Y.; Zhao, G. Crocin Protects PC12 Cells Against MPP$^+$ -Induced Injury through Inhibition of Mitochondrial Dysfunction and ER Stress. *Neurochem. Int.* **2015,** *89,* 101–110.

Plate (1) *Plumbago zeylanica* L.     (2) *Ailanthus excelsa* Roxb.     (3) *Trichosanthes tricuspidata* Laur

(4) *Alangium salvifolium* L.F. Wag.     (5) *Aristolochia indica* L.     (6) *Balanites aegyptiaca* (L.) Del.

(7) *Bridelia retusa* L.     (8) *Cissus quadrangularis* L.     (9) *Madhuca Longifolia* (Koen.) Mac Br.

(10) *Costus speciosa* (Koen.) Sm.     (11) *Gloriosa superba* L.     (12) *Kalanchoe pinnata* (Lam.) Pers.

**FIGURE 5.1**    Plant species of ethnoveterinary importance.

(13) *Macaranga peltata* (Roxb.) Muell-Arg.

(14) *Pergularia daemia* (Forsk.) Chiov.

(15) *Rubia cordifolia* L.

(16) *Strychnos potatorum* L.f.

*Aerva lanata* (L.) Juss. Ex. Sch.

*Amorphophallus paeoniifolius* (Dennst.) Nicolson

*Annona reticulata* L.

*Barringtonia acutangula* (Retz.) Willd.

*Cleistanthus collinus* (Roxb.) Benth. Ex Hook.f.

*Cochlospermum relisiosum* (L.) Alston

*Elephantophus scaber* L.

*Entada pursaetha* DC.

**FIGURE 5.1** *(Continued)*

*Grewia tiliaefolia* Van.

*Holoptelea integrifolia* (Roxb.)
Planch.

*Manilkara hexandra* (Roxb.)
Dubard.

*Pterolobium hexapetalum* (Roth)
Sant. & Wagh.

*Schleichera oleosa* (Lour) Koen

*Semicarpus anacardium* L.f.

**FIGURE 5.1**   *(Continued)*

**FIGURE 5.3**   Author working in the Field.

Plate 1: *Allium sativum*          Plate 2: *Azadirachta indica*          Plate 3: *Curcuma longa*

Plate 4: *Eucalyptus globulus*                          Plate 5: *Pongamia pinnata*

**FIGURE 9.1**    Antimicrobial medicinal plants.

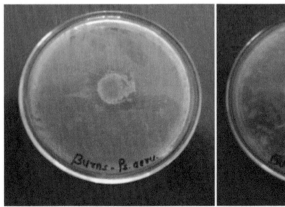

Effect of extract of *P. pinnata* on          Effect of extract of *P. pinnata*
P. aeruginosa from burns.                    on *S. aureus* from burns.

**FIGURE 9.3**    Antimicrobial activity of plant extracts on pathogenic isolates from burn wounds.

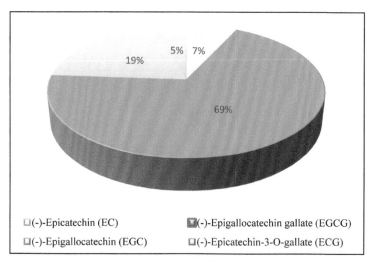

□ (-)-Epicatechin (EC)                    ▼ (-)-Epigallocatechin gallate (EGCG)

□ (-)-Epigallocatechin (EGC)              □ (-)-Epicatechin-3-O-gallate (ECG)

**FIGURE 11.1**    Relative percentage of polyphenol in green tea.

**FIGURE 12.1**    Bark of cinnamon (A) Ceylon cinnamon; (B) Chinese cassia; (C) Indonesian cinnamon; (D) Vietnamese cassia; and (E) Indian cassia.

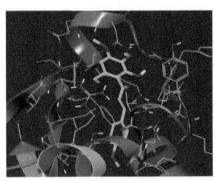

MIEN1_molecule14

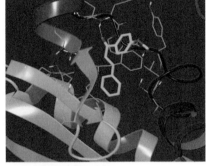

MIEN1_molecule 15

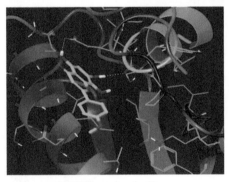

MIEN1_molecule 68

**FIGURE 16.4**   Interaction pattern of migration and invasion enhancer 1 protein and lead methylated flavonoids.

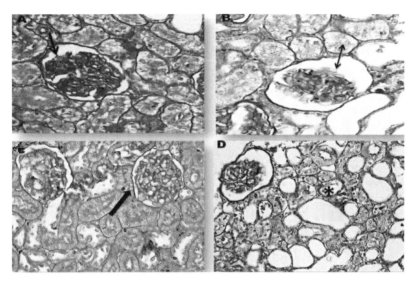

**FIGURE 18.1** Histological investigation in a kidney of diabetic nephropathy (DN): The excised kidney was fixed in neutral formalin solution and subsequently processed for histological studies. The thin micro-sections (5 μ) were cut and stained with periodic acid-Schiff reagent and visualized in a Nikon microscope (original magnification ×400). (A) Extracellular matrix accumulation in glomerular of DN (arrow) (B) Thickening of the glomerular basement membrane (both sided arrow) (C) Sclerotic nodule (Kimmelstiel–Wilson nodules) (thick arrow) (D) Marked glomerulosclerosis along with tubular atrophy and interstitial fibrosis (asterisk).

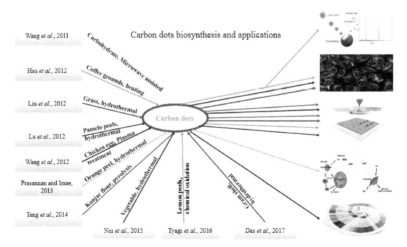

**FIGURE 22.1** Summary of some green approaches for carbon dots (CDs) synthesis and their application. The arrow color indicates the application of those synthesized CDs. The insets on the right hand side, from top to bottom: 1 = SALDI-MS, 2 = bio-imaging, 3 = chemical sensing, 4 = photocatalysis, and 5 = ink printing.

Plate 1: *Atropa belladonna*

Plate 2: *Brugmansia sp*

Plate 3: *Ricinus communis* L.

Plate 4: Seeds of *Ricinus communis* L.

Plate 5: *Dieffenbachia* Schott

Plate 6: *Cicuta maculate*

Plate 4: Seeds of *Ricinus communis* L.

Plate 5: *Dieffenbachia* Schott

Plate 6: *Cicuta maculate*

**FIGURE 23.1**    Some poisonous plants.

## CHAPTER 14

# ROLE OF PHYTOCHEMICALS IN THE TREATMENT OF MALE INFERTILITY

VIJAYKUMAR K. PARMAR[1,3,*] and KETAN VARIYA[2]

[1,2]*Ramanbhai Patel College of Pharmacy, Charotar University of Science and Technology (CHARUSAT), Anand, India*

[2]*Sun Pharma Advanced Research Company Ltd., Vadodara, India*

[3]*Department of Pharmaceutical Sciences, Sardar Patel University, Vallabh Vidyanagar, India*

*\*Corresponding author. E-mail: vijaykparmar@gmail.com*
*\*ORCID: orcid.org/0000-0003-2092-2144.*

## ABSTRACT

The World Health Organization (WHO) estimated that 50–60 million couples worldwide are affected by infertility. Infertility, the inability to conceive after a prolonged period, can be referred to both male and female partner. Impotence or sexual dysfunction is a major cause of male infertility. The cures accessible for male sexual dysfunctions are restricted in present-day pharmaceuticals. Synthetic drugs prescribed for the treatment of erectile dysfunction (ED) are believed to produce unwanted effects, which affects other physiological processes and, ultimately, general health. Aphrodisiacs, defined as drugs, herbs, phytocompounds, or food, which are known for their effects in increasing sexual desire and improving sexual performance, have been used for thousands of years as traditional medicines. The utilization of herbal aphrodisiacs by the society has been exhibited for thousands of years. The in vitro animal and clinical experiments published in the literature showing the effects of phytochemicals in male infertility and sexual dysfunction are summarized in this chapter.

## 14.1   INTRODUCTION

The World Health Organization (WHO) defines infertility as a disease of the reproductive system referred to the failure to carry a clinical pregnancy after one year or more of regular unprotected sexual intercourse. Primary infertility is infertility in a couple who have never had a child. Secondary infertility is a failure to conceive following a previous pregnancy (Mascarenhas et al., 2012). The WHO estimates that 15 to 20% of couples worldwide currently suffer from infertility (WHO report, 2004). The prevalence of infertility in Africa ranges from 9 to 30% (Ombelet et al., 2008). The rate of male infertility is highest in Africa and European countries (Agarwal et al., 2015). The WHO estimates the overall prevalence of primary infertility in India to be between 3.9 and 16.8% (Adamson et al., 2011). Infertility has been relatively neglected as both a health problem and a subject for social science research in South Asia, as in the developing world, more generally (Ganguly and Unisha, 2010). The inability to have children affects both partners across the globe. Infertility can lead to misery and depression, as well as discrimination and exclusion in the society (Cui, 2010; Chachamovich et al., 2010). Whenever a couple is not able to conceive, it is generally assumed in Indian society that the female partner is at fault. Doctors, however, say that in more than 50% of the cases, infertility or inability to conceive can be associated with the male partner. The number has grown extremely in recent times (Gwalani, 2012).

Figure 14.1 depicts the data on causes of infertility reported in a guidance document on fertility services from the Department of Health, UK (2009). The male partner is unable to produce or ejaculate enough normal sperm, in about 30% of cases. Female partner suffers from problems such as failure to ovulate in about 10% of cases or a partial or complete blockage to the passage of the eggs from the ovary to the uterus in about 15% of the cases. In about 10–15% of cases, problems are found with both partners. However, in about one-fourth of the cases, no cause of infertility can be identified (UK, 2009). The trend in masculine sexual dysfunction and impotency, which significantly contributes to male infertility, is increasing at a very alarming rate.

The utilization of herbs for treating various ailments goes back a few centuries. These days, about 80% of the total populace utilizes plant-derived medicines for the treatment of many ailments. Around the world, such medicines make up a one-fourth share of the pharmaceutical arm stockpile (De Smet et al., 2000). The herbal medicine market is growing at a very rapid rate as the consumers are eventually coaxed by the benefits of herbal medicines over synthetic medicines (Yakubu et al., 2007). This business sector for branded nonprescription herbal medicines has grown from $1.5 billion in 1994 to $4.0 billion in 2000 in the US alone with a similar pattern being found in European nations as well (Rosen and Ashton, 1993). Traditional therapeutic approaches of regional significance are found in South and Central America, India (Ayurveda), Traditional Chinese medicine, Africa, Tibet, Indonesia, and the Pacific Islands. Thus, plants and other natural resources have always been of interest to researchers, mainly, for the treatment of numerous medical conditions as well as in the development of new drugs.

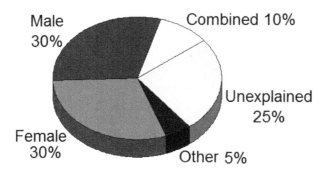

**FIGURE 14.1**    A data from the Department of Health UK, 2009.

## 14.2   CAUSES OF MALE INFERTILITY

The major events in the erection process are (a) release of neurotransmitters from the cavernous nerve terminals, (b) relaxation of the penile arterial system, resulting in increased blood flow in the corpora cavernosa, (c) expansion of sinusoids, (d) compression of the venules against the tunica albuginea, (e) relaxation of the trabecular smooth muscle, (f) increase in pressure of oxygen ($PO_2$), and intracavernous pressure (Dean and Lue, 2005).

The primary reasons for male infertility are sexual dysfunction, inadequate sperm quality, endocrine causes, radiation/chemotherapy, age, environmental factors (pesticides, lead), and factors amenable to personal control, including smoking, exercise, stress, and alcohol consumption. The main areas of sexual dysfunction in men are described in Table 14.1 (Etuk, 2009). Further, the reasons of male infertility may be categorized into pre-testicular, testicular, and post-testicular reasons (Table 14.2; Wiser et al., 2012).

**TABLE 14.1** Main Areas of Sexual Dysfunction in Men.

| Abnormalities | Causes |
| --- | --- |
| **Sperm abnormalities** | Low sperm count (oligospermia), poor sperm motility (asthenospermia), and abnormal sperm morphology (teratospermia) |
| **Sexual problems** | Erection problem, ejaculation problem |
| **Structural abnormalities** | Cryptorchidism, hypospadias, blockage in the tubes that transport sperm |
| **Hormonal problem** | Hypogonadism |
| **Genetic disorders** | Cystic fibrosis, polycystic kidney disease, Klinefelter syndrome, and Kartagener syndrome |

**TABLE 14.2** Causes of male infertility.

| Problems | Causes |
| --- | --- |
| **Pre-testicular** | Hypogonadotropic, hypogonadism, hypergonadotropic, hypogonadism, elevated prolactin, pharmacologic (medications), and Kallmann Syndrome |
| **Testicular** | Varicocele, cryptorchidism, testicular cancer, ionizing radiation, genetic, azoospermia/oligospermia, chemotherapy, environmental factors, testicular injury, primary ciliary dyskinesia, sertoli cell-only syndrome, antisperm antibodies, and DNA damage |
| **Post-testicular** | Absence of the vas deferens, Young's syndrome, seminal vesicle dysfunction, vasectomy, nerve injury, medications, resection of the prostate, and coital |

## 14.3 HERBAL REMEDIES FOR THE TREATMENT OF MALE INFERTILITY

The treatment of male infertility includes drug therapy for any underlying health conditions that contribute to male infertility. For example, hormonal treatment (GnRH and/or gonadotrophin treatment) is used to treat hypogonadism (Howards, 1995; Kamischke and Nieschlag, 1999). Surgery is made to repair any obstructions in the reproductive tract (Young, 1970). Intracytoplasmic sperm injection (ICSI) is commonly used in combination with in vitro fertilization in cases of male factor infertility (Baek, 2007). For reversal of retrograde ejaculation, α-adrenergic agonists (phenylpropanolamine, oxedrine), anticholinergics (brompheniramine maleate and imipramine), which facilitate ejaculation by stimulating peristalsis in the vas deferens and closing the neck of the bladder, have all been used. The modern medicines available for male sexual dysfunction are limited. Allopathic medicines mainly phosphodiesterase-V inhibitors, which include sildenafil and tadalafil, are available for erectile dysfunction (ED). Very few drugs are available in case of lack of sexual desire/decrease in libido, to exemplify, tricyclic anti-depressant in case of lack of sexual desire is due to depression (Burns, 2007). These drugs are believed to produce side effects. The diverse treatment plans are emerging globally for the management of male infertility caused by hormonal imbalance, infections, and other emergencies. Modern treatments, such as assisted reproductive techniques, are costly with low success rates of only 10–30%; however, herbal remedies are gaining more attention as an alternative or supplementary therapeutic mode for male infertility (Sengupta et al., 2017).

The belief that natural medicines are much safer than synthetic medication has picked up fame in recent years and prompted due to the tremendous growth of phytopharmaceuticals utilization. Several plant materials are empirically used to treat distinct conditions of male infertility, namely libido, sperm abnormalities, and erectile and ejaculatory dysfunctions. The biological activities of many of these plants were confirmed by in vitro and/or in vivo animal studies and in humans (Table 14.3).

**TABLE 14.3** Phytochemical Studies on the Male Reproductive System.

| Effect | Plant | Reference |
| --- | --- | --- |
| **In vitro effect in animals tissues/cells** | *Haplopappus rigidus* | Hnatyszyn et al. (2003) |
| | *Huperzia saururus* | Hnatyszyn et al. (2003) |
| | *Lycium barbarum* | Hnatyszyn et al. (2003) |
| | *Satureja parvifolia* | Hnatyszyn et al. (2003) |
| | *Senecio eriophyton* | Hnatyszyn et al. (2003) |
| | *Basella alba* | Moundipa et al. (2005) |
| | *Epimedium brevicornum* | Chiu et al. (2006) |
| **In vitro effect in human tissues/cells** | *Astragalus membranaceus* | Liu et al. (2004) |
| | *Acanthopanacis senticosi* | Liu et al. (2004) |
| | Genistein | Fraser et al. (2006) |
| **In vivo effect in animals** | *Trichopus zeylanicus* | Subramoniam et al. (1997) |
| | *Panax quinquefolium* | Murphy et al. (1998) |
| | *Turnera diffusa* and *Pfaffia paniculata* | Arletti et al. (1999) |

**TABLE 14.3**   *(Continued)*

| Effect | Plant | Reference |
|--------|-------|-----------|
|  | *Terminalia catappa* | Ratnasooriya and Dharmasiri (2000) |
|  | *Vanda tessellata* | Kumar et al. (2000) |
|  | *Lepidium meyenii* | Gonzales et al. (2002) |
|  | *Pentadiplandra brazzeana* | Kamtchouing et al. (2002) |
|  | *Tribulus terrestris* | Gauthaman et al. (2002) |
|  | *Zingiber officinale* | Kamtchouing et al. (2002) |
|  | *Lepidium meyenii* | Gonzales et al. (2003) |
|  | *Butea frondonsa* | Ramachandran et al. (2004) |
|  | *Mondia whitei* | Watcho et al. (2004) |
|  | *Montanoa tomentosa* | Carro-Juarez et al. (2004) |
|  | *Syzygium aromaticum* | Tajuddin et al. (2004) |
|  | *Fadogia agrestis* | Yakubu et al. (2005) |
|  | *Piper guineense* | Mbongue et al. (2005) |
|  | *Ruta chalepensis* | Al-Qarawi (2005) |
|  | *Lycium barbarum* | Luo et al. (2006) |
|  | Meyenii | Rubio et al. (2006) |
|  | *Satureja khuzestanica* | Haeri et al. (2006) |
|  | *Basella alba* | Nantia et al. (2007) |
|  | *Catha edulis* | Abdulwaheb et al. (2007) |
|  | *Cordyceps militaris* | Lin et al. (2007) |
|  | *Croton zambesicus* | Ofusori et al. (2007) |
|  | *Dactylorhiza hatagirea* | Thakur and Dixit (2007) |
|  | *Dracaena arborea* | Watcho et al. (2007) |
|  | *Epimedium koreanum* | Makarova et al. (2007) |
|  | *Myristica fragrans* | Tajuddin et al. (2005) |
|  | *Nigella sativa* | Bashandy (2007) |
|  | *Panax ginseng* | Park et al. (2007) |
|  | *Casimiroa edulis* | Ali and Rakkah (2008) |
|  | *Ginkgo biloba* | Yeh et al. (2008) |
|  | *Massularia acuminate* | Yakubu et al. (2008) |
|  | *Peganum harmala* | Hamden et al. (2008) |
|  | *Punica granatum* | Turk et al. (2008) |
|  | *T. terrestris* | Gauthaman et al. (2008) |

**TABLE 14.3**   *(Continued)*

| Effect | Plant | Reference |
|---|---|---|
| **In vivo effect in humans** | *Eurycoma longifolia* | Ang et al. (2001) |
| | *Phoenix dactylifera* | Bahmanpour et al. (2015) |
| | Y virilin | Rege et al. (1997) |
| | *Lepidium meyenii* | Gonzales et al. (2001) |
| | *Shengjing pill* | Xu et al. (2003) |
| | Formulation of plants *(T. terrestris, Asparagus racemosus, Withania somnifera)* | Devi et al. (2004) |
| | Formulation of Chinese medicinal plants | Tempest et al. (2005) |
| | *Korean Red Ginseng* | De Andrade et al. (2007) |
| | *Mucuna pruriens* | Shukla et al. (2010) |

Aphrodisiacs are the substances which are generally taken to upsurge sexual activity and helps in improving fertility. Aphrodisiacs are defined as food, herbs, or drugs, which are assumed to increase sexual desire and enhance sexual performance. This word is derived from "Aphrodite" the Greek goddess of love and these substances are derived from plants, animals, or minerals (Yakubu et al., 2007). Aphrodisiacs are of two types (a) preparations stimulating psychophysiological actions (aural, visual, olfactory, and tactile) and (b) internal preparations (food and alcoholic drinks) (Rosen and Ashton, 1993). The utilization of aphrodisiacs by the society has been in existence for many years now. Historically, Chinese, Roman, Indian, Greek, and Egyptian cultures have utilized various substances as solutions for enhancing sexual performance. The list of herbal aphrodisiac comprises Chhota Gokhru (*Tribulus terrestris*) (Gauthaman et al., 2002), Fenugreek (*Trigonella foenum-graecum*) (Sakr et al., 2012), Safed musli (*Chlorophytum borivilianum*) (Thakur et al., 2009), Ashwagandha (*Withania somnifera*) (Ilayperuma et al., 2002), Shatavari (*Asparagus racemosus*) (Satyavati et al., 1976), Kapikachhu (*Mucuna pruriens*) (Anantha et al., 1994), Kokilaksha (*Asteracantha longifolia*) (Chauhan et al., 2011), and Nutmeg (*Myristica fragrans*) (Tajuddin et al., 2003). The Ayurvedic system of medicine explicitly addresses the problem of male infertility by treatment with a specialized remedy known as *Vajikarana*. Vajikaran therapy has been prescribed for male impotence, suppressed libido and in debility, especially encountered with advancing age (Sharma et al., 2012).

## 14.4   EFFECTS OF PHYTOCHEMICALS IN ANIMAL MODELS OF MALE INFERTILITY

Phytochemicals are purified extracts from medicinal plants by which the bioactive compounds and other constituents have been well identified and standardized. In past years, there have been several studies using animal models for the efficacy of phytochemicals in male infertility. The animal experiments are designed using sexually active animals administered with phytochemicals. The allopathic medicine such as sildenafil is used as positive control for sexual behavior whereas testosterone as a positive control for androgenic activity. The sexual behavior is monitored with measurement of penile erection index (PEI), monitoring copulatory/mounting activities such as mount latency (ML), mount frequency (MF), intromission latency (IL), intromission frequency (IF), ejaculation latency (EL), and post-ejaculatory latency (PEL). Hormonal estimation (serum testosterone, serum follicle-stimulating hormone (FSH), and serum luteinizing hormone (LH)), weight measurements (body weight, sexual organ weight), histological studies, and sperm analysis (sperm motility, sperm viability, and sperm count) are carried out at the end of treatments.

An aqueous decoction of roots of *Caesalpinia benthamiana* is used in African traditional remedy for the treatment of ED. Zamble et al. studied aphrodisiac properties of aqueous extract of *C. benthamiana* in male rats and confirmed the empirical use of *C. benthamiana* as an aphrodisiac agent (2008). The enhancement of sexual activity is partly supported by its vaso-relaxant properties as a result of an increase in nitric oxide (NO) production in vascular bed and a decline in its destruction (Zamble et al., 2008). Further, Zamble and his group conducted studies on aqueous extract of *Microdesmis keayana* roots to ascertain its aphrodisiac activity (Zamble et al., 2008a). The aphrodisiac activity was linked with its action in NO production and, mainly attributed to the alkaloids, keayanidine B, and keayanine, largely present in *M. keayana* roots. Since ancient times, many Allium species such as garlic, chives, and onion have been used as spices, food and herbal remedies in widespread areas around the world. The Allium genus is a rich source of steroidal saponins, alkaloids, and sulfur-containing compounds. Guohua et al. provided experimental evidence for the aphrodisiac activity of butanolic extract of *Allium tuberosum* in male rats (2009).

Patel et al. summarized 41 plant materials which have been tested for its aphrodisiac activity in vitro or in vivo experimental models (2011). Singh and Gupta have studied the aphrodisiac activity of *T. terrestris* using rat model (2011). Lyophilized powder of the *T. terrestris* dried fruits (100 mg/kg body weight) orally administered has demonstrated anabolic effect confirmed by the weight gain in the body as well as reproductive organs. Enhancement in sexual behavior of male rats was surmised by increased MF, IF, and PEI and decreased ML, IL, and EL. The progression in sexual performance was ascribed to the significant increase in sperm count and serum testosterone. Findings of the study validated the traditional use of *T. terrestris* in Chinese and Indian medicinal systems for its role in the treatment of ED. Similarly, Sharma et al. investigated effects of ethanolic extract of *Pedalium murex* fruits, used as *Vajikaran Rasayana*, on sexual performance and testosterone levels of male rats (2012).

Dhumal et al. formulated a polyherbo-mineral formulation, Afrodet Plus®, for the treatment of male infertility (2013). The major ingredients of the formulation were *M. pruriens*, *W. somnifera*, *A. racemosus*, *C. borivilianum*, *Pueraria tuberosa*, and *Argyreia speciosa*. The efficacy and safety of the formulation were evaluated in Holtzman rats using reverse pharmacology approach. The study shows that the oral administration of Afrodet Plus® (Dose: 90 and 180 mg/kg body weight) lead to significant increase in daily sperm production, sperm motility, epididymal sperm count, serum testosterone level, and weight of epididymis as compared to control group without any adverse effects. Sahin et al. determined the effect of three antioxidant herbals – *M. pruriens*, *T. terrestris*, and *W. somnifera,* on sexual enhancing capacity, serum biochemical parameters and levels of nuclear factor kappa-B (NF-κB), erythroid 2-related factor 2 (Nrf2), and Heme oxygenase (HO-1) in male Sprague-Dawley rats (2016). It was evident from the study that the extracts of these three herbals are potent enhancers of sexual function and behavior by the increased testosterone levels, and regulation of NF-κB and Nrf2/HO–1 pathways. Thus, natural antioxidants and phytochemicals have demonstrated improvement in semen quality.

## 14.5 EFFECTS OF PHYTOCHEMICALS IN IN VITRO MODELS OF MALE INFERTILITY

The reports on in vitro models connecting phytochemicals and male infertility are relatively sparse. Penile erection is essential for effective copulation in males. The NO is assumed to be the main vasoactive nonadrenergic and noncholinergic neurotransmitter of erectile action in the corpora cavernosa (Burnett et al., 1992). The improvement in PEI is generally supported by in vitro NO release using human corpus cavernosum cell line (DS-1). The effect of *P. murex* extracts containing furostanolic glycosides in ED was correlated on inducible NO (iNO) release in vitro with the enhanced erectile function in vivo (Sharma et al., 2012). It is conceivable that phytocompounds present in *P. murex* fruit extract might influence both hypothalamo-pituitary – gonadal axis to increase serum testosterone level and in vitro NO release.

Rakuambo et al. performed in vitro bioassays of *Rhoicissus tridentata, Securidaca longepedunculata,* and *Wrightia natalensis* extracts on the contraction of corpus cavernosal smooth muscle of white New Zealand rabbits using Viagra as the positive control and established a strong relationship between relaxation of this muscle and sexual function (2006). Researchers also analyzed the effect of *Securidaca longepedunculata* extracts on human sperm cells. An in vitro screening of different solvent fractions of *Corchorus depressus* on rabbit corpus cavernosal smooth muscle was conducted to ascertain their aphrodisiac activity.

## 14.6   CLINICAL STUDIES ON PHYTOCHEMICALS FOR THE TREATMENT OF MALE INFERTILITY

Myriad of animal studies is reported on herbal plant materials for the treatment of male infertility. However, randomized controlled clinical trials are scarce. Very few herbal plants are studied in humans (Table 14.4). The quality of most clinical studies is poor. The phytochemicals tested clinically are *T. terrestris* (Garg et al., 2004), *M. pruriens* (Shukla et al., 2010), *W. somnifera* (Mahdi et al., 2011), *Nigella sativa* (Kolahdooz et al., 2014), *Panax ginseng* (Park et al., 2016), and *T. foenum-graecum* (Maheshwari et al., 2017).

## 14.7   CONCLUSION AND FUTURE PERSPECTIVES

Herbal remedies have been widely used in most areas around the world since ancient times. There are a plethora of animal studies reported in the literature to prove the effect of phytochemicals on sexual behavior, sperm parameters, and ED. The in vitro testing supports the screening experiments for evaluation of aphrodisiac properties of plant extracts. Well-defined systematic clinical trials should be exercised to determine efficacy and safety of phytochemicals for the treatment of male infertility. The well-documented studies would be helpful to set phytochemicals as complementary or alternative medicines for the treatment of male infertility.

**TABLE 14.4**  Clinical Studies on Phytochemicals for Treatment of Male Infertility.

| Study | Study design | Infertility criteria | Treatment | Result | Reference |
|---|---|---|---|---|---|
| Evaluation of safety and efficacy of "Tentex Royal," a polyherbal formulation, in patients with erectile dysfunction (ED) | prospective, open, noncomparative, and phase III clinical trial (n=30) | Patient with erectile dysfunction | 2 capsules of "Tentex Royal," once daily, 1 h before the sexual intercourse | Significant improvement in erectile function from the third week onwards | Garg et al. (2004) |
| Role of *M. pruriens* in infertile men under psychological stress | Parallel controlled (n=20) | (i) normozoospermic infertile men, defined as a control group (ii) oligozoospermic infertile men <20 × 106, spermatozoa/ mL, motility >40% and >40% normal morphology and (iii) asthenozoospermic infertile men (>20 × 106 spermatozoa/ mL, <40% motility and >40% normal morphology). | *M. pruriens* seed powder (5 g day1), orally, in a single dose for 3 months with a cup of skimmed milk | Amelioration of psychological stress, improvement in sperm count and motility | Shukla et al. (2010) |
| Effect of *Withania somnifera* in infertile men with normozoospermia either under psychological stress or heavy cigarette smoking habit | Parallel controlled (n = 20) | Not defined | orally, in a single dose (5 g/day) for 3 months | Treatment resulted in a decrease in stress, improved the level of antioxidants, and improved overall semen quality in a significant number of individuals. The treatment resulted in pregnancy in the partners of 14% of the patients. | Mahdi et al. (2011) |

**TABLE 14.4** *(Continued)*

| Study | Study design | Infertility criteria | Treatment | Result | Reference |
|---|---|---|---|---|---|
| Effect of *N. sativa* L. seed oil in infertile men with abnormal semen quality | Randomized, double-blind, placebo-controlled ($n = 34$) | abnormal sperm morphology less than 30% or sperm counts below $20 \times 106$/mL or type A and B motility less than 25and 50%, respectively | 2.5 mL N. sativa oil two times a day orally for 2 months | sperm count, sperm morphology, sperm motility, semen round cells, semen volume and semen pH improved significantly ($p = 0.001$, $p = 0.011$, $p = 0.001$, $p = 0.025$, $p = 0.001$ and $p = 0.001$, respectively) compared with placebo group | Kolhadooz et al. (2014) |
| Effect of protodioscin-enriched fenugreek seed extract (Furosap®) on testosterone level and sperm profile in healthy volunteers | One-arm, open-labeled ($n = 50$) | Healthy male diagnosed with symptomatic hypogonadism | Furosap® (1 capsule of 500 mg each/day after breakfast) over a period of 12 consecutive weeks. | 1.47-fold increase in free testosterone level ($p = 0.0004$), significant increase in sperm count, and sperm motility | Maheshwari et al. (2017) |

## KEYWORDS

- **animal models**
- **aphrodisiacs**
- **male infertility**
- **phytochemicals**
- **sexual dysfunction**

## REFERENCES

Ahmad, S.; Latif, A.; Qasmi I. A. Effect of 50% Ethanolic Extract of *Syzygium aromaticum* (L.) Merr. & Perry.(clove) on Sexual Behaviour of Normal Male Rats. *BMC Complementary And Alternative Medicine*. **2004**, *4,* 17.

Ahmad, S.; Latif, A.; Qasmi I. A.; Amin, K. M. Y. An Experimental Study Of Sexual Function Improving Effect of *Myristica fragrans* Houtt.(nutmeg). *BMC Complement Altern. Med.* **2005**, *5,* 16.

Ahmad, S.; Latif, A.; Qasmi, I. A. Aphrodisiac Activity of 50% Ethanolic Extracts of *Myristica fragrans* Houtt.(nutmeg) and *Syzygium aromaticum* (L) Merr. & Perry.(clove) in Male Mice: A Comparative Study. *BMC Complement Altern Med.* **2003**, *3,* 6.

Dean, R. C.; Lue, T. F. Physiology of Penile Erection and Pathophysiology of Erectile Dysfunction. *Urologic Clin. North Am.* **2005**, *32,* 379.

Department of Health UK. Expert Group on Commissioning Nhs Infertility Provision. Regulated Fertility Services: A Commissioning Aid, 2009.

Sharma, V.; Thakur, M.; Dixit, V. K. A Comparative Study of Ethanolic Extracts of Pedalium Murex Linn. Fruits and Sildenafil Citrate on Sexual Behaviors and Serum Testosterone Level in Male Rats During and After Treatment. *J. Ethnopharmacol.* **2012**, *143,* 201-206.

Thakur, M.; Dixit, V. Aphrodisiac Activity of *Dactylorhiza hatagirea* (D. Don) Soo in Male Albino Rats. *Evidence-Based Complement. Altern. Med. eCAM.* **2007**, *4,* 29–31.

Abdulwaheb, M.; Makonnen, E.; Debella, A.; Abebe, D. Effect of *Catha Edulis* Foresk (khat) Extracts on Male Rat Sexual Behavior. *J. Ethnopharmacol.* **2007**, *110,* 250–256.

Adamson, P. C.; Krupp, K.; Freeman, A. H.; Klausner, J. D.; Reingold, A. L.; Madhivanan, P. Prevalence & and Correlates of Primary Infertility Among Young Women in Mysore, India. *Indian J. Med. Res.* **2011**, *134,* 440–446.

Agarwal, A.,; Mulgund, A.;, Hamada, A.,; Chyatte, M. R. A Unique View on Male Infertility Around the Globe. *Reprod. Biol. Endocrinol.* **2015**, *13*(1), 37.

Ali, S. T.; Rakkah, N. I. Probable Neuro Sexual Mode of Action of *Casimiroa edulis* Seed Extract Verseus Sildenafil Citrate (Viagra (Tmtm)) on Mating Behavior in Normal Male Rats. *Pak. J. Pharm. Sci.* **2008**, *21,* 1–6.

Al-Qarawi, A. A. Stimulatory Effect of the Aqueous Extract of *Ruta Chalepensis* on the Sex Organs and Hormones of Male Rats. *J. Appl. Res.* **2005**, *5,* 206–211.

Anantha K. K.; Srinivasan, K.; Shanbhag, T.; Rao, S. Aphrodisiac Activity of the Seeds of *Mucuna pruriens*. *Indian Drugs*. **1994**, *31,* 321–327.

Ang, H.; Ikeda, S.; Gan, E. Evaluation of the Potency Activity of Aphrodisiac in *Eurycoma Longifolia* Jack. *Phytother. Res.* **2001**, *15*, 435–436.

Arletti, R.; Benelli, A.; Cavazzuti, E.; Scarpetta, G.; Bertolini, A. Stimulating Property of *Turnera diffusa* and *Pfaffia paniculata* Extracts on the Sexual Behavior of Male Rats. *Psychopharmacology* **1999**, *143*, 15–19.

Baek, J. S. Treatment of Male Infertility. *J Korean Med. Assoc.* **2007**, *50*, 424–430.

Bahmanpour, S.; Talaei, T.; Vojdani, Z.; Panjehshahin, M.; Poostpasand, A.; Zareei, S.; Ghaeminia, M. Effect of *Phoenix dactylifera* Pollen on Sperm Parameters and Reproductive System of Adult Male Rats. *Iran. J. Med. Sci.* **2015 2006**, *31*, 208–212.

Bashandy, A. S. Effect of Fixed Oil of Nigella Sativa on Male Fertility in Normal and Hyperlipidemic Rats. *Int. J. Pharm.* **2007**, *3*, 27–33.

Burnett, A. L.; Lowenstein, C. J.; Bredt, D. S.; Chang, T. S.; Snyder, S. H. Nitric Oxide: A Physiologic Mediator of Penile Erection. *Science* **1992**, *257*, 401–403.

Burns, L. H. Psychiatric Aspects of Infertility and Infertility Treatments. *Psychiatr Clin North Am.* **2007**, *30*, 689–716.

Carro-Juarez, M.; Cervantes, E.; Cervantes-Mendez, M.; Rodrıguez-Manzo, G. Aphrodisiac Properties of *Montanoa tomentosa* Aqueous Crude Extract in Male Rats. *Pharmacol. Biochem. Behav.* **2004**, *78*, 129–134.

Chachamovich, J. R.; Chachamovich, E.; Ezer, H.; Fleck, M. P.; Knauth, D.; Passos, E. P. Investigating Quality of Life and Health-Related Quality of Life In Infertility: A Systematic Review. *J. Psychosom. Obstet. Gynaecol.* **2010**, *31*, 101–110.

Chauhan, N. S.; Sharma, V.; Dixit, V. Effect of *Asteracantha longifolia* Seeds on the Sexual Behaviour of Male Rats. *Nat. Prod. Res.* **2011**, *25*, 1423–1431.

Chiu, J.; Chen, K.; Chien, T.; Chiou, W.; Chen, C.; Wang, J.; Lui, W.; Wu, C. *Epimedium brevicornum* Maxim Extract Relaxes Rabbit Corpus Cavernosum Through Multitargets on Nitric Oxide/Cyclic Guanosine Monophosphate Signaling Pathway. *Int. J. Impotence Res.* **2006**, *18*, 335–342.

Cui, W. Mother or Nothing: the Agony of Infertility. *Bull. World Health Organ.* **2010**, *88*, 881–882.

De Andrade, E.; De Mesquita, A. A.; De Almeida Claro, Jde, A.; De Andrade, P. M.; Ortiz, V.; Paranhos, M.; ErdogrunSrougi, M., T. Study of the Efficacy of Korean Red Ginseng in the Treatment of Erectile Dysfunction. *Asian J. Andrology.* **2007**, *9*, 241–244.

De Smet, P. A.; Bonsel, G.; Van der Kuy, A.; Hekster, Y. A.; Pronk, M. H.; Brorens, M. J.; Lockefeer, J. H.; Nuijten, M. J. Introduction to the Pharmacoeconomics of Herbal Medicines. *PharmacoEconomics* **2000**, *18*, 1–7.

Dean, R. C.; Lue, T. F. Physiology of Penile Erection and Pathophysiology of Erectile Dysfunction. *Urol. Clin. North Am.* **2005**, *32*, 379.

Department of Health UK. Expert Group on Commissioning NHS Infertility Provision. Regulated Fertility Services: A Commissioning Aid. 2009.

Devi, P. R., Laxmi, V.; Charulata, C.; Rajyalakshmi, A. "Alternative Medicine"—" – a Right Choice for Male Infertility Management. *Int. Congr. Ser.*; **2004**, *1271*, 67–70.

Dhumal, R.; Vijaykumar, T.; Dighe, V.; Selkar, N.; Chawda, M.; Vahlia, M.; Vanage, G. Efficacy and Safety of a Herbo-Mineral Ayurvedic Formulation 'Afrodet Plus®' in Male Rats. *J. Ayurveda Integr. Med.* **2013**, *4*, 158–164.

Etuk, S. J. Reproductive Health: Global Infertility Trend. *Niger. J. Physiol. Sci.: Off. Publ. Physiol. Soc. Niger.* **2009**, *24*, 85–90.

Fraser, L. R.; Beyret, E.; Milligan, S. R.; Adeoya-Osiguwa, S. A. Effects of Estrogenic Xenobiotics on Human and Mouse Spermatozoa. *Hum. Reprod.* **2006**, *21*, 1184–1193.

Ganguly, S.; Unisa, S. Trends of Infertility and Childlessness in India: Findings from NFHS Data. *Facts, Views Vision ObGyn.* **2010**, *2,* 131–138.

Garg, S. K.; Giri, S.; Kolhapure, S. A. Evaluation of the Efficacy and Safety of "Tentex Royal" in the Management of Erectile Dysfunction. *Medicine* **2004**, *12,* 51–55.

Gauthaman, K.; Adaikan, P.; Prasad, R. Aphrodisiac properties of *Tribulus terrestris* Extract (Protodioscin) in Normal and Castrated Rats. *Life Sci.* **2002**, *71,* 1385–1396.

Gauthaman, K.; Ganesan, Adaikan., P. G.; Prasad, R. N. The Hormonal Effects of *Tribulus Terrestris* and its Role in the Management of Male Erectile Dysfunction – An Evaluation Using Primates, Rabbit and Rat. *Phytomedicine* **2008**, *15,* 44–54.

Gonzales, G. F.; Cordova, A.; Gonzales, C.; Chung, A.; Vega, K.; Villena, A. Lepidium *Meyenii* (Maca) Improved Semen Parameters In Adult Men. *Asian J. Androl.* **2001**, *3,* 301–304.

Gonzales, G. F.; Cordova, A.; Vega, K.; Chung, A.; Villena, A.; Gonez, C.; Castillo, S. Effect of *Lepidium meyenii* (Maca) On Sexual Desire and its Absent Relationship with Serum Testosterone Levels in Adult Healthy Men. *Andrologia* **2002**, *34,* 367–372.

Gonzales, G. F.; Rubio, J.; Chung, A.; Gasco, M.; Villegas, L. Effect of Alcoholic Extract of *Lepidium meyenii* (Maca) on Testicular Function in Male Rats. *Asian J. Androl.* **2003**, *5,* 349.

Guohua, H.; Yanhua, L.; Rengang, M.; Dongzhi, W.; Zhengzhi, M.; Hua, Z. Aphrodisiac Properties of Allium Tuberosum Seeds Extract. *J. Ethnopharmacol.* **2009**, *122,* 579–582.

Gwalani, P. Increasing Male Infertility Worries Doctors. *The Times of India.* 2012 27 July 27, 2012.

Haeri, S.; Minaie, B.; Amin, G.; Nikfar, S.; Khorasani, R.; Esmaily, H.; Salehnia, A.; Abdollahi, M. Effect of *Satureja khuzestanica* Essential Oil on Male Rat Fertility. *Fitoterapia* **2006**, *77,* 495–499.

Hamden, K.; Silandre, D.; Delalande, C.; Elfeki, A.; Carreau, S. Protective Effects of Estrogens and Caloric Restriction During Aging on Various Rat Testis Parameters. *Asian J. Androl.* **2008**, *10,* 837–845.

Hnatyszyn, O.; Moscatelli, V.; Garcia, J.; Rondina, R.; Costa, M.; Arranz, C.; Balaszczuk, A.; Ferraro, G.; Coussio, J. D. Argentinian Plant Extracts with Relaxant Effect on the Smooth Muscle of the Corpus Cavernosum of Guinea Pig. *Phytomedicine* **2003**, *10,* 669–674.

Howards, S. S. Treatment of Male Infertility. *New Engl. J. Med.* **1995**, *332,* 312–317.

Ilayperuma, I.; Ratnasooriya, W.; Weerasooriya, T. Effect of *Withania somnifera* Root Extract on the Sexual Behaviour of Male Rats. *Asian J Androl.* **2002**, *4,* 295–298.

Kamischke, A.; Nieschlag, E. Analysis of Medical Treatment of Male Infertility. *Hum. Reprod.* **1999**, *14,* 1–23.

Kamtchouing, P.; Fandio, G. M.; Dimo, T.; Jatsa, H. B. Evaluation of Androgenic Activity of *Zingiber officinale* and *Pentadiplandra brazzeana* in Male Rats. *Asian J. Androl.* **2002**, *4,* 299–302.

Kolahdooz, M.; Nasri, S.; Modarres, S. Z.; Kianbakht, S.; Huseini, H. F. Effects of Nigella Sativa lL. Seed Oil on Abnormal Semen Quality in Infertile Men: A Randomized, Double-Blind, Placebo-Controlled Clinical Trial. *Phytomedicine* **2014**, *21,* 901–905.

Kumar, P. S.; Subramoniam, A.; Pushpangadan, P. Aphrodisiac Activity of *Vanda tessellata* (ROXB.) Hook Exdon Extract in Male Mice. *Indian J. Pharmacol.* **2000**, *32,* 300–304.

Lin, W. H.; Tsai, M. T.; Chen, Y. S.; Hou, R. C. W, Hung. H. F.; Li, C. H, Wang, H. K.; Lai, M. N.; Jeng, K. C. G. Improvement of Sperm Production in Subfertile Boars by *Cordyceps militaris* Supplement. *The Am. J. Chin. Med.* **2007**, *35,* 631–641.

Liu, J.; Liang, P.; Yin, C.; Wang, T.; Li, H.; Li, Y.; et al. Effects of Several Chinese Herbal Aqueous Extracts on Human Sperm Motility in Vitro. *Andrologia* **2004**, *36*, 78–83.

Luo, Q.; Li, Z.; Huang, X.; Yan, J.; Zhang, S.; Cai, Y. Z. *Lycium barbarum* Polysaccharides: Protective Effects Against Heat-Induced Damage of Rat Testes and $H_2O_2$-induced DNA Damage in Mouse Testicular Cells and Beneficial Effect on Sexual Behavior and Reproductive Function of Hemicastrated Rats. *Life Sci.* **2006**, *79*, 613–621.

Mahdi, A. A.; Shukla, K. K.; Ahmad, M. K.; Rajender, S.; Shankhwar, S. N.; Singh, V.; Dalela, D. *Withania somnifera* Improves Semen Quality in Stress-Related Male Fertility. *Evidence-Based Complementary Altern. Med.* **2011**. Article ID: 576962. http://dx.doi.org/10.1093/ecam/nep138

Maheshwari, A.; Verma, N.; Swaroop, A.; Bagchi, M.; Preuss, H. G.; Tiwari, K.; Bagchi, D. Efficacy of Furosap™, a Novel Trigonella Foenum-Graecum Seed Extract, in Enhancing Testosterone Level and Improving Sperm Profile in Male Volunteers. *Int. J. Med. Sci.* **2017**, *14*, 58.

Makarova, M. N.; Pozharitskaya, O. N.; Shikov, A. N.; Tesakova, S. V.; Makarov, V. G.; Tikhonov, V. P. Effect of Lipid-Based Suspension of *Epimedium Koreanum* Nakai Extract on Sexual Behavior in Rats. *J. Ethnopharmacol.* **2007**, *114*, 412–416.

Mascarenhas, M. N.; Flaxman, S. R.; Boerma, T.; Vanderpoel, S.; Stevens, G. A. National, Regional, and Global Trends in Infertility Prevalence Since 1990: A Systematic Analysis of 277 Health Surveys. *PLoS Med.* **2012**, *9*, e1001356.

Mbongue, F. G.; Kamtchouing, P.; Essame, O. J.; Yewah, P.; Dimo, T.; Lontsi, D. Effect of the Aqueous Extract of Dry Fruits of *Piper Guineense* on the Reproductive Function of Adult Male Rats. *Indian J. Pharmacol.* **2005**, *37*, 30.

Moundipa, P. F.; Beboyl, N. S. E.; Zelefack, F.; Ngouela, S.; Tsamo, E.; Schill, W. B.; Monsees, T. K. Effects of *Basella alba* and Hibiscus Macranthus Extracts on Testosterone Production of Adult Rat and Bull Ley Dig Cells. *Asian J. Androl.* **2005**, *7*, 411–417.

Murphy, L. L.; Cadena, R. S.; Chavez, D.; Ferraro, J. S. Effect of American Ginseng (*Panax quinquefolium*) on Male Copulatory Behavior in the Rat. *Physiology Behav.* **1998**, *64*, 445–450.

Nantia, A.; Moundipa, F.; Beboy, E.; Monsees, K.; Carreau, S. Study of the Androgenic Effect of the Methanol Extract of *Basella alba* L.(Basellaceae) on the Male Rat Reproductive Function. *Andrologie* **2007**, *17*, 129–133.

Ofusori, D.; Oluwayinka, O.; Adelakun, A.; Keji, S.; Oluyemi, K.; Adesanya, O.; Ajeigbe, K. O.; Ayoka, A. O. Evaluation of the Effect of Ethanolic Extract of *Croton zambesicus* on the Testes of Swiss Albino Mice. *Afr. J. Biotechnol.* **2007**, *6*, 2434–2438.

Ombelet, W.; Cooke, I.; Dyer, S.; Serour, G.; Devroey, P. Infertility and the Provision of Infertility Medical Services in Developing Countries. *Human Reproduction Update.* **2008**, *14*, 605–621.

ORC Macro and the World Health Organization. *Infecundity, Infertility, and Childlessness in Developing Countries*. DHS Comparative Reports No 9. World Health OOrganization:, Calverton, Maryland, USA,. 2004; DHS Comparative Reports No 9.

Park, H. J.; Choe, S.; Park, N. C. Effects of Korean Red Ginseng on Semen Parameters in Male Infertility Patients: A Randomized, Placebo-Controlled, Double-Blind Clinical Study. *Chin. J. Integr. Med.* **2016**, *22*, 490–495.

Park, W. S.; Shin, D. Y.; Yang, W. M.; Chang, M. S.; Park, S. K. Korean Ginseng Induces Spermatogenesis in Rats through the Activation of Camp-Responsive Element Modulator (CREM). *Fertil. Steril.* **2007**, *88*, 1000–1002.

Patel, D. K.; Kumar, R.; Prasad, S. K.; Hemalatha, S. Pharmacologically Screened Aphrodisiac Plant-A Review of Current Scientific Literature. *Asian Pac.ific J. Trop. Biomed.* **2011,** *1,* S131–138.

Rakuambo, N. C.; Meyer, J. J.; Hussein, A.; Huyser, C.; Mdlalose, S. P.; Raidani, T. G. In Vitro Effect of Medicinal Plants Used to Treat Erectile Dysfunction on Smooth Muscle Relaxation and Human Sperm. *J. Ethnopharmacol.* **2006,** *105,* 84–88.

Ramachandran, S.; Sridhar, Y.; Sam, S. K. G.; Saravanan M.; Leonard, J. T.; Anbalagan, N.; Sridhar, S. K. Aphrodisiac Activity of *Butea frondosa* Koen. Ex Roxb. Extract in Male Rats. *Phytomedicine* **2004;** *11,* 165–168.

Ratnasooriya, W.; Dharmasiri, M. Effects of *Terminalia catappa* Seeds on Sexual Behaviour and Fertility of Male Rats. *Asian J. Androl.* **2000,** *2,* 213–220.

Rege, N.; Date, J.; Kulkarni, V.; Prem, A.; Punekar, S.; Dahanukar, S. Effect of Y Virilin on Male Infertility. *J. Postgrad. Med.* **1997,** *43,* 64.

Rosen, R. C.; Ashton, A. K. Prosexual Drugs: Empirical Status of the "New Aphrodisiacs". *Arch. Sex. Behav.* **1993,** *22,* 521–543.

Rubio, J.; Riqueros, M. I.; Gasco, M.; Yucra, S.; Miranda, S.; Gonzales, G. F. *Lepidium meyenii* (Maca) Reversed the Lead Acetate Induced— – Damage on Reproductive Function in Male Rats. *Food Chem. Toxicol.* **2006,** *44,* 1114–1122.

Sahin, K.; Orhan, C.; Akdemir, F.; Tuzcu, M.; Gencoglu, H.; Sahin, N.; Turk, G.; Yilmaz, I.; Ozercan, I. H.; Juturu, V. Comparative Evaluation of the Sexual Functions and NF-κB and Nrf2 Pathways of Some Aphrodisiac Herbal Extracts in Male Rats. *BMC Complement. Altern. Med.* **2016,** *16,* 318–328.

Sakr, S. A.; El-Shenawy, S. M.; Al-Shabka A. M. Aqueous Fenugreek Seed Extract Ameliorates Adriamycin-Induced Cytotoxicity and Testicular Alterations in Albino Rats. *Reprod. Sci.* **2012,** *19,* 70–80.

Satyavati, G.; Raina, M.; Sharma, M. *Medicinal Plants of India*: Indian Council of Medical Research: New Delhi; 1976.

Sengupta, P.; Agarwal, A.; Pogrebetskaya, M.; Roychoudhury, S.; Durairajanayagam, D.; Henkel, R. Role of *Withania somnifera* (Ashwagandha) in the Management of Male Infertility. *Reprod. Biomed. Online.* **2017,** *36,* 311–326 Dec 7.

Sharma, V.; Thakur, M.; Dixit, V. K. A Comparative Study of Ethanolic Extracts of *Pedalium murex* linn. Fruits and Sildenafil Citrate on Sexual Behaviors and Serum Testosterone Level in Male Rats During and After Treatment. *J. Ethnopharmacol.* **2012,** *143,* 201–206.

Shukla, K. K.; Mahdi, A. A.; Ahmad, M. K.; Jaiswar, S. P.; Shankwar, S. N.; Tiwari, S. C. *Mucuna Pruriens* Reduces Stress and Improves the Quality of Semen in Infertile Men. *Evidence-Based Complementary and Altern. Med.* **2010,** *7,* 137–144.

Singh, S.; Gupta, Y. Aphrodisiac Activity of *Tribulus terrestris* Linn. in Experimental Models in Rats. *J. Men's Health* **2011,** *8,* S75–S77.

Subramoniam, A.; Madhavachandran, V.; Rajasekharan, S.; Pushpangadan, P. Aphrodisiac Property of *Trichopus zeylanicus* Extract in Male Mice. *J. Ethnopharmacol.* **1997,** *57,* 21–27.

Tajuddin; Ahmad, S.; Latif, A.; Qasmi, I. A. Aphrodisiac Activity of 50% Ethanolic Extracts of *Myristica fragrans* Houtt. (nutmeg) and *Syzygium aromaticum* (L) Merr. & Perry.(clove) in Male Mice: A Comparative Study. *BMC Complementary Altern. Med.* **2003,** *3,* 6.

Tajuddin; Ahmad, S.; Latif, A.; Qasmi, I. A. Effect of 50% Ethanolic Extract of *Syzygium aromaticum* (L.) Merr. and Perry.(clove) on Sexual Behaviour of Normal Male Rats. *BMC Complement. Altern. Med.* **2004,** *4,* 17.

Tajuddin; Ahmad, S.; Latif, A.; Qasmi, I. A.; Amin, K. M. Y. An Experimental Study of Sexual Function Improving Effect of *Myristica fragrans* Houtt. (nutmeg). *BMC Complement. Altern. Med.* **2005,** *5,* 16.

Tempest, H. G.; Homa, S. T.; Zhai, X. P.; Griffin, D. K. Significant Reduction of Sperm Disomy in Six Men: Effect of Traditional Chinese Medicine? *Asian J. Androl.* **2005,** *7,* 419–425.

Thakur, M.; Chauhan, N. S.; Bhargava, S.; Dixit, V. K. A Comparative Study on Aphrodisiac Activity of Some Ayurvedic Herbs in Male Albino Rats. *Arch Sex. Behav.* **2009,** *38,* 1009–1015.

Thakur, M.; Dixit, V. Aphrodisiac Activity of *Dactylorhiza hatagirea* (D. Don) Soo in Male Albino Rats. *Evidence-Based Complement. Altern. Med.: eCAM* **2007,** *4,* 29–31.

Turk, G.; Sonmez, M.; Aydin, M.; Yuce, A.; Gur, S.; Yuksel, M.; Aksu, E. H.; Aksoy, H. Effects of Pomegranate Juice Consumption on Sperm Quality, Spermatogenic Cell Density, Antioxidant Activity and Testosterone Level in Male Rats. *Clin. Nutr.* **2008,** *27,* 289–296.

Watcho, P.; Kamtchouing, P.; Sokeng, S. D.; Moundipa, P. F.; Tantchou, J.; Essame, J. L.; Koueta, N. Androgenic Effect of *Mondia whitei* Roots in Male Rats. *Asian J. Androl.* **2004,** *6,* 269–272.

Watcho, P.; Wankeu-Nya, M.; Nguelefack, T. B.; Tapondjou, L.; Teponno, R.; Kamanyi, A. Pro-Sexual Effects of *Dracaena arborea* (wild) Link (Dracaenaceae) in Sexually Experienced Male Rats. *Pharmacol Online* **2007,** *1,* 400–419.

Wiser, H. J.; Sandlow, J.; Köhler, T. S. Causes of Male Infertility. In *Male Infertility;*, Springer: New York, 2012, (pp 3–14).

Xu, X.; Yin, H.; Tang, D.; Zhang, L.; Gosden, R. G. Application of Traditional Chinese Medicine in the Treatment of Infertility. *Hum. Fertil.* **2003,** *6,* 161–168.

Yakubu, M. T.; Akanji, M. A.; Oladiji, A. T.; Adesokan, A. A. Androgenic Potentials of Aqueous Extract of *Massularia acuminata* (G. Don) Bullock Ex Hoyl. Stem in Male Wistar Rats. *J. Ethnopharmacol* **2008,** *118,* 508–513.

Yakubu, M.; Akanji, M.; Oladiji, A. Aphrodisiac Potentials of the Aqueous Extract of *Fadogia agrestis* (Schweinf. Ex Hiern) Stem in Male Albino Rats. *Asian J. Androl.* **2005,** *7,* 399–404.

Yakubu, M.; Akanji, M.; Oladiji, A. Male Sexual Dysfunction and Methods Used in Assessing Medicinal Plants with Aphrodisiac Potentials. *Pharmacogn Rev.* **2007,** *1,* 49.

Yeh, K. Y.; Pu, H. F.; Kaphle, K.; Lin, S. F.; Wu, L. S.; Lin, J. H.; Tsai, Y. F. *Ginkgo biloba* Extract Enhances Male Copulatory Behavior and Reduces Serum Prolactin Levels in Rats. *Horm. Behav.* **2008,** *53,* 225–231.

Young, D. Surgical Treatment of Male Infertility. *J. Reprod. Fertil.* **1970,** *23,* 541–542.

Zamblé, A.; Martin-Nizard, F.; Sahpaz, S.; Hennebelle, T.; Staels, B.; Bordet, R.; Duriez, P.; Brunet, C.; Bailleul, F. Vasoactivity, Antioxidant and Aphrodisiac Properties of Caesalpinia Benthamiana Roots. *J. Ethnopharmacol.* **2008,** *116,* 112–119.

Zamble, A.; Sahpaz, S.; Brunet, C.; Bailleul, F. Effects Of *Microdesmis Keayana* Roots on Sexual Behavior of Male Rats. *Phytomedicine* **2008a,** *15,* 625–629.

# CHAPTER 15

# ROLES OF MEDICINAL PLANTS IN THE TREATMENT OF CANCER

PRABHAT UPADHYAY[1,4,*], RASHMI SHUKLA[2],
SUNIL KUMAR MISHRA[3], RINKI VERMA[4], SURESH PUROHIT[1], and
G. P. DUBEY[4]

[1]*Department of Pharmacology, Institute of Medical Sciences, Banaras Hindu University, Varanasi, India*

[2]*Department of Medicinal Chemistry, Institute of Medical Sciences, Banaras Hindu University, Varanasi, India*

[3]*Department of Pharmaceutical, Engineering and Technology, Indian Institute of Technology, Banaras Hindu University, Varanasi, India*

[4]*Collaborative program, Institute of Medical Sciences, Banaras Hindu University, Varanasi, India*

*Corresponding author. E-mail: upadhyayprabhat.89@gmail.com
*ORCID: https://orcid.org/0000-0003-1432-6792

## ABSTRACT

This huge group of compounds, now collectively termed as "phytochemicals," found in plants, is responsible for therapeutic effect as well as color, flavor, and aroma of foods. These phytochemicals have different classes (phenol, flavonoid, alkaloid, carotenoid, nitrogen-containing, organosulfur compounds, and so forth). Some of these phytochemicals have proven to exhibit the potentials for the treatment and management of the different type of cancer (breast, lung, prostate, oral, blood, cervical) through antioxidant enzymes, enhancing deoxyribonucleic acid repair pathways, cell cycle arrest, and apoptosis. This chapter enlisted the plant source material with their scientifically proven molecules and possible preclinical and clinical effects, different phytoconstituents sources in the treatment and management of the different type of cancer.

## 15.1   INTRODUCTION

Cancer is the second leading cause of death in the world after cardiovascular diseases (Fitzmaurice et al., 2015). Today, cancer patients can live a fairly normal life once it is identified early and treated (Lodish et al., 2000). Cancer is not a new disease and has afflicted people throughout the world. The word cancer came from Greek words "karkinos" to describe carcinoma tumors by a physician. Cancer develops when normal cells in a particular part of the body begin to grow out of control. Cancer becomes serious if the tumor begins to spread (metastasize) throughout the body. They are named based on where the tumor is located, or where it first started growing in the body. There are different types of cancers; colon, lung, breast, blood, and prostate cancer (Ma and Yu, 2006). Usually, cancer cells continue to grow, divide, and redivide and instead of dying they rather form new abnormal cells. Generally, cancer cells develop from normal cells due to changes in the deoxyribonucleic acid (DNA) structure. It is the normal physiological function of the body to repair such DNA, but unfortunately, in cancer cells, damaged DNA is not repaired.

## 15.2   TREATMENT AND MANAGEMENT APPROACHES

Many approaches have been adopted for the treatment of cancer. They include, (1) religious and spiritual approaches (2) psychosocial approaches (3) nutritional approaches (4) physical approaches (5) traditional medicines from around the world (6) herbal treatments for cancer (7) unconventional pharmacological treatments (8) electromagnetic therapies (9) the unconventional use of conventional or conventional-experimental cancer therapies (10) esoteric therapies and (12) humane approaches alternative diagnostic and treatment instruments, which are used now in these days globally (Allen et al., 2015).

### 15.2.1   HERBAL APPROACHES

Currently, the main treatments for cancer are chemotherapy, radiotherapy, and surgery. Some of the most used chemotherapeutic drugs include antimetabolites (e.g., methotrexate), DNA interactive agents (e.g., cisplatin, doxorubicin), anti-tubulin agents (taxanes), hormones, and molecular-targeting agents. However, clinical uses of these drugs are accompanied by several unwanted effects such as hair loss, suppression of bone marrow, drug resistance, gastrointestinal lesions, neurologic dysfunction, and cardiac toxicity. Moreover, even with the current intensive interventions, a large number of patients suffer from poor prognosis. Therefore, the search for new anticancer agents with better efficacy and lesser side effects has been continued. Experimentally, a number of phytochemicals isolated from medicinal plants have been shown to decrease cell proliferation, induce apoptosis, retard metastasis, and inhibit angiogenesis. Even currently, some of these plant-derived compounds (taxol analogs, vinca alkaloids (vincristine, vinblastine, and podophyllotoxin analogs) are widely used for chemotherapy of cancerous patients (Hosseini and Ghorbani, 2015).

## 15.2.2 MECHANISM OF PHYTOCHEMICALS

The phytochemical of interests involved in the treatment of different types of cancers (breast, lung, prostate, oral, blood, cervical) are phenol, flavonoid, alkaloid, carotenoid, nitrogen-containing compounds and the organosulfur compounds, and so forth. There are different pathways involved in the treatment of cancerous cells to which the major mechanism is the induction of apoptosis and cell cycle arrest. However, the emergence of cancer can be checked through the increase in antioxidant compounds, immunomodulators, and reduced incidence of neoplasia induced by chemical carcinogens. In the field of natural plant product, most of the plant which has the anticancer properties contains phenolic compound, alkaloid, or organosulfur compound (Fig. 15.1 and Table 15.1). The mechanisms to which this phytochemicals act was described by Mollakhalili et al., 2017 and presented in Figure 15.1.

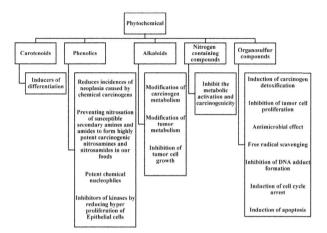

**FIGURE 15.1** Classification of phytochemicals and possible mechanism involved in the treatment and management of different cancer types.

**TABLE 15.1** List of Medicinal Plant with Their Phytochemicals used in Treatment and Management of Different Type of Cancer.

| Name of plant/ family | Phytochemical | Type of cancer | Formula | Molecular Mass | Structure | References |
|---|---|---|---|---|---|---|
| *Catharanthus roseus* (Apocynaceae) | Vinblastine Vincristine | Treatment of lymphomas, leukemia's, breast cancer, testicular cancer, lung cancers, and Kaposi's sarcoma | $C_{46}H_{58}N_4O_9$ | 810.97 | | Kaur et al. (2011) |
| *Agapanthus africanus* (Amaryllidaceae) | Isoliquiritigenin | Endometrial cancer | $C_{15}H_{12}O_4$ | 256.253 | | Wu et al. (2016) |
| *Apium graveolens* (Apiaceae) | Apigenin | Breast *cancer* and skin tumors | $C_{15}H_{10}O_5$ | 270.23 | | Shukla and Gupta (2010) |
| *Aglaia sylvestre* (Meliaceae) | Silvesterol | Prostate *cancer*, Breast *cancer*, orthotopic tumor | $C_{34}H_{38}O_{13}$ | 654.66 | | Kogure et al. (2013) |

**TABLE 15.1** *(Continued)*

| Name of plant/ family | Phytochemical | Type of cancer | Formula | Molecular Mass | Structure | References |
|---|---|---|---|---|---|---|
| *Ailanthus altissima* (Simaroubaceae) | Ailnthone | Prostate cancer | $C_{20}H_{24}O_7$ | 376.40 | | Peng et al. (2017) |
| *Bleckeria vitensis* (Apocynaceae) | Ellipticine | Breast adenocarcinoma | $C_{17}H_{14}N_2$ | 246.3 | | Stiborová et al. (2011) |
| *Brucea anti-dysenterica* (Simaroubaceae) | Bruceantin | Multiple myeloma cancer | $C_{28}H_{36}O_{11}$ | 548.57 | | Issa et al. (2016) |
| *Bursera microphylla* (Burseraceae) | Burseran | Breast cancer | $C_{22}H_{26}O_6$ | 386.43 | | Acevedo et al. (2015) |

**TABLE 15.1** *(Continued)*

| Name of plant/ family | Phytochemical | Type of cancer | Formula | Molecular Mass | Structure | References |
|---|---|---|---|---|---|---|
| *Campotheca acuminata* (Nyssaceae) | Camptothecin | Colon cancer | $C_{22}H_{20}ClN_3O_5$ | 441.86 | | Arango et al. (2003) |
| *Centaurea montana* (Asteraceae) | Montamine | Leukemia's, lymphomas, advanced testicular cancer, breast and lung cancer and kaposi's sarcoma | $C_{40}H_{38}N_4O_8$ | 702.75 | | Mohan and Jeyachandran (2012) |
| *Centaurea schischkinii* (Asteraceae) | Schischkinnin | | $C_{26}H_{26}N_6O_2$ | 454.53 | | Kaur et al. (2011) |

**TABLE 15.1** *(Continued)*

| Name of plant/ family | Phytochemical | Type of cancer | Formula | Molecular Mass | Structure | References |
|---|---|---|---|---|---|---|
| *Cephalotaxus harringtonia* (Cephalotaxaceae) | Homoharringtonine | Lung cancer | $C_{29}H_{39}NO_9$ | 545.62 | | Cao et al. (2015) |
| *Cleistanthus collinus* (Phyllanthaceae) | Cleistanthin | Exhibit cytotoxicity on several cancer cell lines | $C_{28}H_{28}O_{11}$ | 540.51 | | Pinho and Kijjoa (2007) |
| *Croton lechleri* (Euphorbiaceae) | Taspine | Breast cancer | $C_{20}H_{19}NO_6$ | 369.36 | | Zhan et al. (2012) |
| *Daphne mezereum* (Thymelaeaceae) | Mezerein | Prostate cancer | $C_{38}H_{38}O_{10}$ | 654.70 | | Aghajanpour et al. (2017) |

**TABLE 15.1** *(Continued)*

| Name of plant/family | Phytochemical | Type of cancer | Formula | Molecular Mass | Structure | References |
|---|---|---|---|---|---|---|
| *Diphylleia grayi* (Berberidaceae) | Diphyllin | Gastrointestinal carcinomas | $C_{21}H_{16}O_7$ | 380.34 | | Chiurillo (2015) |
| *Dysoxylum binectariferum* (Meliaceae) | Rohitukine | Ovarian, breast, cervical, pancreatic, and prostate cancers | $C_{16}H_{19}NO_5$ | 305.32 | | Ravishankar et al. (2013) |
| *Erythroxylum pervillei* (Erythroxylaceae) | Pervilleine | Colon cancer | $C_{30}H_{37}NO_{11}$ | 587.62 | | Shoeb (2008) |

**TABLE 15.1** *(Continued)*

| Name of plant/family | Phytochemical | Type of cancer | Formula | Molecular Mass | Structure | References |
|---|---|---|---|---|---|---|
| *Euphorbia semiperfoliata* (Euphorbiaceae) | Jatrophane | Lung carcinoma | $C_{39}H_{50}O_{15}$ | 758.81 | | Corea et al. (2009) |
| *Hypericum perforatum* (Hypericaceae) | Hypericin | Bladder cancer | $C_{30}H_{16}O_{8}$ | 504.44 | | Buytaert et al. (2008) |
| *Hypoxis colchicifolia* (Hypoxidaceae) | Hypoxoside | Pancreatic cancer | $C_{29}H_{34}O_{14}$ | 606.57 | | Boukes and van de Venter (2016) |
| *Indigofera tinctoria* (Fabaceae) | Indirubins | Treatment of metastatic renal cell *cancer* | $C_{16}H_{11}N_{3}O_{2}$ | 277.28 | | Perabo et al. (2011) |

**TABLE 15.1** *(Continued)*

| Name of plant/ family | Phytochemical | Type of cancer | Formula | Molecular Mass | Structure | References |
|---|---|---|---|---|---|---|
| *Justicia procumbens* (Acanthaceae) | Justicidin A, B | Colorectal cancer | $C_{22}H_{18}O_7$ | 394.37 | | Lee et al. (2005a) |
| *Lantana camara* (Verbenaceae) | Verbascoside | Colorectal and Gastric cancer | $C_{29}H_{36}O_{15}$ | 624.58 | | Zhou et al. (2014) |
| *Lonicera japonica* (Caprifoliaceae) | Luteolin | Colon cancer | $C_{15}H_{10}O_6$ | 286.23 | | Ashokkumar and Sudhandiran (2008) |

**TABLE 15.1** (Continued)

| Name of plant/ family | Phytochemical | Type of cancer | Formula | Molecular Mass | Structure | References |
|---|---|---|---|---|---|---|
| *Paris polyphylla* (Melanthiaceae) | Polyphyllin | Breast cancer | $C_{44}H_{70}O_{16}$ | 855.02 | | Lee et al. (2005b) |
| *Penstemon deustus* (Plantaginaceae) | Liriodendrin | Gastric cancer | $C_{34}H_{46}O_{18}$ | 742.71 | | Kaur et al. (2011) |
| *Phaleria macrocarpa* (Thymelaeaceae) | Laricinesinol | Abdominal tumors Prostate cancer | $C_{20}H_{24}O_6$ | 360.40 | | Kaur and Goel (2011) |

**TABLE 15.1** *(Continued)*

| Name of plant/family | Phytochemical | Type of cancer | Molecular Structure | Formula | Molecular Mass | References |
|---|---|---|---|---|---|---|
| *Podophyllum emodi* (Berberidaceae) | Epipodophyllotoxin | Leukaemia | | $C_{22}H_{22}O_8$ | 414.40 | Hawkins et al. (1992) |
| *Polygonum cuspidatum* (Knotweed) | Resveratrol | Breast cancer | | $C_{14}H_{12}O_3$ | 228.24 | Scarlatti et al. (2008) |
| *Betula utilis* (Betulaceae) | Betulin | Prostate Cancer | | $C_{30}H_{50}O_2$ | 442.71 | Suresh et al. (2012) |
| *Vitex rotundifolia* (Lamiaceae) | Casticin | Cervical cancer | | $C_{19}H_{18}O_8$ | 374.34 | Cao (2011) |
| *Wikstroemia viridi* (Thymelaeaceae) | Wikstromol | Nasopharyngeal *cancer*; breast cancer | | $C_{20}H_{22}O_7$ | 374.38 | Kaur et al. (2011) |

**TABLE 15.1** *(Continued)*

| Name of plant/ family | Phytochemical | Type of cancer | Formula | Molecular Mass | Structure | References |
|---|---|---|---|---|---|---|
| *Astragalus membranaceus* (Fabaceae) | Swainsonine | Bladder cancer | $C_8H_{15}NO_3$ | 173.21 | | Przybyło et al. (2005) |
| *Fragaria vesca* (Rosaceae) | Borneol | Hepatocellular Carcinoma | $C_{10}H_{18}O$ | 154.24 | | Su et al. (2013) |
| *Echinacea angustifolia* (Asteraceae) | Arabinogalactan | Colon cancer | $C_{11}H_{10}O$ | 158.19 | | Yang et al. (2014) |
| *Annona* species (Annonaceae) | Acetogenins | Prostate cancer | $C_{26}H_{46}O_7$ | 470.64 | | Yang et al. (2015) |
| *Allium sativum* (Amaryllidaceae) | Alicin | Colon cancer | $C_6H_{10}OS_2$ | 162.27 | | Chavan et al. (2013) |

**TABLE 15.1** (Continued)

| Name of plant/ family | Phytochemical | Type of cancer | Formula | Molecular Mass | Structure | References |
|---|---|---|---|---|---|---|
| Pueraria mirifica (Fabaceae) | Coumestans | Breast cancer | $C_{15}H_8O_3$ | 236.22 | | Ziegler (2004) |
| Tinospora cordifolia (Menispermaceae) | Tinosporin | Anticancer activity | $C_{20}H_{22}O_6$ | 358.38 | | Desai et al. (2008) |
| Colchicum luteum (Colchicaceae) | Demecolcine | Breast cancer | $C_{21}H_{25}NO_5$ | 371.42 | | Savel (1966) |
| Gossypium barbadense (Malvaceae) | Gossypol | Prostate cancer | $C_{30}H_{30}O_8$ | 518.55 | | Volate et al. (2010) |

**TABLE 15.1** (Continued)

| Name of plant/ family | Phytochemical | Type of cancer | Formula | Molecular Mass | Structure | References |
|---|---|---|---|---|---|---|
| Berberis vulgaris (Berberidaceae) | Berberine | Breast cancer | $C_{20}H_{18}NO_4$ | 336.36 | | Patil et al. (2010) |
| Azadirecta indica (Meliaceae) | Nimbin | Breast cancer | $C_{30}H_{36}O_9$ | 540.60 | | Alzohairy (2016) |
| Withania somnifera (Solanaceae) | Withanolide | Breast cancer | $C_{28}H_{38}O_6$ | 470.59 | | Wang et al. (2012a) |

**TABLE 15.1** (Continued)

| Name of plant/family | Phytochemical | Type of cancer | Formula | Molecular Mass | Structure | References |
|---|---|---|---|---|---|---|
| *Viscum album* (Santalaceae) | Lectin | Colorectal cancer | $C_{30}H_{48}O_3$ | 456.70 | | Campbell et al. (2001) |
| *Crocus sativus* (Iridaceae) | Crocetin | Hepatocellular carcinoma | $C_{20}H_{24}O_4$ | 328.40 | | Wang et al. (2012b) |
| *Curcuma longa* (Zingiberaceae) | Curcumin | Colon cancer | $C_{21}H_{20}O_6$ | 368.38 | | Singh and Khar (2006) |
| Red berries Buckthorn | Cyanidin | Colon cancer | $C_{15}H_{11}O_6$ | 287.24 | | Cvorovic et al. (2010) |
| *Brassica rapa* (Brassicaceae) | Indole-3-carbinol (I3C) | Prostate cancers | $C_9H_9NO$ | 147.17 | | Chinni et al. (2001) |

**TABLE 15.1** *(Continued)*

| Name of plant/ family | Phytochemical | Type of cancer | Formula | Molecular Mass | Structure | References |
|---|---|---|---|---|---|---|
| *Camellia sinensis* (Theaceae) | Epigallocatechin gallate | Breast and prostate cancer | $C_{22}H_{18}O_{11}$ | 458.37 | | Stuart et al. (2006) |
| *Acacia greggii* (Fabaceae) | Fisetin | Prostate, lung and colon cancer | $C_{15}H_{10}O_6$ | 286.23 | | (Khan et al., 2008) |
| *Flemingia vestita* (Fabaceae) | Genistein | Breast and prostate cancer | $C_{15}H_{10}O_5$ | 270.23 | | Sarkar and Li (2002) |
| *Zingiber officinale* (Zingiberaceae) | Gingerol | Colon cancer | $C_{17}H_{26}O_4$ | 294.38 | | Jeong et al. (2009) |
| *Camellia sinensis* (Fabaceae) | Kaempferol | Colon and lung cancer | $C_{15}H_{10}O_6$ | 286.23 | | Li et al. (2009) |

**TABLE 15.1** *(Continued)*

| Name of plant/ family | Phytochemical | Type of cancer | Formula | Molecular Mass | Structure | References |
|---|---|---|---|---|---|---|
| *Citrus paradisi* (Rutaceae) | Broccoli | Breast cancer | $C_{15}H_{10}O_6$ | 286.23 | | Webb and Mccullough (2017) |
| *Vitis vinifera* (Vitaceae) | Resveratrol | Breast cancer | $C_{14}H_{12}O_3$ | 228.24 | | Scarlatti et al. (2008) |
| *Rosmarinus officinalis* (Lamiaceae) | Rosmarinic | Colon cancer | $C_{18}H_{16}O_8$ | 360.31 | | Huang et al. (2009) |
| *Bras-sica oleracea* (Brassicaceae) | Sulforaphane | Prostate cancer | $C_6H_{11}NOS_2$ | 177.28 | | Herman-Antosiewicz et al. (2006) |
| Mushroom/ Basidiomycota, (Agaricomycetes) | Vitamin D (Cholecalciferol) | Prostate cancer | $C_{27}H_{44}O$ | 384.63 | | Skowronski et al. (1993) |

**TABLE 15.1** *(Continued)*

| Name of plant/ family | Phytochemical | Type of cancer | Formula | Molecular Mass | Structure | References |
|---|---|---|---|---|---|---|
| *Catharanthus roseus* (Apocynaceae) | Naringenin | Colon and lung cancer | $C_{15}H_{12}O_5$ | 272.25 | | Etcheverry et al. (2008) |
| *Pueraria mirifica* (Fabaceae) | Daidzein | Breast cancer | $C_{15}H_{10}O_4$ | 254.23 | | Adlercreutz (1995) |
| *Glycyrrhiza glabra* (Fabaceae) | Glycyrrhizin | leukemia and stomach cancer | $C_{42}H_{62}O_{16}$ | 822.93 | | Hibasami et al. (2006) |
| *Toxicodendron verniciflium* (Anacardiaceae) | Butein | Breast cancer | $C_{15}H_{12}O_5$ | 272.25 | | Samoszuk et al. (2005) |
| *Emblica officinalis* (Phyllanthaceae) | Ellagic acid | Prostate cancer | $C_{14}H_6O_8$ | 302.19 | | Bell and Hawthorne (2008) |

**TABLE 15.1** *(Continued)*

| Name of plant/ family | Phytochemical | Type of cancer | Formula | Molecular Mass | Structure | References |
|---|---|---|---|---|---|---|
| *Arnebia nobilis* (Boraginaceae) | Shikonin | Prostate cancer | $C_{16}H_{16}O_5$ | 288.29 | | Gaddipati et al.(2000) |
| *Aesculus hippocastanum* (Sapindaceae) | Escin IB | Pancreatic Cancer | $C_{55}H_{86}O_{24}$ | 1131.25 | | Rimmon et al. (2013) |
| *Amoora rohituka* (Meliaceae) | Amooranin | Breast and Pancreatic Cancer | $C_{30}H_{46}O_4$ | 470.69 | | Chan et al. (2011) |
| *Aegle marmelos* (Rutaceae) | Lupeol | Prostate cancer | $C_{25}H_{26}O_4$ | 390.47 | | Saleem et al. (2005) |

**TABLE 15.1** *(Continued)*

| Name of plant/ family | Phytochemical | Type of cancer | Formula | Molecular Mass | Structure | References |
|---|---|---|---|---|---|---|
| *Betula utilis* (Betulaceae) | Betulinic acid (3ß)-Hydroxy-lup-20 (29)-en-28-oic acid | Melanoma *cancer* | $C_{30}H_{48}O_3$ | 456.70 | | Suresh et al. (2012) |
| *Cassia fistula* (Fabaceae) | Rhein | Colan cancer | $C_{15}H_8O_6$ | 284.22 | | Duraipandiyan et al. (2012) |
| *Scutellaria* (Lamiaceae) | Baicalin | Prostate cancer | $C_{21}H_{18}O_{11}$ | 446.36 | | Ikezoe et al. (2001) |
| *Cassia tora* (Fabaceae) | Emodin | Colan cancer | $C_{15}H_{10}O_5$ | 270.37 | | Kaczmarczyk et al. (2012) |
| *Chlorella pyrenoidosa* (Chlorellaceae) | Lysine | Bladder cancer | $C_6H_{14}N_2O_2$ | 146.18 | | Kauffman et al. (2011) |

**TABLE 15.1** *(Continued)*

| Name of plant/ family | Phytochemical | Type of cancer | Formula | Molecular Mass | Structure | References |
|---|---|---|---|---|---|---|
| *Picrorrhizia kurroa* (Plantaginaceae) | Picroside I | Prostate cancer | $C_{24}H_{28}O_{11}$ | 492.47 | | Garodia et al. (2007) |
| *Nigella sativa* (Ranunculaceae) | Thymoquinone | Breast cancer | $C_{10}H_{12}O_2$ | 164.20 | | Abukhader (2013) |
| *Taxus brevifolia* (Taxaceae) | Taxane | Breast cancer | $C_{20}H_{36}$ | 276.50 | | Thomas et al. (2007) |
| *Panax ginseng* (Araliaceae) | Ginsenoside | Gastric cancer | $C_{48}H_{82}O_{18}$ | 947.15 | | Hu et al. (2012) |
| *Mylabris phalerlata* (Meloidae) | Magnolol | Prostate cancer | $C_{18}H_{18}O_2$ | 266.33 | | Lee et al. (2009) |

**TABLE 15.1** (Continued)

| Name of plant/ family | Phytochemical | Type of cancer | Molecular Structure | | Formula | Molecular Mass | References |
|---|---|---|---|---|---|---|---|
| Cephalotaxus harringtonia (Cephalotaxaceae) | Harringtonine, | Lung cancer | | | $C_{28}H_{37}NO_9$ | 531.59 | Wang and Liu (1998) |
| Fagopyrum esculentum (Polygonaceae) | Rutin | Colon cancer | | | $C_{27}H_{30}O_{16}$ | 610.51 | Kuo (1996) |
| Ginkgo biloba (Ginkgoaceae) | Ginkgolide-B, A, C, and J | Ovarian cancer | | | $C_{20}H_{24}O_{10}$ | 424.39 | Jiang et al. (2011) |

**TABLE 15.1** *(Continued)*

| Name of plant/ family | Phytochemical | Type of cancer | Formula | Molecular Mass | Structure | References |
|---|---|---|---|---|---|---|
| *Larrea tridentata* (Zygophyllaceae) | Terameprocol | Colon and breast cancer | $C_{22}H_{30}O_4$ | 358.47 | | Lapenna and Giordano (2009) |
| *Linum usitatissimum* (Linaceae) | Secoisolar-iciresinol diglucoside | Breast cancer | $C_{32}H_{46}O_{16}$ | 686.69 | | Mousavi and Adlercreutz (1992) |
| *Ocimum sanctum* (Lamiaceae) | Eugenol | Prostate cancer | $C_{10}H_{12}O_2$ | 164.20 | | Jaganathan & Supriyanto (2012) |
| *Solanum nigrum* (Solanaceae) | Solanine | Pancreatic Cancer Lung cancer | $C_{45}H_{73}NO_{15}$ | 868.05 | | Sun et al. (2014) |

**TABLE 15.1** *(Continued)*

| Name of plant/ family | Phytochemical | Type of cancer | Formula | Molecular Mass | Structure | References |
|---|---|---|---|---|---|---|
| *Psoralea corylifolia* (Fabaceae) | Psoralen | Breast cancer | $C_{11}H_6O_3$ | 186.16 | | Panno and Giordano (2014) |
| *Podophyllum hexandrum* (Berberidaceae) | Astragalin | Colon cancer | $C_{21}H_{20}O_{11}$ | 448.37 | | Huang et al. (2009) |
| *Rubia cordifolia* (Rubiaceae) | Rubidianin | Breast *cancer*, malignant lymphoma, Prostate cancer | $C_{20}H_{18}O_9$ | 402.35 | | Umadevi, et al, (2004) |
| *Pygeum africanum* (Rosaceae) | Amygdalin | Lung cancer | $C_{20}H_{27}NO_{11}$ | 457.42 | | Qian et al. (2015) |
| *Oldenlandia diffusa* (Rubiaceae) | Ursolic acid | Breast cancer | $C_{30}H_{48}O_3$ | 456.70 | | (Yeh et al., 2010) |

**TABLE 15.1** *(Continued)*

| Name of plant/ family | Phytochemical | Type of cancer | Formula | Molecular Mass | Structure | References |
|---|---|---|---|---|---|---|
| *Ochrosia elliptica* (Apocynaceae) | 9-Methoxyellip- ticine | Breast cancer | $C_{18}H_{16}N_2O$ | 276.33 | | (Kuo et al., 2005) |
| *Psoralea coryli- folia* (Fabaceae) | Bavachinin | inhibiting tumor angiogenesis | $C_{21}H_{22}O_4$ | 338.39 | | Nepal et al. (2012) |
| *Phal- eria macrocarpa* (Thymelaeaceae) | Pinoresinol | breast, prostate, and colorec- tal cancer | $C_{20}H_{22}O_6$ | 358.38 | | Adlercreutz (2002) |
| *Pteris multifida* (Pteridaceae) | Pterokaurane | colon, rectum, lung and breast cancer | $C_{20}H_{34}O_3$ | 322.5 | | Nepomuceno (2011) |
| *Saussurea lappa* (Asteraceae) | Cynaropicrin | Leukemia Leukocyte cancer | $C_{19}H_{22}O_6$ | 346.37 | | Cho et al. (2004) |

**TABLE 15.1** (*Continued*)

| Name of plant/ family | Phytochemical | Type of cancer | Formula | Molecular Mass | Structure | References |
|---|---|---|---|---|---|---|
| *Alpinia galanga* (Zingiberaceae) | 1'acetoxychavi-col acetate | Oral Carcinogenesis | $C_{13}H_{14}O_4$ | 234.24 | | Ohnishi et al., (1996) |
| *Bauhinia varie-gata* (Fabaceae) | Malvidin | Colon cancer | $C_{17}H_{15}O_7$ | 331.29 | | Huang et al. (2009) |

## KEYWORDS

- **treatment of cancer**
- **apoptosis**
- **organosulfur compound**
- **cell cycle arrest**
- **medicinal plant**

## REFERENCES

Abukhader, M. M. Thymoquinone in the Clinical Treatment of Cancer: Fact or Fiction? *Pharmacogn. Rev.* **2013,** *7*(14), 117–120.

Acevedo, M.; Nuñez, P.; Gónzalez-Maya, L.; Cardosotaketa, A.; Villarreal, M. L. Cytotoxic and Anti-Inflammatory Activities of Bursera Species from Mexico. *J. Clin. Toxicol.* **2015,** *5*(1), 1–8.

Adlercreutz, H. Phytoestrogens: Epidemiology and a Possible Role in Cancer Protection. *Environ. Health Perspect.* **1995,** *103*(Suppl 7), 103–112.

Adlercreutz, H. Phyto-Oestrogens and Cancer. *Lancet. Oncol.* **2002,** *3*(6), 364–373.

Aghajanpour, M.; Nazer, M. R.; Obeidavi, Z.; Akbari, M.; Ezati, P.; Kor, N. M. Functional Foods and Their Role in Cancer Prevention and Health Promotion: A Comprehensive Review. *Am. J. Cancer Res.* **2017,** *7*(4), 740–769.

Allen, T.; Mba, G. M. N. V; Shoja, G.; Razavi, E. Complementary Medicines and Cancer. *J. Cancer Sci. Res.* **2015,** *1*(13), 17–17.

Alzohairy, M. A. Therapeutics Role of *Azadirachta Indica* (Neem) and Their Active Constituents in Diseases Prevention and Treatment. *Evidence-Based Complement. Altern. Med.* **2016,** *2016*, 1–11.

Arango, D.; Mariadason, J. M.; Wilson, A. J.; Yang, W.; Corner, G. A.; Nicholas, C.; Aranes, M. J.; Augenlicht, L. H. C-Myc Overexpression Sensitises Colon Cancer Cells to Camptothecin-Induced Apoptosis. *Br. J. Cancer* **2003,** *89*(9), 1757–1765.

Ashokkumar, P.; Sudhandiran, G. Protective Role of Luteolin on the Status of Lipid Peroxidation and Antioxidant Defense against Azoxymethane-Induced Experimental Colon Carcinogenesis. *Biomed. Pharmacother.* **2008,** *62*(9), 590–597.

Bell, C.; Hawthorne, S. Ellagic Acid, Pomegranate and Prostate Cancer—a Mini Review. *J. Pharm. Pharmacol.* **2008,** *60*(2), 139–144.

Boukes, G. J.; van de Venter, M. The Apoptotic and Autophagic Properties of Two Natural Occurring Prodrugs, Hyperoside and Hypoxoside, Against Pancreatic Cancer Cell Lines. *Biomed. Pharmacother.* **2016,** *83*, 617–626.

Buytaert, E.; Matroule, J. Y.; Durinck, S.; Close, P.; Kocanova, S.; Vandenheede, J. R.; de Witte, P. A.; Piette, J.; Agostinis, P. Molecular Effectors and Modulators of Hypericin-Mediated Cell Death in Bladder Cancer Cells. *Oncogene* **2008,** *27*(13), 1916–1929.

Campbell, B. J.; Yu, L. G.; Rhodes, J. M. Altered Glycosylation in Inflammatory Bowel Disease: A Possible Role in Cancer Development. *Glycoconj. J.* **2001,** *18*(11–12), 851–858.

Cao, J. Induction of Apoptosis by Casticin in Cervical Cancer Cells Through Reactive Oxygen Species-Mediated Mitochondrial Signaling Pathways. *Oncol. Rep.* **2011**.

Cao, W.; Liu, Y.; Zhang, R.; Zhang, B.; Wang, T.; Zhu, X.; Mei, L.; Chen, H.; Zhang, H.; Ming, P.; et al. Homoharringtonine Induces Apoptosis and Inhibits STAT3 via IL-6/JAK1/STAT3 Signal Pathway in Gefitinib-Resistant Lung Cancer Cells. *Sci. Rep.* **2015**, *5*, 8477.

Chan, L. L.; George, S.; Ahmad, I.; Gosangari, S. L.; Abbasi, A.; Cunningham, B. T.; Watkin, K. L. Cytotoxicity Effects of Amoora Rohituka and Chittagonga on Breast and Pancreatic Cancer Cells. *Evid. Based. Complement. Alternat. Med.* **2011**, *2011*, 860605.

Chavan, S.S. Damale, M.G. Shamkumar, P.B. Pawar, D.K. Traditional Medicinal Plants for Anticancer Activity. *Int J of Current Pharm Res.* **2013**, *5*(4), 50-54.

Chinni, S. R.; Li, Y.; Upadhyay, S.; Koppolu, P. K.; Sarkar, F. H. Indole-3-Carbinol (I3C) Induced Cell Growth Inhibition, G1 Cell Cycle Arrest and Apoptosis in Prostate Cancer Cells. *Oncogene* **2001**, *20*(23), 2927–2936.

Chiurillo, M. A. Role of the Wnt/β-Catenin Pathway in Gastric Cancer: An in-Depth Literature Review. *World J. Exp. Med.* **2015**, *5*(2), 84.

Cho, J. Y.; Kim, A. R.; Jung, J. H.; Chun, T.; Rhee, M. H.; Yoo, E. S. Cytotoxic and Pro-Apoptotic Activities of Cynaropicrin, a Sesquiterpene Lactone, on the Viability of Leukocyte Cancer Cell Lines. *Eur. J. Pharmacol.* **2004**, *492*(2–3), 85–94.

Corea, G.; Di Pietro, A.; Dumontet, C.; Fattorusso, E.; Lanzotti, V. Jatrophane Diterpenes from *Euphorbia* Spp. as Modulators of Multidrug Resistance in Cancer Therapy. *Phytochem. Rev.* **2009**, *8*(2), 431–447.

Cvorovic, J.; Tramer, F.; Granzotto, M.; Candussio, L.; Decorti, G.; Passamonti, S. Oxidative Stress-Based Cytotoxicity of Delphinidin and Cyanidin in Colon Cancer Cells. *Arch. Biochem. Biophys.* **2010**, *501*(1), 151–157.

Desai, A. G.; Qazi, G. N.; Ganju, R. K.; El-Tamer, M.; Singh, J.; Saxena, A. K.; Bedi, Y. S.; Taneja, S. C.; Bhat, H. K. Medicinal Plants and Cancer Chemoprevention. *Curr. Drug Metab.* **2008**, *9*(7), 581–591.

Duraipandiyan, V.; Baskar, A. A.; Ignacimuthu, S.; Muthukumar, C.; Al-Harbi, N. A. Anticancer Activity of Rhein Isolated from *Cassia fistula* L. Flower. *Asian Pac. J. Trop. Dis.* **2012**, 517–523.

Etcheverry, S. B.; Ferrer, E. G.; Naso, L.; Rivadeneira, J.; Salinas, V.; Williams, P. A. M. Antioxidant Effects of the VO(IV) Hesperidin Complex and its Role in Cancer Chemoprevention. *J. Biol. Inorg. Chem.* **2008**, *13*(3), 435–447.

Fitzmaurice, C.; Dicker, D.; Pain, A.; Hamavid, H.; Moradi-Lakeh, M.; MacIntyre, M. F.; Allen, C.; Hansen, G.; Woodbrook, R.; et al. Global, Regional, and National Cancer Incidence, Mortality, Years of Life Lost, Years Lived With Disability, and Disability-Adjusted Life-years for 32 Cancer Groups, 1990–2015. *JAMA Oncol.* **2015**, *1*(4), 505–527.

Gaddipati, J. P.; Mani, H.; Shefali; Raj, K.; Mathad, V. T.; Bhaduri, A. P.; Maheshwari, R. K. Inhibition of Growth and Regulation of IGFs and VEGF in Human Prostate Cancer Cell Lines by Shikonin Analogue 93/637 (SA). *Anticancer Res.* **2000**. *20*(4), 2547–2552.

Garodia, P.; Ichikawa, H.; Malani, N.; Sethi, G.; Aggarwal, B. B. From Ancient Medicine to Modern Medicine: Ayurvedic Concepts of Health and Their Role in Inflammation and Cancer. *J. Soc. Integr. Oncol.* **2007**, *5*(1), 25–37.

Hawkins, M. M.; Wilson, L. M.; Stovall, M. A.; Marsden, H. B.; Potok, M. H.; Kingston, J. E.; Chessells, J. M. Epipodophyllotoxins, Alkylating Agents, and Radiation and Risk of Secondary Leukaemia After Childhood Cancer. *BMJ* **1992**, *304*(6832), 951–958.

Herman-Antosiewicz, A.; Johnson, D. E.; Singh, S. V. Sulforaphane Causes Autophagy to Inhibit Release of Cytochrome *c* and Apoptosis in Human Prostate Cancer Cells. *Cancer Res.* **2006,** *66*(11), 5828–5835.

Hibasami, H.; Iwase, H.; Yoshioka, K.; Takahashi, H. Glycyrrhetic Acid (a Metabolic Substance and Aglycon of Glycyrrhizin) Induces Apoptosis in Human Hepatoma, Promyelotic Leukemia and Stomach Cancer Cells. *Int. J. Mol. Med.* **2006,** *17*(2), 215–219.

Hosseini, A.; Ghorbani, A. Cancer Therapy with Phytochemicals: Evidence from Clinical Studies. *Avicenna J. Phytomed.* **2015,** *5*(2), 84–97.

Hu, C.; Song, G.; Zhang, B.; Liu, Z.; Chen, R.; Zhang, H.; Hu, T. Intestinal Metabolite Compound K of Panaxoside Inhibits the Growth of Gastric Carcinoma by Augmenting Apoptosis via Bid-Mediated Mitochondrial Pathway. *J. Cell. Mol. Med.* **2012,** *16*(1), 96–106.

Huang, W.-Y.; Cai, Y.-Z.; Zhang, Y. Natural Phenolic Compounds From Medicinal Herbs and Dietary Plants: Potential use for Cancer Prevention. *Nutr. Cancer* **2009,** *62*(1), 1–20.

Ikezoe, T.; Chen, S. S.; Heber, D.; Taguchi, H.; Koeffler, H. P. Baicalin is a Major Component of PC-SPES Which Inhibits the Proliferation of Human Cancer Cells via Apoptosis and Cell Cycle Arrest. *Prostate* **2001,** *49*(4), 285–292.

Issa, M. E.; Berndt, S.; Carpentier, G.; Pezzuto, J. M.; Cuendet, M. Bruceantin Inhibits Multiple Myeloma Cancer Stem Cell Proliferation. *Cancer Biol. Ther.* **2016,** *17*(9), 966–975.

Jaganathan, S. K.; Supriyanto, E. Antiproliferative and Molecular Mechanism of Eugenol-Induced Apoptosis in Cancer Cells. *Molecules* **2012,** *17*(12), 6290–6304.

Jeong, C.-H.; Bode, A. M.; Pugliese, A.; Cho, Y.-Y.; Kim, H.-G.; Shim, J.-H.; Jeon, Y.-J.; Li, H.; Jiang, H.; Dong, Z. [6]-Gingerol Suppresses Colon Cancer Growth by Targeting Leukotriene A$_4$ Hydrolase. *Cancer Res.* **2009,** *69*(13), 5584–5591.

Jiang, W.; Qiu, W.; Wang, Y.; Cong, Q.; Edwards, D.; Ye, B.; Xu, C. Ginkgo May Prevent Genetic-Associated Ovarian Cancer Risk. *Eur. J. Cancer Prev.* **2011,** *20*(6), 508–517.

Kaczmarczyk, M. M.; Miller, M. J.; Freund, G. G. The Health Benefits of Dietary Fiber: Beyond the Usual Suspects of Type 2 Diabetes Mellitus, Cardiovascular Disease and Colon Cancer. *Metabolism* **2012,** *61*(8), 1058–1066.

Kauffman, E. C.; Robinson, B. D.; Downes, M. J.; Powell, L. G.; Lee, M. M.; Scherr, D. S.; Gudas, L. J.; Mongan, N. P. Role of Androgen Receptor and Associated Lysine-Demethylase Coregulators, LSD1 and JMJD2A, in Localized and Advanced Human Bladder Cancer. *Mol. Carcinog.* **2011,** *50*(12), 931–944.

Kaur, M.; Goel, R. K. Anti-Convulsant Activity of *Boerhaavia diffusa*: Plausible Role of Calcium Channel Antagonism. *Evid. Based. Complement. Alternat. Med.* **2011,** *2011*, 310420.

Kaur, R.; Singh, J.; Singh, G.; Kaur, H. Anticancer Plants: A Review. *J. Nat. Prod. Plant Resour.* **2011,** *1*(4), 131–136.

Khan, N.; Afaq, F.; Syed, D. N.; Mukhtar, H. Fisetin, a Novel Dietary Flavonoid, Causes Apoptosis and Cell Cycle Arrest in Human Prostate Cancer LNCaP Cells. *Carcinogenesis* **2008,** *29*(5), 1049–1056.

Kogure, T.; Kinghorn, A. D.; Yan, I.; Bolon, B.; Lucas, D. M.; Grever, M. R.; Patel, T. Therapeutic Potential of the Translation Inhibitor Silvestrol in Hepatocellular Cancer. *PLoS One* **2013,** *8*(9), e76136.

Kuo, P.-L.; Hsu, Y.-L.; Chang, C.-H.; Lin, C.-C. The Mechanism of Ellipticine-Induced Apoptosis and Cell Cycle Arrest in Human Breast MCF-7 Cancer Cells. *Cancer Lett.* **2005,** *223*(2), 293–301.

Kuo, S. M. Antiproliferative Potency of Structurally Distinct Dietary Flavonoids on Human Colon Cancer Cells. *Cancer Lett.* **1996,** *110*(1–2), 41–48.

Lapenna, S.; Giordano, A. Cell Cycle Kinases as Therapeutic Targets for Cancer. *Nat. Rev. Drug Discov.* **2009,** *8*(7), 547–566.

Lee, D.-H.; Szczepanski, M.-J.; Lee, Y. J. Magnolol Induces Apoptosis via Inhibiting the EGFR/PI3K/Akt Signaling Pathway in Human Prostate Cancer Cells. *J. Cell. Biochem.* **2009,** *106*(6), 1113–1122.

Lee, J.-C.; Lee, C.-H.; Su, C.-L.; Huang, C.-W.; Liu, H.-S.; Lin, C.-N.; Won, S.-J. Justicidin A Decreases the Level of Cytosolic Ku70 Leading to Apoptosis in Human Colorectal Cancer Cells. *Carcinogenesis* **2005a,** *26*(10), 1716–1730.

Lee, M.-S.; Chan, J. Y.-W.; Kong, S.-K.; Yu, B.; Eng-Choon, V. O.; Nai-Ching, H. W.; Mak Chung-Wai, T.; Fung, K.-P. Effects of Polyphyllin D, a Steroidal Saponin in *Paris Polyphylla*, in Growth Inhibition of Human Breast Cancer Cells and in Xenograft. *Cancer Biol. Ther.* **2005b,** *4*(11), 1248–1254.

Li, W.; Du, B.; Wang, T.; Wang, S.; Zhang, J. Kaempferol Induces Apoptosis in Human HCT116 Colon Cancer Cells via the Ataxia-Telangiectasia Mutated-p53 Pathway with the Involvement of p53 Upregulated Modulator of Apoptosis. *Chem. Biol. Interact.* **2009,** *177*(2), 121–127.

Lodish, H.; Berk, A.; Zipursky, S. L.; Matsudaira, P.; Baltimore, D.; Darnell, J. *DNA Damage and Repair and Their Role in Carcinogenesis*; 4th ed., W. H. Freeman: New York, 2000. https://www.ncbi.nlm.nih.gov/books/NBK21554/ (accessed Jan 8, 2018).

Mohan, K.; Jeyachandran, R.; Alkaloids as Anticancer Agents. *Ann. Phytomed.* **2012,** *1*(1), 46–53.

Mollakhalili Meybodi, N.; Mortazavian, A. M.; Bahadori Monfared, A.; Sohrabvandi, S.; Aghaei Meybodi, F. Phytochemicals in Cancer Prevention: A Review of the Evidence. *Iran. J. Cancer Prev.* **2017,** *10*(1), 1–8.

Mousavi, Y.; Adlercreutz, H. Enterolactone and Estradiol Inhibit Each Other's Proliferative Effect on MCF-7 Breast Cancer Cells in Culture. *J. Steroid Biochem. Mol. Biol.* **1992,** *41*(3–8), 615–619.

Nepal, M.; Jung Choi, H.; Choi, B.-Y.; Lim Kim, S.; Ryu, J.-H.; Hee Kim, D.; Lee, Y.-H.; Soh, Y. Anti-Angiogenic and Anti-Tumor Activity of Bavachinin by Targeting Hypoxia-Inducible Factor-1α. *Eur. J. Pharmacol.* **2012,** *691*(1–3), 28–37.

Nepomuceno, J.C. Antioxidants in Cancer Treatment. In *Current Cancer Treatment—Novel Beyond Conventional Approaches*; InTech, 2011, Pp. 621-650. DOI: 10.5772/23131, ISBN No: 978-953-307-397-2.

Ohnishi, M.; Tanaka, T.; Makita, H.; Kawamori, T.; Mori, H.; Satoh, K.; Hara, A.; Murakami, A.; Ohigashi, H.; Koshimizu, K. Chemopreventive Effect of a Xanthine Oxidase Inhibitor, 1'-Acetoxychavicol Acetate, on Rat Oral Carcinogenesis. *Jpn. J. Cancer Res.* **1996,** *87* (4), 349–356.

Panno, M. L.; Giordano, F. Effects of Psoralens as Anti-Tumoral Agents in Breast Cancer Cells. *World J. Clin. Oncol.* **2014,** *5*(3), 348–358.

Patil, J. B.; Kim, J.; Jayaprakasha, G. K. Berberine Induces Apoptosis in Breast Cancer Cells (MCF-7) Through Mitochondrial-Dependent Pathway. *Eur. J. Pharmacol.* **2010,** *645*(1–3), 70–78.

Peng, S.; Yi, Z.; Liu, M. Ailanthone: A New Potential Drug for Castration-Resistant Prostate Cancer. *Chin. J. Cancer* **2017,** *36*(1), 25.

Perabo, F. G. E.; Landwehrs, G.; Frössler, C.; Schmidt, D. H.; Mueller, S. C. Antiproliferative and Apoptosis Inducing Effects of Indirubin-3′-Monoxime in Renal Cell Cancer Cells. *Urol. Oncol. Semin. Orig. Investig.* **2011,** *29*(6), 815–820.

Pinho, P. M. M.; Kijjoa, A. Chemical Constituents of the Plants of the Genus Cleistanthus and Their Biological Activity. *Phytochem. Rev.* **2007,** *6*(1), 175–182.

Przybyło, M.; Lityńska, A.; Pocheć, E. Different Adhesion and Migration Properties of Human HCV29 Non-Malignant Urothelial and T24 Bladder Cancer Cells: Role of Glycosylation. *Biochimie* **2005,** *87*(2), 133–142.

Qian, L.; Xie, B.; Wang, Y.; Qian, J. Amygdalin-Mediated Inhibition of Non-Small Cell Lung Cancer Cell Invasion in Vitro. *Int. J. Clin. Exp. Pathol.* **2015,** *8*(5), 5363–5370.

Ravishankar, D.; Rajora, A. K.; Greco, F.; Osborn, H. M. I. Flavonoids as Prospective Compounds for Anti-Cancer Therapy. *Int. J. Biochem. Cell Biol.* **2013,** *45* (12), 2821–2831.

Rimmon, A.; Vexler, A.; Berkovich, L.; Earon, G.; Ron, I.; Lev-Ari, S. Escin Chemosensitizes Human Pancreatic Cancer Cells and Inhibits the Nuclear Factor-kappaB Signaling Pathway. *Biochem. Res. Int.* **2013,** *2013*, 251752.

Saleem, M.; Kweon, M.-H.; Yun, J.-M.; Adhami, V. M.; Khan, N.; Syed, D. N.; Mukhtar, H. A Novel Dietary Triterpene Lupeol Induces Fas-Mediated Apoptotic Death of Androgen-Sensitive Prostate Cancer Cells and Inhibits Tumor Growth in a Xenograft Model. *Cancer Res.* **2005,** *65*(23), 11203–11213.

Samoszuk, M.; Tan, J.; Chorn, G. The Chalcone Butein from *Rhus verniciflua* Stokes Inhibits Clonogenic Growth of Human Breast Cancer Cells Co-Cultured with Fibroblasts. *BMC Complement. Altern. Med.* **2005,** *5*(1), 5.

Sarkar, F. H.; Li, Y. Mechanisms of Cancer Chemoprevention by Soy Isoflavone Genistein. *Cancer Metastasis Rev.* **2002,** *21*(3–4), 265–280.

Savel, H. The Metaphase-Arresting Plant Alkaloids and Cancer Chemotherapy. *Prog Exp Tumor Res.* **1966**; *8,* 189–224.

Scarlatti, F.; Maffei, R.; Beau, I.; Codogno, P.; Ghidoni, R. Role of Non-Canonical Beclin 1-Independent Autophagy in Cell Death Induced by Resveratrol in Human Breast Cancer Cells. *Cell Death Differ.* **2008,** *15*(8), 1318–1329.

Shoeb, M. Anticancer Agents from Medicinal Plants. *Bangladesh J. Pharmacol.* **2008,** *1*(2), 35–41.

Shukla, S.; Gupta, S. Apigenin: A Promising Molecule for Cancer Prevention. *Pharm. Res.* **2010,** *27*(6), 962–978.

Singh, S.; Khar, A. Biological Effects of Curcumin and its Role in Cancer Chemoprevention and Therapy. *Anticancer Agents Med. Chem.* **2006,** *6*(3), 259–270.

Skowronski, R. J.; Peehl, D. M.; Feldman, D. Vitamin D and Prostate Cancer: 1,25 Dihydroxyvitamin D3 Receptors and Actions in Human Prostate Cancer Cell Lines. *Endocrinology* **1993,** *132*(5), 1952–1960.

Stiborová, M.; Poljaková, J.; Martínková, E.; Bořek-Dohalská, L.; Eckschlager, T.; Kizek, R.; Frei, E. Ellipticine Cytotoxicity to Cancer Cell Lines—a Comparative Study. *Interdiscip. Toxicol.* **2011,** *4*(2), 98–105.

Stuart, E. C.; Scandlyn, M. J.; Rosengren, R. J. Role of Epigallocatechin Gallate (EGCG) in the Treatment of Breast and Prostate Cancer. *Life Sci.* **2006,** *79*(25), 2329–2336.

Su, J.; Lai, H.; Chen, J.; Li, L.; Wong, Y.-S.; Chen, T.; Li, X. Natural Borneol, a Monoterpenoid Compound, Potentiates Selenocystine-Induced Apoptosis in Human Hepatocellular Carcinoma Cells by Enhancement of Cellular Uptake and Activation of ROS-Mediated DNA Damage. *PLoS One* **2013,** *8*(5), e63502.

Sun, H.; Lv, C.; Yang, L.; Wang, Y.; Zhang, Q.; Yu, S.; Kong, H.; Wang, M.; Xie, J.; Zhang, C.; et al. Solanine Induces Mitochondria-Mediated Apoptosis in Human Pancreatic Cancer Cells. *Biomed. Res. Int.* **2014,** *2014*, 805926.

Suresh, C.; Zhao, H.; Gumbs, A.; Chetty, C. S.; Bose, H. S. New Ionic Derivatives of Betulinic Acid as Highly Potent Anti-Cancer Agents. *Bioorg. Med. Chem. Lett.* **2012,** *22*(4), 1734–1738.

Thomas, E. S.; Gomez, H. L.; Li, R. K.; Chung, H.-C.; Fein, L. E.; Chan, V. F.; Jassem, J.; Pivot, X. B.; Klimovsky, J. V.; de Mendoza, F. H.; et al. Ixabepilone Plus Capecitabine for Metastatic Breast Cancer Progressing After Anthracycline and Taxane Treatment. *J. Clin. Oncol.* **2007,** *25*(33), 5210–5217.

Umadevi, M.; Kumar, K. P. S; Debjit, B. S. D. Traditionally used Anticancer Herbs in India. *J. Med. Plants Studies* **2004,** *1*(3), 56–74.

Ma, X.; Yu, H. Global Burden of Cancer. *Yale J. Biol. Med.* **2006,** *79*(3–4), 85–94.

Volate, S. R.; Kawasaki, B. T.; Hurt, E. M.; Milner, J. A.; Kim, Y. S.; White, J.; Farrar, W. L. Gossypol Induces Apoptosis by Activating p53 in Prostate Cancer Cells and Prostate Tumor-Initiating Cells. *Mol. Cancer Ther.* **2010,** *9*(2), 461–470.

Wang, H.; Khor, T. O.; Shu, L.; Su, Z.-Y.; Fuentes, F.; Lee, J.-H.; Kong, A.-N. T. Plants vs. Cancer: A Review on Natural Phytochemicals in Preventing and Treating Cancers and Their Druggability. *Anticancer Agents Med. Chem.* **2012b,** *12*(10), 1281–1305.

Wang, H.-C.; Tsai, Y.-L.; Wu, Y.-C.; Chang, F.-R.; Liu, M.-H.; Chen, W.-Y.; Wu, C.-C. Withanolides-Induced Breast Cancer Cell Death is Correlated with Their Ability to Inhibit Heat Protein 90. *PLoS One* **2012a,** *7*(5), e37764.

Wang, X. Y.; Liu, H. T. Antisense Expression of Protein Kinase C Alpha Improved Sensitivity to Anticancer Drugs in Human Lung Cancer LTEPa-2 Cells. *Zhongguo Yao Li Xue Bao* **1998,** *19*(3), 265–268.

Webb, A. L.; Mccullough, M. L. Dietary Lignans: Potential Role in Cancer Prevention Dietary Lignans: Potential Role in Cancer Prevention. *Natural and Cancer.* **2009,** *51*(2), 117-131.

Wu, C.-H.; Chen, H.-Y.; Wang, C.-W.; Shieh, T.-M.; Huang, T.-C.; Lin, L.-C.; Wang, K.-L.; Hsia, S.-M. Isoliquiritigenin Induces Apoptosis and Autophagy and Inhibits Endometrial Cancer Growth in Mice. *Oncotarget* **2016,** *7*(45), 73432–73447.

Yang, C.; Gundala, S. R.; Mukkavilli, R.; Vangala, S.; Reid, M. D.; Aneja, R. Synergistic Interactions Among Flavonoids and Acetogenins in Graviola (*Annona muricata*) Leaves Confer Protection against Prostate Cancer. *Carcinogenesis* **2015,** *36*(6), 656–665.

Yang, L.-C.; Hsieh, C.-C.; Lu, T.-J.; Lin, W.-C. Structurally Characterized Arabinogalactan from *Anoectochilus formosanus* as an Immuno-Modulator against CT26 Colon Cancer in BALB/c Mice. *Phytomedicine* **2014,** *21*(5), 647–655.

Yeh, C.-T.; Wu, C.-H.; Yen, G.-C. Ursolic Acid, a Naturally Occurring Triterpenoid, Suppresses Migration and Invasion of Human Breast Cancer Cells by Modulating c-Jun N-Terminal Kinase, Akt and Mammalian Target of Rapamycin Signaling. *Mol. Nutr. Food Res.* **2010,** *54*(9), 1285–1295.

Zhan, Y.; Zhang, Y.; Liu, C.; Zhang, J.; Smith, W. W.; Wang, N.; Chen, Y.; Zheng, L.; He, L. A Novel Taspine Derivative, HMQ1611, Inhibits Breast Cancer Cell Growth via Estrogen Receptor α and EGF Receptor Signaling Pathways. *Cancer Prev. Res. (Phila).* **2012,** *5*(6), 864–873.

Zhou, L.; Feng, Y.; Jin, Y.; Liu, X.; Sui, H.; Chai, N.; Chen, X.; Liu, N.; Ji, Q.; Wang, Y.; et al. Verbascoside Promotes Apoptosis by Regulating HIPK2-p53 Signaling in Human Colorectal Cancer. *BMC Cancer* **2014,** *14,* 747.

Ziegler, R. G. Phytoestrogens and Breast Cancer. *Am. J. Clin. Nutr.* **2004,** *79*(2), 183–184.

# CHAPTER 16

# METHYLATED FLAVONOIDS AS A NOVEL INHIBITOR OF METASTASIS IN THE CANCER CELL

PREM PRAKASH KUSHWAHA[1], PUSHPENDRA SINGH[2], and SHASHANK KUMAR[1,*]

[1]Department of Biochemistry and Microbial Sciences, School of Basic and Applied Sciences, Central University of Punjab, Bathinda, Punjab 151001, India, Mob.: +91 9335647413

[2]National Institute of Pathology, New Delhi, India

*Corresponding author. E-mail: shashankbiochemau@gmail.com; shashank.kumar@cupb.edu.in
*ORCID: https://orcid.org/0000-0002-9622-0512

## ABSTRACT

Flavonoids are an important class of secondary metabolites. During natural synthesis of flavonoids, their basic skeleton gets modified by the different type of chemical group. Various enzymes are involved in this type of modification. Structure–function relationship has been reported for their biological activity. Methylation is a type of natural modification in flavonoids. Methylation of hydroxyl group, where methyl group attached with the oxygen gives rise to O-methylated flavonoids. On the other hand, when methylation occurs at carbon of flavonoid basic structure, it results into synthesis of C-methylated flavonoids. Methylation of flavonoid is known for their various biological activities and has a role in their digestion process also. Metastasis with invasion is a multistep process which accounts for a number of cancer-associated deaths worldwide. This chapter discusses the chemistry and biological activities of methylated flavonoids. We will also discuss the in silico potential of these flavonoids as an anti-metastasis agent.

## 16.1   INTRODUCTION

Flavonoids are a class of secondary metabolites. They have an impact on the color of the flower and other pigmentation tissues of plants. First, the biologically studied flavonoids are the isoflavones. They showed tremendous affinity for estrogen receptors. Another used prime group of flavonoids in the human history are the anthocyanins (as a dye). In the present research era, flavonoids have been associated with anti-inflammatory, anti-allergic, antiviral, and anticarcinogenic properties. Flavonoid skeleton comprise of a typical 15-carbon structure possessing two phenyl rings, namely A and B, and a heterocyclic ring referred as C. Researchers have synthesized a range of novel flavone derivatives to develop effective anticancer drug against various cancers (Liu et al., 2010). Some group of flavonoids (such as flavanones) are not broadly studied for their anticancer activities. Hesperidin (hesperetin 7-rutinoside) has capability to inhibit kinases and phosphodiesterase activity, responsible for cellular signal transduction and activation of inflammation response (Manthey et al., 2001). Flavans from green tea (*Camellia sinensis*) such as (+)-catechin, (-)-epicatechin, (-)-epigallocatechin possess distinct biological properties including anticarcinogenic activity (Crespy and Williamson, 2004; Moore et al., 2009). The anthocyanidins are aglycones of anthocyanins and have flavylium (2-phenylchromenylium) as an ion skeleton. Various studies showed their potential in cancer treatment and human nutrition (Lule and Xia, 2005; Nichenametla et al., 2006). Daidzein, an isoflavone drug showed immunostimulatory consequences which might associate with cancer prevention (Birt et al., 2001). Flavonoids have the potency to combat with different type of cancers like ovarian, breast, cervical, pancreatic, and prostate cancer. *Tecoma stans* flower methanolic extract has significant dose-dependent antitumor activity studied in both in vivo and in vitro (Kameshwaran et al., 2012). Ginger extract also showed remarkably anticancer activity against MCF-7 and MDA-MB-231 breast cancer cell lines, respectively (Rahman et al., 2011). Luteolin-7-methyl ether isolated from *Blumea balsemifera* leaves showed anticancer activity against human lung cancer (potent activity) and oral cavity cancer cell lines (moderate activity) (Saewan et al., 2011). Kaempferol is known to reduce vascular endothelial growth factor (VEGF) expression in cancer cells. Other flavonoids such as genistein, genistin, daidzein, and biochanin A also hamper the growth of murine and human bladder cancer cell lines by inducing cell cycle arrest, apoptosis, and angiogenesis (Luo et al., 2010).

## 16.2    C-METHYLATED AND O-METHYLATED FLAVONOIDS

Natural compounds are valuable sources for the development of nutraceuti-cals and drugs (Brandt et al., 2004; Newman and Cragg, 2007). Flavonoids are one of the important plant secondary metabolites. They are also produced by microorganism and algae. They exert various biological activities, most importantly anti-aging, anticancer, antioxidant, and antimicrobial potential. Due to their pharmaceutical potential various methods for flavonoid extrac-tion from plants have been developed (Tapas et al., 2008). Flavonoid basic structure contains a characteristic aromatic ring. Its biosynthetic pathway starts from tyrosine (aromatic amino acid). Naringenin is one of the impor-tant metabolite of flavonoid biosynthetic process, which further converts into various flavonoid subgroups (Winkel-Shirley 2001). First, tyrosine ammonia lyase converts L-tyrosine to a phenylpropanoic acid, p-coumaric acid. O-methylation, glycosylation, and malonylation are different chemical modifications known during flavonoid biosynthesis process (Winkel-Shirley 2001). Attachment of methyl group with oxygen of hydroxyl moiety in flavonoid skeleton results in O-methylated derivatives of flavonoids (Table 16.1). O-methyltransferase enzyme is responsible for O-methylation of flavonoids, which confers the potential biological activity of these modi-fied flavonoids (Kim et al., 2010). For example, 4′-Omethylnaringenin (ponciretin) shows antibacterial activity against *Helicobacter pylori* (Kim et al., 1999) and 7-O-methylnaringenin (sakuranetin) inhibits germination of the rice blast fungus, *Magnaporthe grisea* (Kodama et al., 1992). Tyrosine biosynthesis pathway in *Escherichia coli* was engineered to produce bioac-tive O-methylated flavonoids such as sakuranetin and ponciretin. Using this approach, the author increased production of these flavonoids up to 40 mg/L (Kim et al., 2013). The C-methylated flavonoids are a category of flavonoids having methylation(s) on carbon. Various C-methylated flavonoids have been reported from plants such as *Pisonia grandis* (roots), Myrtaceae family plants, or *Cleistocalyx operculatus* (Fig. 16.1) (Dao et al., 2010; Koirala et al., 2016).

## 16.3    NATURAL OCCURRENCE OF METHYLATED FLAVONOIDS

*C. operculatus* (Myrtaceae) is distributed widely in tropical Asia. The plant is known for its role in traditional medicine to treat various ailments. Recently some C-methylated flavonoids have been isolated from the plant (Dao et al., 2010). *P. grandis* (Nyctaginaceae), an ornamental plant is well

**TABLE 16.1** List of o-Methylated Flavonoids.

| Flavanones | Flavonols | Flavones | Isoflavones |
|---|---|---|---|
| Hesperetin | Kaempferide | Acacetin | Biochanin A |
| Homoeriodictyol | Annulatin | Chrysoeriol | Calycosin |
| Sakuranetin | Combretol | Diosmetin | Formononetin |
| Isosakuranetin | Europetin | Nepetin | Glycitein |
| Sterubin | Laricitrin (3'-O-Dimethylmyricetin) | Nobiletin | Irigenin |
| | 5-O-methylmyricetin | Oroxylin-A | 5-O-methylgenistein |
| | Syringetin (3',5'-O-imethylmyricetin) | Sinensetin | Pratensein |
| | Ayanin | Tangeritin | Prunetin |
| | Azaleatin | Wogonin | Psi-tectorigenin |
| | Isorhamnetin | | Retusin |
| | Ombuin | | Tectorigenin |
| | Pachypodol | | |
| | Retusin (Quercetin-3,7,3',4'-tetramethyl ether) | | |
| | Rhammazin | | |
| | Rhamnetin | | |
| | Tamarixetin | | |
| | Eupatolitin | | |
| | Natsudaidain | | |

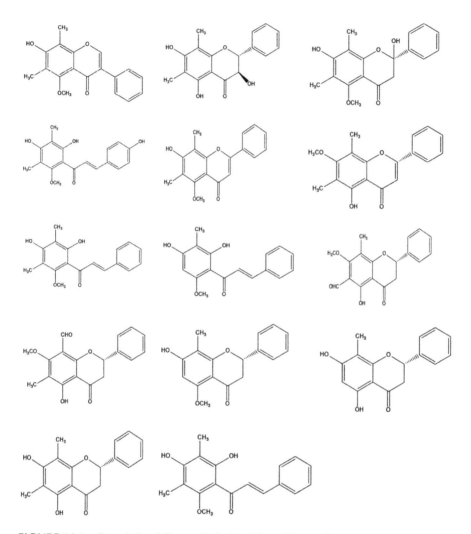

**FIGURE 16.1** C-methylated flavonoids isolated from *Cleistocalyx operculatus.*

documented to possess anti-dysentery, anti-diabetic, anti-inflammatory, and analgesic properties. In another study, isolation of five new C-methylated flavonoids (from *P. grandis*) was reported (Fig. 16.2) (Sutthivaiyakit et al., 2013). Different studies reported the occurrence of C-methylated flavonoids from different parts of the plant such as leaves of *Vellozia* spp., roots of *Talinum triangulare,* and needles of the *Scotch pine.*

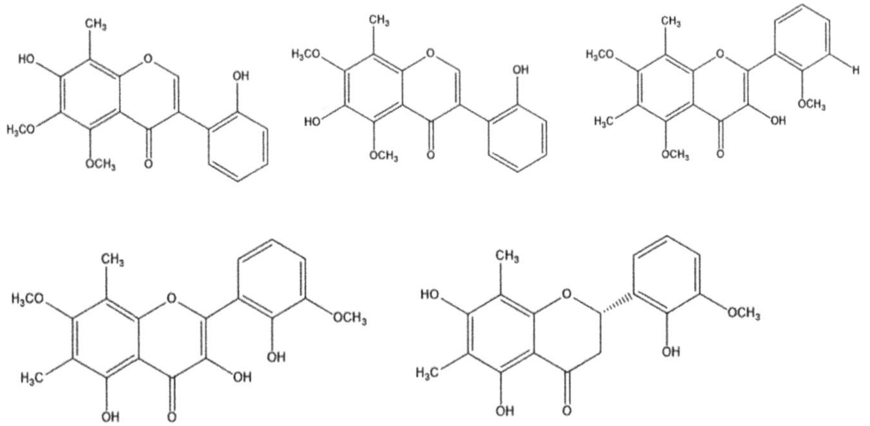

**FIGURE 16.2**   C-methylated flavonoids isolated from *Pisonia grandis.*

## 16.4   BIOLOGICAL ACTIVITIES OF METHYLATED FLAVONOIDS

Different biological activities such as anticancer and cytochrome P450 activities at mRNA and protein level are shown by mono and dimethylated flavones (Table 16.2) (Murakami et al., 2002; Morley et al., 2007). In a different study, methylated flavonoid (7-O-methyl genistein and 7-O-methyl daidzein) showed inhibition of TNF-α-induced invasion of human umbilical vein endothelial cells. At the test concentration, the flavonoids did not show any cellular toxicity in normal cells (Koirala et al., 2015). Several methylated flavonoids such as rhamnetin, sakuranetin, and genkwanin are in the clinical trial phase. C-methylated flavonoid has also been reported as neuraminidase inhibitors and possesses oxidative stress-modulating efficacy (Malterud et al., 1996; Dao et al., 2010).

## 16.5   BIOTECHNOLOGICAL PRODUCTION OF METHYLATED FLAVONOIDS

Methyltransferases are one of the biological tools for synthesis of methylated flavonoid. S-Adenosyl-l-methionine (AdoMet)-dependent O-methyltrans-ferases (OMT) enzymes transfer methyl group to a specific hydroxyl group of an acceptor compound. This methyl transfer resulted in the formation of methyl ether derivative (Ibrahim et al., 1998). OMTs (ubiquitous) can be clas-sified into three classes; Class I, II (methylate phenolic hydroxyl residues),

**TABLE 16.2** Biological Activities of Methylated Flavonoids.

| Plant | Phytochemical | Mode of action | References |
|---|---|---|---|
| Daphene genkwa | Genkwanin (7-O-methylapigenin) | Active against bacteria and plasmodium, showed antioxidant, chemo preventive, and anti-inflammatory effect. | Altinier et al. (2007) |
| Oryza sativa | Sakuranetin (7-Omethylnaringenin) | Inhibition of platelet aggregation,cytotoxic against nasopharyngeal, carcinoma cells | Tuchinda et al. (2002) |
| Rhamnus petiolaris | Rhamnetin (7-O-methylated quercetin) | Anti-melanogenesis, inhibits the formation of β-amyloid, inhibits Notch-1 signaling in cancer cell lines | Igarashi and Ohmuma (1995) |
| Kaempferia galanga | Kaempferide | Inhibit TRP1 expression in cancer cells, pro-coagulant activity in human monocyte, cytotoxic against several cancer cells | Matsuda et al. (2009) |
| Orthosiphon stamineus | Sinensetin | Radical-scavenging activity | Akowuah et al. (2005) |
| Melastomataceae, | Matteucinol | Antitumor activity against several cancer cell lines | Shi et al. (2010) |
| Catharanthus roseus | Hirsutidin | Active against skin cancer | Piovan et al. (1998) |
| Rice bran | Tricin | Antihistaminic activity Antioxidant activity, cyclooxygenase inhibitory potential | Cai et al. (2005) |
| Casimiroa edulis | Zapotin | Anticancer potential against isolated colon cancer cells | Maiti et al. (2007) |
| Myrica serrata | Demethoxymatteucinol | Inhibit HIV-1 replication in host cells, antitumor potential | Gafner et al. (1996); Wu et al. (2003) |

and III enzymes (methylate carboxyl group) (Noel et al., 2003). Methyltransferases can also be categorized as O, N, or C-methyltransferase based on their target attachments such as oxygen, nitrogen, and carbon, respectively (Schubert et al., 2003). Structure–function relationship of different flavonoid methyltransferases such as BcOMT2 (a flavonoid Mg2+ dependent O-methyltransferasehas), TaOMT2 (a tricetin O-methyltransferase), COMT (Medicago truncatula O-methyltransferase), PaMTH1 (Podospora anserina O-methyltransferase) been studied at molecular level. Knowledge about crystal structure of ChOMT (chalcone SAMdependent O-methyltransferase) and IOMT (isoflavone SAMdependent O-methyltransferase) opened new door to understand the substrate specificity. Crystal structure of enzyme and their specific substrate interaction facilitates the bioengineering progress in this field (Zubieta et al., 2001). Detailed information regarding the characterization of C-methyltransferase such as wheat flavone O-methyltransferase, caffeic acid/5-hydroxyferulic acid 3/5-O-methyltransferase is still been awaited (Koirala et al., 2016). Synthesis of methylated flavonoids by in vitro enzymatic reaction is a different technique used nowadays. Merit of this technique includes better tools to study catalytic mechanism, activity, and enzyme kinetics. In this regard, various flavonoids O-methyltransferase have been isolated from bacteria, fungi, and plant and characterized for their substrate reactivity (Koirala et al., 2016). Due to huge size, genome and complex regulatory network in plants, it is difficult to carry out mutagenesis or optimization for production of secondary metabolites in plants. Reconstruction of recombinant plasmids harboring genes and genetically engineered *E. coli* as host can be used to produce various natural and unnatural products biotechnologically. Biotransformation of flavonoids is another mean to produce methylated derivatives of flavonoids (Koirala et al., 2016).

## 16.6 INTESTINAL ABSORPTION AND HEPATIC METABOLISM OF METHYLATED FLAVONOIDS

The potential utility of flavonoids in chemoprevention of various diseases has been investigated in different studies (Wen and Walle, 2006a). Biological efficacies of flavonoids in cell culture studies did not show promising results in in vivo situations, particularly in humans. Very low oral bioavailability of these compounds (chrysin, resveratrol, quercetin, and so forth) is known for the disparity among biological activities in vitro and in vivo (Fig. 16.3). It has been reported that the free hydroxyl groups of most polyphenols involve in glucuronidation and sulfation conjugation reactions (Otake et al., 2002).

Several studies reported that glucuronidation, sulfation, and oxidation of flavonoids increased metabolic stability of the products in the liver. The study also revealed that methylation of flavonoids protects their rapid metabolism in the liver (Wen and Walle, 2006b). Thus, the natural/artificial structural modification in flavonoids provides a promising way to improve the bioavailability of flavonoids. Studies from a group of scientist suggest metabolic stability as a new and effective parameter to assess the human oral bioavailability of methylated flavonoids (Fig. 16.3). This helps to evaluate the chemoprevention potential of these compounds against various human diseases. The study concluded that fully methylated flavones showed higher intestinal permeability and metabolic stability than unmethylated flavones (Wen and Walle, 2006a).

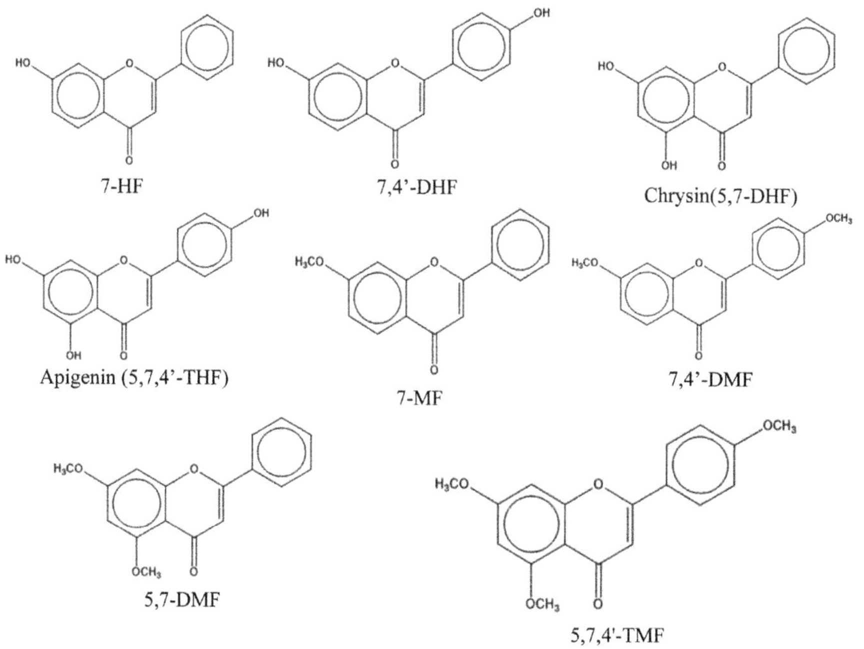

**FIGURE 16.3** Structure of methylated and non-methylated flavonoids used in the study to prove better bioavailability of methylated flavonoids.

## 16.7 METASTASIS

The movement of cancer cells from a primary site to progressively colonize distant organs is known as tumor metastasis. It is a major contributor factor of death in cancer patients. Metastasis is a phenomenon that has been associated with several events in cancer cell such as cancer cell immunity, extracellular matrix (ECM) composition, cell survival signaling pathways, drug resistance, and so forth (Steeg, 2016). The signaling involved in initial phases of tumorigenesis provide metastatic and drug resistance properties to cancer cells. Later on, with the progress of tumorigenesis, the cancer cell acquires strong resistance to anticancer drugs and accelerated metastasis. Mostly cancer therapy has largely targeted to minimize the drug resistance in cancer cells. Targeting metastasis through anticancer agents lag far behind in terms of identification and validation of antimetastatic drugs. The reason behind the clinical validation of antimetastatic drugs is problematic due to their cytostatic and noncytotoxic nature (mostly). Cancer cell metastasis begins with invading the surroundings of primary tissue. Thereafter, the tumor cell enters the bloodstream and gets arrest at the capillary bed encountered. At this colonizing site, various factors such as altered cellular ECM composition, immune status, and blood supply promote the colonization of tumor cell at this new site. Different factors associated with genomic stability such as abnormal chromosomal stability, DNA repair, and altered gene regulation have been known to fuel the metastasis process. Generally, it has been assumed that an antitumor growth drug also inhibit/ target metastasis (Steeg, 2016). But many clinical studies indicate that tumor growth and tumor metastasis both should be targeted separately. Patient getting only antitumor growth drug cannot be relieved by tumor metastasis by the same drug. Several FDA approved anticancer drugs (paclitaxel, cisplatin, anti-androgens, and everolimus, and so forth) are reported for their stimulated metastasis in preclinical studies. These studies substantiate the need for antimetastatic drugs in concurrent with potent antitumor growth remedies. Angiogenesis provide new capillaries to deliver nutrient and oxygen to metastatic colonization. VEGF is one of the important factors in growth and permeability of capillary endothelial cells. Beside targeting angiogenesis there is another antimetastatic target known as angiopoietin 2 (ANGPT2) pathway involved in vessel stabilization. Integrin are receptors that mediate the tumor cells adhesion to ECM. By doing this it affects angiogenesis, viability, invasion, and colonization in tumor cells. Thus the inhibitors of angiogenesis, ANGPT2 pathway, and integrin inhibitors might act through monotherapy and or in combinatorial therapy as antimetastatic agents. But the preclinical studies are disappointing (Steeg, 2016). There is an urgent need to identify and validate a new metastatic protein marker and their novel natural inhibitors.

## 16.8   MIGRATION AND INVASION ENHANCER 1 PROTEIN

Migration and invasion enhancer 1 (MIEN1) protein expression is associated with tumorigenesis and considered as a new tumor-specific target protein (Zhao et al., 2017). The protein is also known as C17orf37 (chromosome 17 open reading frame 37), hepatitis B virus (HBV) XAg-transactivated protein 4, HBV X-transactivated Gene 4 Protein, XTP4, ORB3, RDX12, and C35. It is a membrane-anchored protein (Kauraniemi and Kallioniemi, 2006). MIEN1 gene is located in the "hot spot locus of cancer" on the human chromosome 17q12 (Dasgupta et al., 2010). MIEN1 putative promoter region contains numerous CpG dyads, CpG islands, and a short interspersed nuclear element alu repeat. Various sites in the alu element of MEIN1 promoter remain hypermethylated in the normal cells (Rajendiran et al., 2016). Prenylated proteins belong to a class of protein family, which are post-translationally modified by the addition of isoprenyl groups. It has been reported that a special motif of the prenylated protein serves as a substrate for a series of post-translational modifications and facilitates membrane association. It is believed that MIEN1 is prenylated by a GGTase-I enzyme (Kpetemey et al., 2015). The prenylation of protein helps to stabilize its membrane attachment and enhance signaling (Wang and Casey, 2016). Still, the areas are open to identifying the protein (s) involved in the stabilization of membrane-MIEN1 association. Additionally, MIEN1 and metalloproteases expression on the surface of tumors help to clear the road for the cells to move outward, promoting metastasis. Upregulation of DNp73, an isoform of p53 protein has been observed in a variety of human cancers and showed pro-tumor activities. It has been suggested that DNp73 might in association with MIEN1 promote tumor progression and contribute to cisplatin resistance in cancer cells. Evans et al. reported the overexpression of MIEN1 gene in tumor and normal human mammary cell lines. Immunohistochemical analysis detected robust and frequent expression of the C35 protein in breast and other cancer tissue (Evans et al., 2006). Increased levels of C35 protein are linked with the hallmarks of transformation (colony formation, invasion, and epithelial to mesenchymal transition). Association of MIEN1 overexpression with thick membrane extensions suggests its role in actin cytoskeletal dynamics leading to increased cell motility. MIEN1 have immunoreceptor tyrosine-based activation motif (ITAM), which regulates filopodia generation, migration, and invasion in breast cancer. Mutation in this motif (Y39F/50F) was unable to regulate filopodia generation, migration, and invasion. These facts show a direct link between a MIEN1 protein in metastasis of cancer cells (Kpetemey et al., 2015).

Despite being widely and abundantly expressed, MIEN1 proteins have not been extensively studied, and their functional contributions are not readily categorized. It may be possible that MIEN1 proteins can directly or indirectly interact with soluble or cellular ligands. How importantly MIEN1 functions, is unclear. Within the context of the sITAM motif, MIEN1 can strongly influence cell adhesion, migration, invasion, and signaling. These functions are all relevant during multiple stages of cancer in different types of cancer development. The development of small molecules specifically to target MIEN1 in cancer has not yet been reported. However, as structural information becomes available, and as relevant molecular interactions are identified, this approach should become more feasible. Keeping this in our mind we model the MIEN1 protein structure and performed in the silico study to find natural inhibitors of the protein. In conclusion, advances in our understanding of MIEN1 biochemistry and MIEN1 tumor biology, together with technological advances in therapeutic targeting approaches, should lead to further in vivo validation of the anticancer benefits of targeting MIEN1 and associated proteins.

We at this moment first time performed in the silico study to find out some methylated flavonoids active against the MIEN1 protein. Protein crystal of the protein has not been reported yet. Thus we performed protein homology modeling of MIEN1 protein. Various offline and online tools were used for docking study (protein, ligand, and structure preparation). The best doc score of three methylated flavonoids molecule 14 (5-hydroxy-7,8-dimethoxy-2-(4-methoxyphenyl)-4H-chromen-4-one), molecule 15 (5-hydroxy-2-(4-hydroxyphenyl)-6,7,8-trimethoxy-4H-chromen-4-one), and molecule 68 (2-(3,4-dimethylphenyl)-5,7-dimethyl-4H-chromen-4-one) are −5.56, −6.84 and −5.14, respectively (Figs. 16.3 and 16.4). Interaction pattern of these methylated flavonoids and the target protein is depicted in Figure 16.4. We also predicted the druglikeness and properties and physiochemical parameters of test methylated flavonoids. The result of these parameters is shown in Table 16.3.

Molecule 14 = 5-hydroxy-7,8-dimethoxy-2-(4-methoxyphenyl)-4H-chromen-4-one; Molecule 15 = 5-hydroxy-2-(4-hydroxyphenyl)-6,7,8-trimethoxy-4H-chromen-4-one and Molecule 68 = 2-(3,4-dimethylphenyl)-5,7-dimethyl-4H-chromen-4-one.

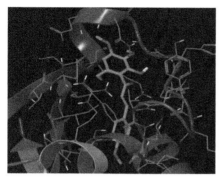

MIEN1_molecule14

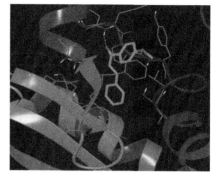

MIEN1_molecule 15

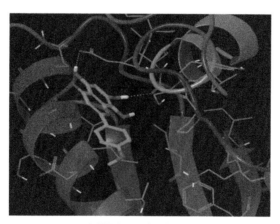

MIEN1_molecule 68

**FIGURE 16.4** **(See color insert.)** Interaction pattern of migration and invasion enhancer 1 protein and lead methylated flavonoids.

**TABLE 16.3** Drug Likeness Properties and Physiochemical Parameters of Test Methylated Flavonoids.

| M | Gscore | Mol_MW | Volume | A | B | C | D | E | F |
|---|---|---|---|---|---|---|---|---|---|
| M1 | −2.23 | 274.235 | 915.826 | 3.956 | 13.946 | 5.077 | −4.959 | 0 | 2.5 |
| M2 | −2.54 | - | - | - | - | - | - | - | - |
| M3 | −2.93 | - | - | - | - | - | - | - | - |
| M5 | −4.85 | 290.188 | 836.272 | −0.971 | 19.416 | 12.427 | −0.766 | 0 | 9 |
| M11 | −3.78 | - | - | - | - | - | - | - | - |
| M12 | −3.96 | - | - | - | - | - | - | - | - |
| M13 | −1.72 | 380.226 | 1105.523 | 2.778 | 14.432 | 7.407 | −2.812 | 0 | 7 |

**TABLE 16.3**    *(Continued)*

| M | Gscore | Mol_MW | Volume | A | B | C | D | E | F |
|---|--------|--------|--------|---|---|---|---|---|---|
| M14 | −5.56 | - | - | - | - | - | - | - | - |
| M15 | −6.84 | 230.222 | 797.12 | 3.308 | 13.412 | 5.251 | −3.529 | 0 | 2.5 |
| M28 | −3.6 | - | - | - | - | - | - | - | - |
| M36 | −1.92 | 356.247 | 1036.522 | 3.524 | 13.309 | 6.426 | −3.721 | 0 | 5 |
| M39 | −0.86 | - | - | - | - | - | - | - | - |
| M40 | −3.36 | 344.236 | 997.878 | 2.227 | 20.022 | 8.469 | −2.877 | 0 | 6.5 |
| M41 | −1.18 | 332.225 | 952.771 | 0.833 | 31.886 | 10.542 | −1.909 | 0 | 8 |
| M45 | −3.49 | 344.236 | 985.05 | 2.215 | 16.438 | 8.385 | −2.647 | 0 | 6.5 |
| M46 | −4.07 | 344.236 | 983.68 | 2.125 | 16.227 | 8.404 | −2.621 | 0 | 6.5 |
| M48 | −3.72 | - | - | - | - | - | - | - | - |
| M49 | −4.07 | - | - | - | - | - | - | - | - |
| M50 | −4.52 | 362.208 | 972.873 | −1.635 | 18.968 | 14.509 | 0.169 | 0 | 12 |
| M51 | −3.13 | - | - | - | - | - | - | - | |
| M52 | −3.6 | - | - | - | - | - | - | - | - |
| M53 | −3.45 | - | - | - | - | - | - | - | - |
| M54 | −3.56 | - | - | - | - | - | - | - | - |
| M55 | −2.79 | 332.225 | 962.246 | 0.869 | 17.684 | 10.583 | −2.017 | 0 | 8 |
| M56 | 1.09 | - | - | - | - | - | - | - | - |
| M57 | −4.08 | - | - | - | - | - | - | - | - |
| M58 | −0.86 | - | - | - | - | - | - | - | - |
| M60 | −4.12 | 302.242 | 909.351 | 3.288 | 14.712 | 6.291 | −3.54 | 0 | 4 |
| M61 | −4.82 | - | - | - | - | - | - | - | - |
| M62 | −3.36 | - | - | - | - | - | - | - | - |
| M63 | −3.95 | 320.214 | 904.52 | −0.486 | 29.636 | 12.42 | −0.826 | 0 | 9.5 |
| M64 | −3.33 | 320.214 | 907.641 | −0.568 | 19.085 | 12.509 | −0.859 | 0 | 9.5 |
| M65 | −3.17 | 316.226 | 931.924 | 0.966 | 15.205 | 10.28 | −2.16 | 0 | 7.5 |
| M66 | −3.31 | 290.231 | 871.204 | 1.9 | 16.442 | 8.39 | −2.561 | 0 | 5.5 |
| M67 | −4.58 | 308.203 | 873.247 | −2.042 | 23.137 | 14.579 | −0.069 | 0 | 11 |
| M68 | −5.14 | 308.203 | 864.455 | −1.984 | 26.595 | 14.498 | 0.062 | 0 | 11 |
| M69 | −4.9 | - | - | - | - | - | - | - | - |

**TABLE 16.3**  *(Continued)*

M=molecule; A= QPlogPo/w; B= QPlogPoct; C= QPlogPw; D= QPlogS; E= donorHB;
F= accptHB Molecule 1=5,7-dihydroxy-2-(3-hydroxy-4-methoxyphenyl)-4H-chromen-4-one; Molecule 2=5,7,8-trihydroxy-2-(4-methoxyphenyl)-4H-chromen-4-one; Molecule 3=2-(3,4-dihydroxyphenyl)-3,5-dihydroxy-7-methoxy-4H-chromen-4-one; Molecule 5=8-hydroxy-7-methoxy-2-(2-methoxyphenyl)-4H-chromen-4-one;Molecule11=5,7-dimethoxy-2-(4-methoxyphenyl)-4H-chromen-4-one; Molecule 12=5-hydroxy-6,7-dimethoxy-2-(4-methoxyphenyl)-4H-chromen-4-one;Molecule13=8-hydroxy-5,7-dimethoxy-2-(4-methoxyphenyl)-4H-chromen-4-one;Molecule14=5-hydroxy-7,8-dimethoxy-2-(4-methoxyphenyl)-4H-chromen-4-one; Molecule 15=5-hydroxy-2-(4-hydroxyphenyl)-6,7,8-trimethoxy-4H-chromen-4-one; Molecule 28=2-(4-hydroxyphenyl)-5,6,7,8-tetramethoxy-4H-chromen-4-one; Molecule 36=5-hydroxy-3,6,7,8-tetramethoxy-2-(4-methoxyphenyl)-4H-chromen-4-one; Molecule 39=2-(4-hydroxy-3-methoxyphenyl)-5,6,7,8-tetramethoxy-4H-chromen-4-one; Molecule 40=2-(3,4-dimethoxyphenyl)-5,6,7,8-tetramethoxy-4H-chromen-4-one; Molecule 41=3-hydroxy-2-(3-hydroxy-4-methoxyphenyl)-5,6,7,8-tetramethoxy-4H-chromen-4-one; Molecule 45=2-(3,4-dimethoxyphenyl)-3,5,7,8-tetramethoxy-4H-chromen-4-one; Molecule 46=2-(3,4-dimethoxyphenyl)-5-hydroxy-3,6,7,8-tetramethoxy-4H-chromen-4-one; Molecule 48=2-(2,5-dimethoxyphenyl)-3,6,7,8-tetramethoxy-4H-chromen-4-one; Molecule 49=2-(3,4-dimethoxyphenyl)-3-hydroxy-5,6,7,8-tetramethoxy-4H-chromen-4-one; Molecule 50=6-hydroxy-3,5,7,8-tetramethoxy-2-(3,4,5-trimethoxyphenyl)-4H-chromen-4-one; Molecule 51=2-(3,4-dimethoxyphenyl)-3,5,6,8-tetramethoxy-4-methylidene-4H-chromene; Molecule 52=2-(3,4-dimethoxyphenyl)-8-hydroxy-3,5,6,7-tetramethoxy-4H-chromen-4-one; Molecule 53=2-(3,4-diethylphenyl)-8-hydroxy-3,5,6,7-tetramethoxy-4H-chromen-4-one; Molecule 54=2-(3,4-diethylphenyl)-8-hydroxy-3,5,6,7-tetramethoxy-4H-chromen-4-one; Molecule 55= (2E)-3-(3,4-dimethoxyphenyl)-1-(6-ethyl-2-hydroxy-3,4-dimethoxyphenyl)prop-2-en-1-one; Molecule 56= (2E)-3-(3,4-dimethoxyphenyl)-1-(2-hydroxy-3,4,5,6-tetramethoxyphenyl) prop-2-en-1-one; Molecule 57=2E)-3-(3,4-dimethoxyphenyl)-1-(6-hydroxy-2,3,4-trimethoxyphenyl) prop-2-en-1-one; Molecule 58= (2E)-1-(4-hydroxy-2,3,6-trimethoxyphenyl)-3-(4-methoxyphenyl) prop-2-en-1-one; Molecule 60=(2S)-2-(3,4-dimethoxyphenyl)-5,6,7,8-tetramethoxy-3,4-dihydro-2H-1-benzopyran-4-one; Molecule 61=2S)-2-(3,4-dimethoxyphenyl)-5,6,7-trimethoxy-3,4-dihydro-2H-1-benzopyran-4-one; Molecule 62=(2S)-2-(3,4-dimethoxyphenyl)-5,7,8-trimethoxy-3,4-dihydro-2H-1-benzopyran-4-one; Molecule 63=(2S)-5,6,7,8-tetramethoxy-2-(4-methoxyphenyl)-3,4-dihydro-2H-1-benzopyran-4-one; Molecule 64=(2S)-2-(3,4-dimethoxyphenyl)-5,6,7-trimethoxy-3,4-dihydro-2H-1-benzopyran-4-one; Molecule 65=(2S)-5,7-dihydroxy-2-(3-hydroxy-4-methoxyphenyl)-3,4-dihydro-2H-1-benzopyran-4-one; Molecule 66=2-(3,4-dimethylphenyl)-5,6,7,8-tetramethyl-4H-chromen-4-one; Molecule 67=2-(3,4-dimethylphenyl)-3,5,7,8-tetramethyl-4H-chromen-4-one; Molecule 68=2-(3,4-dimethylphenyl)-5,7-dimethyl-4H-chromen-4-one; Molecule 69=(2S)-2-(3,4-dimethylphenyl)-5,6,7-trimethyl-3,4-dihydro-2H-1-benzopyran-4-one.

## ACKNOWLEDGMENT

Shashank Kumar acknowledges Central University of Punjab, Bathinda and University Grants Commission, India for providing necessary infrastructure facility and financial support in the form of UGC-BSR Research Start-Up-Grant, GP: 87 [No. F.30–372/2017 (BSR)], respectively. Prem Prakash Kushwaha acknowledges financial support from University Grants Commission, India in the form of CSIR-UGC Junior Research fellowship.

## KEYWORDS

- **flavonoid**
- **metastasis**
- **cancer**
- **inhibitor**
- **methylation**

## REFERENCES

Akowuah, G. A.; Ismail, Z.; Norhayati, I.; Sadikun, A. The Effects of Different Extraction Solvents of Varying Polarities on Polyphenols of Orthosiphon Stamineus and Evaluation of the Free Radical-Scavenging Activity. *Food Chem.* **2005**, *93*(2), 311–317.

Altinier, G.; Sosa, S.; Aquino, R. P.; Mencherini, T.; Loggia, R. Della; Tubaro, A.; Della Loggia, R.; Tubaro, A. Characterization of Topical Anti-Inflammatory Compounds in *Rosmarinus officinalis* L. *J. Agric. Food Chem.* **2007**, *55*(5), 1718–1723.

Birt, D. F.; Hendrich, S.; Wang, W. Dietary Agents in Cancer Prevention: Flavonoids and Isoflavonoids. *Pharmacol Ther.* **2001**, *90*(2–3), 157–177.

Brandt, K.; Christensen, L.; Hansen-Møller, J.; Hansen, S.; Haraldsdottir, J.; Jespersen, L.; Purup, S.; Kharazmi, A.; Barkholt, V.; Frøkiær, H.; Kobæk-Larsen, Health-Promoting Compounds in Vegetables and Fruits: a Systematic Approach for Identifying Plant Components with Impact on Human Health. *Trends Food Sci. Technol.* **2004**, *15*(7–8), 384–393.

Cai, H. The Rice Bran Constituent Tricin Potently Inhibits Cyclooxygenase Enzymes and Interferes with Intestinal Carcinogenesis in Apcmin Mice. *Mol. Cancer Ther.* **2005**, *4*(9), 1287–1292.

Crespy, V.; Williamson, G. A Review of the Health Effects of Green Tea Catechins in Vivo Animal Models. *J. Nutr.* **2004**, *134*(12), 3431S–3440S.

Dao, T. T.; Tung, B. T.; Nguyen, P. H.; Thuong, P. T.; Yoo, S. S.; Kim, E. H.; Kim, S. K.; Oh, W. K. C-Methylated Flavonoids from *Cleistocalyx operculatus* and Their Inhibitory Effects on Novel Influenza a (H1N1) Neuraminidase, *J. Nat. Prod.* **2010**, *73*(10), 1636–1642.

Dasgupta, S.; Wasson, L. M.; Rauniyar, N.; Prokai, L.; Borejdo, J.; Vishwanatha, J. K. Novel Gene C17orf37 in 17q12 Amplicon Promotes Migration and Invasion of Prostate Cancer Cells. *Oncogene.* **2010,** *28*(32), 2860–2872.

Evans, E. E.; Henn, A. D.; Jonason, A.; Paris, M. J.; Schiffhauer, L. M.; Borrello, M. A.; Smith, E. S.; Sahasrabudhe, D. M.; Zauderer, M C35 (C17orf37) is a Novel Tumor Biomarker Abundantly Expressed in Breast Cancer. *Mol. Cancer Ther.* **2006,** *5*(11), 2919–2930.

Gafner, S.; Wolfender, J. L.; Mavi, S.; Hostettmann, K. Antifungal and Antibacterial Chalcones from Myrica Serrata. *Planta Med.* **1996,** *62*(1), 67–69.

Ibrahim, R. K.; Bruneau, A.; Bantignies, B. Plant O-Methyltransferases: Molecular Analysis, Common Signature, and Classification. *Plant Mol. Biol.* **1998,** *36*, 1–10.

Igarashi, K.; Ohmuma, M. Effects of Isorhamnetin, Rhamnetin, and Quercetin on the Concentrations of Cholesterol and Lipoperoxide in the Serum and Liver and on the Blood and Liver Antioxidative Enzyme Activities of Rats. *Biosci. Biotechnol. Biochem.* **1995,** *59*(4), 595–601.

Kameshwaran, S.; Suresh, V. In Vitro and in Vivo Anticancer Activity of Methanolic Extract of *Tecoma stans* Flowers. *Int. Res. J. Pharm.* **2012,** *3* (3), 246–251.

Kauraniemi, P.; Kallioniemi, A. Activation of Multiple Cancer-Associated Genes At the Erbb2 Amplicon in Breast Cancer. *Endocr. Relat. Cancer* **2006,** *13*(1), 39–49.

Kim, D. H.; Bae, E. A.; Han, M. J. Anti-*Helicobacter pylori* Activity of the Metabolites of Poncirin from Poncirus Trifoliate by Human Intestinal Bacteria. *Biol. Pharm. Bull.* **1999,** *22*, 422–424.

Kim, B. G.; Sung, S. H.; Chong, Y.; Lim, Y.; Ahn, J. H. Plant Flavonoid O-Methyltransferases: Substrate Specificity and Application. *J. Plant Biol.* **2010,** *53*(5), 321–329.

Kim, M. J.; Kim, B. G.; Ahn, J. H. Biosynthesis of Bioactive O-Methylated Flavonoids in *Escherichia coli. Appl. Microbiol. Biotechnol.* **2013,** *97*(16), 7195–7204.

Kodama, O.; Miyakawa, J.; Akatsuka, T.; Kiyosawa, S. Sakuranetin, A. Flavanone Phytoalexin Flavanone Phytoalexin from Ultraviolet-Irradiated Rice Leaves. *Phytochemistry* **1992,** *31*(11), 3807–3809.

Koirala, N.; Thuan, N. H.; Ghimire, G. P.; Jung, H. J.; Oh, T.-J.; Sohng, J. K. Metabolic Engineering of *E. coli* for the Production of Isoflavonoid-7-O-Methoxides and Their Biological Activities. *Biotechnol. Appl. Biochem.* **2015,** 1–34. https://doi.org/10.1002/bab.1452.

Koirala, N.; Thuan, N. H.; Ghimire, G. P.; Thang, D. Van; Sohng, J. K. Methylation of Flavonoids: Chemical Structures, Bioactivities, Progress and Perspectives for Biotechnological Production. *Enzyme Microb. Technol.* **2016,** *86*, 103–116.

Kpetemey, M.; Dasgupta, S.; Rajendiran, S.; Das, S.; Gibbs, L. D.; Shetty, P.; Gryczynski, Z.; Vishwanatha, J. K. MIEN1, a Novel Interactor of Annexin A2, Promotes Tumor Cell Migration by Enhancing Anxa2 Cell Surface Expression, *Mol. Cancer* **2015,** *14*(1), 156.

Liu, X.-H.; Liu, H-F.; Chen, J.; Yang, Y.; Song, B-A.; Bai, L-S.; Liu, J-X.; Zhu, H-L.; Qi, X-B. Synthesis and Molecular Docking Studies of Novel 2-Chloro-Pyridine Derivatives Containing Flavone Moieties as Potential Antitumor Agents. *Bioorg. Med. Chem. Lett.* **2010,** *20*(19), 5705–5708.

Lule, S. U.; Xia, W. Food Phenolics, Pros and Cons: a Review. *Food Rev. Int.* **2005,** *21*(4), 367–388.

Luo, H.; Daddysman, M. K.; Rankin, G. O.; Jiang, B-H.; Chen, Y. C. Kaempferol Enhances Cisplatin's Effect on Ovarian Cancer Cells Through Promoting Apoptosis Caused by Down Regulation of Cmyc. *Cancer Cell Int.* **2010,** *10*(1), 16.

Maiti, A.; Cuendet, M.; Kondratyuk, T.; Croy, V. L.; Pezzuto, J. M.; Cushman, M. Synthesis and Cancer Chemopreventive Activity of Zapotin, a Natural Product from Casimiroa Edulis. *J. Med. Chem.* **2007,** *50*(2), 350–355.

Malterud, K. E.; Diep, O. H.; Sund, R. B. C-Methylated Dihydrochalcones from Myrica Gale L: Effects as Antioxidants and as Scavengers of 1,1-Diphenyl-2-Picrylhydrazyl, *Pharmacol. Toxicol.* **1996,** *78*, 111–116.

Manthey, J. A.; Guthrie, N.; Grohmann, K. Biological Properties of Citrus Flavonoids Pertaining to Cancer and Inflammation. *Curr. Med. Chem.* **2001,** *8*(2), 135–153.

Matsuda, H.; Nakashima, S.; Oda, Y.; Nakamura, S.; Yoshikawa, M. Melanogenesis Inhibitors From the Rhizomes of Alpinia of Ficinarum in B16 Melanoma Cells. *Bioorganic Med. Chem.* **2009,** *17*(16), 6048–6053.

Moore, R. J.; Jackson, K. G.; Minihane, A. M. Green Tea (*Camellia sinensis*) Catechins and Vascular Function. *Br. J. Nutr.* **2009,** *102*(12), 1790–1802.

Morley, K. L.; Ferguson, P. J. Koropatnick, Tangeretin and Nobiletin Induce G1 Cell Cycle Arrest But Not Apoptosis in Human Breast and Colon Cancer Cells, *J. Cancer Lett.* **2007,** *251*(1), 168–178.

Murakami, A.; Koshimizu, K.; Ohigashi, H.; Kuwahara, S.; Kuki, W.; Takahashi, Y.; Hosotani, K.; Kawahara, S.; Matsuoka, Y. Characteristic Rat Tissue Accumulation of Nobiletin, A Chemopreventive Polymethoxyflavonoid, In Comparison with Luteolin. *Biofactors* (Oxford Engl.) **2002,** *16*(3–4), 73–82.

Newman, D. J.; Cragg, G. M. Natural Products as Sources of New Drugs Over the Last 25 Years. *J. Nat. Prod.* **2007,** *70*(3), 461–477.

Nichenametla, S. N.; Taruscio, T. G.; Barney, D. L.; Exon, J. H. A Review of the Effects and Mechanisms of Polyphenolics in Cancer. *Crit. Rev. Food Sci. Nutr.* **2006,** *46*(2), 161–183.

Noel, J. P.; Dixon, R. A.; Pichersky, E.; Zubieta, C.; Ferrer, J. L. *Integrative Phytochemistry: from Ethnobotany to Molecular Ecology.* Pergamon, U.S.A **2003,** *37*, 37–58.

Otake, Y.; Hsieh, F.; Walle, T. Glucuronidation Versus Oxidation of the Flavonoid Galangin by Human Liver Microsomes and Hepatocytes. *Drug Metab. Dispos.* **2002,** *30*(5), 576–581.

Piovan, A.; Filippini, R.; Favretto, D. Characterization of the Anthocyanins of Catharanthus Roseus (L.) G. Donin Vivo Andin Vitro by Electrospray Ionization Ion Trap Mass Spectrometry. *Rapid Commun. Mass Spectrom.* **1998,** *12*, 361–367.

Rahman, S.; Salehin, F.; Iqbal, A. Retracted Article: in Vitro Antioxidant and Anticancer Activity of Young Zingiber of ficinale Against Human Breast Carcinoma Cell Lines. *BMC Complement Altern. Med.* **2011,** 11, 76.

Rajendiran, S.; Gibbs, L. D.; Treuren, T. Van; Klinkebiel, D. L.; Vishwanatha, J. K. MIEN1 is Tightly Regulated by SINE Alu Methylation in Its Promoter. *Oncotarget.* **2016,** *7*(40), 65307–65319.

Saewan, N.; Koysomboon, S.; Chantrapromma, K. Anti-Tyrosinase and Anti-Cancer Activities of Flavonoids from Blumea Balsamifera Dc. *J. Med. Plants Res.* **2011,** *5*(6), 1018–1025.

Schubert, H. L.; Blumenthal, R. M.; Cheng, X. Many Paths to Methyl Transfer: A Chronicle of Convergence, *Trends Biochem. Sci.* **2003,** *28*(6), 329–335.

Shi, L.; Feng, X. E.; Cui, J. R.; Fang, L. H.; Du, G. H.; Li, Q. S. Synthesis and Biological Activity of Flavanone Derivatives, *Bioorganic Med. Chem. Lett.* **2010,** *20*(18), 5466–5468.

Steeg, P. S. Targeting Metastasis. *Nat. Rev. Cancer.* **2016,** *16*(4), 201–218.

Sutthivaiyakit, S.; Seeka, C.; Wetprasit, N.; Sutthivaiyakit, P. C-Methylated Flavonoids from *Pisonia grandis* Roots. *Phytochem. Lett.* **2013,** *6*(3), 407–411.

Tapas, A.; Sakarkar, D.; Kakde, R. B. Flavonoids as Nutraceuticals: A Review. *Trop. J. Pharm. Res.* **2008,** *7*(3), 1089–1099.

Tuchinda, P.; Reutrakul, V.; Claeson, P.; Pongprayoon, U.; Sematong, T.; Santisuk, T.; Taylor, W. C. Anti-inflammatory cyclohexenyl chalcone derivativesAnti-Inflammatory Cyclohexenyl Chalcone Derivatives in Boesenbergia Pandurata. *Phytochemistry* **2002,** *59*(2), 169–173.

Wang, M.; Casey, P. J. Protein Prenylation: Unique Fats Make Their Mark on Biology. *Nat. Rev. Mol. Cell Biol.* **2016,** *17*(2), 110–122.

Wen, X.; Walle, T. Methylated Flavonoids Have Greatly Improved Intestinal Absorption and Metabolic Stability. *Drug Metab. Dispos.* **2006a,** *34*(10), 1786–1792.

Wen, X.; Walle, T. Methylation Protects Dietary Flavonoids from Rapid Hepatic Metabolism. *Xenobiotica.* **2006b,** *36*(5), 387–397.

Winkel-Shirley, B. Flavonoid Biosynthesis. A Colorful Model for Genetics, Biochemistry, Cell Biology, and Biotechnology. *Plant Physiol.* **2001,** *126*(2), 485–493.

Wu, J. H.; Wang, X. H.; Yi, Y. H.; Lee, K. H. Anti-Aids Agents 54. A Potent Anti-Hiv Chalcone and Flavonoids from Genus Desmos. *Bioorganic Med. Chem. Lett.* **2003,** *13*(10), 1813–1815.

Zhao, H. B.; Zhang, X. F.; Wang, H. B.; Zhang, M. Z. Migration and Invasion Enhancer 1 (MIEN1) is Overexpressed in Breast Cancer and is A Potential New Therapeutic Molecular Target *Genet. Mol. Res.* **2017,** *16*(1), 1–9.

Zubieta, C.; He, X. Z.; Dixon, R. A.; Noel, J. P. Structures of Two Natural Product Methyltransferases Reveal the Basis for Substrate Specificity in Plant O-Methyltransferases, *Nat. Struct. Biol.* **2001,** *8*(3), 271–279.

# CHAPTER 17

# MEDICINAL ROLES OF PHYTOMOLECULES IN THE TREATMENT AND MANAGEMENT OF DIABETES MELLITUS

MARIA ASLAM*, SIDRA KHALID, and HAFSA KAMRAN

*University Institute of Diet and Nutritional Sciences, University of Lahore, Pakistan, Tel.: +92 3224300729*

*Corresponding author. E-mail: mnarz.aslam@gmail.com
*ORCID: https://orcid.org/0000-0002-5681-4260

## ABSTRACT

Diabetes mellitus (DM) influences over 150 million people globally. It is predicted to be the seventh foremost reason of casualty in 2030, with a projected increase to 300 million sufferers. Phytochemicals, for example, tannins, alkaloids, carbohydrates, terpenoids, steroids, and flavonoids that are found in vegetables, fruits, beverages, and medicinal herbs have a role in the treatment and management of diabetics through blood glucose reduction, homeostasis regulation, and by other mechanisms. This chapter discusses the roles of phytomolecules in the prevention and treatment of DM with emphasis on the various plants reported to possess antidiabetic properties. Other comorbidities (disease linked to diabetes) such as obesity and Alzheimer's diseases were also presented.

## 17.1 INTRODUCTION

Diabetes mellitus (DM) is a prevalent systemic disease affecting a significant section of the populace around the globe. The effects of diabetes are devastating and well distinguished (Rahimi et al., 2005). Type 2 diabetes (T2D) is altogether a complicated, multifactorial disease condition and an uncertain, escalating public-health difficulty (Janero, 2014). It has been projected that the cases of diabetics will raise from 150 million persons to 300 million through 2025–2030 (Figueiredo-González et al., 2016; Sen et al., 2016). DM is one of the key prevailing metabolic diseases caused by elevated blood glucose level (Table 17.1), and an abnormality in insulin discharge, insulin mechanism, or both (Firdous, 2014). DM is among the widespread disease in the endocrine gland system through an escalating prevalence in the human community. Type 1 diabetes is caused by insulin discharge scarcity, whereas T2D is by a progressive rate of insulin resilience in the liver and peripheral tissues. DM brings about acute metabolic after effects involving ketoacidosis, hyperosmolar coma together with chronic diseases and long-standing, unpleasant reactions, for example, renal failure, neuropathy, retinopathy, skin complexities, in addition to growing cardiovascular complication threats. Moreover, ordinary signs of diabetes are repeated urine, overeating, and thirsty (Bahmani et al., 2014).

**TABLE 17.1**   Glucose Concentration Value (mg/dL).

| Condition | Fasting value | 2 h after meals |
| --- | --- | --- |
| Normal | 80–115 | <180 |
| Impaired glucose tolerance | 120–150 | 120–180 |
| Diabetes mellitus | ≥120 | ≥180 |

## 17.2   DIABETES PATHOPHYSIOLOGY

There are increasing concerns that in certain pathologic conditions, the increased production and/or incomplete clearance of reactive oxygen species (ROS) may possibly cause a major problem. Highly ROS are known to chemically alter cellular components leading to lipid peroxidation. The production of ROS and altered antioxidant resistance in diabetic subjects have been stated. It has been observed that increased creation of free radicals and oxidative stress is vital to the development of diabetic complications. This idea has been sustained by expression of elevated altitudes of indicators of oxidative stress in diabetic persons suffering from obstacles. There is significant information on the positive function of antioxidants in diabetes management. The use of antioxidants diminishes oxidative stress and subsides complications due to diabetes (Rahimi et al., 2005).

## 17.3 DIABETES MANAGEMENT

The fundamental and efficient medicines for DM are insulin injection and hypoglycemic factors. However, these composites have numerous unpleasant effects and have no effects on diabetes complications in continuing. Concerning the human improved awareness with reference to DM and its problems, it is essential to get efficient compounds having lower after effects in curing diabetes. Medicinal plants are good sources as replacing or balancing managements for this, along with other diseases. Though varieties of plants have been used all through history to diminish blood glucose and improve diabetes complications, there is not sufficient scientific knowledge regarding few of them (Bahmani et al., 2014).

### 17.3.1 HERBAL DRUGS

The helpful effects of several herbal medicines in lowering blood glucose and its obstacles have been previously documented. A study conducted in Iran revealed that antidiabetic plants (Lamiaceae family, *Citrullus colocynthis*; traditionally being used in treating diabetes) have potential active ingredients to influence diabetes. Iran has a precious established structure of medicine called traditional Persian medicine. In early medieval time, Persian physicians such as Rhazes, Akhawayni, Haly Abbas, Avicenna, Jorjani, and so forth have developed medical sciences, whereas few of their documents like *Canon of Medicine* were included in chief textbooks in eastern and western universities till the 18th century. Twenty-three Persian physicians knew and identified diabetes for centuries and had potentially good medicines (Bahmani et al., 2014)

### 17.3.2 THERAPEUTIC PLANTS

Therapeutic plants are the chief origin of organic compounds, for example, polyphenols, tannins, alkaloids, carbohydrates, terpenoids, steroids, and flavonoids. These natural compounds are the basis for the breakthrough and formation of novel antidiabetic molecules (Firdous, 2014). Recent phytochemicals have evolved from traditional herbs. Phytochemicals focus their mechanism of a remedial feat to the root origin, that is, the incapability to manage the appropriate task of the entire body system (Adimoelja, 2000).

## 17.3.3   DIETARY PATTERN

The Mediterranean diet is attributed among the healthiest diets in the world. This is frequently recognized as a high vegetable, low-saturated fat, and moderate wine consumption. Though, herbs and spices consumed in these diets may also likely to play a beneficial role in the quality of this diet (Bower et al., 2016). Recently, dietary pattern studies have emerged as an alternative and complementary strategy to investigate the association between food and the threat of chronic diseases. Rather than exploring the role of single nutrients or foods, pattern analysis inspects the consequences of food in general. Dietary patterns conceptually represent a wider image of food and nutrient utilization, thus more predictive of disease risk than individual foods or nutrients (Hu, 2002). The association between chief dietary patterns and the threat of type 2 DM in a cohort of women acknowledged two major dietary patterns: "prudent" and "western." The prudent pattern was distinguished by higher consumption of whole grains, fruits, vegetables, legumes, fish, and poultry while the western pattern included higher intakes of refined grains, red and processed meats, sweets and desserts, and french fries. Positive associations were observed between red meat and other processed meats and risk of T2D (Fung et al., 2004).

## 17.4   ROLES OF PHYTOMOLECULES IN DIABETES MANAGEMENT

Phytochemicals, particularly polyphenols in vegetables, fruits, beverages, berries, and herbal medications, might alter glucose and lipid homeostasis, in this manner lowering the effect of the T2D and metabolic syndrome complications (Dembinska-Kiec et al., 2008). As lignin, flavonoids, carotenoids, salicylic acid, and protection response substances (tannins and phytoalexins) are phenolic compounds which play an important role as an antioxidant. In the human body, these substances have been found to be involved in decreasing the occurrence of certain chronic diseases, for example, diabetes, tumors, and heart diseases by lowering oxidative stress. Additionally, food items with low amylase level and high-glucosidase inhibitor like berries prevent initial stages of high blood glucose linked with T2D (Lin et al., 2016). Phenolic acids, proanthocyanidins, and flavonoids are phenolic compounds which play a defensive role in biotic and abiotic stresses, which are commonly spread in plants. The richest resources of dietary polyphenols are herbs, spices, vegetables, fruits, and grains. Oxidative stress is the origin of the majority of the widespread chronic, degenerative diseases. High intake of these foods has been linked to decreasing possibilities of these diseases (Zhang and Tsao, 2016).

Dietary polyphenols might play a role in suppressing α-amylase and α-glucosidase, inhibiting glucose incorporation in the intestine by sodium-reliant glucose transporter 1, regulating insulin discharge and lessening hepatic glucose production. Polyphenols might as well activate 5′ adenosine monophosphate-activated protein kinase, boost insulin-reliant glucose uptake, amend the microbiome, and have anti-inflammatory properties.

### 17.4.1   FLAVONOIDS

For management and prevention of diabetes and its long-standing complications, naturally occurring flavonoids including flavones, flavonols, flavanones, flavonols, isoflavones, and anthocyanidins have been projected to be a valuable addition based on animal and human studies. Enormous studies have revealed that dietary flavonoids could protect humans against diseases like obesity, diabetes, and their complications by regulating glucose metabolism, lipid profile, and hormones and enzymes in the human body (Vinayagam and Xu, 2015).

### 17.4.2   CAROTENOIDS

Available in numerous foods, predominantly fruits, and vegetables, an assembly of fat-soluble pigments; carotenoids have plentiful conjugated double bonds to disrupt the series reaction of lipid oxidation and to quench peroxyl radicals. Against the progression of T2D, the powerful antioxidant capability of carotenoids might offer a defense. Among the chief carotenoids identified in human tissues, lycopene has shown to possess the most powerful antioxidant properties(Wang et al., 2006).

## 17.5   ANTIHYPERGLYCEMIC ROLES OF FOOD-BASED MATERIALS

Dietary materials have remarkable potential as effective antidiabetic agents. The increasing prevalence of diabetes globally demands cost-effective antidiabetic strategies. In different animal models, edible oils and a number of vegetables, fruits, cereals, beverages, legumes, spices, and oilseeds have shown antidiabetic effects (Wang and Zhu, 2016).

## 17.5.1  CEREALS AND GRAINS

Dietary therapy plays important role in the prevention and treatment of diabetes. Cereal grains especially whole grains are known as the staple diet for humans in most of the countries having an antidiabetic role also (Singhal and Kaushik, 2016). Cereals, millets, and beans are main sources of phenolic and other bioactive compounds and they also contain carbohydrates complex food with higher levels of dietary fiber which are health benefits for a human being. Such as they have health-promoting properties in such a way that whole-grain consumption due to the presence of phenolic compounds, help to lower risk of cardiovascular disease (CVD), ischemic, stroke, T2D, metabolic syndrome, and gastrointestinal cancer (Kumar and Kaur, 2017).

## 17.5.2  FRUITS AND VEGETABLES

Dietary flavonoids present as glycosides in fruits and vegetables are considered as bioactive food components potentially capable of various health benefits (Zhu et al., 2017). The fruits' low glycemic indices, strong antioxidant properties, and inhibition of α-amylase and α-glucosidase activities might be probable mechanisms for their usage in the cure and deterrence of DM. Certain fruits and vegetables are functional foods and their ingestion decreases the occurrence of T2D. Hypoglycemic effects of fruits and vegetables (Tables 17.2 and 17.3) may be due to their encouraging nature on pancreatic β-cells for insulin discharge or bioactive compounds such as flavonoids, alkaloids, and anthocyanins, which act as insulin-like molecules or insulin secretagogues (Beidokhti and Jäger, 2017).

### 17.5.2.1  BITTER MELON

Bitter melon, Momordica charantia, is commonly referred to as the bitter gourd, karela, and balsam pear. It is one of the plants that has received the most attention for its antidiabetic properties. Its fruit is also used for the treatment of diabetes and related conditions (Joseph and Jini, 2013). Bitter gourd is a highly nutritious plant as its aril contains high contents of phytochemicals, especially lycopene and β-carotene (Bootprom et al., 2015). Bitter melon has been widely used as a traditional medicine treatment for diabetic

**TABLE 17.2** Phytomolecules in Vegetables.

| S/ No. | Food items | | Phytochemicals | Plant part used | Effects | References |
|---|---|---|---|---|---|---|
| | Common name | Scientific name | | | | |
| 1. | Bitter melon | Momordica charantia L. | Lycopene, β-carotene, isoflavones, terpenes, anthroquinones, glucosinolates | Fruit | Hypoglycemic activity, reduce fructosamine level | Fuangchan et al. (2011) |
| 2. | Okra | Abelmoschus esculentus | Myricetin, catechin, epicatechin (EC), quercetin, rutin | Fruit | Diabetes mellitus-associated; nephropathy, glaucoma, cataract | Prabhune et al. (2017) |
| 3. | Fenugreek seeds | Trigonella foenum-graecum | Galactomannan, 4-Hydroxyisoleucin (4-OH-Ile), diosgenin, trigonelline | Leaves, seeds | β-cells regeneration, increases insulin secretion, decreases insulin resistance and glucose resorption from the gastrointestinal tract | Baliga et al. (2017) |
| 4. | Moringa leaves | Moringa oleifera | Phenolic acids (gallic and chlorogenic acids) and flavonoids (rutin, luteolin, quercetin, apigenin, and kaempferol | Fruit, leaves | Hypoglycemic, antioxidative | Kumar (2017) |
| 5. | Ridyegourd | Luffa acutnagula Linn. | Flavonoids, saponins, luffanguli, sapogenin, oleanolic acid, and cucurbitacin B. | Fruit | Antidiabetic, α-glucosidase inhibitory effect | Pimple et al. (2011) |
| 6. | Celery | Apium graveolens L. Apiaceae family | Phenolic acids | Fruit | Antioxidative | Al-Sa'aidi et al. (2012) |

**TABLE 17.3** Phytomolecules in Fruits.

| Sl No. | Food Items Common name | Scientific name | Phytochemicals | Plant part used | Effects | References |
|---|---|---|---|---|---|---|
| 1. | Apple | *Malus domestica* | Phloretin, quercetin, phloridzin, chlorogenic acid | Fruit | Antioxidative, hinder intestinal glucose absorption, prevent α-amylase and α-glucosidase | Manzano et al. (2016) |
| 2. | Mango | Mangifera indica L. | Mangiferin, gallic acid, gallotannins, quercetin, isoquercetin, ellagic acid, and β-glucogallin | Fruit | Decline in blood glucose level, increased plasma insulin level, decreased levels of fructosamine and glycated hemoglobin | Lauricella et al. (2017) |
| 3. | Guava | Psidium guajava Linn. | Phenolic, flavonoid, carotenoid, terpenoid, and triterpene | Fruit, leaves | Decrease blood glucose, hypoglycemia, hypoinsulinemia | Deguchi and Miyazaki (2010) |
| 4. | Papaya | Carica papaya Linn. | Alkaloids, saponins, glycosides, tannins, and anthraquinones | Fruit | Inhibit pancreatic lipase, antidiabetic | Gironés-Vilaplana et al. (2014) |
| 5. | Kala jamun | *Eugenia jambolana* | Jamboline, jambosine, anthroquinones, tannins, and cardiac glycosides | Fruit, seeds | Hypoglycemic, hypocholesterolemia, antioxidant | Banu and Jyothi (2016) Jagetia (2017) |
| 6. | Wolf fruit | *Solanum lycocarpum* St.-Hil. | Glycoalkaloids, solamargine, and solasonine | Fruit | Increased release of gastrointestinal hormones lowers blood glucose | Dall'Agnol and von Poser (2000) |

patients in Asia. It shows hypoglycemic activity in vitro in both animal and human studies. Bitter melon has a modest hypoglycemic effect and helps to reduced fructosamine levels among patients with T2D (Fuangchan et al., 2011). Medicinal value of bitter melon has also been increased due to its high antioxidant properties (Thakur and Sharma, 2016).

### 17.5.2.2   OKRA

*Abelmoschus esculentus* commonly known as okra or lady finger is popular all over the world as a vegetable due to its nutritional values and health benefits. Traditionally, it has been used as an alternative treatment for diabetes when taken regularly as a part of a diet. Okra has been shown to have a preventive effect in chronic diseases due to antioxidant effect exerted by high contents of natural flavonoids, Myricetin. It can prevent and help manage hyperglycemia when ingested regularly as a dietary supplement (Prabhune et al., 2017). Okra is a widely accepted vegetable in subtropical and tropical regions due to its good palatability and source of dietary medicine.

### 17.5.2.3   MANGO

Mangifera indica L. is a real functional food containing a large variety of antioxidants, pigments, and vitamins that are present in all part of the plant. Mango is a rich source of polyphenols, a diverse group of organic micronutrients found in plants which exert specific fitness benefits for their ability to scavenge free radicals. Polyphenols found in mango include mangiferin, gallic acid, gallotannins, rhamnetin, quercetin, isoquercetin, ellagic acid, β-glucogallin, and kaempferol. Gallic acid has been identified as the major polyphenol present in this fraction. Mango exocarp has been found to be a good source of polyphenols, carotenoids, dietary fibers, and vitamin E. The bioactive compounds in mango have been reported to exert antidiabetic effects (Lauricella et al., 2017).

### 17.5.2.4   JAMUN (BLACK PLUM)

*Eugenia jambolana* (kala jamun) seeds contain a glucoside jamboline and alkaloid jambosine. These two compounds have the ability to check the pathological conversion of starch into sugar. Clinical trials showed the hypoglycemic and hypocholesterolemic effect of *E. jambolana*. It is also found to relieve certain symptoms like polyuria, tiredness, and fatigue (Banu and Jyothi, 2016). Jamun has been found to contain phytochemicals including anthraquinones, alkaloids, catechins, flavonoids, glycosides, steroids, phenols, tannins, saponins, and cardiac glycosides. The diverse activities of jamun may be due to its abilities to scavenge free radicals, increase the antioxidant status of cells by increasing glutathione, glutathione peroxidase, catalase, and/or superoxide dismutase, and to attenuate lipid peroxidation (Jagetia, 2017).

### 17.5.2.5   WOLF FRUIT

In the Brazilian cerrado, a mixture acquired from the fruits of *Solanum lycocarpum* St.-Hil. (Solanaceae), generally identified as "fruta-de-lobo" (wolf fruit), have been broadly enlisted for diabetes management, obesity, and reduction of cholesterol intensities. The remedial composition of the green fruits are crushed in the presence of water and strained. The white "gum" dumped is poured and gradually desiccated, giving a powder which is advertised in capsules by the name of "polvilho-de-lobeira." By phytochemical investigation of this phytomedicine and the fruit of *S. lycocarpum* polysaccharides as the chief constituent has been found. Some polysaccharides delay gastric emptying and work in the endocrine system influencing the release of gastrointestinal hormones, lowering blood glucose levels (Dall'Agnol and von Poser, 2000).

### 17.5.3   SPICES AND HERBS

Spices are bionutrient supplements that boost the taste, flavor, and aroma of food and helps in managing numerous illnesses (Singh et al., 2017). Spices are the usual nutritional condiment that add taste and flavor to the foods. Spices possess some advantageous physiological effects against diabetics (Table 17.4). Among the spices, fenugreek seeds (*Trigonella foenum-graecum*), onion (*Allium cepa*), garlic (*Allium sativum*), turmeric (*Curcuma longa*) have been acknowledged to retain antidiabetic properties. Studies have reported cumin seeds (*Cuminum cyminum*), curry leaves (*Murraya koenigii*), coriander (*Coriandrum sativum*), mustard (*Brassica nigra*), and ginger (*Zingiber officinale*) to possess hypoglycemic potentials (Srinivasan, 2005).

**TABLE 17.4** Phytomolecules in Herbs and Spices.

| S/ No. | Food Items Common name | Scientific name | Phytochemicals | Plant Part Used | Effects | References |
|---|---|---|---|---|---|---|
| 1. | Onion | Allium cepa | Phenolics and flavonoids | Roots | Antidiabetic, hypoglycaemic | Jevas (2011) |
| 2. | Garlic | Allium sativum | Phenolics and flavonoids | Roots | Antidiabetic | Corzo-Martínez et al. (2007) |
| 3. | Ginger | Zingiber officinale | Gingerols, terpenoids, sesquiterpenoids, monoterpenoids; curcumene, and geraniol | Roots | Fasting blood sugar and HbA1c decline, insulin resistance indices, enhances insulin sensitivity and lipid profile | Mozaffari-Khosravi et al. (2014). Arablou et al. (2014) |
| 4. | Cumin seeds | Cuminum cyminum | Cuminaldehyde, cuminol, pinene, oleoresin, cymene, terpinene, and thymol | Seeds | Insulinotropic constituents | Al-Snafi, 2016; Singh et al. (2017) |
| 5. | Coriander | Coriandrum sativum | Alkaloids, flavonoids, terpenoids, sterols, carbohydrates, saponins, and phenolic compounds | Leave, seeds | Antihyperglycemic, insulin-releasing, insulin-like activity | Gray and Flatt, 1999 |
| 6. | Turmeric | Curcuma longa | Curcumin | Roots | Antioxidative, lowers glucose | Zanzer et al. (2017) |
| 7. | Curry leaves | Murraya koenigii | Muconicine, mahanimbidine, isomahanimbine, koenine mahanimbine, koenimbine, koenigine, and koenidine | Leaves | Antioxidative, antidiabetic | Nouman et al. (2015) |
| 8. | Cinnamon | Cinnamtannin trans-cinnamic acid, catechin, procyanidin, and flavones | Stem | Decrease serum HbA1C, blood glucose, low-density lipoprotein cholesterol, total cholesterol, and triglyceride | Cinnamtannin trans-cinnamic acid, catechin, procyanidin, and flavones | Allen et al., (2013) |

**TABLE 17.4** (Continued)

| Sl No. | Food Items Common name | Scientific name | Phytochemicals | Plant Part Used | Effects | References |
|---|---|---|---|---|---|---|
| 9. | Kasni/chicory | *Cichorium intybus* L. | Inulin, sesquiterpene lactones, coumarins, flavonoids, saponins tannins, chicoric acid, and caffeic acid | Leaves | Stimulate insulin secretion, insulin-secreting properties | Katiyar et al. (2015) |
| 10. | Kalongi | *Nigella sativa* | Alkaloids, phenols, phytosterols, saponins, sterols, tannins, flavonoids, and terpenoids | Seeds | Insulinotropic action, antioxidant properties | Bamosa et al. (2010) |
| 11. | Leaves of tomato | *Lycopersicon esculentum* Mill., Solanaceae | Phenolics, alkaloids, carotenoids | Leaves | Antidiabetic | Figueiredo-González et al. (2016) |
| 12. | Olive leaves | *Olea europaea* L. | Flavones, flavonols, catechin, oleuropein, | Fruit, leaves | Antioxidative, glucose-induced insulin release, increases peripheral glucose uptake | Boaz et al. (2011) |
| 13. | Aloe vera | *Aloe vera* L. | Aloin, glucomannans, salicylic acid | Leaves; gel | Hypoglycemic, antioxidative, insulin resistance | John (2017) |
| 14. | Green tea | *Camellia sinensis* | Epigallocatechin gallate, epigallocatechin, EC gallate, and EC | Leaves | Enhances insulin sensitivity, lowers blood glucose level, antioxidative | Salim (2014) |

### 17.5.3.1 ONION AND GARLIC

Onion (*A. cepa*) and garlic (*A. sativum*) have been reported to have effects against CVDs due to their hypocholesterolemic, antihypertensive hypolipidemic, antithrombotic, antidiabetic, and anti-hyperhomocysteinemia effects. They possess biological activities together with antioxidant, antimicrobial, antimutagenic, anticarcinogenic, immunomodulatory, antiasthmatic, and prebiotic activities (Corzo-Martínez et al., 2007). *A. cepa* (onions) have a hypoglycemic and hypolipidemic effect that might exemplify a defensive process compared to the growth of hyperglycemia and hyperlipidemia characteristic of DM (Jevas, 2011).

### 17.5.3.2 GINGER

Ginger plants contain volatile oil 1–4% and about a 100 different compounds. The predominant are the terpenoids and sesquiterpenoids (zingiberine, zingibrol, and bisabolene) and lesser quantities of monoterpenoids (camphene, curcumene, borneol, cineole, and geraniol). The active principles, the gingerols (4–7.5%), are a homologous sequence of phenols. Throughout drying and storage, gingerols are incompletely dehydrated to the corresponding shogaols which might go through further decline to make paradols. Other components are starch (about 50%) and the others include lipids, proteins, and inorganic compounds. Numerous animal and clinical studies recommended that ginger, or its constituents, has hypoglycemic properties and reduces the problems of DM (Salim, 2014). Daily consumption of ginger showed to be valuable for sufferers of type 2 DM because of fasting blood sugar and HbA1c decline and improvement of insulin resistance indices (Mozaffari-Khosravi et al., 2014). Additionally, it enhanced insulin sensitivity and lipid profile. Thus ginger intake may be a preventive measure against DM complications (Arablou et al., 2014).

### 17.5.3.3 TURMERIC

Turmeric (*C. longa*) is a type of herb belonging to the ginger family, which is widely grown in southern and southwestern tropical Asia region. Turmeric, which has an importance place in the cuisines of Iran, Malaysia, India, China, Polynesia, and Thailand is often used as a spice and has an effect on the nature, color, and taste of foods. Turmeric is also known to have been used for centuries in India and China for the medical treatments of diseases such as dermatologic diseases, infection, stress, and depression. Turmeric effects on health are generally centered upon an orange-yellow colored, lipophilic polyphenol substance called "curcumin," which is acquired from the rhizomes of the herb. Curcumin is known recently to have antioxidant, anti-inflammatory, anticancer, and antidiabetic effects (Zanzer et al., 2017).

### 17.5.3.4 CINNAMON

Polyphenols such as cinnamtannin trans-cinnamic acid, catechin, procyanidin, and flavones (cinnamaldehyde and trans-cinnamaldehyde) are present in cinnamon (Kim et al., 2016). In randomized controlled trials, cinnamon has been considered for its glycemic-lowering effects. A noteworthy reduction in intensities of fasting plasma glucose, total cholesterol, triglyceride levels, and low-density lipoprotein (LDL) and a rise in high-density lipoproteins levels is linked by using cinnamon (Allen et al., 2013). Consuming cinnamon is helpful for decreasing serum HbA1C in T2D (Crawford, 2009). Clinical trials revealed that in type 2 diabetic people, the level of blood glucose, LDL cholesterol, total cholesterol, and triglyceride can be decreased by consuming cinnamon daily, so decreasing the chances of CVDs and diabetes (Khan et al., 2003).

### 17.5.3.5 NIGELLA SATIVA

*Nigella sativa* seeds or kalonji are used as an adjuvant treatment in sufferers of DM. Kalonji has been found to decrease HbA1c and insulin resistance and improved β-cell role. The evident antidiabetic effect of *N. sativa* is attributed to its insulinotropic action and the antioxidant properties which decrease oxidative stress and preserve pancreatic β-cell integrity. The glycemic control obtained by *N. sativa* was also attributed to its extrapancreatic actions, mainly the inhibition of hepatic gluconeogenesis (Bamosa et al., 2010).

## 17.6 COMORBIDITIES ASSOCIATED WITH DIABETES

### 17.6.1 DIABESITY

The association of obesity and diabetes, termed "diabesity," defines a combination of primarily metabolic disorders with insulin resistance as the underlying common pathophysiology (Potenza et al., 2017). DM is linked with alterations in the insulin and glucose metabolism in adipose tissues, muscles, and liver which occur by lessening insulin sensitivity. In the development of obesity-linked insulin resistance, oxidative stress and subclinical grade inflammation might play an important part. Thus numerous antioxidant supplied in foods have potentials in the amelioration of obesity-related diseases. Diet plays an important role in the prevention of most cases of diabetes and CVDs and several obesity-linked chronic diseases (van Dam et al., 2002).

## 17.6.2  HEPATIC INSULIN RESISTANCE

Hepatic insulin resistance is linked to T2D and metabolic syndromes which progressively lead towards fibrosis and nonalcoholic fatty liver. Berries, grapes, tea, olives, legumes, and nuts are the richest source of phytochemical and they play an important role in the treatment of liver disorders (Basu et al., 2016).

## 17.6.3  ALZHEIMER'S DISEASE

Diabetes is a significant risk factor for mental illnesses and a few investigations refer to type 3 diabetes, a condition resulting due to insulin resistance in the brain. Alzheimer's disease, the common kind of dementia and diabetes, surprisingly, share fundamental pathological routes, harmony in threat reasons, plus pathways for interference. Antioxidant compounds commonly found in tea such as flavonoids, caffeine, tannins, behenic acid, polyphenols, theobromine, theophylline, anthocyanins, gallic acid, and epigallocatechin-3-gallate decrease the threat of diabetes and Alzheimer's illness. The results of various studies also express that catechins may stop the creation of amyloid-β plaques and boost cerebral functions and might be beneficial in the treatment of Alzheimer's disease or mental disorder. Additionally, other biological active compounds present in tea play an important role in stimulating the neurons within the cell by signal transduction paths and mitochondrial function (Binosha et al., 2017).

## 17.7  CONCLUSION

The increasing prevalence of diabetes demands cost-effective antidiabetic strategies. Phytomolecules, particularly polyphenols in vegetables, fruits, beverages, whole-grain cereals, and herbal medications has proven to be useful in the management of diabetes. Dietary flavonoids present as glycosides in fruits and vegetables are considered as bioactive food components potentially capable of various health benefits. The fruits' low glycemic indices, strong antioxidant properties, and inhibition of α-amylase and α-glucosidase activities might be probable mechanisms for their usage in the cure and prevention of DM.

## KEYWORDS

- diabetes
- diabetes mellitus
- insulin
- phytochemicals
- diseases

## REFERENCES

Adimoelja, A. Phytochemicals and the Breakthrough of Traditional Herbs in the Management of Sexual Dysfunctions. *Int. J. Andrology* **2000**, *23*(S2), 82–84.

Al-Sa'aidi, J. A.; Alrodhan, M. N.; Ismael, A. K. Antioxidant Activity of N-Butanol Extract of Celery (*Apium Graveolens*) Seed in Streptozotocin-Induced Diabetic Male Rats. *Res. Pharm. Biotechnol.* **2012**, *4*(2), 24–29.

Al-Snafi, A. E. The Pharmacological Activities of *Cuminum cyminum* – A Review. *IOSR J. Pharm.* **2016**, *6*(6), 46–65.

Allen, R. W.; Schwartzman, E.; Baker, W. L.; Coleman, C. I.; Phung, O. J. Cinnamon Use in Type 2 Diabetes: An Updated Systematic Review and Meta-Analysis. *Annals Family Med.* **2013**, *11*(5), 452–459.

Arablou, T.; Aryaeian, N.; Valizadeh, M.; Sharifi, F.; Hosseini, A.; Djalali, M. The Effect of Ginger Consumption on Glycemic Status, Lipid Profile and Some Inflammatory Markers in Patients with Type 2 Diabetes Mellitus. *Int. J. Food Sci. Nutr.* **2014**, *65*(4), 515–520.

Bahmani, M.; Zargaran, A.; Rafieian-Kopaei, M.; Saki, K. Ethnobotanical Study of Medicinal Plants Used in the Management of Diabetes Mellitus in the Urmia, Northwest Iran. *Asian Pac. J. Tropical Med.* **2014**, *7*, S348–S354.

Baliga, M.; Palatty, P.; Adnan, M.; Naik, T.; Kamble, P.; et al. Anti-Diabetic Effects of Leaves of *Trigonella foenum-graecum* L. (Fenugreek): Leads from Preclinical Studies. *J. Food Chem. Nanotechnol.* **2017**, *3*(2), 67–71.

Bamosa, A. O.; Kaatabi, H.; Lebda, F. M.; Elq, A. M. A.; Al-Sultan, A. Effect of *Nigella sativa* Seeds on the Glycemic Control of Patients with Type 2 Diabetes Mellitus. *Indian J Physiol Pharmacol.* **2010**, *54*(4), 344-54.

Banu, H.; Jyothi, A. Hypoglycemic and Hypo Cholesterolemic Effect of *Eugenia jambolana* (Kala Jamun) Spicy Mix on Type II Diabetic Subjects. *Imp. J. Interdiscip. Res.* **2016**, *2*(4), 850–857.

Basu, A., Basu, P., Lyons T.J. Hepatic Biomarkers in Diabetes as Modulated by Dietary Phytochemicals. In *Biomarkers in Liver Disease. Biomarkers in Disease: Methods, Discoveries and Applications;* Patel, V., Preedy, V., Eds.; Springer: Dordrecht, 2007, pp 957–975.

Beidokhti, M. N.; Jäger, A. K. Review of Antidiabetic Fruits, Vegetables, Beverages, Oils and Spices Commonly Consumed in the Diet. *J. Ethnopharmacol.* **2017**, *201,* 26–41.

Binosha, F. W.; Somaratne, G.; Williams, S.; Goozee, K. G.; Singh, H.; Martins, R. N. Diabetes and Alzheimer's Disease: Can Tea Phytochemicals Play a Role in Prevention? *J. Alzheimer's Dis.* **2017**, *59*(2), 481–501.

Boaz, M.; Leibovitz, E.; Dayan, Y. B.; Wainstein, J. Functional Foods in the Treatment of Type 2 Diabetes: Olive Leaf Extract, Turmeric and Fenugreek, A Qualitative Review. *Funct. Foods Health Dis.* **2011**, *1*(11), 472–481.

Bootprom, N.; Songsri, P.; Suriharn, B.; Lomthaisong, K.; Lertrat, K. Genetics Diversity Based on Agricultural Traits and Phytochemical Contents in Spiny Bitter Gourd (*Momordica cochinchinensis* (Lour.) Spreng). *SABRAO J. Breed. Genet.* **2015**, *47*(3), 278–290.

Bower, A.; Marquez, S.; de Mejia, E. G. The Health Benefits of Selected Culinary Herbs and Spices Found in the Traditional Mediterranean Diet. *Crit. Rev. Food Sci. Nutr.* **2016**, *56*(16), 2728–2746.

Corzo-Martínez, M.; Corzo, N.; Villamiel, M. Biological Properties of Onions and Garlic. *Trends Food Sci. Technol.* **2007**, *18*(12), 609–625.

Crawford, P. Effectiveness of Cinnamon for Lowering Hemoglobin A1C in Patients with Type 2 Diabetes: A Randomized, Controlled Trial. *J. Am. Board Family Med.* **2009**, *22*(5), 507–512.

Dall'Agnol, R.; von Poser, G. L. The use of Complex Polysaccharides in the Management of Metabolic Diseases: The Case of *Solanum lycocarpum* Fruits. *J. Ethnopharm.* **2000**, *71*(1), 337–341.

Deguchi, Y.; Miyazaki, K. Anti-Hyperglycemic and Anti-Hyperlipidemic Effects of Guava Leaf Extract. *Nutr. Metab.* **2010**, *7*(9), 1–10.

Dembinska-Kiec, A.; Mykkänen, O.; Kiec-Wilk, B.; Mykkänen, H. Antioxidant Phytochemicals against Type 2 Diabetes. *Br. J. Nutr.* **2008**, *99*(E-S1), ES109–ES117.

Figueiredo-González, M.; Valentão, P.; Andrade, P. B. Tomato Plant Leaves: From By-Products to the Management of Enzymes in Chronic Diseases. *Ind. Crops Prod.* **2016**, *94,* 621–629.

Firdous, S. Phytochemicals for Treatment of Diabetes. *EXCLI J.* **2014**, *13,* 451–453.

Fuangchan, A.; Sonthisombat, P.; Seubnukarn, T.; Chanouan, R.; Chotchaisuwat, P.; Sirigulsatien, V.; Ingkaninan, K.; Plianbangchang, P.; Haines, S. T. Hypoglycemic Effect of Bitter Melon Compared with Metformin in Newly Diagnosed Type 2 Diabetes Patients. *J. Ethnopharmacol.* **2011**, *134*(2), 422–428.

Fung, T. T.; Schulze, M.; Manson, J. E.; Willett, W. C.; Hu, F. B. Dietary Patterns, Meat Intake, and the Risk of Type 2 Diabetes in Women. *Arch. Intern. Med.* **2004**, *164*(20), 2235–2240.

Gironés-Vilaplana, A.; Baenas, N.; Villaño, D.; Speisky, H.; García-Viguera, C.; Moreno, D. A. Evaluation of Latin-American Fruits Rich in Phytochemicals with Biological Effects. *J. Funct. Foods* **2014,** *7,* 599–608.

Gray, A. M.; Flatt, P. R. Insulin-Releasing and Insulin-Like Activity of the Traditional Anti-Diabetic Plant *Coriandrum sativum* (Coriander). *Br. J. Nutr.* **1999,** *81*(3), 203–209.

Hu, F. B. Dietary Pattern Analysis: A New Direction in Nutritional Epidemiology. *Curr. Opin. Lipidol.* **2002,** *13*(1), 3–9.

Jagetia, G. C. Phytochemical Composition and Pleotropic Pharmacological Properties of Jamun, *Syzygium Cumini* Skeels. *J. Explor. Res. Pharmacol.* **2017,** *2*(2), 54–66.

Janero, D. R. Relieving the Cardiometabolic Disease Burden: A Perspective on Phytometabolite Functional and Chemical Annotation for Diabetes Management. *Expert Opin Pharmacother.* **2014,** *15*(1), 5–10.

Jevas, C. Anti-Diabetic Effects of *Allium cepa* (Onions) Aqueous Extracts on Alloxan-Induced Diabetic Rattus Novergicus. *J. Med. Plants Res.* **2011,** *5*(7), 1134–1139.

John, J. Evaluation of Hypoglycemic Effect of Aloe Vera on Allaxon Induced Diabetic Rats. **2017.**

Joseph, B.; Jini, D. Antidiabetic Effects of *Momordica Charantia* (Bitter Melon) and its Medicinal Potency. *Asian Pac. J. Trop. Dis.* **2013,** *3*(2), 93–102.

Katiyar, P.; Kumar, A.; Mishra, A. K.; Dixit, R. K.; Kumar, R.; Gupta, A. Kasni (Chicorium Intybus L.) A Propitious Traditional Medicinal Herb. *Int. J. Pharmacogn.* **2015,** *2*(8), 368–380.

Khan, A.; Safdar, M.; Khan, M. M. A.; Khattak, K. N.; Anderson, R. A. Cinnamon Improves Glucose and Lipids of People with Type 2 Diabetes. *Diabetes Care* **2003,** *26*(12), 3215–3218.

Kim, Y.; Keogh, J. B.; Clifton, P. M. Polyphenols and Glycemic Control. *Nutrients* **2016,** *8*(1), E17.

Kumar, H.; Kaur, C. A Comprehensive Evaluation of Total Phenolics, Flavonoids Content and In-Vitro Antioxidant Capacity of Selected 18 Cereal Crops. *Int. J. Pure App. Biosci.* **2017,** *5*(2), 569–574.

Kumar, S. Medicinal Importance of *Moringa Oleifera*: Drumstick Plant. *Indian J. Sci. Res.* **2017,** *16*(1), 129–132.

Lauricella, M.; Emanuele, S.; Calvaruso, G.; Giuliano, M.; D'Anneo, A. Multifaceted Health Benefits of *Mangifera indica* L. (Mango): The Inestimable Value of an Orchard Recently Rooted in Sicilian Rural Areas. *Nutrients.* **2017,** *9*(5), 525.

Lin, D.; Xiao, M.; Zhao, J.; Li, Z.; Xing, B.; Li, X.; Kong, M.; Li, L.; Zhang, Q.; Liu, Y.; Chen, H. An Overview of Plant Phenolic Compounds and their Importance in Human Nutrition and Management of Type 2 Diabetes. *Molecules* **2016,** *21*(10), 1374.

Manzano, M.; Giron, M. D.; Vilchez, J. D.; Sevillano, N.; El-Azem, N.; Rueda, R.; Salto, R.; Lopez-Pedrosa, J. M. Apple Polyphenol Extract Improves Insulin Sensitivity in Vitro and in Vivo in Animal Models of Insulin Resistance. *Nutr. Metab.* **2016,** *13*(1), 32.

Mozaffari-Khosravi, H.; Talaei, B.; Jalali, B. A.; Najarzadeh, A.; Mozayan, M. R. The Effect of Ginger Powder Supplementation on Insulin Resistance and Glycemic Indices in Patients with Type 2 Diabetes: A Randomized, Double-Blind, Placebo-Controlled Trial. *Complementary Ther. Med.* **2014,** *22*(1), 9–16.

Nouman, S. M.; Shehzad, A.; Butt, M. S.; Khan, M. I.; Tanveer, M. Phytochemical Profiling of Curry (*Murraya koenijii*) Leaves and Its Health Benefits. *Pak. J. Food Sci.* **2015,** *25*(4), 204–215.

Pimple, B.; Kadam, P.; Patil, M. Antidiabetic and Antihyperlipidemic Activity of Luffa Acutangula Fruit Extracts in Streptozotocin Induced Niddm Rats. *Asian J. Pharm. Clin. Res.* **2011**, *4*(Suppl 2), 156–163.

Potenza, M. A.; Nacci, C.; De Salvia, M. A.; Sgarra, L.; Collino, M.; Montagnani, M. Targeting Endothelial Metaflammation to Counteract Diabesity Cardiovascular Risk: Current and Perspective Therapeutic Options. *Pharmacol. Res.* **2017**, *120*, 226–241.

Prabhune, A.; Sharma, M.; Ojha, B. *Abelmoschus Esculentus* (Okra) Potential Natural Compound for Prevention and Management of Diabetes and Diabetic Induced Hyperglycemia. *Int. J. Herb. Med.* **2017**, *5*, 65–68.

Rahimi, R.; Nikfar, S.; Larijani, B.; Abdollahi, M. A Review on the Role of Antioxidants in the Management of Diabetes and Its Complications. *Biomed. Pharmacother.* **2005**, *59*(7), 365–373.

Salim, K. S. Hypoglycemic Property of Ginger and Green Tea and their Possible Mechanisms in Diabetes Mellitus. *The Open Conference Proceedings Journal,* **2014**, *5,* 13–19.

Sen, S.; Chakraborty, R.; De, B. *Diabetes Mellitus in 21st Century;* Springer: Berlin, Germany, 2016.

Singh, R. P.; Gangadharappa, H.; Mruthunjaya, K. *Cuminum cyminum* – A Popular Spice: An Updated Review. *Pharmacogn. J.* **2017**, *9*(3), 292–301.

Singhal, P.; Kaushik, G. Therapeutic Effect of Cereal Grains: A Review. *Crit. Rev. Food Sci. Nutr.* **2016**, *56*(5), 748–759.

Srinivasan, K. Plant Foods in the Management of Diabetes Mellitus: Spices as Beneficial Antidiabetic Food Adjuncts. *Int. J. Food Sci. Nutr.* **2005**, *56*(6), 399–414.

Thakur, M.; Sharma, R. Bitter Gourd: Health Properties and Value Addition at Farm Scale. *Marumegh.* **2016**, *1,* 17–21.

van Dam, R. M.; Rimm, E. B.; Willett, W. C.; Stampfer, M. J.; Hu, F. B. Dietary Patterns and Risk for Type 2 Diabetes Mellitus in U. S. Men. *Ann. Intern. Med.* **2002**, *136*(3), 201–209.

Vinayagam, R.; Xu, B. Antidiabetic Properties of Dietary Flavonoids: A Cellular Mechanism Review. *Nutr. Meta.* **2015**, *12,* 60.

Wang, L.; Liu, S.; Pradhan, A. D.; Manson, J. E.; Buring, J. E.; Gaziano, J. M.; Sesso, H. D. Plasma Lycopene, Other Carotenoids, and the Risk of Type 2 Diabetes in Women. *Am. J. Epidemiol.* **2006**, *164*(6), 576–585.

Wang, S.; Zhu, F. Antidiabetic Dietary Materials and Animal Models. *Food Res. Int.* **2016**, *85,* 315–331.

Zhang, H.; Tsao, R. Dietary Polyphenols, Oxidative Stress and Antioxidant and Anti-Inflammatory Effects. *Curr. Opin. Food Sci.* **2016**, *8,* 33–42.

Zhu, X.; Ouyang, W.; Miao, J.; Xiong, P.; Feng, K.; Li, M.; Cao, Y.; Xiao, H., Dietary Avicularin Alleviated Type 2 Diabetes in Mice. *FASEB J.* **2017**, *31*(1 Suppl), 46.7–46.7.

Zanzer, Y. C.; Plaza, M.; Dougkas, A.; Turner, C.; Björck, I.; Östman, E. Polyphenol-Rich Spice-Based Beverages Modulated Postprandial Early Glycaemia, Appetite and PYY after Breakfast Challenge in Healthy Subjects: a Randomized, Single Blind, Crossover Study. *J. Funct. Foods.* **2017**, *35,* 574–583.

## CHAPTER 18

# ROLES OF PHYTOMOLECULES IN THE TREATMENT OF DIABETIC NEPHROPATHY

RASHMI SHUKLA[1,*], PRABHAT UPADHYAY[2], and YAMINI B. TRIPATHI[1]

[1]Department of Medicinal Chemistry, Institute of Medical Sciences, Banaras Hindu University, Varanasi, India

[2]Department of Pharmacology, Institute of Medical Sciences, Banaras Hindu University, Varanasi, India

*Corresponding author. E-mail: rashmishukla561@gmail.com
*ORCID: https://orcid.org/0000-0002-7825-6212

## ABSTRACT

Diabetic nephropathy (DN) is a major microvascular complication of diabetes which leads to end-stage renal disease. It is a progressive disease, assessed by a time-dependent rise in urinary albumin excretion and the decline in renal functions along with extracellular matrix (ECM) accumulation, glomerular basement thickening, glomerulosclerosis, and tubulointerstitial fibrosis. It has multi-etiological pathogenesis such as oxidative stress, inflammation, production of advanced glycation end products and protein kinase C activation, which are attributed to hyperglycemia, hyperlipidemia, and hypertensive state. Herbal medicines are being progressively used all over the world. Bioactive phytomolecules are secondary metabolites having pharmacological effects. The enormous cost and side effects of allopathic drugs have increased the need for newer therapeutic agents that have the potential to prevent the primary mechanisms contributing to the pathogenesis of DN with less cost and minimum toxicity. This chapter explains the role of various herbal phytomolecules in signaling pathways associated with DN.

## 18.1   INTRODUCTION

The prevalence of diabetes is increasing globally as a result of urbanization, human aging, and change in lifestyles. Diabetic nephropathy (DN)
is one of the important microvascular complications in both type I and
type II diabetes and is a leading cause of end-stage renal disease (ESRD)
throughout the world and associated risks of cardiovascular disease
(Kanwar et al., 2008). It is reported that approximately 35% of insulin-
dependent diabetes mellitus (IDDM) and 15% of non-insulin dependent
diabetes mellitus (NIDDM) patients have nephropathy with diabetes. The
prevalence of diabetes is expected to rise from 6 to over 10% in the next
decade and thus DN will become a significant clinical problem associated with increased morbidity and mortality (The Diabetes Control and
Complications Trial Research group, 1993). It is a progressive disease
that takes several years to develop (Bretzel, 1997). It is defined as
diabetes with albuminuria (ratio of urine albumin to creatinine $\geq 30$ mg/g),
impaired glomerular filtration rate (GFR) (<60 mL/min/1.73m$^2$), or
both and is the single strongest predictor of mortality in patients with
diabetes (Palatini, 2012). The activation of the various cellular events
and signaling pathways occur during its progression. Such cellular events
include uncontrolled hyperglycemia intermediaries into various metabolic pathways with activation of protein kinase C (PKC), generation
of advanced glycation products, generation of reactive oxygen species
(ROS), and increased expression of transforming growth factor $\beta$ (TGF-$\beta$)
and guanosine triphosphate-binding proteins. In addition to the deregulation of these biochemical and metabolic pathways, there are changes in
the intra-glomerular hemodynamics, modulated by local activation of the
renin–angiotensin system. In the kidney, various intersecting pathways
occur in its most cell types (Giunti et al., 2006). Several experimental and
clinical researches have been conducted to understand the pathogenesis of
DN, yet little is known about its mechanism. Thus, more knowledge about
the mechanism is necessary to expand new horizon in the development of
therapeutic drugs for the treatment of DN.

## 18.2   ETIOLOGY

Not all of the diabetics develop to DN but those who develop it, the progression varies with time. Various factors appeared in diabetic condition to the

renal damage. Hyperglycemia is the major etiological factors in the patient with DN but smoking and genetic susceptibility also contribute. The major lifestyle risk factors include hypertension, hyperglycemia, and dyslipidemia. Hypertension is also one of the major risk factor involved in the progression of DN, before and after the presence of the microalbuminuric condition. Since the patients with diabetes and a positive family history of hypertension are at higher risk of nephropathy, age, race, and genetic profile are modifiable risk factors.

## 18.3 IDENTIFICATION MARKERS OF DIABETIC NEPHROPATHY (DN)

Clinically, early characterization of DN are hyperfiltration and microalbuminuria, followed by a progressive decline in the GFR, as well as a marked increase in albumin excretion, which over years or decades leads to ESRD in many patients with type I or II diabetes mellitus (Kanwar et al., 2008; Najafian et al., 2011). The early pathophysiological changes of DN are hypertrophy of glomerular and tubular cells associated with thickening of glomerular and tubular basement membranes, mesangial expansion, and tubulointerstitial in compartments accompanied by glomerular podocyte damage and loss, inexorable scarring of the renal glomerulus (Vujičić et al., 2012). The most promising biomarker currently is serum tumor necrosis factor $\alpha$ (TNF-$\alpha$) receptor level, which may predict progression of chronic kidney disease (CKD) and ESRD CKD and ESRD, in type I and type II diabetics. Different tubular and glomerular markers were identified to manifest early DN (Lee and Choi, 2015). Tubular markers are kidney injury molecule-1, secreted in the blood and urine and it is associated with the decline in GFR. Neutrophil gelatinase-associated lipocalin, found in serum/urine, is elevated before the onset of microalbuminuria in urine. Liver-type fatty acids binding protein, found in urine, helps to predict the progression of albuminuria. Cystatin C is found in urine and ratio of C/Creatinine is associated with increased estimated glomerular filtration rate (eGFR). Some glomerular markers include podocalyxin, a major constituent of glycocalyx of podocyte in the glomerulus. Nephrin is a structural component of the podocyte and slit diaphragm and required for maintenance of glomerular permeability and glomerular filtration barrier. Leakage of nephrin protein in urine was found in case of DN.

## 18.4   HISTOPATHOLOGICAL CHANGES IN DN

The histopathological changes found in case of DN are mesangial expansion caused by increased extracellular matrix (ECM) accumulation which is a characteristic feature of DN, glomerular basement thickening (GBM) and enlargement of tubules, Kimmelstiel-Wilson nodules, and diffuse diabetic glomerulosclerosis along with tubular atrophy and interstitial fibrosis are seen in later stages of disease on light microscopy (Fig. 18.1A–D). In literature, it has been found that the electron microscopy demonstrates the glomerular basement membrane thickening and podocyte effacement. Podocyte hypertrophy cause effacement of foot process which leads to excess excretion of urinary albumin (Dronavalli and Duka, 2008; Gaballa and Farag, 2013).

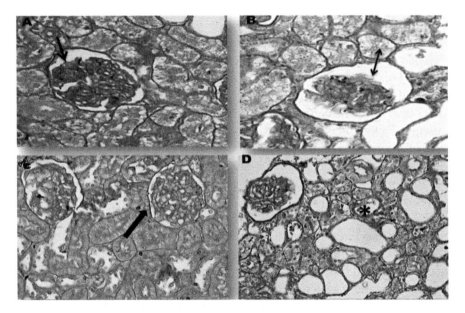

**FIGURE 18.1   (See color insert.)** Histological investigation in a kidney of diabetic nephropathy (DN): The excised kidney was fixed in neutral formalin solution and subsequently processed for histological studies. The thin micro-sections (5 μ) were cut and stained with periodic acid-Schiff reagent and visualized in a Nikon microscope (original magnification ×400). (A) Extracellular matrix accumulation in glomerular of DN (arrow) (B) Thickening of the glomerular basement membrane (both sided arrow) (C) Sclerotic nodule (Kimmelstiel– Wilson nodules) (thick arrow) (D) Marked glomerulosclerosis along with tubular atrophy and interstitial fibrosis (asterisk).

## 18.5   STAGES OF DN

The various stages of DN progression are presented in Figure 18.2.

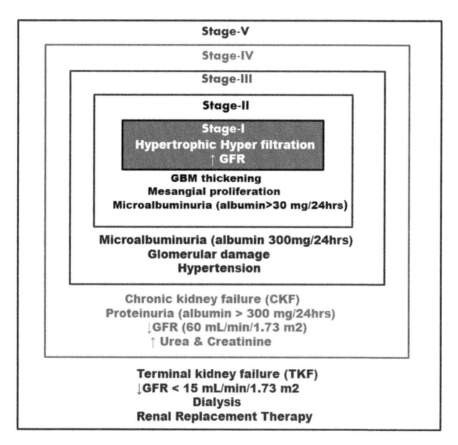

**FIGURE 18.2**   Different stages of DN: DN progresses with the five stages (I–V) to reach end-stage renal disease. GFR – glomerular filtration rate, GBM – glomerular basement thickening, CKF – chronic kidney failure, TFK – terminal kidney failure.

**Stage 1** is characterized by early hypertrophy and hyperfunction, diagnosed before insulin treatment. Increased excretion of urinary albumin and aggravation during physical exercise is also found at this stage. These changes can be easily reversed by treatments.

**Stage 2** is a silent phase which takes many years to develop. It shows morphological lesions without any clinical symptom. A number of patients continue in stage 2 throughout their lives.

**Stage 3** is also called as incipient DN. It is characterized by abnormal excretion of albumin in urine. A slow and gradual progression over the years along with hypertension occurs. The increased rate of albumin excretion is directly proportional to increased blood pressure in a patient of DN.

**Stage 4** is overt DN, chronic kidney failure is the irreversible stage and characterized by persistent proteinuria (greater than 0.5 g/24 h) and declined GFR.

**Stage 5** is an end-stage renal failure with uremia due to DN and terminal kidney failure (GFR < 15 mL/min/1.73 m$^2$). The only option for its treatment is renal replacement therapy (hemodialysis, peritoneal dialysis, kidney transplantation).

## 18.6  POSSIBLE MECHANISM INVOLVED IN DN

The pathogenesis of DN is multifactorial (Cao and Cooper, 2011). The Diabetes Control and Complication Trial Research Group and the United Kingdom Prospective Diabetes Study strongly implicates that hyperglycemia is a crucial factor in the development of the pathogenesis and progression of microvascular complications in both type I and type II diabetes (King et al., 1999). Hyperglycemia acts through a variety of mechanisms to cause long-term diabetic kidney complications (Kanwar et al., 2008). It is associated with an increase in mesangial cell proliferation and hypertrophy as well as increased matrix production and basement membrane thickening (Giunti et al., 2006; Najafian et al., 2011; Lim, 2014). It leads to end-stage renal damage in diabetes through both metabolic and hemodynamics pathways including (1) polyol pathway (increased intracellular sorbitol and fructose) (2) auto-oxidative glycosylation and nonenzymatic glycation, (3) formation of advanced glycation end products (AGEs), (4) oxidative stress, (5) activation of PKC, (6) intracellular renin–angiotensin system activation, and nitric oxide release. All these pathways operate simultaneously or independently ultimately leading to renal failure (Kanwar et al., 2008; Vujičić et al., 2012) as shown in (Fig. 18.3). A hemodynamic factor involves the deregulation of the renin–angiotensin system. The metabolic and hemodynamic factors interact with each other through different molecular and signaling pathways such as PKC with associated ROS generation and nuclear factor kappa-light-chain-enhancer of activated B cells (NF-kB) (Cooper, 2001). These changes contribute to pathological damage to glomerulus, podocyte, and tubulointerstitium which leads to renal damage.

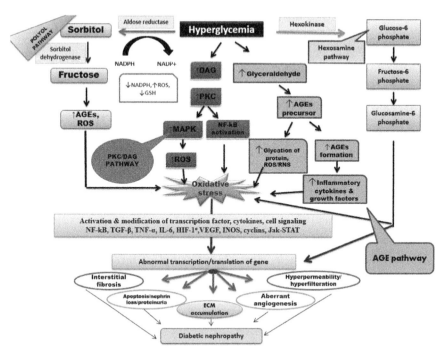

**FIGURE 18.3** Different mechanisms involved in progression of DN. DAG – diacylglycerol, PKC – protein kinase C, MAPK – mitogen-activated protein kinase, NF-kB – nuclear factor kappa light chain enhancer of activated B-cells, ROS – reactive oxygen species, AGE – advance glycation end product, TGF-β – transforming growth factor β, TNF-α – tumor necrosis factor α, IL-6 – interleukin-6, HIF-1α – hypoxia-inducing factor 1α, VEGF – vascular endothelial growth factor, iNOS – inducible nitric oxide synthase, ECM – extracellular matrix.

## 18.7 METABOLIC PATHWAY

Metabolic factors involve glucose-dependent pathways, such as polyol pathway, AGEs, and their receptors. Hyperglycemia induces vascular injury through complex overlapping pathways, including the formation of AGEs, activation of PKC and generation of ROS. The excess generation of ROS which leads to oxidative stress has been implicated in the pathogenesis of DN.

### 18.7.1 POLYOL PATHWAY

The polyol pathway is up-regulated as a result of increased continuous and persistent hyperglycemic condition. In general, glucose is converted into sorbitol via aldose reductase enzyme NADPH-dependent enzyme, in which sorbitol is then converted into fructose (Brownlee, 2001). The reduction in the conversion of glucose to sorbitol results in decreased level of intracellular NADPH which acts as a cofactor for regeneration of an antioxidant, reduced glutathione (GSH). The decreased levels of GSH are thought to contribute to increased intracellular oxidative stress which in turn causes increased cell stress and apoptosis. An increased ratio of $NADH:NAD^+$, increases the formation of diacylglycerol (DAG) and methylglyoxal which are precursors of PKC and AGE pathways.

## 18.7.2   ADVANCED GLYCATION END PRODUCTS PATHWAY

Chronic hyperglycemia leads to accumulation of AGEs (Singh et al., 2014). AGEs are the outcome of irreversible glycation of proteins that occur in the presence of the intracellular hyperglycemic condition. Increase in the uptake of glucose to the proximal tubular cells via sodium glucose cotransporter 2 (SGLT2) stimulates free radical generation and receptor for advanced glycation end products (RAGE) expression; These AGEs can modulate various intracellular events, such as the activation of PKC, mitogen-activated protein kinase, and transcription factors such as NF-κB (Lim, 2014). These events, in turn, regulate the expression of diverse growth factors and cytokines such as TGF-β, which inevitably influence the synthesis of ECM proteins.

## 18.7.3   HEXOAMINASE PATHWAY

This pathway originates from third step of glycolysis in which fructose-6-phosphate is converted to glucosamine-6-phosphate by the enzyme glutamine fructose-6-phosphate amidotransferase. The glucosamine-6-phosphate is then used as a substrate to increase transcription of inflammatory cytokines like TNF-α and TGF-β. An increased concentration of TGF-β levels is known to promote renal cell hypertrophy and increase mesangial matrix components which are characteristics of DN, whereas TNF-α is an inflammatory cytokine which significantly is involved in the progression of DN (Dronavalli and Duka, 2008; Cao and Cooper, 2011).

## 18.7.4   PROTEIN KINASE C PATHWAY

PKC is a multifunctional, serine/threonine-related protein kinase that involved in many cellular functions and affects many signal transduction pathways (Newton, 2003). Increased concentration of DAG in turn activates the PKC. The activation of PKCs is considered as an important signaling molecule in diabetic complication. Many stimuli responsible for its activation are hyperglycemia, excess ROS generation, and various growth factors (Maurya and Vinayak, 2015). Several studies indicated that the activation of PKC is one of the important mediators for the DN as it is attributed to an abnormality in vascular and cellular processes causing vascular permeability, endothelial dysfunction, angiogenesis, ECM accumulation, basement thickening and alteration in various enzymatic activities, and transcription factors. It also mediates the expression vascular endothelial growth factor, TGF-β, NF-kB, and connective tissue growth factor (CTGF) as well as the production of fibronectin, type IV and type VI collagen which contributes to GBM thickening and ECM accumulation (Craven *et al.*, 1995)

## 18.7.5   OXIDATIVE STRESS: CONNECTING LINK

Excess generation of ROS leads to oxidative stress and a decreased antioxidant defense systems (Rahal et al., 2014). The oxidative stress is considered as a major risk factor in the progression of DN which couples hyperglycemia and vascular complications via different mechanisms such as the alterations in the renal hemodynamics and metabolic modifications of target tissue molecules. In the clinical and experimental diabetes, the enhanced ROS production has been linked to vasoconstriction, endothelial dysfunction, modification of ECM proteins with progression to fibrosis and ESRD (Lee *et al.*, 2003; Selim et al., 2017). Studies suggested that ROS production increase in podocytes induces glomerular filtration barrier dysfunction and increases albuminuria (Agnieszka et al., 2011). The excess generation of ROS promotes podocyte apoptosis in DN (Wagener et al., 2009). The prevention of the generation of ROS may be a promising therapeutic to ameliorating renal damage from DN.

## 18.8   MANAGEMENT AND TREATMENT OF DN

### 18.8.1   ALLOPATHIC APPROACH FOR DN

The main therapeutic targets of drugs are in glycemic control and renin–angiotensin–aldosterone system (RAAS) inhibition with angiotensin-converting enzyme inhibitors and angiotensin receptor blockers, anti-inflammatory, PKC inhibitors, phosphodiesterase inhibitors, and SGLT2 inhibitors (Table 18.1). These drugs work by different mechanisms. Despite the tremendous effect of these drugs on the pathways that contribute to DN, clinicians are still a long way away from having a new drug in their prescribing arsenal because of reported side effects of drugs (Phillips et al., 2007).

### 18.8.2   HERBAL APPROACH TO TREAT DN

The established drugs are already in clinical use and their therapeutic and side effects are well studied, so only phase three clinical trials would be enough to establish the newer claim of old drugs (Bailey and Day, 1989). The herbal medicines could also contribute significantly to this mission because of their long history of clinical use in the traditional system of medicine. Once validated, they can also be used as a food supplement in the early stage of the disease to prevent its further propagation. To achieve this goal, the biochemical mechanism behind its therapeutic claims and associated safety and toxicity measures must also be studied and documented. World Health Organization (WHO) also currently recommending and promoting the traditional/herbal remedies in national health care systems because such drugs are easily available at low cost, are comparatively safe and the people have faith in such remedies (Oluyemisi et al., 2012). According to WHO, the estimated world market for herbal products is 62 billion US dollars which is projected to grow to 5 trillion US dollar by 2050 (Express, 2004).

Some of the active constituents from plant product have been identified to cure DN which is shown in Table 18.2. The WHO endorses the evaluation of the potential of plants as effective therapeutic agents, especially in areas where there is a lack of safe modern drugs. Clinical use of such phytochemicals would also reduce the financial burden of health care in any country, which has high projected value.

**TABLE 18.1** Showing The Different Class of Compounds Which are Currently Used to Prevent The Progression of DN.

| Category | Drugs | Mechanism of action | Effects | References |
|---|---|---|---|---|
| Angiotensin-converting enzyme inhibitor | Ramipril, captopril, perindopril | Produce vasodilation by inhibiting the formation of angiotensin II | Reduce the level of albuminuria | Langham et al. (2002) |
| Angiotensin II receptor blockers | Losartan, telmisartan, valsartan | Block the action of angiotensin II by preventing angiotensin II from binding to angiotensin II receptors | Decreased the accumulation of extracellular matrix (ECM) | Wencheng et al. (2012) |
| Anti-inflammatory | Pioglitazone thiazolidinedione | Decrease expression of transforming growth factor β (TGF-β), type IV collagen, and ICAM-1 | Reduced albuminuria and glomerular hypertrophy | Jia et al. (2014) |
| β-blockers | Carvedilol, atenolol, and metoprolol | Lowering arterial pressure | Reduce hypertension | Bakris (2003) |
| Aldol reductase inhibitors | Alrestatin, epalrestat | Reduces sorbitol concentration through polyol pathway | Reduced mesangial expansion | Chung and Chung (2005) |
| Protein kinase C (PKC) inhibitor | Ruboxistaurin | Inhibit activation of PKC-β, TGF-β | Decrease in albuminuria while maintaining a stable eGFR and urinary TGF-α level | AL-Onazi et al. (2016) |
| Phosphodiesterase inhibitor | Cilostazol, pentoxifylline | Decreased expression of tumor necrosis factor α (TNF-α), interleukin (IL)-1, IL-6, and interferon | Decreased glomerular basement thickening thickening, podocyte flattening, loss of fenestration in the endothelial cell layer, and albuminuria | He and Cooper (2014) |
| Sodium-glucose cotransporter 2 inhibitors | Dapagliflozin, gliflozins, canagliflozin | Inhibit the reabsorption of glucose in the proximal tubule, reduce HbA1c levels by 0.5–1% | Reduced hyperglycemia, albuminuria, expression of inflammatory cytokines and oxidative stress, glomerular mesangial expansion, and interstitial fibrosis | Paola et al. (2016) |

**TABLE 18.2** Showing The Active Constituents Present in Plant and Their Mechanism of Action in The Prevention of Progression of DN.

| Plant name/family | Active constituent | Class | Mechanism of actions | References |
|---|---|---|---|---|
| Averrhoe carambola/ Oxalidaceae | 2-Dodecyl-6-methoxy-cyclohexa-2,5-diene-1,4-dione | Quinone | ↓Hyperglycemia, ↓ Nuclear factor kappa-light-chain-enhancer of activated B cells (NF-κB), advanced glycation end products (AGE), TGF-β1 expression ↑antioxidant enzyme, ↓ECM accumulation | Zheng et al. (2013) |
| Anacardium occidentale/ Anacardiaceae | 2-Hydroxy- 6-pentadecyl-benzoic acid | Phenolic derivative | Attenuates renal function | Tedong et al. (2004) |
| Andrographis paniculata/ Acanthaceae | Andrographolide | Diterpenoid | Attenuated hyperglycemia-induced oxidative stress via AKT/NF-kB pathway | Lan et al. (2013) |
| Acorus tatarinowii/ Araceae | Acortatarin A | Spiroalkaloid | ↓ ECM accumulation | Zhao et al. (2013) |
| Astragalus membranaceus/Fabaceae | Astragalosides IV | Saponin | Restores mitochondrial quality control network | Liu et al. (2017) |
| Brachystem-ma calycinum/ Caryophyllaceae | Duanbanhuain | Cyclic peptides | ↓ MCP-1, IL-6, collagen IV and reactive oxygen species (ROS) production in mesangial cell | Cheng et al. (2011) |
| Benincasa cerifera/ Cucurbitaceae |  |  | ↓ ROS production, ↑ concentration of antioxidant enzyme | Bhalodia et al. (2009) |
| Camellia sinensis/ Theaceae | Catechins | Polyphenolic | Trap methylglyoxal, ↓ AGEs formation, IL-1β, TNF-α | Zhu et al. (2014) |
| Cinnamomum zeylanicum/Lauraceae | Cinnamon oil |  | ↓ Mesangial expansion, hyline cast, and tubular dilation. | Mishra et al. (2010) |
| Cinnamomum cassia/ Lauraceae | Cinnamoids | Sesquiterpenoid | Inhibit fibronectin, MCP-1, IL-6 | Yan et al. (2015) |

**TABLE 18.2** (Continued)

| Plant name/family | Active constituent | Class | Mechanism of actions | References |
|---|---|---|---|---|
| Curcuma longa/ Zingiberaceae | Curcumin | Polyphenol | ↓PKC activation, AGEs formation, NF-kB, ↑ renal AMPK, Nrf2, and so forth (Soetikno et al., 2013) | Soetikno et al. (2013) |
| Cassia tora (seeds)/ Caesalpiniaceae | Naphthopyrone | Glucosides | ↓ AGEs formation | Lee et al. (2006) |
| Dioscorea hypoglaucae/ Dioscoreaceae | Diosgenin | Phytosteroid sapogenin | ↓ Oxidative stress, advanced glycation end product & TGF-β | Guo et al. (2016) |
| Ganoderma lucidum/ Ganodermataceae | Polysaccharide peptide | | ↓ Oxidative stress | He et al. (2006) |
| Ginkgo biloba/ Ginkgoaceae | Ginkgolides, bilobalide, ascorbic acid, catechin | Flavonol and flavone glycosides, lactone derivatives | Suppress AKT/mTOR signaling | Lu et al. (2015) |
| Glycine max/Fabaceae | β-Conglycinin, genistein, | Isoflavones | Restore expression of nephrin | Yang et al. (2014) |
| Euryale ferox/ Euryalaceae | Buddlenol E | Triterpenoic | ↓ ROS production in mesangial cells | Song et al. (2011) |
| Linum usitatissimum/ Linaceae | Secoisolariciresinol | Diglycoside | ↓ AGEs production & oxidative stress | Sherif (2014) |
| Rosa laevigata/Rosaceae | Rosalaevin A | Lignan | ↓ Expression of NF-kB & MCP-1 | He et al. (2012) |
| Panax quinquefolius/ Araliaceae | Ginsenosides | Steroid glycosides, and triterpene saponins | ↓TNF-α, TGF-β, MCP-1 | Zhang et al. (2009); Ma et al. (2010) |

**TABLE 18.2** (Continued)

| Plant name/family | Active constituent | Class | Mechanism of actions | References |
|---|---|---|---|---|
| Polygonatum odoratum/ Asparagaceae | 3-(4'-Hydroxybenzyl)-5,7- Dihydroxy-6-methyl-8-methoxychroman-4-one) | Homoisoflavanones | ↓ AGEs formation | Dong et al. (2010) |
| Pterocarpus santalinus/ Pterocarpus | Santalin, pterolinus L | Anthracene | ↓Oxidative stress | Guo et al. (2016) |
| Rheum officinale/ Polygonaceae | Rhein | Anthraquinone | ↓Proinflammatory cytokines, NF-kB, AP-1, Vascular cell adhesion protein 1 (VCAM-1), inhibit epithelial mesenchymal transition (EMT) progression. | Zeng et al. (2014) |
| Salvia miltiorrhiza/ Lamiaceae | Tanshinone IIA | Quinone | ↓ Renal hypertrophy, TGF-β, monocyte/ macrophage (ED-1) | Kim et al. (2009) |
| Terminalia chebula/ Combretaceae | Chebulic acid | Tannin | Increase G6PDH, glutathione, superoxide dismutase, glomerular filtration rate | Silawat, N. and Gupta, V. B. (2014) |
| Vitis vinifera/Vitaceae | Resveratrol | Polyphenol | ↓ Vascular endothelial growth factor (VEGF), hypoxia-inducing factor-1α (HIF-1α) | Wen et al. (2013) |
| Murraya paniculata/ Rutaceae | 5,7,3',4'-Tetramethoxyflavone and 5,7,3',4',5'-pentamethoxyflavone | Total flavonoid | ↓ TGF-β, CTGF | Zou et al. (2014) |
| Pueraria tuberosa/ Fabaceae | Puerarin, genistein, tuberosin | Flavonoids | ↓ HIF-1α, VEGF, ECM, ↑MMP-9 | Tripathi et al. (2016); Shukla et al. (2017) |
| Tribulus terrestrial/ Zygophyllaceae | Protodioscin | Saponin | ↓TGF-β, TNF- α, VEGF, serum cystatin-c and β 2 microglobulin | Gandhi (2012) |

**TABLE 18.2** *(Continued)*

| Plant name/family | Active constituent | Class | Mechanism of actions | References |
|---|---|---|---|---|
| Salvia tomentosa/ Lamiaceae | Luteolin | Flavonoids | Increasing HO-1 expression and elevating antioxidant | Wang et al. (2011) |
| Smilax glabraRoxb/ Lamiaceae | Astilbin | Flavanonol | ↓ CTGF, TGF-β1 | Li et al. (2009) |
| Cornus officinalis/ Cornaceae | Loganin | Glycosides | ↓Collagen IV, fibronectin, IL-6 | Ma et al. (2014) |
| Herba epimedii/ Berberidaceae | Icariin | Flavonoid | ↓ TGF-β1, collagen IV | Qi et al. (2011) |
| Silybum marianum/ Asteraceae | Silibinin | Flavanone | ↓Oxidative stress | Jain and Somani (2015) |

## KEYWORDS

- diabetic nephropathy
- diabetes
- phytomolecules
- herbal medicine
- renal disease

## REFERENCES

Agnieszka, P.; Dorota, R.; Irena, A.; Maciej, J.; Stefan, A. High Glucose Concentration Affects the Oxidant-Antioxidant Balance in Cultured Mouse Podocytes. *J. Cell. Biochem.* **2011,** *112*(6), 1661–1672.

AL-Onazi, A. S.; AL-Rasheed, N. M.; Attia, H. A.; AL-Rasheed, N. M.; Ahmed, R. M.; AL-Amin, M. A.; Poizat, C. Ruboxistaurin Attenuates Diabetic Nephropathy via Modulation of TGF-β1/Smad and GRAP Pathways. *J. Pharm. Pharmacol.* **2016,** *68*(2), 219–232.

Bailey, C. J.; Day, C. Traditional Plant Medicines as Treatments for Diabetes. *Diabetes Care* **1989,** *12*(8), 553–564.

Bakris, G. Role for β-Blockers in the Management of Diabetic Kidney Disease. *Am. J. Hypertens.* **2003,** *16*(9), 7–12.

Bhalodia, Y.; Kanzariya, N.; Patel, R.; Patel, N.; Vaghasiya, J.; Jivani, N.; Raval, H. Renoprotective Activity of Benincasa Cerifera Fruit Extract on Ischemia/Reperfusion-Induced Renal Damage in Rat. *Iran. J. Kidney Dis. IJKD* **2009,** *33*(2), 80–85.

Bretzel, R. G. Prevention and Slowing down the Progression of the Diabetic Nephropathy through Antihypertensive Therapy. *J. Diabetes Complications* **1997,** *11*(2), 112–122.

Brownlee, M. Biochemistry and Molecular Cell Biology of Diabetic Complications. *Nature* **2001,** *414*(6865), 813–820.

Cao, Z.; Cooper, M. E. Pathogenesis of Diabetic Nephropathy. *J. Diabetes Investig.* **2011,** *2*(4), 243–247.

Cheng, Y. X.; Li-Li, Z.; Yong-Ming, Y.; Ke-Xin, C.; Fan-Fan, H. Diabetic Nephropathy-Related Active Cyclic Peptides from the Roots of Brachystemma Calycinum. Bioorg. Med. Chem. Lett. **2011,** 21(24), 7434–7439.

Chung, S. S.; Chung, S. K. Aldose Reductase in Diabetic Microvascular Complications. *Curr. Drug Targets* **2005,** *6*(4), 475–486.

Cooper, M. E. Interaction of Metabolic and Haemodynamic Factors in Mediating Experimental Diabetic Nephropathy. *Diabetologia* **2001,** *44*(11), 1957–1972.

Craven, P. A.; Studer, R. K.; Negrete, H.; DeRubertis, F. R. Protein Kinase C in Diabetic Nephropathy. *J. Diabetes Complications* **1995,** *9*(4), 241–245.

Dong, W.; Shi, H. B.; Ma, H.; Miao, Y. B.; Liu, T. J.; Wang, W. Homoisoflavanones from Polygonatum Odoratum Rhizomes Inhibit Advanced Glycation End Product Formation. *Arch. Pharm. Res.* **2010,** *33*(5), 669–674.

Dronavalli, S.; Duka, I. The Pathogenesis of Diabetic Nephropathy. *Nat. Clin. Pract. Endocrinol. Metab.* **2008**, *4*(8), 1–13.

Gaballa, M.; Farag, Y. Predictors of Diabetic Nephropathy. *Open Med.* **2013**, *8*(3), 287–296.

Gandhi, S. Effect of Tribulus Terrestris in Diabetic Nephropathy in STZ Induced NIDDM in Rats, 2012. http://shodhganga.inflibnet.ac.in/bitstream/10603/31457/14/14_part%203.pdf (accessed Jan 10, 2018).

Giunti, S.; Barit, D.; Cooper, M. E. Mechanisms of Diabetic Nephropathy. *Hypertension* **2006**, *48*(4), 519–526.

Guo, C.; Ding, G.; Huang, W.; Wang, Z.; Meng, Z.; Xiao, W. Total Saponin of Dioscoreae Hypoglaucae Rhizoma Ameliorates Streptozotocin-Induced Diabetic Nephropathy. *Drug Des. Devel. Ther.* **2016**, *10*, 799–810.

He, C-Y.; Li, W-D.; Guo, S-X.; Lin, S-Q.; Lin, Z-B. Effect of Polysaccharides from Ganoderma Lucidum on Streptozotocin-Induced Diabetic Nephropathy in Mice. *J. Asian Nat. Prod. Res.* **2006**, *8*(8), 705–711.

He, G.; Liao, Q.; Luo, Y.; Qing, Z.; Zhang, Q.; He, G. Renal Protective Effect of Rosa Laevigata Michx. by the Inhibition of Oxidative Stress in Streptozotocin-Induced Diabetic Rats. *Mol. Med. Rep.* **2012**, *5*(6), 1548–1554.

He, T.; Cooper, M. E. Diabetic Nephropathy: Renoprotective Effects of Pentoxifylline in the PREDIAN Trial. *Nat. Rev. Nephrol.* **2014**, *10*(10), 547–548.

Jain, D. P.; Somani, R. S. Silibinin, A Bioactive Flavanone, Prevents the Progression of Early Diabetic Nephropathy in Experimental Type – 2 Diabetic Rats. *Int. J. Green Pharm.* **2015**, *9*(2), 118–124.

Jia, Z.; Sun, Y.; Yang, G.; Zhang, A.; Huang, S.; Heiney, K. M.; Zhang, Y. New Insights into the PPAR γ Agonists for the Treatment of Diabetic Nephropathy. *PPAR Res.* **2014**, Article ID 818530, 1–17. http://dx.doi.org/10.1155/2014/818530.

Kanwar, Y.; Wada, J.; Sun, L.; Xie, P.; Wallner, E.; Chen, S.; Chugh, S.; Danesh, F. Diabetic Nephropathy: Mechanisms of Renal Disease Progression. *Exp. Biol. Med.* **2008**, *233*(1), 4–11.

Kim, S. K.; Jung, K-H.; Lee, B-C. Protective Effect of Tanshinone IIA on the Early Stage of Experimental Diabetic Nephropathy. *Biol. Pharm. Bull.* **2009**, *32*(2), 220–224.

King, P.; Peacock, I.; Donnelly, R. The UK Prospective Diabetes Study(UKPDS): Clinical and Therapeutic Implications for Type 2 Diabetes. *Br. J. Clin. Pharmacol.* **1999**, *48*(5), 643–648.

Lan, T.; Wu, T.; Gou, H.; Zhang, Q.; Li, J.; Qi, C.; He, X.; Wu, P.; Wang, L. Andrographolide Suppresses High Glucose-Induced Fibronectin Expression in Mesangial Cells via Inhibiting the AP-1 Pathway. *J. Cell. Biochem.* **2013**, *114*(11), 2562–2568.

Langham, R.; Kelly, D.; Cox, A.; Thomson, N.; Holth, H.; Zaoui, P.; Pinel, N.; Cordonnier, D.; Gilbert, R. Proteinuria and the Expression of the Podocyte Slit Diaphragm Protein, Nephrin, in Diabetic Nephropathy: Effects of Angiotensin Converting Enzyme Inhibition. *Diabetologia* **2002**, *45*(11), 1572–1576.

Lee, H. B.; Ha, H.; King, G. L. *Reactive Oxygen Species and Diabetic Nephropathy*. Proceedings of the Hyonam Kidney Laboratory, Soon Chun Hyang University International Diabetes Symposium. Seoul, Korea, January 18–19, 2003. *J. Am. Soc. Nephrol.* **2003**, *14*(8 Suppl 3), S209-96.

Lee, G. Y.; Jang, D. S.; Lee, Y. M.; Kim, J. M.; Kim, J. S. Naphthopyrone Glucosides from the Seeds ofCassia Tora with Inhibitory Activity on Advanced Glycation End Products (AGEs) Formation. *Arch. Pharm. Res.* **2006**, *29*(7), 587.

Lee, S-Y.; Choi, M. E. Urinary Biomarkers for Early Diabetic Nephropathy: Beyond Albuminuria. *Pediatr. Nephrol.* **2015**, *30*(7), 1063–1075.

Li, G-S.; Jiang, W-L.; Yue, X-D.; Qu, G-W.; Tian, J-W.; Wu, J.; Fu, F-H. Effect of Astilbin on Experimental Diabetic Nephropathy in Vivo and in Vitro. *Planta Med.* **2009**, *75*(14), 1470–1475.

Lim, A. K. Diabetic Nephropathy – Complications and Treatment. *Int. J. Nephrol. Renovasc. Dis.* **2014**, *7*, 361–381.

Liu, X.; Wang, W.; Song, G.; Wei, X.; Zeng, Y.; Han, P.; Wang, D.; Shao, M.; Wu, J.; Sun, H.; *et al.* Astragaloside IV Ameliorates Diabetic Nephropathy by Modulating the Mitochondrial Quality Control Network. *PLoS One* **2017**, *12*(8), e0182558.

Lu, Q.; Zuo, W-Z.; Ji, X-J.; Zhou, Y-X.; Liu, Y-Q.; Yao, X-Q.; Zhou, X-Y.; Liu, Y-W.; Zhang, F.; Yin, X-X. Ethanolic Ginkgo Biloba Leaf Extract Prevents Renal Fibrosis through Akt/mTOR Signaling in Diabetic Nephropathy. *Phytomedicine* **2015**, *22*(12), 1071–1078.

Ma, X.; Xie, X.; Zuo, C.; Fan, J. Effects of Ginsenoside Rg1 on Streptozocin-Induced Diabetic Nephropathy in Rats. *J. Biomed. Eng.* **2010**, *27*(2), 342–347.

Ma, W.; Wang, K-J.; Cheng, C-S.; Yan, G.; Lu, W-L.; Ge, J-F.; Cheng, Y-X.; Li, N. Bioactive Compounds from Cornus Officinalis Fruits and Their Effects on Diabetic Nephropathy. *J. Ethnopharmacol.* **2014**, *153*(3), 840–845.

Maurya, A. K.; Vinayak, M. Modulation of PKC Signaling and Induction of Apoptosis through Suppression of Reactive Oxygen Species and Tumor Necrosis Factor Receptor 1 (TNFR1): Key Role of Quercetin in Cancer Prevention. *Tumour Biol.* **2015**, *36*(11), 8913–8924.

Mishra, A.; Bhatti, R.; Singh, A.; Singh Ishar, M. Ameliorative Effect of the Cinnamon Oil from Cinnamomum Zeylanicum upon Early Stage Diabetic Nephropathy. *Planta Med.* **2010**, *76*(5), 412–417.

Najafian, B.; Alpers, C. E.; Fogo, A. B. Pathology of Human Diabetic Nephropathy. *Contrib. Nephrol.* **2011**, *170*, 36–47.

Newton, A. C. Regulation of the ABC Kinases by Phosphorylation: Protein Kinase C as a Paradigm. *Biochem. J.* **2003**, *370*(2), 361–371.

Oluyemisi, F.; Henry, O.; Ochogu, P. Standardization of Herbal Medicines - A Review. *Int. J. Biodivers. Conserv.* **2012**, *4*(3), 101–112.

Palatini, P. Glomerular Hyperfiltration: A Marker of Early Renal Damage in Pre-Diabetes and Pre-Hypertension. *Nephrol. Dial. Transplant.* **2012**, *27*(5), 1708–1714.

Paola, F.; Alberto, Z.; Marco, R.; Luca, B.; Roberto, V. SGLT2 Inhibitors and the Diabetic Kidney. *Diabetes Care* **2016**, *39*, 165–171.

Phillips, C. O.; Kashani, A.; Ko, D. K.; Francis, G.; Krumholz, H. M. Adverse Effects of Combination Angiotensin II Receptor Blockers Plus Angiotensin-Converting Enzyme Inhibitors for Left Ventricular Dysfunction. *Arch. Intern. Med.* **2007**, *167*(18), 1930.

Qi, M-Y.; Kai-Chen; Liu, H-R.; Su, Y.; Yu, S-Q. Protective Effect of Icariin on the Early Stage of Experimental Diabetic Nephropathy Induced by Streptozotocin via Modulating Transforming Growth Factor β1 and Type IV Collagen Expression in Rats. *J. Ethnopharmacol.* **2011**, *138*(3), 731–736.

Rahal, A.; Kumar, A.; Singh, V.; Yadav, B.; Tiwari, R.; Chakraborty, S.; Dhama, K. Oxidative Stress, Prooxidants, and Antioxidants: The Interplay. *Biomed Res. Int.* **2014**, *2014*, 761264.

Selim, F.; Wael, A.; KeithE, J. Diabetes-Induced Reactive Oxygen Species: Mechanism of Their Generation and Role in Renal Injury. *J. Diabetes Res.* **2017**, *2017*, 1–30.

Sherif, I. O. Secoisolariciresinol Diglucoside in High-Fat Diet and Streptozotocin-Induced Diabetic Nephropathy in Rats: A Possible Renoprotective Effect. *J. Physiol. Biochem.* **2014**, *70*(4), 961–969.

Shukla, R.; Pandey, N.; Banerjee, S.; Tripathi, Y. B. Effect of Extract of Pueraria Tuberosa on Expression of Hypoxia Inducible Factor-1α and Vascular Endothelial Growth Factor in Kidney of Diabetic Rats. *Biomed. Pharmacother.* **2017,** *93,* 276–285.

Singh, V. P.; Bali, A.; Singh, N.; Jaggi, A. S. Advanced Glycation End Products and Diabetic Complications. *Korean J. Physiol. Pharmacol.* **2014,** *18*(1), 1.

Soetikno, V.; Suzuki, K.; Veeraveedu, P. T.; Arumugam, S.; Lakshmanan, A. P.; Sone, H.; Watanabe, K. Molecular Understanding of Curcumin in Diabetic Nephropathy. *Drug Discov. Today* **2013,** *18*(15–16), 756–763.

Song, C-W.; Wang, S-M.; Zhou, L-L.; Hou, F-F.; Wang, K-J.; Han, Q-B.; Li, N.; Cheng, Y-X. Isolation and Identification of Compounds Responsible for Antioxidant Capacity of Euryale Ferox Seeds. *J. Agric. Food Chem.* **2011,** *59*(4), 1199–1204.

Silawat, N.; Gupta, V. B. Chebulic Acid, Tannin from Terminalia chebula, Attenuates Diabetic Nephropathy in Rats. *Ethnopharmacology,* **2014,** *4,*100–108.

Tedong, L.; Dimo, T.; Desire, P.; Dzeufiet, D.; Asongalem, A. E.; Sokeng, D. S.; Callard, P.; Flejou, J–F.; Kamtchouing, P. Antihyperglycemic and Renal Protective Activities of *Anacardium Occidentale* (Anacardiaceae) Leaves in Streptozotocin Induced Diabetic Rats. *African J. Tradit. Complement. Altern. Med.* **2004,** *3,* 23–35.

The Diabetes Control and Complications Trial Research group. The Effect of Intensive Treatment of Diabetes on the Development and Progression of Long-Term Complications in Insulin-Dependent Diabetes Mellitus. *N. Engl. J. Med.* **1993,** *329*(14), 977–986.

Tripathi, Y. B.; Shukla, R.; Pandey, N.; Pandey, V.; Kumar, M. An Extract of Pueraria Tuberosa Tubers Attenuates Diabetic Nephropathy by Upregulating Matrix Metalloproteinase-9 Expression in the Kidney of Diabetic Rats. *J. Diabetes* **2016,** *9,* 123–132.

Vujičić, B.; Turk, T.; Crnčević-Orlić, Ž.; Đorđević, G.; Rački, S. Diabetic Nephropathy. In *Pathophysiology and Complications of Diabetes Mellitus*; 2012; pp 71–96.

Wagener, F. A.; Dekker, D.; Berden, J. H.; Scharstuhl, A.; van der Vlag, J. The Role of Reactive Oxygen Species in Apoptosis of the Diabetic Kidney. *Apoptosis* **2009,** *14*(12), 1451–1458.

Wang, G. G.; Lu, X. H.; Li, W.; Zhao, X.; Zhang, C. Protective Effects of Luteolin on Diabetic Nephropathy in STZ-Induced Diabetic Rats. *Evidence-Based Complement. Altern. Med.* **2011,** *2011,* 1–7.

Wen, D.; Huang, X.; Zhang, M.; Zhang, L.; Chen, J.; Gu, Y.; Hao, C-M. Resveratrol Attenuates Diabetic Nephropathy via Modulating Angiogenesis. *PLoS One* **2013,** *8*(12), e82336.

Wencheng, F.; Yunman, W.; Zhouhui, J.; Wang, H.; Cheng, W.; Zhou, H.; Yin, P.; Peng, W. Losartan Alleviates Renal Fibrosis by down-Regulating HIF-1α and up-Regulating MMP-9/TIMP-1 in Rats with 5/6 Nephrectomy. *Ren. Fail.* **2012,** *34*(10), 1297–1304.

Yan, Y-M.; Fang, P.; Yang, M-T.; Li, N.; Lu, Q.; Cheng, Y-X. Anti-Diabetic Nephropathy Compounds from Cinnamomum Cassia. *J. Ethnopharmacol.* **2015,** *165,* 141–147.

Yang, H-Y.; Wu, L-Y.; Yeh, W-J.; Chen, J-R. Beneficial Effects of β-Conglycinin on Renal Function and Nephrin Expression in Early Streptozotocin-Induced Diabetic Nephropathy Rats. *Br. J. Nutr.* **2014,** *111*(1), 78–85.

Zeng, C-C.; Liu, X.; Chen, G-R.; Wu, Q-J.; Liu, W-W.; Luo, H-Y.; Cheng, J-G. The Molecular Mechanism of Rhein in Diabetic Nephropathy. *Evid. Based. Complement. Alternat. Med.* **2014,** *2014,* 487097.

Zhang, L.; Xie, X.; Zuo, C.; Fan, J. Effect of Ginsenoside Rgl on the Expression of TNF-Alpha and MCP-1 in Rats with Diabetic Nephropathy. *J. Sichuan Univ. Med. Sci. Ed.* **2009,** *40*(3), 466–471.

Zhao, Z.-F.; Zhou, L-L.; Chen, X.; Cheng, Y-X.; Hou, F-F.; Nie, J. Acortatarin A Inhibits High Glucose-Induced Extracellular Matrix Production in Mesangial Cells. *Chin. Med. J. (Engl).* **2013,** *126*(7), 1230–1235.

Zheng, N.; Lin, X.; Wen, Q.; Kintoko; Zhang, S.; Huang, J.; Xu, X.; Huang, R. Effect of 2-Dodecyl-6-Methoxycyclohexa-2,5-Diene-1,4-Dione, Isolated from Averrhoa Carambola L. (Oxalidaceae) Roots, on Advanced Glycation End-Product-Mediated Renal Injury in Type 2 Diabetic KKAy Mice. *Toxicol. Lett.* **2013,** *219*(1), 77–84.

Zhu, D.; Wang, L.; Zhou, Q.; Yan, S.; Li, Z.; Sheng, J.; Zhang, W. (+)-Catechin Ameliorates Diabetic Nephropathy by Trapping Methylglyoxal in Type 2 Diabetic Mice. *Mol. Nutr. Food Res.* **2014,** *58*(12), 2249–2260.

Zou, J.; Yu, X.; Qu, S.; Li, X.; Jin, Y.; Sui, D. Protective Effect of Total Flavonoids Extracted from the Leaves of Murraya Paniculata (L.) Jack on Diabetic Nephropathy in Rats. *Food Chem. Toxicol.* **2014,** *64*, 231–237.

# CHAPTER 19

# PHYTOCHEMICALS AS PROTAGONIST FOR THE TREATMENT AND MANAGEMENT OF AUTOIMMUNE DISEASES

PRAGYA MISHRA[1,*], PARJANYA KUMAR SHUKLA[2], and RAGHVENDRA RAMAN MISHRA[3]

[1]Centre of Food Technology, University of Allahabad, Allahabad 211002, India, Mob.: +91 9452096368

[2]Department of Pharmaceutical Sciences, Faculty of Health Science, Sam Higginbottom Institute of Agriculture Technology and Sciences – Deemed University, Allahabad, India

[3]Medical Laboratory Technology, DDU Kaushal Kendra, Banaras Hindu University, Varanasi 221005, Uttar Pradesh, India

*Corresponding author. E-mail: 23raksha@gmail.com
*ORCID: https://orcid.org/0000-0003-1612-1424

## ABSTRACT

The basis of good health is connected to a well-functioning immune system. An overactive immune response as a consequence of immune dysfunction that results in a disease condition is called autoimmune disease. Autoimmune disease is generally associated with factors such as genetics, infections, and/or environment causing a condition which is triggered by the immune system initiating an attack on self-molecules due to the deterioration of immunologic tolerance to auto reactive immune cells. Phytochemicals are good sources which can influence our response to drugs as many of them function as antioxidants, influence hormonal function, have antibacterial, antidiuretic, memory enhancing, adaptogenic property, anticancer, antiviral abilities with potential immunostimulating and immunomodulatory activity. This chapter is a study on the role of phytochemicals for the management and treatment of autoimmune diseases.

## 19.1    INTRODUCTION

Immunity conveyed by the body's immune system has been considered as a sophisticated defense system. The sign of good health is noticeable when there is the occurrence of balanced biological defenses against the various types of infections, diseases, or invasion of foreign particles that may be unwanted biological material, however, possessing tolerance to avoid allergy as well as autoimmune diseases (Venkatalakshmi et al., 2016). The autoimmune disease has proven to be an overwhelming clinical challenge to medical practitioners as well as to the complementary health practitioners. Over the last decades, the continual toxic exposure of drugs, chemicals, or environmental pollutants (collectively referred to as xenobiotics) due to increased urbanization, industrialization, and unhealthy eating habits which results in cellular imbalance are followed by immune toxic alterations that have a major negative impact on the host defense mechanisms. This may further lead to a low or altered immunity against invading pathogens or neoplasia alongside dysregulated immune response which ultimately causes allergy, hypersensitivity, and autoimmune reactions. The main disorders related to immunity are allergy, infectious diseases, cancers, and autoimmune diseases (Venkatalakshmi et al., 2016). Generally, the occurrence of autoimmune diseases are being relatively uncommon (prevalence of autoimmunity is approximately 3–5% in the general population), but due to their significant ill effects on mortality and morbidity, are of matter of great concern (Wang et al., 2015).

The immune system is of importance and is assured to include specific self-components due to a random expression of a specific action of the B- and T-cell populations. As a consequence, the human body must establish self-tolerance mechanisms to distinguish between self and nonself molecules, which the immune system employs to avoid autoreactivity. However, no mechanism has 100% efficiency and is at a risk of breakdown. The self-recognition mechanisms are no exception, and a number of diseases have been identified in which there is autoimmunity, due to mimicking copies of autoantibodies and autoreactive T cells. The main cause of autoimmune diseases is tolerance to self-antigens and consequently due to failure in immune recognition of self and injury of self-tissues (autoimmunity) which may be acquired by various mechanisms. It was also reported that an individual may suffer from more than one autoimmune disease at the same time. A diverse array of synthetic, natural, and recombinant agents (levamisole, isoprinosine, pentoxifylline, and thalidomide) is available. However, these therapeutic agents should be specific for treating patients with autoimmunity that will completely reverse, if not cure, the disease. The conventional method for the treatment of autoimmunity usually is based on the same principle that may be an immune response to self-antigens. This can be achieved by immunosuppression, immunostimulation, removal of the thymus gland, plasmapheresis, T-cell vaccination, block major histocompatibility complex with similar peptide, and so forth.

In earlier studies, supplementation with phytochemical-rich foods emerged as one of the most accepted preventative strategies to promote health endogenous defenses to fight against disease conditions. Previously, it was suggested that nutritional intervention by investigating the role of bioactive dietary molecules such as phytochemicals and antioxidants along with dietary control will be able to influence cell metabolism and sequentially help to improve the overall wellness of autoimmune patients. The potential role of phytochemicals as an immunoadjuvant, immunosuppressant, immunostimulant against both types of autoimmune diseases (Systemic and organ-specific) is given in Figure 19.1. Although the use of phytochemicals as nutraceutical intervention and its potential role on account of the molecular basis for prevention of autoimmune diseases are of crucial importance, it is projected to be a major area for future research.

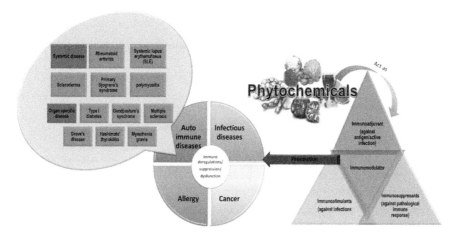

**FIGURE 19.1**   Role of phytochemicals in the prevention of autoimmune diseases.

## 19.2   AUTOIMMUNITY: AN OVERVIEW

Autoimmunity is a kind of immune dysfunction or dysregulation of immune processes and pathways that are involved in normal immunity due to the overreaction of immune responses in which immune system fails to recognize and distinguish between self and nonself molecules. The main factors affecting the immune system are age, sex, genetic variability, drug/alcohol consumption, malnutrition, environmental pollution, and an unhealthy lifestyle with poor eating habits. There are two major areas related to the autoimmunity dysfunction or dysregulation; first is the immune deficiency syndromes (inability of one or more components of the immune system to respond in a protective fashion to a pathogen) and second is the autoimmune diseases (failure to distinguish among self from nonself antigen more often leading to immune intolerance) (Wang et al., 2015). There are mainly two types of autoimmune diseases which are given below:

### 19.2.1   SYSTEMIC AUTOIMMUNE DISEASES

These kinds of diseases are associated with autoantibodies to antigens, which are generally not tissue specific meaning that the incriminating antigens and the autoimmunity are distributed in many tissues. Below is a list of different systemic autoimmune diseases with associated tissue.

1.   Rheumatoid arthritis – joints
2.   Systemic lupus erythematosus – kidney
3.   Scleroderma – skin
4.   Primary Sjogrens's syndrome
5.   Polymyositis – muscles

### 19.2.2   ORGAN-SPECIFIC OR LOCALIZED AUTOIMMUNE DISEASES

As the name implies, organ-specific or localized autoimmune diseases occur in a particular organ. In an organ-specific autoimmune disease, the invading antigens and the autoimmunity are restricted to specific organs in the body. Below is a list of different organ-specific autoimmune diseases with the associated organ given.

1. Type I diabetes (T1D) – pancreas
2. Good pasture's syndrome
3. Multiple sclerosis – brain
4. Grave's disease
5. Hashimoto thyroiditis – thyroid
6. Myasthenia gravis

Among all autoimmune diseases, the most common autoimmune conditions are thyroid disease and T1D (Wang et al., 2015). A short enumeration for causes of autoimmunity includes lymphocytes abnormalities and polyclonal lymphocytes activation, changes in lymphocytes causing cytokine dysregulation, failure of central tolerance and peripheral tolerance, antigen modified/generated by molecular changes, alteration in the aging process, molecular mimicry, infection, and genetic factors. Eventually, to avoid the occurrence of autoimmune conditions, the concept of immune tolerance came into existence which may be defined as an ability of the immune system to prevent itself from targeting self-molecules, cells, or tissues (Wang et al., 2014). To achieve the balanced immune system homeostasis, thymus and bone marrow plays a crucial role in immune tolerance.

## 19.3   IMMUNE TOLERANCE/IMMUNOMODULATION

To prevent an autoimmune condition, there are a variety of specific and non-specific approaches which may include synthetic and natural agents that have potential stimulatory, suppressive, and regulatory activity. Any changes in the immune response which may involve induction, expression, amplification, or inhibition of any part or phase in the immune response has been termed as immunomodulation. The control of autoimmune disease by immunologic means due to immunomodulation comprising mainly two objectives; first is the development and maintenance of immunity and other is the prevention of undesired immune reactions. Plant as a source of food has positive impacts on human health which stake on harmony of a wide range of metabolites derived from the plant prospectively for introducing various novel phytochemicals with therapeutic potentials such as pharmaceuticals, multicomponent drugs, nutritional supplements and functional foods, and so forth, (Mishra and Prasad, 2016). Phytochemicals literally mean plant chemicals inclusive of plant secondary metabolites (bioactive components) such as alkaloids, tannins, flavonoids, and phenolic compounds that may aid the maintenance of health status of organisms, but are not essential nutrients.

Phytochemicals of nutraceuticals importance are bioactive constituents that may sustain or promote health that occur at the intersection of food and pharmaceutical industries. The complex nature of plant secondary metabolism and diverse array of secondary metabolites demands a holistic approach to understand the combined biochemical, molecular, cellular, and physiological perspective to order to appreciate the true nature of their occurrence and significance. The nutritional intervention by investigating the role of bioactive dietary molecules such as phytochemicals and antioxidants along with dietary control will be able to influence cell metabolism and sequentially help to improve the overall wellness of autoimmune patients. Additionally, the importance of secondary metabolites is increasing rapidly as these compounds are utilized as raw materials for various industries and as therapeutic agents (Shah et al., 2010). Generally, phytochemicals are categorized as phytoestrogens, terpenoids, carotenoids, limonoids, phytosterols, glucosinolates, polyphenols, flavonoids, isoflavonoids, and anthocyanidins as given in Table 19.1.

All these phytochemicals have a broad spectrum of potential pharmacological properties with special reference to the human health such as anti-inflammatory, antiallergic, antioxidants, antibacterial, antifungal, antispasmodic, chemopreventive, hepatoprotective, hypolipidemic, neuroprotective, hypotensive, antiaging, diabetes, osteoporosis, DNA damage, cancer and heart diseases, induce apoptosis, diuretic, central nervous system stimulant, analgesic, inhibition of ultraviolet B-induced carcinogenesis, immunomodulator, and carminative (Prakash et al., 2012).

## 19.4   NOVEL APPROACHES FOR TREATMENT AND PREVENTION

Among all immunity dysfunction, the autoimmunity has emerged as the major challenge for all clinicians due to their negative impact on morbidity and mortality. Wang et al. (2015) also conscripted the more commonly used therapeutic agents along with their side effects. There has always been the search for novel nutraceuticals of plant origin and phytochemicals in different food (complex matrix) which may evolve as a vital compound with suitable dietary intake for prevention of autoimmune disease. It was also suggested that the objective of treating patients with autoimmunity should be target-specific agent that will completely have no side or ill effect if unable in curing the disease. Over the last five decades, antibiotics have been applied for the treatment of autoimmune diseases (Rosman et al., 2014). Further, due to the fact that autoimmunity is a very complicated condition as

**TABLE 19.1** List of Major Phytochemicals, Food Sources, Main Constituents, and Health Benefits.

| Major phytochemicals | Chemical structure | Food sources | Representative major constituents | Health benefits | References |
|---|---|---|---|---|---|
| Carotenoids | | Red, orange, and green fruits and vegetables including broccoli, carrots, cooked tomatoes, leafy greens, sweet potatoes, winter squash, apricots, cantaloupe, oranges, and watermelon and umbelliferous vegetables (carrots, celery, cilantro, parsley, parsnips) | β-carotene, lycopene, lutein, zeaxanthin | Inhibition of cancer cell growth, also act as antioxidants and helps to improve immune response | Prakash et al. (2012) |
| Flavonoids (anthocyanidins) | | Blueberries, blackberries, cranberry, raspberry, blackcurrant, black grape, strawberries, cherries, plums, pomegranate juice, red wine | Cyanidin 3-glycosides, malvidin, delphimidin, pelargonidin | Inhibition of inflammation and tumor growth; help to boost immunity and also increases production of detoxifying enzymes | Prakash et al. (2012) |

**TABLE 19.1** *(Continued)*

| Major phytochemicals | Chemical structure | Food sources | Representative major constituents | Health benefits | References |
|---|---|---|---|---|---|
| Flavones | | Celery hearts, celery, olives, peppers, fresh parsley, dry parsley, oregano, rosemary, thyme | Apigenin, chrysin, luteolin | Antidiabetic, anti-inflammatory, antioxidative, hepatoprotective, antiviral, antianxiety, antitumor, and antihypertensive effects | Prakash et al. (2012) |
| Flavanones | | Orange, orange juice, lemon, grapefruit, tangerine juice | Hesperitin, erodictyol, naringenin | Antidiabetic, anti-inflammatory, anti-oxidative, antitumor, and anticancerous activity | Prakash et al. (2012) |

**TABLE 19.1** *(Continued)*

| Major phytochemicals | Chemical structure | Food sources | Representative major constituents | Health benefits | References |
|---|---|---|---|---|---|
| Flavanols | | Grapes, citrus fruits, and their juices, tangerine juice, peppermint | Morin, procyanidins, prodelphinidins, catechin, epicatechin, and their gallates | Antidiabetic, anti-inflammatory, α-fetoprotein level decreased, and anticancerous activity | Prakash et al. (2012) |
| Anthoxanthins (flavonols) | | Apricots, apples, grapes, peaches, pears, plums, raisins, berries, cherries, red wine, tea, chocolate | Myricetin, fisetin, quercetin, kaempferol, isorhamnetin | Antioxidant protectants for human beings and play beneficial role in reducing the risk of coronary heart disease, diabetes, hypertension, and some types of cancer | Prakash et al. (2012) |

**TABLE 19.1** (Continued)

| Major phytochemicals | Chemical structure | Food sources | Representative major constituents | Health benefits | References |
|---|---|---|---|---|---|
| Isoflavones | | Cherry, tomatoes, spinach, celery, onions, peppers, sweet potato, lettuce, broccoli, kale, buckwheat, beans, apples, apricots, grapes, plums, berries, currants, cherries, juices, ginkgo biloba, red wine, tea, cocoa | Genistein, daidzein, equol soybean, soy products, soy cheese and sauces, grape seeds/skin | Inhibition of tumor growth, limit production of cancer-related hormones and have antioxidant potential | Prakash et al. (2012) |
| Phenolic acids | | Lemon, peach, lettuce, coffee beans, tea, coffee, cider, strawberry, raspberry grape juice, pomegranate juice, blueberry, cranberry, pear, cherry, cherry juice, apple, apple juice, orange, grapefruit | Caffeic acid, chlorogenic acid, ferulic acid, P-coumaric acid, sinapic acid, ellagic acid, gallic acid | Prevention of cancer, prevention of inflammation, and antioxidant activity | Prakash et al. (2012) |

**TABLE 19.1** *(Continued)*

| Major phytochemicals | Chemical structure | Food sources | Representative major constituents | Health benefits | References |
|---|---|---|---|---|---|
| Tannins | | Pomegranate, walnuts, peach, olive, plum, chickpea, peas, grape seeds and skin, apple juice, strawberries, raspberries, blackberry, lentils, haricot bean, red wine, cocoa, chocolate, tea, coffee, immature fruits, sorghum, and corn | Catechin, epicatechin polymers, ellagitannins, proanthocyanidin, tannic acids | act as an anti-inflammatory, antiseptic, antioxidant, and hemostatic agent, also have | Prakash et al. (2012); Saxena et al. (2014) |
| Diferuloylmethane | | Turmeric | Curcuminoids | Have medicinal properties against biliary disorders, anorexia, coryza, cough, diabetic wounds, hepatic disorder, rheumatism, and sinusitis | Prakash et al. (2012); Raina et al. (2014) |

**TABLE 19.1** *(Continued)*

| Major phytochemicals | Chemical structure | Food sources | Representative major constituents | Health benefits | References |
|---|---|---|---|---|---|
| Alkaloids | | Tobacco, opium, cinchona tree | Piperidine, caffine, atropine, morphine, cholchicine, nicotine, coniine | Used as local anesthetic and stimulant as cocaine, antiarrhythmic effect (due to quinidine, spareien), antibacterial, antifungal, and antihypertensive activities | Saxena et al. (2014) |

its prevalence includes exposure of numerous infectious agents rather than a single inciter; antibiotic agents gained their position more comprehensive than assumed because of their target specificity. Presently, no such kind of (target-specific agent with completely reverse effect, if not cure, the disease) therapeutic strategy exists for any autoimmune disease. However, during the past decade, the prognosis for patients with these diseases has dramatically improved due to the existence of more effective therapeutic and preventive strategies, which could be possible with the use of natural compounds such as phytochemicals obtained from different food source as given in Table 19.1. In recent years the protective roles of phytochemicals with balanced dietary intake for boosting up the human health is well described (Saxena et al., 2013). A wide range of dietary phytochemicals (more than 4000) has been obtained from fruits, vegetables, legumes, whole grains, nuts, seeds, fungi, herbs, and spices. The most common sources may be broccoli, cabbage, carrots, onions, garlic, whole wheat bread, tomatoes, grapes, cherries, strawberries, raspberries, beans, legumes, and soy foods. Although phytochemicals reported to have health benefits nevertheless historically have been considered as anti-nutrients by nutritionists, for example, tannins, have adverse effects such as decreasing the activities of digestive enzymes, energy, protein and amino acid availabilities, mineral uptake, and having other toxic effects. It was also envisaged that phytochemicals obtained from various fruits, vegetables, herbs, and spices may exert relevant negative immunoregulatory activities with special reference to brain aging. Hence, it is essential to determine effective physiological concentrations or exact dose for the dietary intake to explore the actual impact of dietary phytochemicals for prevention of early commencement of symptoms leading to cognitive decline and inflammatory neurodegeneration. Some studies regarding the significant dose of dietary intake of phytochemicals with their consecutive health befits are summarized in Table 19.2.

In addition, cautious studies are necessary for the dietary intake of various phytochemicals for their roles in the prevention of chronic degenerative diseases as well as immunity-related diseases.

## 19.5   CONCLUSION

This association of plant-derived phytochemicals in the prevention and management of autoimmune diseases need further research and experimentation in order to develop and design possible newer potential therapeutic strategy of natural origin. Therefore, this chapter summarized information

**TABLE 19.2** Prognosis of Different Phytochemicals with their Health Benefits on Different Target Groups.

| Phytochemical | Food source | Dose | Target group | Health benefits | References |
|---|---|---|---|---|---|
| Anthocyanins | - | 300 mg/day | Healthy adults, n = 120, (age = 40–74 years) for 3 weeks | To decrease the plasma concentrations of several NF-κB-regulated pro-inflammatory mediators | Karlsen et al. (2007) |
| | Maqui berry extract | 486 mg/day | Healthy, overweight, and smoker adult, n = 42 (age = 45–65 years) for 4 weeks | Improve oxidative status (Oxidized low-density lipoprotein [LDL] and F2-isoprostanes) | Davinelli et al. (2015) |
| | Blueberry juice | 877 mg/L | n = 9, (mean age 76.2 years) | Improved memory function in older adults with early memory decline | Krikorian et al. (2010) |
| | Anthocyanin-rich juice | 200 ml/day | Older adults, n = 49, (age = +70 years) over 12 weeks | Prevention of mild-to-moderate dementia, improvements in verbal fluency, short-term memory, and long-term memory | Kent et al. (2015) |
| Curcumin | Turmeric | (2 or 4 g/day) | n = 36 subjects | In cognitive function, in CSF Aβ or tau, between placebo and intervention groups | Ringman et al. (2012) |
| | | 500 mg, twice daily | n = 50 | Antidepressant action | Lopresti et al. (2015) |
| Flavanols | Cocoa | 990 mg (high), 520 mg (intermediate), or 45 mg (low) | n = 90 for 3 weeks | Modulating cognitive function | Desideri et al. (2012) |
| | | 494 mg total flavanols; epicatechin 89 mg | In healthy older adults n = 18, (mean age 61 years) | Improves regional cerebral perfusion and showed cognitive performance | Lamport et al. (2015) |
| Phenolic contents | Olive oils | 25 mL/day, high (366 mg/kg), medium (164 mg/kg), and low (2.7 mg/kg) | Healthy men n = 200, for 3 week | Increased the level of oxidized LDL autoantibodies which have protective role in atherosclerosis | Castañer et al. (2011) |

about plant phytochemicals and their potential role against various autoimmune conditions.

## KEYWORDS

- **phytochemicals**
- **autoimmune disease**
- **immunostimulating**
- **immunomodulatory activity**
- **nutritional intervention**

## REFERENCES

Castañer, O.; Fitó, M.; López-Sabater, M. C.; Poulsen, H. E.; Nyyssönen, K.; Schröder, H.; Salonen, J. T.; De la Torre-Carbot, K.; Zunft, H. F.; De la Torre, R.; Bäumler, H.; Gaddi, V. A.; Saez, G. T.; Tomás, M.; Covas, M. I. The Effect of Olive Oil Polyphenols on Antibodies Against Oxidized LDL. A Randomized Clinical Trial. *Clinic. Nutri.* **2011**, *30*(4), 490–493.

Davinelli, S.; Bertoglio, J. C.; Zarrelli, A.; Pina, R.; Scapagnini, G. A Randomized Clinical Trial Evaluating The Efficacy of An Anthocyanin-Maqui Berry Extract (Delphinol®) on Oxidative Stress Biomarkers. *J. Am. Coll. Nutr.* **2015**, *34*(1), 28–33.

Desideri, G.; Kwik-Uribe, C.; Grassi, D.; Necozione, S.; Ghiadoni, L.; Mastroiacovo, D.; Raffaele, A.; Ferri, L.; Bocale, R.; Lechiara, M. C.; Marini, C.; Ferri, C. Benefits in Cognitive Function, Blood Pressure, and Insulin Resistance Through Cocoa Flavanol Consumption in Elderly Subjects With Mild Cognitive Impairment: The Cocoa, Cognition, and Aging (CoCoA) Study. *Hypertension* **2012,** *60*, 794–801.

Karlsen, A.; Retterstøl, L.; Laakc, P.; Paur, I.; Bøhn, S. K.; Sandvik, L.; Blomhoff, R. Anthocyanins Inhibit Nuclear Factor-kappab Activation in Monocytes and Reduce Plasma Concentrations of Pro-Inflammatory Mediators in Healthy Adults. *J. Nutr.* **2007,** *137*, 1951–1954.

Kent, K.; Charlton, K.; Roodenrys, S.; Batterham, M.; Potter, J.; Traynor, V.; Gilbert, H.; Morgan, O.; Richards, R. Consumption of Anthocyanin-Rich Cherry Juice For 12 Weeks Improves Memory and Cognition in Older Adults with Mild-To-Moderate Dementia. *Eur. J. Nutr.* **2017,** *56,* 333–341.

Krikorian, R.; Shidler, M. D.; Nash, T. A.; Kalt, W.; Vinqvist-Tymchuk, M. R, Shukitt-Hale, B.; Joseph, J. A. Blueberry Supplementation Improves Memory in Older Adults. *J. Agric. Food Chem.* **2010,** *58*, 3996–4000.

Lamport, D. J.; Pal, D.; Moutsiana, C.; Field, D. T.; Williams, C. M.; Spencer, J. P.; Butler, L. T. The Effect of Flavanol-Rich Cocoa on Cerebral Perfusion in Healthy Older Adults During Conscious Resting State: A Placebo-Controlled, Crossover, Acute Trial. *Psychopharmacology (Berl).* **2015,** *232*, 3227–3234.

Lopresti, A. L.; Maes, M.; Meddens, M. J.; Maker, G. L.; Arnoldussen, E.; Drummond, P. D. Curcumin and Major Depression: A Randomised, Double-Blind, Placebo Controlled Trial Investigating The Potential Of Peripheral Biomarkers to Predict Treatment Response and Antidepressant Mechanisms of Change. *Eur. Neuropsycho. Pharmacol.* **2015,** *25*, 38–50.

Mishra, P. and Prasad, S. M. Mounting Insights over Human Wellness by Utilizing Plant's Primed Defense against Precise/Mild Oxidative Stress. *Crop Res.* **2016,** *51*(1), 1–10.

Prakash, D.; Gupta, C.; Sharma, G. Importance of Phytochemicals in Nutraceuticals. *J. Chinese Med. Res. Develop.* **2012,** *1*(3), 70–78.

Raina, H.; Soni, G.; Jauhari, N.; Sharma, N.; Bharadvaja, N. Phytochemical Importance of Medicinal Plants as Potential Sources of Anticancer Agents. *Turk. J. Bot.* **2014,** *38*, 1027–1035.

Ringman, J. M.; Frautschy, S. A.; Teng. E.; Begum, A. N.; Bardens, J.; Beigi, M. et al. Oral Curcumin for Alzheimer's Disease: Tolerability and Efficacy in A 24-Week Randomized, Double Blind, Placebo-Controlled Study. *Alzheimers Res. Ther.* **2012,** *4*, 43.

Rosman, Y.; Lidar, M.; Shoenfeld, Y. Antibiotic therapy in autoimmune disorders. *Clin. Pract.* **2014,** *11*(1), 91–103.

Saxena, M.; Saxena, J.; Nema, R.; Singh, D.; Gupta, A. Phytochemistry of Medicinal Plants *J. Pharmacog. Phytochem.* **2013,** *1*(6), 168–182.

Shah, S.; Saravanan, R.; Gajbhiye, N. A. Phytochemical and Physiological Changes in Ashwagandha (*Withania somnifera Dunal*) Under Soil Moisture Stress. *Braz. J. Plant Physiol.* **2010,** *22*(4), 255–261.

Venkatalakshmi, P.; Vadivel, V.; Brindha, P. Role of phytochemicals as immunomodulatory agents: A review. *Int. J. Green Pharma.* **2016,** *10*(1), 1.

Wang, L.; Wang, F. S.; Chang, C. Gershwin, M. E. Breach of tolerance: primary biliary cirrhosis. *Semin. Liver Dis.* **2014,** *34*, 297–317.

Wang, L.; Wang, F. S.; Gershwin, M. E. Human autoimmune diseases: a comprehensive update. *J. Intern. Med.* **2015,** *278*, 369–395.

# PART III

# Nanoparticle Biosynthesis and Its Biomedical Applications

# CHAPTER 20

# GREEN BIOSYNTHESIS OF METALLIC NANOPARTICLES

SESHU VARDHAN POTHABATHULA[1],
PREM PRAKASH KUSHWAHA[2] and SHASHANK KUMAR[2,*]

[1]*School of Biotechnology, JNTUK, Kakinada,
Andhra Pradesh 533003, India*

[2]*School of Basic and Applied Sciences, Department of Biochemistry
and Microbial Sciences, Central University of Punjab, Bathinda,
Punjab 151001, India*

*Corresponding author. E-mail: shashankbiochemau@gmail.com;
shashank.kumar@cupb.edu.in; Mob.: +91 9335647413.
*ORCID: https://orcid.org/0000-0002-9622-0512.*

## ABSTRACT

The green biosynthesis of metallic nanoparticles is an eco-friendly method which involves the use of cell-free extracts of plants, microorganisms, macrofungi, macroalgae and whole organisms such as plants, mushrooms, seaweeds, and so forth. This chapter presents the nitty-gritty knowledge required for a successful biosynthesis of various metallic nanoparticles. The choice of a biosynthetic approach is necessitated due to the quest for cheap, easy, safe and environment-friendly nanoparticles. To ensure successful biosynthesis, research in identifying the plant extract sources and optimized conditions for the biosynthesis of nanoparticles has shown better results. The most popular metallic particles are silver, gold, palladium, selenium, copper, and indium oxide. Their analysis is usually by Ultraviolet–visible spectra, X-ray diffractometer, energy-dispersive X-ray spectroscopy, transmission electron microscope, high-performance liquid chromatography (HPLC)-diode array detector (DAD) chromatogram and Fourier transform infrared spectroscopy (FTIR). Biosynthesized nanoparticles are less toxic and can be helpful to mankind for treating cancers and other deadly diseases.

## 20.1   INTRODUCTION

The term green biosynthesis itself clearly states that the process is eco-friendly. The studies on phytochemicals have helped corroborate nanotechnology with plant biotechnology. The research on nanosized metallic particles is creating a huge impact in the field of Materials Science and Technology. Nanoparticles are synthesized by both physical and chemical methods but these methods are creating a negative impact on the environment. The quest for an eco-friendly method has evolved the understanding of the nature of plants and alternative conventional methods are developed for the production of nanoparticles. The application-oriented nanoparticles synthesized from multiple methods are gold, silver, selenium, platinum, zinc oxide, iron, copper, palladium, indium oxide, and so forth. The advantages of green biosynthesis of nanoparticles are:

i.   Minimal waste and low energy consumption
ii.  Renewable raw materials
iii. Cost-effective process
iv.  Low-risk methods
v.   Environment-friendly
vii. Nontoxic stabilizer for the resultant nanoparticles.

## 20.2   MAJOR NANOPARTICLES SYNTHESIZED FROM PLANT EXTRACTS

### 20.2.1   SILVER NANOPARTICLE (AgNPs)

The most metallic nanoparticles reported to date are silver nanoparticles (AgNPs). The biosynthesis of AgNPs can be accomplished using plant extracts or those of marine sources such as the red algae, *Acanthophora spicifera* (Ibraheem et al., 2016). The use of plants for the production of nanoparticles has yielded lots of nanoparticles. However, some of the most efficient plant extracts involved in producing fine size 5–60nM AgNPs are *Ananas comosus* (Ahmad et al., 2012), *Rumex hymenosepalus, Cochlospermum religiosum, Bergenia ciliata, Clitoria ternatea, Pinus eldarica* extracts (Sorescu et al., 2016), and so forth. The major applications of AgNPs are as an anticancer, antimicrobial and antiviral agent.

### 20.2.2   GOLD NANOPARTICLE (AuNPs)

For gold nanoparticles (AuNPs) biosynthesis, experimental data reviews that the following plants: *Mangifera indica, Coriander spp., Sesbania grandiflora, Gymnocladus assamicus, Eucommia ulmoides, Nerium oleander, Hibiscus cannabinus, Cacumen platycladi, Pogostemon benghalensis, Salix alba, Galaxaura elongata, Solanum nigrum, Ocimum sanctum, Pistacia integerrima, Morinda citrifolia* have been utilized for the production of fine nano-sized gold particles ranging from 2 to 50 nM (Ahmed et al., 2015). The Major applications of AuNPs are in DNA labeling.

### 20.2.3   PALLADIUM NANOPARTICLE (PdNPs)

The palladium nanoparticle (PdNPs) biosynthesis has been reported from *Solanum trilobatum* leaf extract which yielded nanoparticles in the range of 60–70 nM size (Kanchana et al., 2010). Also, *Cinnamomum camphora* was reported to yield 3.2–20 nm (Yang et al., 2010) sized nanoparticles. An application of PdNP is as a biocatalyst.

### 20.2.4   SELENIUM NANOPARTICLE (SeNPs)

Plants containing selenium (Se) may be grouped into two broad categories, (i) those that accumulate Se in direct proportion to the amount of Se available from the soil (e.g. wheat) and (ii) those that actively accumulate Se in orders of magnitude greater than the Se concentrations in the soil (e.g., *Astragalus spp.*). In plants, broccoli is used for the higher biosynthesis of SeNPs with the size of 50–150 nm (Kapur et al., 2017).

### 20.2.5   COPPER NANOPARTICLE (CuNPs)

*Magnolia kobus* leaf extract acts as a better reducing agent for biosynthesis of copper nanoparticles (CuNPs). With a combination of CuSO4·5H$_2$O and *M. Kobus* produces fine nanoparticles in the size range of 40–100 nm, which is analyzed using X-ray spectroscopy (Lee et al., 2011).

## 20.2.6   INDIUM OXIDE NANOPARTICLE (IN$_2$O$_3$)

*Aloe barbadensis* miller, commonly called *Aloe vera*, is identified as a better reducing agent compared with other plants' aqueous solutions in combination with indium acetylacetonate, indium oxide, an amphoteric oxide. Indium oxide is used as an n-type semiconductor with wideband gaps of 3.55–3.75 eV (Maensiri et al., 2008). So this makes indium oxide important. For biosynthesis of Indium oxide nanoparticles (In$_2$O$_3$), the broth of *Aloe vera* (*A. barbadensis* miller) plays a key role and produces 5–50 nm sized nanoparticles. In$_2$O$_3$ particles have multiple applications such as in solar cells, panel displays, architectural glasses, semiconductor gas sensors, photocatalysts, and field emission.

## 20.2.7   PLATINUM NANOPARTICLE

*Diospyros kaki* is a well-known plant commonly used for the biosynthesis of platinum nanoparticles and also a better reducing agent. With a chemical combination of an aqueous H(2)PtCl(6).6H(2)O solution and *D. kaki* leaf extract yields fine-sized nanoparticles. Taking a broth concentration of >10% gives 90% of platinum nanoparticles at 95°C yielding an average particle size range of 2–12 nm (Song et al., 2010). There are multiple methods used to characterize and confirm the size of platinum nanoparticles, for example, X-ray photoelectron spectroscopy and Fourier-transform infrared spectroscopy.

## 20.3   PRINCIPLES INVOLVED IN NANOPARTICLES BIOSYNTHESIS

**Bio-reduction**: The cell-free extracts of plants, microorganisms, macro-fungi, macroalgae and whole organisms (plant, mushrooms, seaweeds, microbial cells), can be used for the biosynthesis of metallic nanoparticles (Figs. 20.1 and 20.2). However, the biochemical reactions such as conversion of NADPH to NADP involved in the plant extracts, help to convert metal salts to metallic nanoparticles. Natural availability of chemicals in one source such as a chemical factory and the conversions and reactions are optimized by plants. Based on their natural intensity and capability, they produce less toxic nanoparticles. Size of the formed nanoparticles varies from method to method.

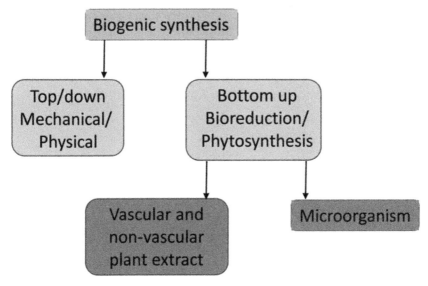

**FIGURE 20.1**   Green biosynthesis method.

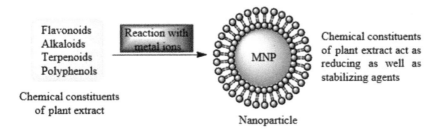

**FIGURE 20.2**   Bioreduction of metallic salts to nanoparticles.

## 20.4   METHODS AND CHARACTERIZATION IN NANOPARTICLES BIOSYNTHESIS

### 20.3.1   METHODS INVOLVED IN NANOPARTICLE BIOSYNTHESIS

The process of nanoparticle synthesis may be classified into the following two methods:

A.   Top-down methods for size reduction
  i.   Mechanical milling/ball milling

ii.   Chemical etching
iii.  Thermal ablation/laser ablation
iv.   Explosion process
v.    Sputtering

B.   Bottom-up methods for buildup from smaller entities
i.    Chemical/electrochemical precipitation
ii.   Vapor deposition
iii.  Atomic/molecular condensation
iv.   Sol-gel processes
v.    Spray Pyrolysis
vi.   Laser pyrolysis
vii.  Aerosol processes

## 20.3.2   CHARACTERIZATION OF NANOPARTICLE

Nanoparticles are generally characterized by their size, shape, surface area, and dispersity. Homogeneity of these properties is important in many applications (Jiang et al., 2009). The common techniques of characterizing nanoparticles are Ultraviolet-visible spectrophotometry, dynamic light scattering (DLS), scanning electron microscopy (SEM), transmission electron microscopy (TEM), Fourier transform infrared spectroscopy (FTIR), powder X-ray diffraction (XRD) and energy-Preparegold i spectroscopy (EDS) are some of the instruments to study the physical conditions of nanoparticles (Table 20.1; Dubey et al., 2010).

**TABLE 20.1**   Various Characterization Tools and Methods for Nanoparticles.

| Parameter | Characterization method |
| --- | --- |
| Carrier-drug interaction | Differential scanning calorimetry |
| Surface hydrophobicity | Rose Bengal (dye) binding, water contact angle measurement, X-ray photoelectron spectroscopy. |
| Drug stability | Bioassay of drug extracted from nanoparticles chemical analysis of drug |
| Release profile | In vitro release characteristics under physiologic and sink conditions |
| Charge determination | Laser doppler anemometry, zeta potentiometer |
| Nanoparticle dispersion stability | Critical flocculation temperature (CFT) |

**TABLE 20.1**  *(Continued)*

| Parameter | Characterization method |
| --- | --- |
| Chemical analysis of surface | Static secondary ion mass spectrometry, sorptometer |
| Particle size and distribution | Atomic force microscopy (AFM), laser diffractometry (LD), photon correlation spectroscopy (PCS), scanning electron microscopy (SEM), transmission electron microscopy (TEM), UV-visible spectroscopy, dynamic light scattering (DLS), Brewster angle microscope (BAM), 2-photon fluorescence microscopy (2PFM), confocal laser scanning microscopy (CLSM), scanning near-field optical microscopy (SNOM), Fourier transform infrared spectroscopy (FTIR), powder X-ray diffraction (XRD), energy-dispersive spectroscopy (EDS) |

## 20.5   FACTORS AFFECTING BIOSYNTHESIS OF METALLIC NANOPARTICLES

Some important factors to consider during the process of biosynthesis of nanoparticles that can affect the size and shape of the nanoparticles are temperature, pH, reactant concentration and reaction time. Many experimental reports suggest that these factors are considered as major reasons for changes in size and shape of nanoparticles. Some of the examples are given in the following sections.

### 20.5.1   INFLUENCE OF PH

In the conditioned medium, the value of pH affects the reaction rate in the conversion of metallic salts into nanoparticles (Sathishkumar et al., 2010). In the comparative studies to determine the effects of pH in the production of nanoparticles, it was found that a higher particle size was yielded at a lower acidic pH condition while smaller sized nanoparticles were reported in a higher acidic condition (Armendariz et al., 2004; Dubey et al., 2010). For example, the size of AuNPs found under a lower acidic condition falls within the range 25–85 nm and under a higher acidic condition, shows a lesser size (5–20 nm) (Huang et al., 2007). The study suggested that between pH 3 and 4, there is more access to functional groups contained within the extract.

## 20.5.2   INFLUENCE OF REACTANT CONCENTRATION

The concentration changes of plant extract can affect the particle size or shape. A study by Huang et al. reported that the leaf extract of *C. camphora* (camphor) shows a difference in the shape of Au nanoparticles due to changes in the concentration (Huang et al., 2007). Other studies reported that the concentration variation of carbonyl compounds in *Plectranthus amboinicus* leaf extract yielded 50 and 350 nm sized metallic nanoparticles and different shapes were obtained (decahedral, hexagonal, triangular).

## 20.5.3   INFLUENCE OF REACTION TIME

Recent studies revealed that there is an influence in the yield of nanoparticles when the reaction time is varied. For instance, the *Chenopodium album* leaf extract used for gold and AgNPs shows differences in the synthesis of nanoparticles at varied reaction time of 15 min to 2 h. It was observed that at 15 mins, there is more synthesis which gradually decreased at maximum reaction time (Chandran et al., 2006; Dwivedi et al., 2010).

## 20.5.4   INFLUENCE OF REACTION TEMPERATURE

Reaction temperature plays a major role in the biosynthesis of nanoparticles. It has been found that temperature is also an important factor in determining the size, shape, and yield of nanoparticles synthesized via plant extracts (Kaviya et al., 2011). A study conducted on *Citrus sinensis* (sweet orange) peel for AgNP biosynthesis at a reaction temperature of 25°C yielded 35 nm scaled particles. But when the temperature was increased to 60°C, it yielded 10 nm sized nanoparticles (Narayanan et al., 2010).

## 20.6   BIOSYNTHESIS OF Ag, Au, Pd, Se, Cu, AND IN NANOPARTICLES

### 20.6.1   BIOSYNTHESIS OF AgNPs

In the biosynthesis of nanoparticles, similar procedures are usually observed. As an example, *Capparis spinosa* should be collected and washed with distilled water to remove dust and other impurities. Take the best leaves and weigh up to 20 g, then boil in 100 mL of distilled water. After 10 min, grind the leaves and boil again for 5 min. Filter the solution using Whatman filter paper to gain pure extract of leaves. Then store the solution at 4°C.

Prepare 30 mL of 0.01 M aqueous solution of $AgNO_3$ taken into the flask. Then take different volumes such as 2.0, 2.5, 3.0, 4.0 of *C. spinosa* leaf extract suspended in $AgNO_3$ solution separately at room temperature for 15 min. Then the solution turns dark brown in color, which indicates the formation of AgNPs. Then centrifuge the whole sample at 9500 rpm for 20 min and collect the pellet containing AgNPs and dry in a hot air oven (Benakashani et al., 2016).

### 20.1.6.1 ULTRAVIOLET-VISIBLE SPECTRAL STUDIES

The leaf extract suspended in the silver nitrate solution turns from green color to brown color due to the reduction of silver ions, which is a result of the surface plasmon vibrations in the AgNPs that allow them to get excited and exhibit yellowish color in the aqueous solution. The determination of nanoparticles present in the solution can be identified by the spectrum in the visible range of 300–600 nm using UV-Vis spectrophotometer. By reading the absorbance, a peak at 400 nm shows the presence of AgNPs. The changes in the concentration of leaf extract affect the absorbance peak.

### 20.6.2 BIOSYNTHESIS OF AuNPs

The leaves of the following plants namely, *Lippia citrodora, Salvia officinalis, Pelargonium graveolens* can be used for the production of AuNPs. Collect leaves and wash twice thoroughly with double-distilled water and take 10 g of leaf source chopped and suspended in 30 mL of double-distilled water and homogenize the mixture and make the solution volume up to 50 mL. Then centrifuge the mixture for 2 min for 2000 rpm. Take the solution and filter using 0.45 μm filter. Do this for all samples and keep all plant extracts at -18°C.

Solution of gold ion preparation starts with diluting 16.6 μL of 30 wt% HAuCl4 solution in 50 mL of distilled water to form a 0.1 g/L solution. After that, 10 mL of the gold ion solution (1.0 mg HAuCl4) is mixed with 0.75 mL of plant extract which leads to the formation of gold nanoparticles. Observe as the mixture turns into deep purple or red color within a few seconds due to the gold nanoparticle formation. Take the readings at ultraviolet-visible absorption of 530 nm. Some plant extract mixture and gold ions require heating at 30–40°C to get the solution color turn to purple. An example is the plant extracts of *L. citrodora* and *S. officinalis*. They require some reaction temperature to synthesize AuNPs due to their lower reducing activity. Finally, the dried AuNPs obtained can be analyzed using energy-dispersive X-ray spectroscopy (EDXS) to identify the particle size and purity of AuNPs (Elia et al., 2014).

### 20.6.3  BIOSYNTHESIS OF PDNPS

Biosynthesis of PdNPs can be done by collecting the plant leaves of *Hippophae rhamnoides Linn* followed by washing with double-distilled water and drying for some days. The dried leaves (around 100 g) are ground into powder. Suspend the fine powder in 500 mL hydroalcoholic solution (30% methanol). Mix the solution thoroughly and boil for 30 min. Now centrifuge the solution at 6500 rpm for 5 min and filter the solution with Whatman filter paper and keep at 4°C for further use. Perform high-performance liquid chromatography (HPLC) analysis for confirming the presence of flavonoids and antioxidants in the plant extract.

Now take 10 mL of the plant leaf extract and add 50 mL of 0.003 M aqueous solution of $PdCl_2$ with continuous stirring at 80°C. Observe the color of the solution as it turns from transparent yellow to dark brown color within 25 min at 80°C. Centrifuge the colored solution at 7000 rpm for 30 min.

Analysis of plant extract using the reverse phase-HPLC-diode array detector (DAD) instrument consisting of the auto-injector, sample cooler, pumps and column oven and diode array detector could help in the identification of the required antioxidants for the synthesis of nanoparticles in the selected plant extract (Nasrollahzadeh et al., 2015).

### 20.6.4  BIOSYNTHESIS OF SENPS

The leaves of *Leucas lavandulifolia* should be identified and collected. Wash leaves multiple times with distilled water. Then shade-dry them, after a few days collect the dried leaves and make their fine pieces. Take 10 g of the sample in 400 mL double-distilled water. The mixture should be boiled for 5 min until the color of the solution turns from colorless to yellow. Cool the mixture at room temperature and filter the solution using Whatman filter paper. Centrifuge the filtered solution at 1200 rpm for 2 min. Now store the solution at room temperature for future use.

Take 2 mL of the plant extract and add to 10 mL of 50 mM selenous acid solution along with 200 µL of 40 mM ascorbic acid used as an initiator of the reduction reaction. For positive standard, take 10 mL selenous acid added to 200 µL of 40 mM ascorbic acid for SeNP synthesis while only plant extract with 200 µL of 40 mM ascorbic acid is used as a negative control. The color change to ruby red represents the biosynthesis of SeNPss (Kirupagaran et al., 2016).

### 20.6.4.1   FTIR ANALYSIS OF SeNPs

A previously experimented *L. lavandulifolia* leaf extract contains polyphenols, heterocyclic components such as fatty acids, alkaloid, flavonoids, and flavones. Their capping and reducing agent's activity leads to the conversion of selenium ions to SeNPs. FTIR analysis results show the absorbance range of 724–3420 cm$^{-1}$ between locating strong band due to the presence of alcohols and phenols and their bond interactions.

## 20.6.5   BIOSYNTHESIS OF CuNPs S

Biosynthesis of CuNPs comprises collection of *Euphorbia prostrata* leaves *and washed thoroughly under tap water. Take* 10 g of leaves and wash with distilled water so that the surface of leaves is free from accumulated dust which may contaminate the extract. Then add 10 g of leaves in 100 mL double-distilled water and boil at 80°C for 30 min. The extract was filtered through Whatman No. 1 filter paper and used for the subsequent analyses (Chung et al., 2017).

Take 100 mL of 3 mM copper acetate solution (Cu(OAc)$_2$) in an Erlenmeyer flask and stir for 2 h. Then add 20 mL of plant extract of *E. prostrate* to 80 mL of 3 mM Cu(OAc)$_2$ and keep the solution at room temperature for 24 h.

### 20.6.5.1   SCANNING ELECTRON MICROSCOPY (SEM) ANALYSIS

SEM micrographs of the previously synthesized CuNPs by the reduction of copper acetate revealed spherical, hexagonal and cubical nanoparticles ranging between 28 nm and105 nm, with an average size of $41 \pm 0.8$ nm due to Cu ions. It was observed that they were approximately spherical in shape with a smooth surface. The EDXS of the synthesized CuNPs showed strong copper signals along with P and C peaks, which may originate from the biomolecules that were bound to the surface of the CuNPs.

## 22.6.6  BIOSYNTHESIS OF INDIUM OXIDE

Take *Aloe vera* leaves and wash with double-distilled water. Then take 35 g of the thoroughly washed *Aloe vera* leaves, cut and boil in 100 mL of deionized water. The resulting extract is used as *Aloe vera* extract solution. Then take 3 g of indium (III) acetylacetonate dissolved in 30 mL *Aloe vera* extract solution. Keep the solution at 60°C for several hours in a stirring condition until the solution is dry. The dried sample should be calcined to characterize the $In_2O_3$ in crystal phase and analyze using powder (XRD) with CuKα radiation (λ = 0.15406 nm) (Chung et al., 2017).

## ACKNOWLEDGMENT

SK acknowledges Central University of Punjab, Bathinda, India for providing necessary infrastructure facility. PPK acknowledges financial support from University Grants Commission, India in the form of CSIR-UGC Junior Research fellowship.

## KEYWORDS

- green biosynthesis
- metallic nanoparticles
- gold nanoparticles
- silver nanoparticles
- scanning electron microscope

## REFERENCES

Ahmad, N.; Sharma, S. Green Synthesis of Silver Nanoparticles Using Extracts of *Ananas Comosus*. *Gr. Sustain. Chem.* **2012,** *2*, 141–147.

Ahmed, S.; Ikram, S. Synthesis of Gold Nanoparticles Using Plant Extract: an Overview. *Nano Res. Appl.* **2015,** *1*, 1–6.

Armendariz, V.; Herrera, I.; Jose-yacaman, M.; Troiani, H.; Santiago, P.; Gardea-Torresdey, J. L. Size Controlled Gold Nanoparticle Formation By *Avena Sativa* Biomass: Use of Plants in Nanobiotechnology. *J. Nanopart. Res.* **2004,** *6*(4), 377–382.

Benakashani, F.; Allafchian, A. R.; Jalali, S. A. H. Biosynthesis of Silver Nanoparticles using *Capparis Spinosa* L. Leaf Extract and their Antibacterial Activity. *Karbala Int. J. Mod. Sci.* **2016,** *2*(4), 251–258.

Chandran, S. P.; Chaudhary, M.; Pasricha, R.; Ahmad, A.; Sastry, M. Synthesis of Gold Nanotriangles and Silver Nanoparticles Using *Aloe Vera* Plant Extract. *Biotechnol. Prog.* **2006,** *22*, 577–583.

Chung, I. M.; Abdul Rahuman, A.; Marimuthu, S.; Vishnu Kirthi, A.; Anbarasan, K.; Padmini, P.; Rajakumar, G. Green Synthesis of Copper Nanoparticles Using *Eclipta Prostrata* Leaves Extract and Their Antioxidant and Cytotoxic Activities. *Exp. Ther. Med.* **2017,** *14*, 18–24.

Dubey, S. P.; Lahtinen, M.; Sillanpaa, M. Tansy Fruit Mediated Greener Synthesis of Silver and Gold Nanoparticles. *Process Biochem.* **2010,** *45*, 1065–1071.

Dwivedi, A. D.; Gopal, K. Biosynthesis of Silver and Gold Nanoparticles Using *Chenopodium Album* Leaf Extract. *Colloids Surf. A.* **2010,** *369*, 27–33.

Elia, P.; Zach, R.; Hazan, S.; Kolusheva, S.; Porat, Z. E.; Zeiri, Y. Green Synthesis of Gold Nanoparticles Using Plant Extracts as Reducing Agents. *Int. J. Nanomedicine.* **2014,** *9*, 4007–4021.

Huang, J., Li, Q., Sun, D., Lu, Y., Su, Y., Yang, X., Wang, H.; Wang, Y.; Shao, W.; He, N.; Hong, J.; Chen, C. Biosynthesis of Silver and Gold Nanoparticles by Novel Sundried *Cinnamomum camphora* Leaf. *Nanotechnology* **2007,** *18*(10), 1–11.

Ibraheem, I. B. M.; Abd Elaziz B. E. E.; Saad, W. F.; Fathy, W. A. Green Biosynthesis of Silver Nanoparticles Using Marine Red Algae *Acanthophora Specifera* and Its Antimicrobial Activity. *J. Nanomed. Nanotechnol.* **2016,** *7*, 1–4.

Jiang, J.; Oberdorster, G.; Biswas, P. Characterization of Size, Surface Charge, and Agglomeration State of Nanoparticle Dispersions for Toxicological Studies. *J. Nanopart. Res.* **2009,** *11*, 77–89.

Kanchana, A.; Devarajan, S.; Ayyappan, S. R. Green Synthesis and Characterization of Palladium Nanoparticles and its Conjugates from *Solanum Trilobatum* Leaf Extract. *Nano-micro Lett.* **2010,** *2*(3), 169–176.

Kapur, M.; Soni, K.; Kohli, K. Green Synthesis of Selenium Nanoparticles from Broccoli, Characterization, Application and Toxicity. *Adv. Tech. Biol. Med.* **2017,** *5*, 1–7.

Kaviya, S.; Santhanalakshmi, J.; Viswanathan, B.; Muthumary, J.; Srinivasan, K. Biosynthesis of Silver Nanoparticles Using *Citrus sinensis* peel Extract and its Antibacterial Activity. *Spectrochim. Acta A Mol. Biomol. Spectrosc.* **2011,** *79*, 594–598.

Kirupagaran, R.; Saritha, A.; Bhuvaneswari, S. Green Synthesis of Selenium Nanoparticles from Leaf and Stem Extract of *Leucas Lavandulifolia* Sm. and Their Application. *J. Nanosci. Nanotechnol.* **2016,** *2*(5), 224–226.

Lee, H. J.; Lee, G.; Jang, N. R.; Yun, J. H.; Song, J. Y.; Kim, B. S. Biological Synthesis of Copper Nanoparticles Using Plant Extract. *Nanotechnology* **2011,** *1*(1), 371–374.

Maensiri, S.; Laokul, P.; Klinkaewnarong, J.; Phokha, S.; Promarak, V.; Seraphin, S. Indium Oxide (In$_2$O$_3$) Nanoparticles Using *Aloe Vera* Plant Extract: Synthesis and Optical Properties. *J. Optoelectron. Adv. Mater.* **2008,** *10*, 161–165.

Narayanan, K. B.; Sakthivel, N. Phytosynthesis of Gold Nanoparticles Using Leaf Extract of *Coleus amboinicus* Lour. *Mater. Charact.* **2010,** *61*(11), 1232–1238.

Nasrollahzadeh, M.; Sajadi, S. M.; Maham, M. Green Synthesis of Palladium Nanoparticles Using *Hippophae rhamnoides* Linn Leaf Extract and Their Catalytic Activity for the Suzuki–Miyaura Coupling in Water. *J. Mol. Catal. A: Chem.* **2015,** *396,* 297–303.

Sathishkumar, M.; Sneha, K.; Yun, Y. S. Immobilization of Silver Nanoparticles Synthesized Using *Curcuma Longa* Tuber Powder and Extract on Cotton Cloth for Bactericidal Activity. *Bioresour. Technol.* **2010,** *101*(20), 7958–7965.

Song, J. Y.; Kwon, E. Y.; Kim, B. S. Biological Synthesis of Platinum Nanoparticles Using Diopyros Kaki Leaf Extract. *Bioprocess Biosyst. Eng.* **2010,** *33*(1), 159.

Sorescu, A. A.; Nuţa, A.; Ion, R. M.; Ioana-Raluca, S. B. In *Green Synthesis of Silver Nanoparticles Using Plant Extracts,* The 4th International Virtual Conference on Advanced Scientific Results, National Institute of Research and Development for Chemistry and Petrochemistry - ICECHIM, 202 Splaiul Independentei Bucharest, Romania, June 6–10, 2016; ISBN: 978-80-554-1234-4.

Yang, X.; Li, Q.; Wang, H.; Huang, J.; Lin, L.; Wang, W.; Green Synthesis of Palladium Nanoparticles Using Broth of *Cinnamomum Camphora* Leaf. *J. Nanopart. Res.* **2010,** *12,* 1589–1598.

# CHAPTER 21

# CYTOTOXICITY AND BIOMEDICAL APPLICATIONS OF METAL OXIDE NANOPARTICLES SYNTHESIZED FROM PLANTS

YIIK SIANG HII[1], JAISON JEEVANANDAM[1*], YEN SAN CHAN[1], and MICHAEL K. DANQUAH[2]

[1]*Department of Chemical Engineering, Faculty of Engineering and Science, Curtin University CDT 250, 98009, Miri, Sarawak, Malaysia*

[2]*Department of Civil & Chemical Engineering, University of Tennessee, Chattanooga, TN 37403, United States.*

*Corresponding author.*
*E-mail: jaison.jeevanandam@postgrad.curtin.edu.my*

## ABSTRACT

Metal oxide nanoparticles have attracted the interest of scientists in the drugs industry and other health-related applications. Hence, the preparation method of metal oxide nanoparticles plays an essential role in developing highly safe and biocompatible nanoparticles for healthcare applications. Green synthesis is among one of the promising methods for nanoparticles synthesis because it is rapid, reproducible, cost-effective, biocompatible, and safe. Phytochemical extraction is the first step in the green synthesis of nanoparticles. The selection of phytochemical extraction method plays a crucial role in determining the physicochemical characteristics of nanoparticles which eventually help in enhancing their properties for biomedical applications. This chapter emphasizes on the recent technologies in phytochemical extraction methods, the preparation methods of metal oxide nanoparticles from various phytochemicals, as well as the cytotoxicity of green-synthesized metal oxide nanoparticles. Biomedical applications of metal oxide nanoparticles synthesized through phytochemicals are also considered.

## 21.1 INTRODUCTION

For centuries, metal oxides have been used widely in the chemistry, material science, and medical field (Rodriguez et al., 2007) due to their strong anti-oxidant and antibacterial action (Gupta et al., 2014). It is also the primary ingredient for sunscreens with broad-spectrum protection (Bartholomey et al., 2016). Metal and metal oxide nanoparticles possess unique characteristics in terms of optical, chemical, physical, and thermal properties (Astruc, 2008) compared to their bulk counterparts. The unique properties of nanoparticles are due to their extraordinary surface area to volume ratio, new quantum effects (Alagarasi, 2011), and surface effects (Duan et al., 2013). These exclusive properties make nanoparticles favorable in applications such as catalysis (Ba-Abbad et al., 2012), biosensing (Bao et al., 2008), medicine (Laurent et al., 2008), bioimaging, and biomedical therapeutics (Andreescu et al., 2012). Remarkably, metal nanoparticles are thermodynamically unstable and highly chemically reactive. Thus, metal nanoparticles need to be embedded in a matrix for stabilization. Some of the matrices include polypropylene, polyethylene, Teflon (Meech et al., 2005), and nanocellulose (Kaushik et al., 2016). Metal oxide nanoparticles have attracted substantial importance from researchers as they have higher mechanical stability and are biocompatible (Andreescu et al., 2012).

The two major approaches in nanoparticle synthesis are top-down and bottom-up. The top-down approach produces nanoparticles from bulk materials by physical methods, whereas bottom-up approach synthesizes nanoparticles from atoms by chemical or biological method(s). A major disadvantage of the top-down method is the surface structure imperfection, while the bottom-up approach is able to overcome the defects and synthesize homogeneous nanoparticles (Thakkar et al., 2010). Generally, the main drawback of physicochemical methods is the involvement of highly toxic and hazardous chemicals such as sodium borohydride (Li et al., 1999) or hydrazine (Tan et al., 2003) during the synthesis. Besides, the nanoparticles produced from these methods are not appropriate for oral administration (Ramanujam et al., 2014). Hence, the biological method has drawn the consideration of researchers as biological synthesis uses environmentally benign materials. Biological synthesis employs the bottom-up approach whereby the nanoparticles are prepared with the support of biological capping and reducing agents.

Biological methods use reducing agents from natural sources that include metabolites from microorganisms and phytochemical extracts from plant sources for nanoparticles synthesis (Thakkar et al., 2010; Jeevana-ndam et al., 2017a). Bacteria have the capabilities to convert and reduce heavy metal ions and are potential candidates for nanoparticles synthesis (Pereira et al., 2015). Notably, these bacteria belong largely to genera such as *Bacillus* (Sunkar et al., 2012) and *Pseudomonas* (Narayanan et al., 2010). As for fungi, their diverse intracellular enzymes are significant in the nanoparticle synthesis (Hulkoti et al., 2014). It has been reported by various authors that fungi belonging to *Aspergillus* sp. have superior poten-tial in nanoparticle synthesis (Jeevanandam et al., 2016). However, micro-organism-mediated nanoparticle synthesis is time-consuming (Mukherjee et al., 2001) and the physicochemical characteristics such as size, shape, and crystallinity of nanoparticles are difficult to control (Narayanan et al., 2010). On the other hand, phytochemicals act as a better reducing agent and also as a stabilizing agent (Rathod et al., 2011) in plant extract-mediated green synthesis of nanoparticle. The nanoparticles synthesized through plant extracts are less toxic and more biocompatible compared to chemical and physical synthesized nanoparticles (Moulton et al., 2010). The main phytochemicals involved in the nanoparticles synthesis are ketones, aldehydes, amides, flavones, terpenoids, and carboxylic acids (Prathna et al., 2010). These phytochemicals decrease the nanoparticle agglomeration and thereby stabilize and control the morphology formation of nanoparticles (Malik et al., 2014). Thus, recent researches have focused on phytochemicals extracted from plant sources as a preferred method for nanoparticle synthesis due to the enormous potential benefits in several biomedical applications such as preparation of enhanced antibiotics and drug delivery systems.

## 21.2  BIOSYNTHESIS OF METAL OXIDE NANOPARTICLES

In recent years, phytochemical-mediated metal oxide nanoparticle synthesis has attracted great attention from researchers. The presence of phytochemi-cals such as flavonoids, tannins, alkaloids, and other phenolic compounds in plants acts as good reducing and capping agent (Rathod et al., 2011). Capping agent minimizes the agglomeration of nanoparticles, thereby controlling the morphology and helps to protect and stabilize the nanoparticles (Jothi et al., 2012). Several factors such as reaction time, the concentration of precursor and reducing agents, pH, and temperature can affect the synthesis of nanoparticles.

### 21.2.1   ZINC OXIDE

Zinc oxide (ZnO) nanoparticles are used in numerous applications such as biosensors, medical diagnostics, catalysis, bioimaging, DNA labeling, and drug delivery (Wang et al., 2004). ZnO nanoparticles display remarkable semiconducting properties due to their large band gap (3.37 eV) and high exciton binding energy (60 meV). This unique semiconducting property allows high UV filtering, making it a suitable candidate in the cosmetology industry for the manufacture of sunscreen lotions (Agarwal et al., 2017). Besides, ZnO nanoparticles show eclectic antibacterial effect even at a very low concentration (Stoimenov et al., 2002) and towards highly heat-resistant spores (Azam et al., 2012). Various plant leaf extracts were used for the green synthesis of ZnO nanoparticles which include extracts of *Aloe barbadensis* Miller (Sangeetha et al., 2011), *Solanum nigrum* (Ramesh et al., 2015), *Calotropis gigantea* (Vidya et al., 2013), *Vitex trifolia* (Elumalai et al., 2015), *Azadirachta indica* (Bhuyan et al., 2015), and *Artocarpus heterophyllus* (Vidya et al., 2016). Generally, smaller sized ZnO nanoparticles in the range of 9–40 nm were obtained by using phytochemical-based green synthesis method as listed in Table 21.1.

Ramesh et al. (2015) synthesized ZnO nanoparticles using the leaves extract of *S. nigrum* as the reducing agent and zinc nitrate as the precursor. The green synthesis was carried out by mixing the extract and zinc nitrate, and the mixture was boiled at 60°C. Later on, the obtained paste was subjected to calcination in the air furnace at 400°C to obtain the ZnO nanoparticles. The transmission electron microscopy (TEM) images revealed that the ZnO nanoparticles were spherical in shape with size of 29.79 nm. Fourier-transform infrared (FTIR) analysis was performed to identify the phytochemicals that act as the capping agent and responsible for the reduction of ZnO. It was observed that terpenoids were the phytochemicals that prevented the agglomeration of nanoparticles by stabilizing it. Similarly, Elumalai et al. (2015) synthesized ZnO nanoparticles by using the leaf extracts of *V. trifolia* and zinc nitrate hexahydrate. The effect of the leaf extract volume on the size of ZnO nanoparticles was investigated. It was revealed that an increase in the volume of leaf extract leads to an increase in the crystalline size of ZnO nanoparticles. The authors suggested that the increment was due to the higher nucleation rate.

Although ZnO nanoparticles are widely used for targeted drug delivery applications, the studies on the cytotoxicity of green synthesized ZnO nanoparticles are limited (Ma et al., 2013). Recently, Mahendiran et al. (2017) studied the in vitro cytotoxicity of the green synthesized ZnO

**TABLE 21.1** Plant Species, Synthesis Parameters and Their Physicochemical Characteristics Used for the Green Synthesis of Metal Oxide Nanoparticles.

| Metal oxide nanoparticles | Plant species | Plant part | Reaction temperature °C | Size and shape of nanoparticles | References |
|---|---|---|---|---|---|
| ZnO | *Aloe barbadensis* Miller | Leaves | 150 | 25–40 nm, spherical | Sangeetha et al. (2011) |
| | *Solanum nigrum* | Leaves | 60–80 | 20–30 nm, spherical | Ramesh et al. (2015) |
| | *Calotropis gigantea* | Leaves | 60 | 11–25 nm, spherical | Vidya et al. (2013) |
| | *Citrus aurantifolia* | Fruits | 90 | 50–200 nm, hexagonal | Samat et al. (2013) |
| | *Vitex trifolia* | Leaves | 60–80 | 30 nm, hexagonal | Elumalai et al. (2015) |
| | *Aloe vera* and *Hibiscus sabdariffa* | Gel and leaves | 80 | 9–18 nm, hexagonal | Mahendiran et al. (2017) |
| | *Azadirachta indica* | Leaves | Room temperature | 9.6–25.5 nm, spherical | Bhuyan et al. (2015) |
| | *Ocimum basilicum* | Leaves | 60 | 9–50 nm, hexagonal | Parthasarathy et al. (2017) |
| | *Artocarpus heterophyllus* | Leaves | Not reported | 15–25 nm, spherical | Vidya et al. (2016) |
| | *Laurus nobilis* | Leaves | Room temperature | 47.27 nm, hexagonal | Vijayakumar et al. (2016) |
| | *Passiflora caerulea* | Leaves | Room temperature | 70 nm, spherical | Santhoshkumar et al. (2017) |
| MgO | *Nephelium lappaceum* L. | Peels | 80 | 100 nm, spherical | Suresh et al. (2014) |
| | *Clitoria ternatea* | Whole plant | Room temperature | 50–400 nm, irregular | Sushma et al. (2016) |
| | *Emblica officinalis* | Fruits | Room temperature | 21–31 nm, spherical | Ramanujam et al. (2014) |
| | *Amaranthus tricolor and blitum,* *Andrographis paniculata* | Leaves | 60 | 18–28 nm, spherical | Jeevanandam et al. (2017b) |
| | *Amaranthus tricolor* | Leaves | 60 | 50–80 nm, hexagonal | |
| | *A. paniculata* | Leaves | Microwave (1 kW, 2.45 GHz) | 46 nm (diameter) and 185 nm (length), rods | Karthik et al. (2017) |

**TABLE 21.1** (Continued)

| Metal oxide nanoparticles | Plant species | Plant part | Reaction temperature °C | Size and shape of nanoparticles | References |
|---|---|---|---|---|---|
| | Citrus limon | Leaves | Room temperature | 12–80 nm, flakes | Awwad et al. (2014) |
| | Rosmarinus officinalis | Leaves | Room temperature | 73 nm, irregular | Ghashang et al. (2016) |
| NiO | N. lappaceum L. | Peels | 80 | 50 nm, irregular | Yuvakkumar et al. (2014) |
| | Moringa oleifera | Leaves | 90 | 9.69 nm, spherical | Ezhilarasi et al. (2016) |
| CeO₂ | Gloriosa superba L. | Leaves | 80 | 5 nm, spherical | Arumugam et al. (2015) |
| | Acalypha indica | Leaves | 80 | 25–30 nm, irregular | Kannan et al. (2014) |
| | A. barbadensis Miller | Leaves | 80 | 63.6 nm, spherical | Priya et al. (2014) |
| | H. sabdariffa | Flowers | Room temperature | 3.9 nm, spherical | Thovhogi et al. (2015) |
| CuO | Syzygium alternifolium | Stem | 50 | 5–13 nm, spherical | Yugandhar et al. (2017) |
| | Phaseolus vulgaris L. | Beans | 120 | 26.6 nm, spherical | Nagajyothi et al. (2017a) |
| | Punica granatum | Peels | Room temperature | 10–100 nm, spherical | Ghidan et al. (2016) |
| | Gundelia tournefortii | Aerial parts | 60 | 40–60 nm, spherical | Nasrollahzadeh et al. (2015a) |
| | Thymus vulgaris L. | Leaves | 60 | 30 nm, spherical | Nasrollahzadeh et al. (2016c) |
| | G. superba L. | Leaves | 400 | 5–10 nm, spherical | Naika et al. (2015) |
| | C. gigantea | Leaves | 60–80 | 20 nm, spherical | Sharma et al. (2015) |
| | Thymbra spicata | Leaves | 80 | 10–20 nm, spherical | Veisi et al. (2017) |
| PbO | Sageretia thea | Leaves | 80 | 15–35 nm, spherical | Khalil et al. (2017a) |
| TiO₂ | M. oleifera | Leaves | 50 | 12 nm, not reported | Patidar et al. (2017) |
| | Nyctanthes arbor-tristis | Leaves | 50 | 100–150 nm, spherical | Sundrarajan et al. (2011) |
| | Psidium guajava | Leaves | Room temperature | 32.58 nm, spherical | Santhoshkumar et al. (2014) |

**TABLE 21.1** (Continued)

| Metal oxide nanoparticles | Plant species | Plant part | Reaction temperature °C | Size and shape of nanoparticles | References |
|---|---|---|---|---|---|
| | *Euphorbia prostrata* | Leaves | Room temperature | 83.22 nm, spherical | Zahir et al. (2015) |
| | *Mangifera indica* L. | Leaves | Room temperature | 30–35 nm, spherical | Rajakumar et al. (2015) |
| | *Euphorbia heteradena* Jaub | Roots | 60 | 17–45 nm, spherical | Nasrollahzadeh et al. (2015b) |
| | *Solanum trilobatum* | Leaves | Room temperature | 70 nm, spherical | Rajakumar et al. (2014) |
| | *Eclipta prostrata* | Leaves | Room temperature | 36–68 nm, spherical | Rajakumar et al. (2012) |
| | *Annona squamosa* | Peels | Room temperature | 23 nm, spherical | Roopan et al. (2012) |
| | *Origanum vulgare* | Leaves | 50 | 341 nm, spherical | Sankar et al. (2014a) |
| HgO | *Callistemon viminalis* | Flowers | 60 | 2–4 nm, spherical | Das et al. (2015) |
| CuO/ZnO | *Melissa officinalis* L. | Leaves | 75 | 10–20 nm, spherical | Bordbar et al. (2018) |

nanoparticles and a standard drug cisplatin was tested using 3-(4,5-dimethylthiazol-2-yl)-2, 5-diphenyltetrazolium bromide (MTT) assay against three cancer cell lines, such as human breast adenocarcinoma (MCF-7), cervical (HeLa), and epithelioma (Hep-2), and one normal human dermal fibroblasts (NHDF) cell line. ZnO nanoparticles showed prominent cytotoxicity activity on MCF-7 cells followed by HeLa and Hep-2. The cytotoxicity of ZnO nanoparticles on NHDF is much lower compared to the three cancer cell lines. Notably, ZnO nanoparticles presented a higher cytotoxic activity than cisplatin. Thus, it can be deduced that phytochemical-mediated synthesis of ZnO nanoparticles has higher anticancer properties than the standard drug which was chemically synthesized. Moreover, it was observed that the $IC_{50}$ values of the ZnO nanoparticles were two times lower than cisplatin against all the tested cell lines.

### 21.2.2   COPPER OXIDE

Copper oxide (CuO) nanoparticles have attracted substantial attention as it can be used as an antimicrobial, antibiotic, and antifungal agent (Dizaj et al., 2014). In recent years, numerous articles on the phytochemical-mediated synthesis of CuO nanoparticles have been published. Recently, stem extracts of *Syzygium alternifolium* (Yugandhar et al., 2017), peel extracts of *Punica granatum* (Ghidan et al., 2016), and leaves extract of *Thymus vulgaris* L. (Nasrollahzadeh et al., 2016c) and *Thymbra spicata* (Veisi et al., 2017) were used in synthesizing CuO nanoparticles. For instance, Yugandhar et al. (2017) synthesized CuO nanoparticles by using stem extracts of *S. alternifolium* as a reducing agent and copper sulfate pentahydrate as the precursor at 50°C. The acquired CuO nanoparticles were spherical in shape with particle size ranging from 5 to 13 nm. FTIR spectral results indicated that phytochemicals such as phenols and primary amines of plant extracts were mainly accountable for capping and stabilizing the synthesized nanoparticles. Similarly, proteins were also found to act as a reducing and stabilizing agent for the formation of CuO nanoparticles when peel extracts of *P. granatum* were used for green synthesis by Ghidan et al. (2016).

The cytotoxic studies of phytochemical-mediated synthesis nanoparticles were very limited. Recently, Yugandhar et al. (2017) studied the cytotoxicity of green synthesized CuO nanoparticles from *S. alternifolium* stem extract. The cytotoxicity was tested against human breast cancer (MDA-MB-231) cell lines by using MTT assay. The results obtained showed that an increase in the concentration of CuO nanoparticles from 10 to 100 µg/mL reduces the cell viability significantly. Nagajyothi et al. (2017a) prepared CuO nanoparticles using the bean extracts of *Phaseolus vulgaris* L. and tested their cytotoxicity against cervical carcinoma (HeLa) cell lines by sulforhodamine-B assay. It was observed that 1 mg/mL of CuO nanoparticles was sufficient to inhibit the growth of HeLa. Additionally, clonogenic survival assay was performed to identify the potential of HeLa cells to proliferate under the presence of CuO nanoparticles. The results showed that the CuO nanoparticles significantly reduce the multiplication of HeLa cells. These studies showed that the green synthesized CuO nanoparticles are highly toxic to cancer cells. Many literatures showed that copper nanoparticles are cytotoxic and genotoxic towards cell lines such as airway epithelial (HEp-2) cells (Fahmy et al., 2009), human pulmonary epithelial cells (A549) (Ahamed et al., 2010), and mouse neuroblastoma Neuro-2A (Perreault et al., 2012). However, the stabilization of copper oxide nanoparticles helps in reducing their cytotoxicity. The stabilization of copper oxide nanoparticles by a carbon layer was proven to reduce their cytotoxicity in various cells (Studer et al., 2010). Thus, surface-stabilized copper oxide nanoparticles can be used in biomedical applications compared to non-stabilized CuO nanoparticles.

### 21.2.3 MAGNESIUM OXIDE

MgO nanoparticles are extensively used in numerous applications such as catalysis, bioremediation, paints, and superconductors (Bhattacharya et al., 2005). Particularly, MgO is nontoxic in nature and thus attracted substantial attention in biomedical applications (Staiger et al., 2006). Ghashang et al. (2016) synthesized MgO nanopowders recently by using magnesium chloride as precursor along with leaf extract of *Rosmarinus officinalis*. The field emission scanning electron microscopy (FE-SEM) micrographs showed that the average size of nanoparticles was approximately 73 nm. Notably, microwave-assisted synthesis was employed by Karthik et al. (2017) in the synthesis of MgO nanoparticles with leaves extracts of *Andrographis paniculata* and magnesium acetate tetrahydrate. The SEM images displayed that the MgO nanoparticles were rod-shaped with a diameter of 46 nm and length of 185 nm. It was also described in the FTIR spectral analysis that phenols were involved in reducing and stabilizing the nanoparticles. The synthesized MgO nanorods were subjected to in vitro hemolysis assay using human red blood cells at different concentrations (20 and 40 mg/mL). The results revealed that the hemolytic activity at 20 and 40 mg/mL were 1.17 and 2.86%, respectively. According to the International Organization for

Standardization Technical Report 7406, the admissible level for biomaterials is less than 5%. Similarly, MgO nanoparticles have been proven to be less toxic towards human umbilical vein endothelial cells (Ge et al., 2011), human cardiac microvascular endothelial cells (Sun et al., 2011), liver (HepG2), kidney (NRK-52E), intestine (Caco-2), lung (A549) cell lines (Mahmoud et al., 2016), and human intestinal (INT 407) cells (Patel et al., 2013). Thus, all these studies proved that the less toxic green synthesized MgO nanoparticles are appropriate for biomedical applications.

### 21.2.4   TITANIUM DIOXIDE

Titanium dioxide ($TiO_2$) nanoparticles gained focus from scientists due to their wide range of applications and their unique properties such as strong oxidizing power, nontoxic, long-term thermodynamic stability, and high photocatalytic activity (Nasrollahzadeh et al., 2015a). A vast number of articles on the green synthesis of $TiO_2$ nanoparticles have been published in recent years. Recently, the leaf extracts of *Moringa oleifera, Psidium guajava, Mangifera indica, Eclipta prostrata,* and *Annona squamosa* were used in the synthesis of $TiO_2$ nanoparticles. Nasrollahzadeh et al. (2015b) synthesized spherical $TiO_2$ nanoparticles using the root extracts of *Euphorbia heteradena* Jaub and meta-titanium acid as the precursor at 60°C. The average size of nanoparticles was obtained in the range of 17–45 nm and the results revealed that polyphenolics compounds act as the reducing and capping agent during the synthesis. Similarly, Rajakumar et al. (2012) synthesized $TiO_2$ nanoparticles by employing leaf extract of *Eclipta prostrata* and meta-titanium acid at room temperature. The nanoparticles were spherical in shape and their size ranges from 36 to 68 nm. It was suggested that the functional groups of beta-amyrin, stigmasterol, triterpenoids, wedelolactone, luteolin-7-O-glucoside, flavonoids, L-terthienyl methanol, and flavones were responsible for reducing and stabilizing the nanoparticles. The cytotoxicity of $TiO_2$ is evaluated in various cell lines such as human astrocytes-like astrocytoma U87 cells (Lai et al., 2008), adult human skin HaCaT cells (Lopes et al., 2016), adipose-derived stromal cells (Xu et al., 2017), and adenocarcinomic human alveolar basal epithelial A549 cells (Armand et al., 2016). All these studies revealed that the titanium oxide nanoparticles possess doses and concentration-dependent toxicity towards normal cells. Thus, it is recommended that in the formulation with biocompatible polymers and surface stabilization of titanium oxide nanoparticles, these nanoparticles ought to be used in biomedical applications.

## 21.2.5  METAL OXIDE NANOCOMPOSITES

Nanocomposites are the composite materials in which at least one of their dimensions are in nanometer range. Composites are the combinations of several materials that consist of matrix with fillers that are made up of fibers, particles, or sheets. Nanocomposites are materials with high performance, exclusive opportunities for unique design, and peculiar properties (Ebrahimi, 2012; Peng et al., 2012). Metal oxide nanocomposites are generally categorized into magnetic metal oxide and nonmagnetic metal oxide nanocomposites. Magnetic metal oxide nanocomposites include iron oxide–metal (Li et al., 2014; Sood et al., 2016), iron oxide–carbon allotrope (Vermisoglou et al., 2014), and iron oxide–polymer nanocomposites (Oh et al., 2011), whereas nonmagnetic metal oxide nanocomposites include metal–metal oxide (Kochuveedu et al., 2013), metal oxide–carbon allotrope (Liu et al., 2008), and metal oxide–polymer nanocomposites (Cai et al., 2013; Bhagavathula, 2017). Green synthesis has been employed to reduce the toxicity of these nanocomposites as chemical-synthesized nanocomposites are toxic to cells. Recently, silver-reduced graphene oxide–titanium dioxide nanocomposites using *Euphorbia helioscopia* L. leaf extract (Nasrollahzadeh et al., 2016a), palladium-reduced graphene oxide–iron oxide nanocomposites through *Withania coagulans* leaf extract (Atarod et al., 2016a), copper–iron oxide nanocomposites using *Morinda morindoides* leaf extract (Nasrollahzadeh et al., 2016b), silver–titanium oxide nanocomposites through *Euphorbia heterophylla* leaf extract (Atarod et al., 2016b), and copper–iron oxide nanocomposites using *Silybum marianum* L. seed extracts (Sajadi et al., 2016) were synthesized through green method. These studies show that phytochemicals help not only in nanoparticle formation but also in nanocomposite formation. However, there are limited cytotoxic studies available for green synthesized nanocomposites towards normal cells. Recently, novel zinc oxide–silver core-shell nanocomposites were synthesized using essential oil from wild ginger and their cytotoxic analysis towards VERO cells shows a dose-dependent toxicity with less toxicity below 100 μg/mL concentration (Azizi et al., 2016). This shows that the green synthesis of nanocomposites will help in enhancing their property as well as reducing their cytotoxicity.

## 21.3  BIOMEDICAL APPLICATIONS

Metal oxide nanoparticles have recently gained importance in many biomedical applications such as in the diagnosis and treatment of cancer, neurodegenerative diseases, biosensing applications, antibacterial activity, and in treating other rare diseases such as Lafora, Marasmus, and Kwashiorkor.

### 21.3.1  CANCER DIAGNOSIS AND TREATMENT

Iron oxide nanoparticles are one of the important metal oxide nanoparticles that is significantly used in cancer diagnosis and treatment (Huang et al., 2016). Superparamagnetic iron oxide nanoparticles possess favorable biodegradability, superparamagnetic, and surface properties which make them a novel magnetic resonance imaging contrast agent to diagnose cancer (Shabestari et al., 2017). Recently, Nagajyothi et al. (2017b) synthesized 40-nm-sized iron oxide nanoparticles using an aqueous extract of *Psoralea corylifolia* seeds. The results revealed that the magnetic nanoparticles are good growth inhibitor of renal carcinoma cells (Caki-2 cells) at lower concentrations (Nagajyothi et al., 2017b). These magnetic nanoparticles can also be used in hyperthermia and drug delivery application for the diagnosis as well as in the efficient treatment of cancer (Hola et al., 2015). Other than iron oxide nanoparticles, copper oxides in nanoform show better anticancer activity against several cancer cells. Copper oxide nanoparticles have been synthesized using *Ficus religiosa* leaf extract and they showed anticancer activity against A549 human lung cancer cells (Sankar et al., 2014b). Recently, copper oxide nanoparticles were synthesized by using aqueous black bean extract and they showed enhanced in vitro anticancer activity against human cervical carcinoma cells (Nagajyothi et al., 2017a).

Zinc oxide (ZnO) and titanium oxide (TiO$_2$) nanoparticles also showed anticancer ability towards several cancer cells. In 2014, Vimala et al. prepared ZnO nanoparticles using *Borassus flabellifer* fruit extract and their in vitro studies revealed that the green synthesized ZnO nanoparticles regulate expression of Bcl-2 gene, which is encoded to a Bcl-2 protein that regulates cell death (apoptosis). The green synthesized ZnO nanoparticles were also loaded with doxorubicin (DOX) and the results disclosed that DOX-ZnO nanoparticles have low toxicity and high therapeutic efficacy. Further characterization and in vivo tests will provide convincing evidence to use these nanoparticles as a promising drug delivery system and to regulate cancer-causing genes (Vimala et al., 2014). Similarly, DOX-loaded zinc oxide nanoparticles were proved to inhibit breast cancer cell explosion rate through E2F3/Akt signaling circuit with higher efficacy and decreased side effects (Vimala et al., 2016). Likewise, titanium dioxide nanoparticles are produced using aqueous leaf extracts of *P. guajava* and the scavenging radical results, estimated by 2,2-diphenyl-1-picrylhydrazyl method, showed that the green synthesized nanoparticles possess higher antioxidant activity than phytochemicals in the leaf extract. These antioxidant nanoparticles are highly beneficial in the treatment of early-stage cancers (Santhoshkumar et al., 2014). Recently, novel gadolinium-doped titanium oxide nanoparticles were prepared by using *Piper betel* leaf and the results revealed that they possess enhanced antioxidant ability which will be highly helpful in cancer treatment (Hunagund et al., 2017). Other than these usual nanoparticles, cerium oxide (Charbgoo et al., 2017), tin oxide (Tammina et al., 2017), magnesium oxide (Karthik et al., 2017), and manganese oxide (Kim et al., 2011) nanoparticles are also extensively prepared through green synthesis method and are studied for detection and treatment of cancer applications. Moreover, metal oxide nanocomposites produced through green synthesis are gaining popularity in cancer diagnosis and treatment due to their enhanced composite properties (Khan et al., 2016; Ali et al., 2017).

## 21.3.2   NEURODEGENERATIVE DISEASES

Metal oxide nanoparticles, such as cerium oxides and iron oxides, were extensively studied to prove their potential in the treatment of neurodegenerative diseases. Increased oxidative stress due to enormous oxygen usage, lower antioxidant systems, and lipid peroxidation are the causative factors for neurodegenerative disorders (Mariani et al., 2005). Trauma, Parkinson's disease, aging, Alzheimer's disease, and ischemic stroke are some of the diseases associated with neurodegenerative disorders (Emerit et al., 2004). Cerium oxide nanoparticles are highly helpful in the removal of reactive oxygen species and in reducing their formation, which is valuable in treating neurodegenerative diseases (Xu et al., 2014). They are also proved to protect neurons from free radical-mediated damage (Das et al., 2007) and exogenous oxidants (Schubert et al., 2006). Recently, it was found that cerium oxide nanoparticles have the ability to promote neurogenesis and 5'-adenine monophosphate-activated protein kinase-protein kinase C-cyclic adenosine monophosphate response element-binding protein binding (AMPK-PKC-CBP) signaling cascade-mediated hypoxia-induced memory impairment (Arya et al., 2016). Iron oxide nanoparticles have also been employed in the detection and treatment of neurodegenerative diseases. Zhang et al. (2016) reported that iron oxide nanoparticles intake through diet delay aging and ameliorate neurodegeneration in *Drosophila* (Zhang et al., 2016). Many literatures reported that magnetic iron oxide nanoparticles (superparamagnetic and ultra-small superparamagnetic) are excellent tools that can help in Alzheimer's disease imaging (Zhou et al., 2014). Moreover, Migliore et al. (2015) reported the benefits of several other metal oxide nanoparticles such as $TiO_2$, silicon dioxide, and zinc oxide nanoparticles in the diagnosis and treatment of neurodegenerative diseases (Migliore et al., 2015). All these metal oxide nanoparticles can be synthesized extensively through phytochemical-mediated green synthesis methods, which enhance their bioavailability and bioactivity with less cytotoxicity, in order to be used as a potential curative agent against neuro-degenerative diseases.

### 21.3.3 BIOSENSING APPLICATIONS

Biosensing is one of the important aspects of metal oxide nanoparticles as they possess a band gap between their valence and conduction band which gives a signal when there is a slight modification in their band gap (Comini, 2006). Several metal oxide nanoparticles are proved to possess biosensing capability and are employed in efficient biosensing applications. For instance, metal oxide nanoparticles such as ZnO, $TiO_2$, and iron oxide nanoparticles were reported to possess enhanced enzymatic biosensing ability (Shi et al., 2014). Iron oxide nanoparticles showed potential to be used as uric acid biosensor (Devi et al., 2014), electrochemical nanobiosensors to detect enzyme, DNA, immuno- and aptamers (Hasanzadeh et al., 2015), and magnetic bead-based biosensor to detect glucose, cholesterol, lactate, urea, creatinine, and creatine (Wu et al., 2015). Similarly, cerium oxide nanoparticles also possess electrochemical, fluorometric, and colorimetric sensing ability to detect hydrogen peroxide in cells, which helps to identify and sense cell death (Patil, 2006; Liu et al., 2015; Sardesai et al., 2015; Charbgoo et al., 2017). Likewise, multiwalled carbon nanotubes with cobalt oxide nanoparticles are reported to be highly useful in the fabrication of enzyme-free biosensor to detect hydrogen peroxide (Heli et al., 2014). Other than these metal oxides, nanocomposites prepared by using metal oxides also display efficient biosensing property. Graphene–tungsten oxide–gold nanohybrid membranes can be used as a glucose sensor (Devadoss et al., 2014), graphene oxide–silver nanocomposites for the simultaneous detection of quercetin and morin (Yola et al., 2014), and iron oxide–chitosan nano-composite as an electrochemical biosensor to detect carbofuran (Jeyapra-gasam et al., 2014) are some of the examples of metal oxide nanocomposites that are favorable for biosensing applications. Phytochemical-based green synthesized metal oxide nanoparticles will be highly useful in biosensing application, provided that they are highly stable and similar to chemically synthesized nanoparticles.

## 21.3.4  ANTIMICROBIAL ACTIVITY

It can be noted from the literatures that almost all the green synthesized metal oxide nanoparticles show antimicrobial activity towards several organisms. However, the enhanced antimicrobial activity can be observed in a few metal oxide nanoparticles. Recent literatures suggested that zinc oxide nanoparticles prepared by using milky latex of *C. gigantea* (Panda et al., 2017), silver oxide nanoparticles synthesized by *Ficus benghalensis* prop root extract (Manikandan et al., 2017), zinc oxide nanoparticles prepared through microwave-assisted aqueous extraction of *Suaeda aegyptiaca* plant (Rajabi et al., 2017), and copper oxide nanoparticles produced by *Abutilon indicum* plant extract possess enhanced antibacterial activities. Similarly, efficient antifungal activity of zinc oxide nanoparticles prepared through *Amaranthus caudatus* aqueous extract (Jeyabharathi et al., 2017) and *Sageretia thea* aqueous leaf extract (Khalil et al., 2017b), undoped ZnO and copper-doped ZnO nanoparticles synthesized through *A. indicum*, *Clerodendrum infortunatum* and *Clerodendrum inerme* aqueous extracts (Khan et al., 2018), copper oxide nanoparticles produced using *Cissus quadrangularis* (Devipriya et al., 2017), and titanium oxide nanoparticles prepared by hydrothermal leaf extracts of *Morinda citrifolia* (Sundrarajan et al., 2017) against several harmful fungal species has recently been reported. Recent research has focused on using metal oxide nanoparticles that are synthesized by phytochemicals as antiviral agents (Yadavalli et al., 2017; Sathishkumar et al., 2018). Many studies revealed that individual phytochemical also possesses extreme antimicrobial property (Lal et al., 2017; Racowski et al., 2017). All these studies clearly show that the combined antimicrobial activity of metal oxide nanoparticles prepared using phytochemicals helps them to be effectual antimicrobial agents.

## 21.4  FUTURE BIOMEDICAL APPLICATIONS

Since green synthesized metal oxide nanoparticles are less toxic towards human cells and possess enormous biomedical importance, they can be further explored to be used as therapeutic agents against rare diseases. In diabetes, metal oxide nanoparticles are employed for the controlled delivery of insulin (Turcheniuk et al., 2014) and as glucose sensors (Fang et al., 2017; Nor et al., 2017). Recently, magnesium oxide nanoparticles were proposed to have potential in reversing insulin resistance; this reversal can be an alternative to insulin therapy in type 2 diabetes. Similarly, several metal oxide nanoparticles are extensively used in drug delivery (Sharma et al., 2016; Mirzaei et al., 2017) and gene delivery applications (Riley et al., 2017; Yadavalli et al., 2017). Thus, less toxic, green synthesized metal oxide nanoparticles will be a crucial asset in treating rare diseases such as Lafora, Progeria, Marasmus, and Kwashiorkor. Lafora is an autosomal recessive genetic disorder in which there will be Lafora body inclusions with the cytoplasm of heart, muscle, and liver cells (Wrone et al., 2001). It is a neurodegenerative disease in which the patients do not live past 25 years of age (Minassian, 2001). Green synthesized metal oxide nanoparticles can be used to deliver antisense oligonucleotides, clustered regularly interspaced short palindromic repeats/Cas9 gene, and small molecule glycogen synthase inhibitors (Tagliabracci et al., 2008) due to their enhanced drug and gene delivery capability, which can help in controlling and treating Lafora which is a severest form of epilepsies. Similarly, metal oxide nanoparticles can also be helpful in delivering sirolimus, a mechanistic target of rapamycin inhibitor (Cao et al., 2011), in patients suffering from Progeria disease. Progeria is a rare genetic disorder which includes several progeroid syndromes, such as aging at a very early age (Sinha et al., 2014). This disorder is due to the mutation in LMNA gene that encodes lamin A/C protein which is involved in nuclear stability, gene expression, and chromatin structure in cells (Novelli et al., 2002). Kwashiorkor is a type of malnutrition in children, caused by the insufficient intake of protein, whereas Marasmus is caused by the insufficient intake of carbohydrates, fats, and proteins (Boyne et al., 2017). As metal oxide nanoparticles possess superior drug and gene delivery property, in future, they can be used for the controlled and targeted delivery of protein in patients with protein deficiency diseases.

## 21.5 CONCLUSION

Phytochemical-based green synthesized metal oxide nanoparticles possess extensive biomedical applications due to their less toxic nature towards human cells. However, cytotoxicity of certain green synthesized metal oxide nanoparticles has to be controlled to effectively use them in the diagnosis and treatment of diseases. Moreover, very few studies are available that show that phytochemical-based green synthesized metal oxide nanoparticles are toxic to normal human cells. Surface functionalization and formulation with biocompatible polymers help in reducing their cytotoxicity and in making them biocompatible. In future, extensive research in the field of nanoparticle synthesis through phytochemicals will also help to reduce their drawbacks such as stability and dissociation in body fluids. Metal oxide nanoparticles that are prepared using phytochemicals will be the future of nanomedicines due to their potential in large-scale production, less cytotoxicity, and superior biomedical applicational importance.

## KEYWORDS

- **cytotoxicity**
- **green synthesis**
- **metal oxide nanoparticles**
- **phytochemicals**
- **plant extracts**

## REFERENCES

Agarwal, H.; Kumar, S. V.; Rajeshkumar, S.A Review on Green Synthesis of Zinc Oxide Nanoparticles – An Eco-Friendly Approach. *Res. Efficient Technolog.* **2017,** *3*(4), 406–413.

Ahamed, M.; Siddiqui, M. A.; Akhtar, M. J.; Ahmad, I.; Pant, A. B.; Alhadlaq, H. Genotoxic Potential of Copper Oxide Nanoparticles in Human Lung Epithelial Cells. *Biochem. Biophys. Res. Commun.* **2010,** *396*(2), 578–583.

Alagarasi, A. *Introduction to Nanomaterials.* National Center for Environmental Research, USA, 2011.

Ali, M. A.; Singh, C.; Srivastava, S.; Admane, P.; Agrawal, V. V.; Sumana, G.; John, R.; Panda, A.; Dong, L.; Malhotra, B. D. Graphene Oxide–Metal Nanocomposites for Cancer Biomarker Detection. *RSC Adv.* **2017,** *7*(57), 35982–35991.

Andreescu, S.; Ornatska, M.; Erlichman, J. S.; Estevez, A.; Leiter, J. Biomedical Applications of Metal Oxide Nanoparticles. In *Fine Particles in Medicine and Pharmacy;* Matijevic, E., Ed.; Springer: USA, 2012, pp 57–100.

Armand, L.; Tarantini, A.; Beal, D.; Biola-Clier, M.; Bobyk, L.; Sorieul, S.; Pernet-Gallay, K.; Marie-Desvergne, C.; Lynch, I.; Herlin-Boime, N. Long-Term Exposure of A549 Cells to Titanium Dioxide Nanoparticles Induces DNA Damage and Sensitizes Cells Towards Genotoxic Agents. *Nanotoxicology* **2016,** *10*(7), 913–923.

Arumugam, A.; Karthikeyan, C.; Hameed, A. S. H.; Gopinath, K.; Gowri, S.; Karthika, V. Synthesis of Cerium Oxide Nanoparticles Using *Gloriosa superba* L. Leaf Extract and Their Structural, Optical and Antibacterial Properties. *Mater Sci. Eng. C.* **2015,** *49,* 408–415.

Arya, A.; Gangwar, A.; Singh, S. K.; Roy, M.; Das, M.; Sethy, N. K. Bhargava, K. Cerium Oxide Nanoparticles Promote Neurogenesis and Abrogate Hypoxia-Induced Memory Impairment Through AMPK–PKC–CBP Signaling Cascade. *Int. J. Nanomed.* **2016,** *11,* 1159–73.

Astruc, D. *Nanoparticles and Catalysis;* Wiley, 2008.

Atarod, M.; Nasrollahzadeh, M.; Mohammad Sajadi, S. Green Synthesis of Pd/RGO/Fe$_3$O$_4$ Nanocomposite Using *Withania coagulans* Leaf Extract and its Application as Magnetically Separable and Reusable Catalyst for the Reduction of 4-Nitrophenol. *J. Colloid Inter. Sci.* **2016a,** *465*(Supplement C), 249–258.

Atarod, M.; Nasrollahzadeh, M.; Mohammad Sajadi, S. *Euphorbia heterophylla* Leaf Extract Mediated Green Synthesis of Ag/TiO$_2$ Nanocomposite and Investigation of its Excellent Catalytic Activity for Reduction of Variety of Dyes in Water. *J. Colloid. Inter. Sci.* **2016b,** *462*(Supplement C), 272–279.

Awwad, A. M.; Ahmad, A. L. Biosynthesis, Characterization, and Optical Properties of Magnesium Hydroxide and Oxide Nanoflakes Using *Citrus limon* Leaf Extract. *Arab. J. Phys. Chem.* **2014,** *1*(2), 66.

Azam, A.; Ahmed, A. S.; Oves, M.; Khan, M. S.; Habib, S. S.; Memic, A. Antimicrobial Activity of Metal Oxide Nanoparticles Against Gram-Positive and Gram-Negative Bacteria: a Comparative Study. *Int. J. Nanomed.* **2012,** *7,* 6003–6009.

Azizi, S.; Mohamad, R.; Rahim, R. A.; Moghaddam, A. B.; Moniri, M.; Ariff, A.; Saad, W. Z.; Namvab, F. ZnO-Ag Core Shell Nanocomposite Formed by Green Method Using Essential Oil of Wild Ginger and Their Bactericidal and Cytotoxic Effects. *Appl. Surf. Sci.* **2016,** *384*(Supplement C), 517–524.

Ba-Abbad, M. M.; Kadhum, A. A. H.; Mohamad, A. B.; Takriff, M. S.; Sopian, K. Synthesis and Catalytic Activity of TiO$_2$ Nanoparticles for Photochemical Oxidation of Concentrated Chlorophenols Under Direct Solar Radiation. *Int. J. Electrochem. Sci.* **2012,** *7,* 4871–4888.

Bao, S. J.; Li, C. M.; Zang, J. F.; Cui, X. Q.; Qiao, Y.; Guo, J. New Nanostructured TiO$_2$ for Direct Electrochemistry and Glucose Sensor Applications. *Adv. Funct. Mater.* **2008,** *18*(4), 591–599.

Bartholomey, E.; House, S.; Ortiz, F. A. Balanced Approach for Formulating Sunscreen Products Using Zinc Oxide. *SOFW J.* **2016,** *14,* 203.

Bhagavathula, S. Metal-oxide Polymer Nanocomposite Films from Disposable Scrap Tire Powder/Poly-Îµ-Caprolactone for Advanced Electrical Energy (Capacitor) Applications. *J. Cleaner Prod.* **2017.**

Bhattacharya, D.; Gupta, R. K. Nanotechnology and Potential of Microorganisms. *Crit. Rev. Biotechnol.* **2005,** *25*(4), 199–204.

Bhuyan, T.; Mishra, K.; Khanuja, M.; Prasad, R.; Varma, A. Biosynthesis of Zinc Oxide Nanoparticles from *Azadirachta indica* for Antibacterial and Photocatalytic Applications. *Mater. Sci. Semicond. Process.* **2015,** *32,* 3255–3261.

Bordbar, M.; Negahdar, N.; Nasrollahzadeh, M. *Melissa officinalis* L. Leaf Extract Assisted Green Synthesis of CuO/ZnO Nanocomposite for the Reduction of 4-Nitrophenol and Rhodamine B. *Sep. Purif Technol.* **2018,** *191,* 295–300.

Boyne, M. S.; Francis-Emmanuel, P.; Tennant, I. A.; Thompson, D. S.; Forrester, T. E. Cardiometabolic Risk in Marasmus and Kwashiorkor Survivors. In *Handbook of Famine,*

*Starvation, and Nutrient Deprivation: from Biology to Policy;* Preedy, V.R., Patel, V.B. Eds.; Springer: New York, 2017; pp 1–23.

Cai, W.; Chen, Q.; Cherepy, N.; Dooraghi, A.; Kishpaugh, D.; Chatziioannou, A.; Payne, S.; Xiang, W.; Pei, Q. Synthesis of Bulk-Size Transparent Gadolinium Oxide–Polymer Nanocomposites for Gamma Ray Spectroscopy. *J. Mater. Chem. C.* **2013,** *1*(10), 1970–1976.

Cao, K.; Graziotto, J. J.; Blair, C. D.; Mazzulli, J. R.; Erdos, M. R.; Krainc, D.; Collins, F. S. Rapamycin Reverses Cellular Phenotypes and Enhances Mutant Protein Clearance in Hutchinson-Gilford Progeria Syndrome Cells. *Sci. Transl. Med.* **2011,** *3*(89), 89ra58.

Charbgoo, F.; Ahmad, M. B.; Darroudi, M. Cerium Oxide Nanoparticles: Green Synthesis and Biological Applications. *Int. J. Nanomed.* **2017,** *12,* 1401–1413.

Comini, E. Metal Oxide Nano-Crystals for Gas Sensing. *Anal. Chim. Acta* **2006,** *568*(1), 28–40.

Das, M.; Patil, S.; Bhargava, N.; Kang, J-F.; Riedel, L. M.; Seal, S.; Hickman, J. J. Auto-Catalytic Ceria Nanoparticles Offer Neuroprotection to Adult Rat Spinal Cord Neurons. *Biomaterials.* **2007,** *28*(10), 1918–1925.

Das, A. K.; Marwal, A.; Sain, D.; Pareek, V. One-Step Green Synthesis and Characterization of Plant Protein-Coated Mercuric Oxide (HgO) Nanoparticles: Antimicrobial Studies. *Int. Nano Lett.* **2015,** *5*(3), 125–132.

Devadoss, A.; Sudhagar, P.; Das, S.; Lee, S. Y.; Terashima, C.; Nakata, K.; Fujishima, A.; Choi, W.; Kang, Y. S.; Paik, U. Synergistic Metal–Metal Oxide Nanoparticles Supported Electrocatalytic Graphene for Improved Photoelectrochemical Glucose Oxidation. *ACS Appl. Mater. Interfaces.* **2014,** *6*(7), 4864–4871.

Devi, R.; Pundir, C. Construction and Application of an Amperometric Uric Acid Biosensor Based on Covalent Immobilization of Uricase on Iron Oxide Nanoparticles/Chitosan-*g*-Polyaniline Composite Film Electrodeposited on Pt Electrode. *Sens. Actuators, B.* **2014,** *193,* 608–615.

Devipriya, D.; Roopan, S. M. *Cissus quadrangularis* Mediated Ecofriendly Synthesis of Copper Oxide Nanoparticles and its Antifungal Studies Against *Aspergillus niger, Aspergillus flavus. Mater. Sci. Eng. C.* **2017,** *80*(Supplement C), 38–44.

Dizaj, S. M.; Lotfipour, F.; Barzegar-Jalali, M.; Zarrintan, M. H.; Adibkia, K. Antimicrobial Activity of the Metals and Metal Oxide Nanoparticles. *Mater. Sci. Eng. C.* **2014,** *44,* 278–284.

Duan, S.; Wang, R. Bimetallic Nanostructures with Magnetic and Noble Metals and Their Physicochemical Applications. *Prog. Nat. Sci.: Mater. Int.* **2013,** *23*(2), 113–126.

Ebrahimi, F. *Nanocomposites – New Trends and Developments;* 2012. DOI: 10.5772/3389; ISBN: 978-953-51-0762-0.

Elumalai, K.; Velmurugan, S.; Ravi, S.; Kathiravan, V.; Raj, G. A. Bio-approach: Plant Mediated Synthesis of ZnO Nanoparticles and Their Catalytic Reduction of Methylene Blue and Antimicrobial Activity. *Adv. Powder Technol.* **2015,** *26*(6), 1639–1651.

Emerit, J.; Edeas, M.; Bricaire, F. Neurodegenerative Diseases and Oxidative Stress. *Biomed. Pharmacother.* **2004,** *58*(1), 39–46.

Ezhilarasi, A. A.; Vijaya, J. J.; Kaviyarasu, K.; Maaza, M.; Ayeshamariam, A.; Kennedy, L. J. Green Synthesis of NiO Nanoparticles Using *Moringa oleifera* Extract and Their Biomedical Applications: Cytotoxicity Effect of Nanoparticles against HT-29 cancer cells. *J. Photochem. Photobiol. B.* **2016,** *164,* 352–360.

Fahmy, B.; Cormier, S. A. Copper Oxide Nanoparticles Induce Oxidative Stress and Cytotoxicity in Airway Epithelial Cells. *Toxicol. In Vitro.* **2009,** *23*(7), 1365–1371.

Fang, L.; Wang, F.; Chen, Z.; Qiu, Y.; Zhai, T.; Hu, M.; Zhang, C.; Huang, K. Flower-Like MoS$_2$ Decorated with Cu$_2$O Nanoparticles for Non-enzymatic Amperometric Sensing of Glucose. *Talanta* **2017**, *167,* 593–599.

Ge, S.; Wang, G.; Shen, Y.; Zhang, Q.; Jia, D.; Wang, H.; Dong, Q.; Yin, T. Cytotoxic Effects of MgO Nanoparticles on Human Umbilical Vein Endothelial Cells in Vitro. *IET Nanobiotechnol.* **2011**, *5*(2), 36–40.

Ghashang, M.; Mansoor, S. S.; Mohammad Shafiee, M. R.; Kargar, M.; Najafi Biregan, M.; Azimi, F.; Taghrir, H. Green Chemistry Preparation of MgO Nanopowders: Efficient Catalyst for the Synthesis of Thiochromeno[4, 3-B]pyran and Thiopyrano[4, 3-B]pyran Derivatives. *J. Sulfur. Chem.* **2016**, *37*(4), 377–390.

Ghidan, A. Y.; Al-Antary, T. M.; Awwad, A. M. Green Synthesis of Copper Oxide Nanoparticles Using *Punica granatum* Peels Extract: Effect on Green Peach Aphid. *Environ. Nanotechnol. Monit. Manag.* **2016**, *6,* 95–98.

Gupta, M.; Mahajan, V. K.; Mehta, K. S.; Chauhan, P. S. Zinc Therapy in Dermatology: A Review. *Dermatol. Res. Pract.* **2014**, Article ID 709152, 1–11. http://dx.doi.org/10.1155/2014/709152

Hasanzadeh, M.; Shadjou, N.; de la Guardia, M. Iron and Iron-Oxide Magnetic Nanoparticles as Signal-Amplification Elements in Electrochemical Biosensing. *TrAC Trends Anal. Chem.* **2015**, *72*(Supplement C), 1–9.

Heli, H.; Pishahang, J. Cobalt Oxide Nanoparticles Anchored to Multiwalled Carbon Nanotubes: Synthesis and Application for Enhanced Electrocatalytic Reaction and Highly Sensitive Nonenzymatic Detection of Hydrogen Peroxide. *Electrochim. Acta* **2014**, *123*(Supplement C), 518–526.

Hola, K.; Markova, Z.; Zoppellaro, G.; Tucek, J.; Zboril, R. Tailored Functionalization of Iron Oxide Nanoparticles for MRI, Drug Delivery, Magnetic Separation and Immobilization of Biosubstances. *Biotechnol Adv.* **2015**, *33*(6), 1162–1176.

Huang, C-C.; Liao, Z-X.; Lu, H-M.; Pan, W-Y.; Wan, W-L.; Chen, C-C.; Sung, H-W. Cellular Organelle-Dependent Cytotoxicity of Iron Oxide Nanoparticles and Its Implications for Cancer Diagnosis and Treatment: a Mechanistic Investigation. *Chem. Mater.* **2016**, *28*(24), 9017–9025.

Hulkoti, N. I.; Taranath, T. Biosynthesis of Nanoparticles Using Microbes—A Review. *Colloids Surf B* **2014**, *121,* 474–483.

Hunagund, S. M.; Desai, V. R.; Barretto, D. A.; Pujar, M. S.; Kadadevarmath, J. S.; Vootla, S.; Sidarai, A. H. Photocatalysis Effect of a Novel Green Synthesis Gadolinium Doped Titanium Dioxide Nanoparticles on Their Biological Activities. *J. Photochem. Photobiol. A.* **2017**, *346*(Supplement C), 159–167.

Jeevanandam, J.; Chan, Y. S.; Danquah, M. K. Biosynthesis of Metal and Metal Oxide Nanoparticles. *Chem. Bio. Eng. Rev.* **2016**, *3*(2), 55–67.

Jeevanandam, J.; Aing, Y. S.; Chan, Y. S.; Pan, S.; Danquah, M. K. Nanoformulation and Application of Phytochemicals as Antimicrobial Agents. In *Antimicrobial Nanoarchitectonics*; Elsevier, 2017a; pp 61–82.

Jeevanandam, J.; Chan, Y. S.; Danquah, MK Biosynthesis and Characterization of MgO Nanoparticles from Plant Extracts via Induced Molecular Nucleation. *New J. Chem.* **2017b**, *41*(7), 2800–2814.

Jeyabharathi, S.; Kalishwaralal, K.; Sundar, K.; Muthukumaran, A. Synthesis of Zinc Oxide Nanoparticles (ZnONPs) by Aqueous Extract of *Amaranthus caudatus* and Evaluation of Their Toxicity and Antimicrobial Activity. *Mater. Lett.* **2017c,** *209*(Supplement C), 295–298.

Jeyapragasam, T.; Saraswathi, R. Electrochemical Biosensing of Carbofuran Based on Acetylcholinesterase Immobilized onto Iron Oxide–Chitosan Nanocomposite. *Sens. Actuators B* **2014**, *191,* 681–687.

Jothi, N.; Sagayaraj, P. The Influence of Capping by TGA and PVP in Modifying the Structural, Morphological, Optical and Thermal Properties of ZnS Nanoparticles. *Arch. Appl. Sci. Res.* **2012**, *4,* 1079–1090.

Kannan, S.; Sundrarajan, M. A Green Approach for the Synthesis of a Cerium Oxide Nanoparticle: Characterization and Antibacterial Activity. *Int. J. Nanosci.* **2014**, *13*(03), 1450018.

Karthik, K.; Dhanuskodi, S.; Kumar, S. P.; Gobinath, C.; Sivaramakrishnan, S. Microwave Assisted Green Synthesis of MgO Nanorods and Their Antibacterial and Anti-Breast Cancer Activities. *Mater. Lett.* **2017**, *206,* 217–220.

Kaushik, M.; Moores, A. Nanocelluloses as Versatile Supports for Metal Nanoparticles and Their Applications in Catalysis. *Green Chem.* **2016**, *18*(3), 622–637.

Khalil, A. T.; Ovais, M.; Ullah, I.; Ali, M.; Jan, S. A.; Shinwari, Z. K.; Maaza, M. Bioinspired Synthesis Of Pure Massicot Phase Lead Oxide Nanoparticles And Assessment Of Their Biocompatibility, Cytotoxicity And In -Vitro Biological Properties. *Arabian J. Chem.* **2017a**. https://doi.org/10.1016/j.arabjc.2017.08.009.

Khalil, A. T.; Ovais, M.; Ullah, I.; Ali, M.; Jan, S. A.; Shinwari, Z. K.; Maaza, M. *Sageretia thea* (Osbeck.) Mediated Synthesis of Zinc Oxide Nanoparticles and its Biological Applications. *Nanomedicine.* **2017b**, *12*(15), 1767–1789.

Khan, M.; Khan, M.; Al-Marri, A. H.; Al-Warthan, A.; Alkhathlan, H. Z.; Siddiqui, M. R. H.; Nayak, V. L.; Kamal, A.; Adil, S. F. Apoptosis Inducing Ability of Silver Decorated Highly Reduced Graphene Oxide Nanocomposites in A549 Lung Cancer. *Int. J. Nanomed.* **2016**, *11,* 873–883.

Khan, S. A.; Noreen, F.; Kanwal, S.; Iqbal, A.; Hussain, G. Green Synthesis of ZnO and Cu-Doped ZnO Nanoparticles from Leaf Extracts of *Abutilon indicum, Clerodendrum infortunatum, Clerodendrum inerme* and Investigation of Their Biological and Photocatalytic Activities. *Mater. Sci. Eng. C.* **2018**, *82*(Supplement C), 46–59.

Kim, T.; Momin, E.; Choi, J.; Yuan, K.; Zaidi, H.; Kim, J.; Park, M.; Lee, N.; McMahon, M. T. Quinones-Hinojosa, A. Mesoporous Silica-Coated Hollow Manganese Oxide Nanoparticles as Positive $T_1$ Contrast Agents for Labeling and MRI Tracking of Adipose-Derived Mesenchymal Stem Cells. *J. Am. Chem. Soc.* **2011**, *133*(9), 2955–2961.

Kochuveedu, S. T.; Jang, Y. H.; Kim, D. H. A Study on the Mechanism for the Interaction of Light with Noble Metal-Metal Oxide Semiconductor Nanostructures for Various Photophysical Applications. *Chem. Soc. Rev.* **2013**, *42*(21), 8467–8493.

Lai, J. C.; Lai, M. B.; Jandhyam, S.; Dukhande, V. V.; Bhushan, A.; Daniels, C. K.; Leung, S. W. Exposure to Titanium Dioxide and Other Metallic Oxide Nanoparticles Induces Cytotoxicity on Human Neural Cells and Fibroblasts. *Int. J. Nanomed.* **2008**, *3*(4), 533.

Lal, K.; Ahmed, N.; Mathur, A. Study of the Antimicrobial Profile and Phytochemical Composition of Solvent Extracts of Leaves and Female Cones of *Cycas revoluta. Int. J. Curr. Microbiol. App. Sci.* **2017**, *6*(4), 2514–2522.

Laurent, S.; Forge, D.; Port, M.; Roch, A.; Robic, C.; Vander Elst, L.; Muller, R. N. Magnetic Iron Oxide Nanoparticles: Synthesis, Stabilization, Vectorization, Physicochemical Characterizations, and Biological Applications. *Chem. Rev.* **2008**, *108*(6), 2064–2110.

Li, Y.; Duan, X.; Qian, Y.; Yang, L.; Liao, H. Nanocrystalline Silver Particles: Synthesis, Agglomeration, and Sputtering Induced by Electron Beam. *J. Colloid Interface Sci.* **1999**, *209*(2), 347–349.

Li, L.; Mak, K.; Leung, C.; Leung, C.; Ruotolo, A.; Chan, K.; Chan, W.; Pong, P. Synthesis and Morphology Control of Gold/Iron Oxide Magnetic Nanocomposites via a Simple Aqueous Method. *IEEE Trans. Magn.* **2014**, *50*(1), 1–5.

Liu, D.; Yang, S.; Lee, S-T. Preparation of Novel Cuprous Oxide–Fullerene [60] Core–Shell Nanowires and Nanoparticles via a Copper (I)-Assisted Fullerene-Polymerization Reaction. *J. Phys. Chem. C.* **2008**, *112*(18), 7110–7118.

Liu, B.; Sun, Z.; Huang, P-J. J.; Liu, J. Hydrogen Peroxide Displacing DNA from Nanoceria: Mechanism and Detection of Glucose in Serum. *J. Am. Chem. Soc.* **2015**, *137*(3), 1290–1295.

Lopes, V. R.; Loitto, V.; Audinot, J-N.; Bayat, N.; Gutleb, A. C.; Cristobal, S. Dose-Dependent Autophagic Effect of Titanium Dioxide Nanoparticles in Human Hacat Cells at Non-cytotoxic Levels. *J. Nanobiotechnol.* **2016**, *14*(1), 22.

Ma, H.; Williams, P. L.; Diamond, S. A. Ecotoxicity of Manufactured ZnO Nanoparticles – A Review. *Environ. Pollut.* **2013**, *17,* 276–285.

Mahendiran, D.; Subash, G.; Selvan, D. A.; Rehana, D.; Kumar, R. S.; Rahiman, A. K. Biosynthesis of Zinc Oxide Nanoparticles Using Plant Extracts of *Aloe vera* and *Hibiscus sabdariffa*: Phytochemical, Antibacterial, Antioxidant and Anti-Proliferative Studies. *J. Bionanosci.* **2017**, *7*(3), 530–545.

Mahmoud, A.; Ezgi, Ö.; Merve, A.; Özhan, G. In Vitro Toxicological Assessment of Magnesium Oxide Nanoparticle Exposure in Several Mammalian Cell Types. *Int. J. Toxicol.* **2016**, *35*(4), 429–437.

Malik, P.; Shankar, R.; Malik, V.; Sharma, N.; Mukherjee, T. K. Green Chemistry Based Benign Routes for Nanoparticle Synthesis. *J. Nanopart.* **2014**. Article ID: 302429, 1–14. http://dx.doi.org/10.1155/2014/302429

Manikandan, V.; Velmurugan, P.; Park, J.-H.; Chang, W.-S.; Park, Y.-J.; Jayanthi, P.; Cho, M.; Oh, B.-T. Green Synthesis of Silver Oxide Nanoparticles and its Antibacterial Activity Against Dental Pathogens. *3 Biotech.* **2017**, *7*(1), 72.

Mariani, E.; Polidori, M.; Cherubini, A.; Mecocci, P. Oxidative Stress in Brain Aging, Neurodegenerative and Vascular Diseases: an Overview. *J. Chromatogr. B* **2005**, *827*(1), 65–75.

Meech, J. A.; Kawazoe, Y.; Kumar, V.; Maguire, J. F. *Intelligence in a Small Materials World*; DEStech Publications, 2005.

Migliore, L.; Uboldi, C.; Di Bucchianico, S.; Coppedè, F. Nanomaterials and Neurodegeneration. *Environ. Mol. Mutagen.* **2015**, *56*(2), 149–170.

Minassian, B. A. Lafora'S Disease: Towards a Clinical, Pathologic, and Molecular Synthesis. *Pediatr. Neurol.* **2001**, *25*(1), 21–29.

Mirzaei, H.; Darroudi, M. Zinc Oxide Nanoparticles: Biological Synthesis and Biomedical Applications. *Ceram. Int.* **2017**, *43*(1), 907–914.

Moulton, M. C.; Braydich-Stolle, L. K.; Nadagouda, M. N.; Kunzelman, S.; Hussain, S. M.; Varma, R. S. Synthesis, Characterization and Biocompatibility of "Green" Synthesized Silver Nanoparticles Using Tea Polyphenols. *Nanoscale* **2010**, *2*(5), 763–770.

Mukherjee, P.; Ahmad, A.; Mandal, D.; Senapati, S.; Sainkar, S. R.; Khan, M. I.; Parishcha, R.; Ajaykumar, P.; Alam, M. Kumar, R. Fungus-Mediated Synthesis of Silver Nanoparticles and Their Immobilization in the Mycelial Matrix: a Novel Biological Approach to Nanoparticle Synthesis. *Nano Lett.* **2001**, *1*(10), 515–519.

Nagajyothi, P. C.; Muthuraman, P.; Sreekanth, T.; Kim, D. H.; Shim, J. Green Synthesis: in Vitro Anticancer Activity of Copper Oxide Nanoparticles Against Human Cervical Carcinoma Cells. *Arabian J. Chem.* **2017a**, *10*(2), 215–225.

Nagajyothi, P. C.; Pandurangan, M.; Kim, D. H.; Sreekanth, T.; Shim, J. Green Synthesis of Iron Oxide Nanoparticles and Their Catalytic and in Vitro Anticancer Activities. *J. Cluster Sci.* **2017b,** *28*(1), 245–257.

Naika, H. R.; Lingaraju, K.; Manjunath, K.; Kumar, D.; Nagaraju, G.; Suresh, D.; Nagabhushana, H. Green Synthesis of CuO Nanoparticles Using *Gloriosa superba* L. Extract and Their Antibacterial Activity. *J. Taibah Uni. Sci.* **2015,** *9*(1), 7–12.

Narayanan, K. B.; Sakthivel, N. Biological Synthesis of Metal Nanoparticles by Microbes. *Adv. Colloid Interface Sci.* **2010,** *156*(1), 1–13.

Nasrollahzadeh, M.; Maham, M.; Sajadi, S. M. Green Synthesis of CuO Nanoparticles by Aqueous Extract of *Gundelia tournefortii* and Evaluation of Their Catalytic Activity for the Synthesis of N-Monosubstituted Ureas and Reduction of 4-Nitrophenol. *J. Colloid Interface Sci.* **2015a,** *455,* 245–253.

Nasrollahzadeh, M.; Sajadi, S. M. Synthesis and Characterization of Titanium Dioxide Nanoparticles Using *Euphorbia heteradena* Jaub Root Extract and Evaluation of Their Stability. *Ceram. Int.* **2015b,** *41*(10), 14435–14439.

Nasrollahzadeh, M.; Atarod, M.; Jaleh, B.; Gandomirouzbahani, M. In Situ Green Synthesis of Ag Nanoparticles on Graphene Oxide/TiO$_2$ Nanocomposite and Their Catalytic Activity for the Reduction of 4-Nitrophenol, Congo Red and Methylene Blue. *Ceram. Int.* **2016a,** *42*(7), 8587–8596.

Nasrollahzadeh, M.; Atarod, M.; Sajadi, S. M. Green Synthesis of the Cu/Fe$_3$O$_4$ Nanoparticles Using *Morinda morindoides* Leaf Aqueous Extract: A Highly Efficient Magnetically Separable Catalyst for the Reduction of Organic Dyes in Aqueous Medium at Room Temperature. *Appl. Surf. Sci.* **2016b,** *364*(Supplement C), 636–644.

Nasrollahzadeh, M.; Sajadi, S. M.; Rostami-Vartooni, A.; Hussin, S. M. Green Synthesis of CuO Nanoparticles Using Aqueous Extract of *Thymus vulgaris* L. Leaves and Their Catalytic Performance for N-Arylation of Indoles and Amines. *J. Colloid Interface Sci.* **2016c,** *466,* 113–119.

Nor, N. M.; Razak, K. A.; Lockman, Z. Physical and Electrochemical Properties of Iron Oxide Nanoparticles-Modified Electrode for Amperometric Glucose Detection. *Electrochim. Acta* **2017,** *248,* 160–168.

Novelli, G.; Muchir, A.; Sangiuolo, F.; Helbling-Leclerc, A.; D'Apice, M. R.; Massart, C.; Capon, F.; Sbraccia, P.; Federici, M.; Lauro, R. Mandibuloacral Dysplasia is Caused by a Mutation in LMNA-Encoding Lamin A/C. *Am. J. Hum. Genet.* **2002,** *71*(2), 426–431.

Oh, J. K.; Park, J. M. Iron Oxide-Based Superparamagnetic Polymeric Nanomaterials: Design, Preparation, and Biomedical Application. *Prog. Polym. Sci.* **2011,** *36*(1), 168–189.

Panda, K. K.; Golari, D.; Venugopal, A.; Achary, V. M. M.; Phaomei, G.; Parinandi, N. L.; Sahu, H. K.; Panda, B. B. Green Synthesized Zinc Oxide (ZnO) Nanoparticles Induce Oxidative Stress and DNA Damage in *Lathyrus sativus* L. Root Bioassay System. *Antioxidants.* **2017,** *6*(2), 35.

Parthasarathy, G.; Saroja, M.; Venkatachalam, M.; Evanjelene, V. Characterization and Antibacterial activity of Green Synthesized ZnO Nanoparticles from *Ocimum basilicum* Leaf Extract. *Adv. Bio. Res.* **2017,** *8*(3), 29–35.

Patel, M. K.; Zafaryab, M.; Rizvi, M.; Agrawal, V. V.; Ansari, Z.; Malhotra, B.; Ansari, S. Antibacterial and Cytotoxic Effect of Magnesium Oxide Nanoparticles on Bacterial and Human Cells. *J. Nanoeng. Nanomanuf.* **2013,** *3*(2), 162–166.

Patidar, V.; Jain, P. Green Synthesis of TiO$_2$ Nanoparticle Using *Moringa oleifera* Leaf Extract. *Int. Res. J. Eng. Technol. (IRJET)* **2017,** *4*(3), 470–473.

Patil, S. Fundamental Aspects of Regenerative Cerium Oxide Nanoparticles and Their Applications in Nanobiotechnology. PhD Thesis, College of Engineering and Computer Science, Department of Mechanical, Materials and Aerospace Engineering, University of Central Florida, Advisor-Sudipta Seal, 2006.

Peng, H.; Cui, B.; Li, L.; Wang, Y. A Simple Approach for the Synthesis of Bifunctional $Fe_3O_4@Gd_2O_3$: Eu3+ Core–Shell Nanocomposites. *J. Alloys Compd.* **2012,** *531,* 30–33.

Pereira, L.; Mehboob, F.; Stams, A. J.; Mota, M. M.; Rijnaarts, H. H.; Alves, M. M. Metallic Nanoparticles: Microbial Synthesis and Unique Properties for Biotechnological Applications, Bioavailability and Biotransformation. *Crit. Rev. Biotechnol.* **2015,** *35*(1), 114–128.

Perreault, F.; Melegari, S. P.; da Costa, C. H.; de Oliveira Franco Rossetto, A. L.; Popovic, R. Matias, W. G. Genotoxic Effects of Copper Oxide Nanoparticles in Neuro 2A Cell Cultures. *Sci. Total Environ.* **2012,** *441*(Supplement C), 117–124.

Prathna, T.; Mathew, L.; Chandrasekaran, N.; Raichur, A. M.; Mukherjee, A. Biomimetic Synthesis of Nanoparticles: Science, Technology and Applicability. In *Biomimetics learning from Nature;* InTechOpen: London. 2010, pp 1–20.

Priya, G. S.; Kanneganti, A.; Kumar, K. A.; Rao, K. V.; Bykkam, S. Biosynthesis of Cerium Oxide Nanoparticles Using *Aloe barbadensis* Miller. Gel. *Int. J. Sci. Res. Publ.* **2014,** *4*(6), 199–224.

Racowski, I.; Piotto, J.; Procópio, V.; Freire, V. Evaluation of Antimicrobial Activity and Phytochemical Analysis of Thaiti Lemon Peels (*Citrus latifolia* Tanaka). *J. Microbiol. Res.* **2017,** *7*(2), 39–44.

Rajabi, H. R.; Naghiha, R.; Kheirizadeh, M.; Sadatfaraji, H.; Mirzaei, A. Alvand, Z. M. Microwave Assisted Extraction as an Efficient Approach for Biosynthesis of Zinc Oxide Nanoparticles: Synthesis, Characterization, and Biological Properties. *Mater. Sci. Eng. C.* **2017,** *78*(Supplement C), 1109–1118.

Rajakumar, G.; Rahuman, A. A.; Priyamvada, B.; Khanna, V. G.; Kumar, D. K.; Sujin, P. J. *Eclipta prostrata* Leaf Aqueous Extract Mediated Synthesis of Titanium Dioxide Nanoparticles. *Mater. Lett.* **2012,** *68*(Supplement C), 115–117.

Rajakumar, G.; Rahuman, A. A.; Jayaseelan, C.; Santhoshkumar, T.; Marimuthu, S.; Kamaraj, C.; Bagavan, A.; Zahir, A. A.; Kirthi, A. V.; Elango, G. *Solanum trilobatum* Extract-Mediated Synthesis of Titanium Dioxide Nanoparticles to Control *Pediculus humanus* Capitis, *Hyalomma anatolicum* Anatolicum and *Anopheles subpictus. Parasitol Res.* **2014,** *113*(2), 469–479.

Rajakumar, G.; Rahuman, A. A.; Roopan, S. M.; Chung, I.-M.; Anbarasan, K.; Karthikeyan, V. Efficacy of Larvicidal Activity of Green Synthesized Titanium Dioxide Nanoparticles Using *Mangifera indica* Extract Against Blood-Feeding Parasites. *Parasitol Res.* **2015,** *114*(2), 571–581.

Ramanujam, K.; Sundrarajan, M. Antibacterial Effects of Biosynthesized MgO Nanoparticles Using Ethanolic Fruit Extract of *Emblica officinalis. J. Photochem. Photobiol. B.* **2014,** *141,* 296–300.

Ramesh, M.; Anbuvannan, M.; Viruthagiri, G. Green synthesis of ZnO nanoparticles using *Solanum nigrum* leaf extract and their antibacterial activity. *Spectrochim. Acta, Part A* **2015,** 136864–136870.

Rathod, N. R.; Chitme, H. R.; Irchhaiya, R.; Chandra, R. Hypoglycemic Effect of *Calotropis gigantea* Linn. Leaves and Flowers in Streptozotocin-Induced Diabetic Rats. *Oman Med. J.* **2011,** *26*(2), 104.

Riley, M. K.; Vermerris, W. Recent Advances in Nanomaterials for Gene Delivery—A Review. *Nanomaterials* **2017,** *7*(5), 94.

Rodriguez, J. A.; Garcia, M. *The World of Oxide Nanomaterials.* John Wiley & Sons, Inc.: Hoboken, NJ, 2007, pp 1–733.

Roopan, S. M.; Bharathi, A.; Prabhakarn, A.; Abdul Rahuman, A.; Velayutham, K.; Rajakumar, G.; Padmaja, R. D.; Lekshmi, M.; Madhumitha, G. Efficient Phyto-Synthesis and Structural Characterization of Rutile $TiO_2$ Nanoparticles Using *Annona squamosa* Peel Extract. *Spectrochim. Acta, Part A* **2012,** *98*(Supplement C), 86–90.

Sajadi, S. M.; Nasrollahzadeh, M.; Maham, M. Aqueous Extract from Seeds of *Silybum marianum* l. As a Green Material for Preparation of the $Cu/Fe_3O_4$ Nanoparticles: A Magnetically Recoverable and Reusable Catalyst for the Reduction of Nitroarenes. *J. Colloid Interface Sci.* **2016,** *469*(Supplement C), 93–98.

Samat, N. A.; Nor, R. M. Sol–Gel Synthesis Of Zinc Oxide Nanoparticles Using *Citrus aurantifolia* Extracts. *Ceram. Int.* **2013,** *39,* S545–548.

Sangeetha, G.; Rajeshwari, S.; Venckatesh, R. Green Synthesis of Zinc Oxide Nanoparticles by *Aloe barbadensis* Miller Leaf Extract: Structure and Optical Properties. *Mater. Res. Bull.* **2011,** *46*(12), 2560–2566.

Sankar, R.; Dhivya, R.; Shivashangari, K. S.; Ravikumar, V. Wound Healing Activity of *Origanum vulgare* Engineered Titanium Dioxide Nanoparticles in Wistar Albino Rats. *J. Mater. Sci. Mater. Med.* **2014a,** *25*(7), 1701–1708.

Sankar, R.; Maheswari, R.; Karthik, S.; Shivashangari, K. S.; Ravikumar, V. Anticancer Activity of *Ficus religiosa* Engineered Copper Oxide Nanoparticles. *Mater. Sci. Eng. C* **2014b,** *44*(Supplement C), 234–239.

Santhoshkumar, T.; Rahuman, A. A.; Jayaseelan, C.; Rajakumar, G.; Marimuthu, S.; Kirthi, A. V.; Velayutham, K.; Thomas, J.; Venkatesan, J. Kim, S.-K. Green Synthesis of Titanium Dioxide Nanoparticles Using *Psidium guajava* Extract and Its Antibacterial and Antioxidant Properties. *Asian Pac. J. Trop. Med.* **2014,** *7*(12), 968–976.

Santhoshkumar, J.; Kumar, S. V.; Rajeshkumar, S. Synthesis of Zinc Oxide Nanoparticles Using Plant Leaf Extract Against Urinary Tract Infection Pathogen. *Res. Efficient Technolog.* **2017,** *3,* 459–465.

Sardesai, N. P.; Ganesana, M.; Karimi, A.; Leiter, J. C.; Andreescu, S. Platinum-Doped Ceria Based Biosensor for in Vitro and in Vivo Monitoring of Lactate During Hypoxia. *Anal Chem.* **2015,** *87*(5), 2996–3003.

Sathishkumar, P.; Gu, F. L.; Zhan, Q.; Palvannan, T.; Mohd Yusoff, A. R. Flavonoids Mediated 'Green' Nanomaterials: A Novel Nanomedicine System to Treat Various Diseases – Current Trends and Future Perspective. *Mater. Lett.* **2018,** *210*(Supplement C), 26–30.

Schubert, D.; Dargusch, R.; Raitano, J.; Chan, S-W. Cerium and Yttrium Oxide Nanoparticles are Neuroprotective. *Biochem. Biophys. Res. Commun.* **2006,** *342*(1), 86–91.

Shabestari, K. S.; Farshbaf, M.; Akbarzadeh, A.; Davaran, S. Magnetic Nanoparticles: Preparation Methods, Applications in Cancer Diagnosis and Cancer Therapy. *Artif. Cells Nanomed., Biotechnol.* **2017,** *45*(1), 6–17.

Sharma, J. K.; Akhtar, M. S.; Ameen, S.; Srivastava, P.; Singh, G. Green Synthesis of CuO Nanoparticles with Leaf Extract of *Calotropis gigantea* and Its Dye-Sensitized Solar Cells Applications. *J. Alloys Compd.* **2015,** *632,* 321–325.

Sharma, H.; Kumar, K.; Choudhary, C.; Mishra, P. K.; Vaidya, B. Development and Characterization of Metal Oxide Nanoparticles for the Delivery of Anticancer Drug. *Artif. Cells Nanomed. Biotechnol.* **2016,** *44*(2), 672–679.

Shi, X.; Gu, W.; Li, B.; Chen, N.; Zhao, K.; Xian, Y. Enzymatic Biosensors Based on the Use of Metal Oxide Nanoparticles. *Microchim. Acta* **2014**, *181*(1), 1–22.

Sinha, J. K.; Ghosh, S.; Raghunath, M. Progeria: a Rare Genetic Premature Ageing Disorder. *Indian J. Med. Res.* **2014**, *139*(5), 667.

Sood, A.; Arora, V.; Shah, J.; Kotnala, R.; Jain, T. K. Ascorbic Acid-Mediated Synthesis and Characterisation of Iron Oxide/Gold Core–Shell Nanoparticles. *J. Exp. Nanosci.* **2016**, *11*(5), 370–382.

Staiger, M. P.; Pietak, A. M.; Huadmai, J.; Dias, G. Magnesium and Its Alloys as Orthopedic Biomaterials: a Review. *Biomaterials* **2006**, *27*(9), 1728–1734.

Stoimenov, P. K.; Klinger, R. L.; Marchin, G. L.; Klabunde, K. J. Metal Oxide Nanoparticles as Bactericidal Agents. *Langmuir* **2002**, *18*(17), 6679–6686.

Studer, A. M.; Limbach, L. K.; Van Duc, L.; Krumeich, F.; Athanassiou, E. K.; Gerber, L. C.; Moch, H.; Stark, W. J. Nanoparticle Cytotoxicity Depends on Intracellular Solubility: Comparison of Stabilized Copper Metal and Degradable Copper Oxide Nanoparticles. *Toxicol. Lett.* **2010**, *197*(3), 169–174.

Sun, J.; Wang, S.; Zhao, D.; Hun, F. H.; Weng, L.; Liu, H. Cytotoxicity, Permeability, and Inflammation of Metal Oxide Nanoparticles in Human Cardiac Microvascular Endothelial Cells. *Cell Biol. Toxicol.* **2011**, *27*(5), 333–342.

Sundrarajan, M.; Gowri, S. Green Synthesis of Titanium Dioxide Nanoparticles by *Nyctanthes arbor-tristis* Leaves Extract. *Chalcogenide Lett.* **2011**, *8*(8), 447–451.

Sundrarajan, M.; Bama, K.; Bhavani, M.; Jegatheeswaran, S.; Ambika, S.; Sangili, A.; Nithya P.; Sumathi R. Obtaining Titanium Dioxide Nanoparticles with Spherical Shape and Antimicrobial Properties Using *M. citrifolia* Leaves Extract by Hydrothermal Method. *J. Photochem. Photobiol. B* **2017**, *171*(Supplement C), 117–124.

Sunkar, S.; Nachiyar, C. V. Biogenesis of Antibacterial Silver Nanoparticles Using the Endophytic Bacterium *Bacillus cereus* Isolated from *Garcinia xanthochymus*. *Asian Pac. J. Trop. Biomed.* **2012**, *2*(12), 953–959.

Suresh, J.; Yuvakkumar, R.; Sundrarajan, M.; Hong, S. I. *Green Synthesis of Magnesium Oxide Nanoparticles*; Advanced Materials Research; Trans Tech Publ. 2014.

Sushma, N. J.; Prathyusha, D.; Swathi, G.; Madhavi, T.; Raju, B. D. P.; Mallikarjuna, K; Kim, H-S. Facile Approach to Synthesize Magnesium Oxide Nanoparticles by Using *Clitoria ternatea*—Characterization and in Vitro Antioxidant Studies. *Appl. Nanosci.* **2016**, *6*(3), 437–444.

Tagliabracci, V. S.; Girard, J. M.; Segvich, D.; Meyer, C.; Turnbull, J.; Zhao, X.; Minassian, B. A.; Depaoli-Roach, A. A.; Roach, P. J. Abnormal Metabolism of Glycogen Phosphate as a Cause for Lafora Disease. *J. Biol. Chem.* **2008**, *283*(49), 33816–33825.

Tammina, S. K.; Mandal, B. K.; Ranjan, S.; Dasgupta, N. Cytotoxicity Study of *Piper nigrum* Seed Mediated Synthesized $SnO_2$ Nanoparticles Towards Colorectal (HCT116) and Lung Cancer (A549) Cell Lines. *J. Photochem. Photobiol., B* **2017**, *166*, 158–168.

Tan, Y.; Dai, X.; Li, Y.; Zhu, D. Preparation of Gold, Platinum, Palladium and Silver Nanoparticles by the Reduction of Their Salts with a Weak Reductant – Potassium Bitartrate. *J. Mater. Chem.* **2003**, *13*(5), 1069–1075.

Thakkar, K. N.; Mhatre, S. S.; Parikh, R. Y. Biological Synthesis of Metallic Nanoparticles. *Nanomedicine NBM.* **2010**, *6*(2), 257–262.

Thovhogi, N.; Diallo, A.; Gurib-Fakim, A.; Maaza, M. Nanoparticles Green Synthesis by *Hibiscus sabdariffa* Flower Extract: Main Physical Properties. *J. Alloys. Compd.* **2015**, *647*, 392–396.

Turcheniuk, K.; Khanal, M.; Motorina, A.; Subramanian, P.; Barras, A.; Zaitsev, V.; Kuncser, V.; Leca, A.; Martoriati, A.; Cailliau, K. Insulin Loaded Iron Magnetic Nanoparticle–Graphene Oxide Composites: Synthesis, Characterization and Application for in Vivo Delivery of Insulin. *RSC Adv.* **2014,** *4*(2), 865–875.

Veisi, H.; Hemmati, S.; Javaheri, H. N-Arylation of Indole and Aniline by a Green Synthesized CuO Nanoparticles Mediated by *Thymbra spicata* Leaves Extract as a Recyclable and Heterogeneous Nanocatalyst. *Tetrahedron Lett.* **2017,** *58*(32), 3155–3159.

Vermisoglou, E.; Devlin, E.; Giannakopoulou, T.; Romanos, G.; Boukos, N.; Psycharis, V.; Lei, C.; Lekakou, C.; Petridis, D.; Trapalis, C. Reduced Graphene Oxide/Iron Carbide Nanocomposites for Magnetic and Supercapacitor Applications. *J. Alloys Compd.* **2014,** *590,* 102–109.

Vidya, C.; Hiremath, S.; Chandraprabha, M.; Antonyraj, I.; Gopal, I. V.; Jain, A.; Bansal K Green Synthesis of ZnO Nanoparticles by *Calotropis gigantea. Int. J. Curr. Eng. Technol.* **2013,** *11,* 18–20.

Vidya, C.; Prabha, M. C.; Raj, M. A. Green Mediated Synthesis of Zinc Oxide Nanoparticles for the Photocatalytic Degradation of Rose Bengal Dye. *Environ. Nanotechnol. Monit. Manag.* **2016,** *61,* 34–38.

Vijayakumar, S.; Vaseeharan, B.; Malaikozhundan, B.; Shobiya, M. *Laurus nobilis* Leaf Extract Mediated Green Synthesis of ZnO Nanoparticles: Characterization and Biomedical Applications. *Biomed. Pharmacother.* **2016,** *84,* 1213–1222.

Vimala, K.; Soundarapandian, K. 129p Inhibition of Breast Cancer Cell Explosion Rate by Irinotecan Loaded Zinc Oxide Nanoparticles Through E2F3/Akt Signaling Circuits: A Milestone In Cancer Gene Therapy. *Ann. Oncol.* **2016,** *27*(Suppl. 9), mdw577.013.

Vimala, K.; Sundarraj, S.; Paulpandi, M.; Vengatesan, S.; Kannan, S. Green Synthesized Doxorubicin Loaded Zinc Oxide Nanoparticles Regulates the Bax and Bcl-2 Expression in Breast and Colon Carcinoma. Process *Biochem.* **2014,** *49*(1), 160–172.

Wang, R.; Xin, J. H.; Tao, X. M.; Daoud, W. A. ZnO Nanorods Grown on Cotton Fabrics At Low Temperature. *Chem. Phys. Lett.* **2004,** *398*(1), 250–255.

Wrone, D. A.; Kwon, N. J. Andrews' Diseases of the Skin: Clinical Dermatology. *Arch Dermatol.* **2001,** *137*(4), 518–518.

Wu, W.; Wu, Z.; Yu, T.; Jiang, C.; Kim, W-S. Recent Progress on Magnetic Iron Oxide Nanoparticles: Synthesis, Surface Functional Strategies and Biomedical Applications. *Sci. Technol. Adv. Mater.* **2015,** *16*(2), 023501.

Xu, C.; Qu, X. Cerium Oxide Nanoparticle: a Remarkably Versatile Rare Earth Nanomaterial for Biological Applications. *NPG Asia Mater.* **2014,** *6*(3), e90.

Xu, Y.; Hadjiargyrou, M.; Rafailovich, M.; Mironava, T. Cell-Based Cytotoxicity Assays for Engineered Nanomaterials Safety Screening: Exposure of Adipose Derived Stromal Cells to Titanium Dioxide Nanoparticles. *J. Nanobiotechnol.* **2017,** *15*(1), 50.

Yadavalli, T.; Shukla, D. Role of Metal and Metal Oxide Nanoparticles as Diagnostic and Therapeutic Tools for Highly Prevalent Viral Infections. *Nanomedicine NBM.* **2017,** *13*(1), 219–230.

Yola, M. L.; Gupta, V. K.; Eren, T.; Şen, A. E.; Atar, N. A Novel Electro Analytical Nanosensor Based on Graphene Oxide/Silver Nanoparticles for Simultaneous Determination of Quercetin and Morin. *Electrochim. Acta* **2014,** *120*(Supplement C), 204–211.

Yugandhar, P.; Vasavi, T.; Devi, P. U. M.; Savithramma, N. Bioinspired Green Synthesis of Copper Oxide Nanoparticles from *Syzygium alternifolium* (Wt.) Walp: Characterization and Evaluation of Its Synergistic Antimicrobial and Anticancer Activity. *Appl. Nanosci.* **2017,** *7*(7), 417–427.

Yuvakkumar, R.; Suresh, J.; Nathanael, A. J.; Sundrarajan, M.; Hong, S. I. Rambutan (*Nephelium lappaceum* L.) Peel Extract Assisted Biomimetic Synthesis of Nickel Oxide Nanocrystals. *Mater. Lett.* **2014,** *128*(Supplement C), 170–174.

Zahir, A. A.; Chauhan, I. S.; Bagavan, A.; Kamaraj, C.; Elango, G.; Shankar, J.; Arjaria, N.; Roopan, S. M.; Rahuman, A. A.; Singh, N. Green Synthesis of Silver and Titanium Dioxide Nanoparticles Using *Euphorbia prostrata* Extract Shows Shift from Apoptosis to G0/G1 Arrest Followed by Necrotic Cell Death in *Leishmania donovani*. *Antimicrob. Agents Chemother.* **2015,** *59*(8), 4782–4799.

Zhang, Y.; Wang, Z.; Li, X.; Wang, L.; Yin, M.; Wang, L.; Chen, N.; Fan, C.; Song, H. Dietary Iron Oxide Nanoparticles Delay Aging and Ameliorate Neurodegeneration in *Drosophila*. *Adv. Mater.* **2016,** *28*(7), 1387–1393.

Zhou, J.; Fa, H.; Yin, W.; Zhang, J.; Hou, C.; Huo, D.; Zhang, D.; Zhang, H. Synthesis of Superparamagnetic Iron Oxide Nanoparticles Coated with a DDNP-Carboxyl Derivative for in Vitro Magnetic Resonance Imaging of Alzheimer's Disease. *Mater. Sci. Eng. C.* **2014,** *37,* 348–355.

## CHAPTER 22

# GREEN SYNTHETIC APPROACHES AND PRECURSORS FOR CARBON DOT NANOPARTICLES

HAMEED SHAH[1,2,*] and ASHFAQ AHMAD KHAN[3]

[1]*CAS Key Laboratory for Biomedical Effects of Nanomaterials and Nanosafety, National Center for Nanoscience and Technology, Beijing, China, Tel.: 008615201044821*

[2]*University of Chinese Academy of Science, 100049 Beijing, China*

[3]*Women University of Azad Jammu and Kashmir, Bagh, Pakistan*

*Corresponding author. E-mail: hameed@nanoctr.cn;
hameed_shah2002@yahoo.com
*ORCID: https://orcid.org/0000-0003-1348-5193.*

## ABSTRACT

Carbon dot (CD) nanoparticles (NPs) are small diameter particles, up to 10 nm in range, having extensive use in bio-imaging, bio-labeling, chemical sensing, bio-sensing, photocatalysis, electrocatalysis, and nano-medicines and so forth, owing to its large surface area, more exposed surface atoms, presenting chemical inertness, low cytotoxicity, low cost, and easy fabrication methods. Carbon NPs are composed of carbon core skeleton, and their surface functional groups, in varying composition as per the methods and materials used. The choice for phytochemicals as precursors for CDs synthesis is due to its cheap, easy fabricating, and environmentally friendly, widely available and applicable sources, which will boost the fields of phytomedicine and nanotechnology simultaneously. This chapter discusses the biosynthesis of CD NPs and materials involved along with the new proposed plants-based precursors suitable for both top-down and bottom-up methods, their applications, their photoluminescent details and their phytotoxic effects.

## 22.1  INTRODUCTION

Since their accidental discovery in 2004 during the electrophoretic purification of one-dimensional single-walled carbon nanotubes (Xu et al., 2004), carbon nanodots or carbon dots (CDs), as the first 0D fluorescent materials of the low-dimensional two-dimensional graphene family, remained a hot topic for researchers including academicians and industrialists, mainly due to their simple chemistry, diverse electrical, magnetic and optical intrinsic properties, along with their humble synthetic methods, both methodologically and in context to their starting material, and their promising applications, for which the scientists were eager that these materials will soon replace semiconductor quantum dots and organic dyes. The dream came true soon with the efforts of researchers, when CDs truly replaced semiconductor quantum dots and organic dyes, especially in the context of their quantum yield, with additional qualities of almost no cell toxicity, resistant to photo-bleaching and photoblinking, as well as good water solubility and biocompatibility, along with their prominent tunable band gap. These properties make them widely applicable ranging for their use in vivo and in vitro metal detection, bio-sensing, bio-imaging, nano-medicine, photocatalysis, electrocatalysis, and photovoltaic devices to solar cells, and so forth. (Baptista et al., 2015; Lim et al., 2015). Their biocompatibility, achievable chemical, colloidal and thermal stability primarily by their passivation, photostability, and broad fluorescence, their solubility in a wide range of solvents and so forth are other factors enhancing their wide applications. These days, or very soon in near future, CDs will be applicable in almost every field of life. The fascinating optical and electro-optical properties of CDs are attributed to their quantum confinement and edge effects, in other words, pointing to their quantum particle nature (Zhang et al., 2012). Currently, the scientists are exploring their new applications in different fields of science and technology, with special emphasis on improving their quantum yield and near infrared emission for deep tissue imaging (Li et al., 2013a; Baptista et al., 2015; Lim et al., 2015; Chen and Zhang, 2017; Hill and Galan, 2017). Since last decade, different material carbon sources have been employed for CD's synthesis either in their isolated form or in combination with other chemicals taking part in reaction with carbon precursor or as an exfoliating agent of this carbon precursor. These include primarily graphite, coal, citric acid, chitosan, and so forth, for the successful synthesis of CD's precursors. In these approaches, much attention has been given to the CD's synthesis with a focus on their cheap sources such as coal, graphene, graphene oxide, or graphite, citric acid, urea (Fan et al., 2017) and so forth. However, most of these techniques involve complex fabrication methods with much laborious work, as well as in many cases the involvement of expensive and hazardous chemicals, not only increase their cost but also made them extremely menacing for the environment. Recently, natural sources such as coffee grounds were reported for CD's synthesis by Hsu et al. (2012), effectively followed by Liu et al. (2012), Lu et al. (2012), and Zhou et al. (2012) whom readily embarked on the use of green approaches for CD's synthesis because these methods will not only utilize the huge natural resources but will also lead to simple, eco-friendly, less toxic side products and so on. Amidst the other natural sources, the plants and their isolated natural products are the prominent figures in this context. Natural products, as factories of natural phytochemicals, are being an emerging source of CD's synthesis because plants as a whole or their isolated phytochemicals individually can easily provide the basic carbon core skeleton as well as the surface functional groups of these entities.

A number of organic and biomolecules have been reported for the synthesis of CDs including some plants sources, such as grass (Liu et al., 2012), coffee grounds (Hsu et al., 2012) and so forth, with some fruits such as mango fruit (Jeong et al., 2014), watermelon peels (Zhou et al., 2012), date molasses (Das et al., 2014), pomelo peels (Lu et al., 2012), orange juice (Prasannan and Imae, 2013), strawberry juice (Huang et al., 2013), sugarcane juice (Huang et al., 2017), apple juice (Xu et al., 2014) and animal sources such as chicken egg (Wang et al., 2012), chitosan (Verpoorte and Schripsema, 1994; Chowdhury et al., 2012; Konwar et al., 2015; Pinto et al., 2017), gelatin (Liang et al., 2013) hair fiber (Sun et al., 2013), and so forth.

Plants, both in terms of their whole material as well as isolated and purified chemicals, no doubt are the best options for the CD's synthesis. In a broader context, as the plants have biocompatible nature, hence in any form their vast use will never harm the environment, as well as the organism on which their synthesized materials are to be applied. Moreover, these are the factories of a large number of important chemicals and hence can serve their CDs as potential sources of antimicrobial, antitumor agents, and so forth. Plants and their phytochemicals, basically are composed of carbon, hence they provide the basic carbon skeleton for CD's synthesis. On the other hand, a large number of important molecules in plants can provide useful functional groups to be replaced on the CD's surface during their synthetic conversion, which will help in enhancing the properties of CDs, both as photoluminescent materials, as well as in making them more useful antimicrobial and antitumor agents.

## 22.2  METHODS OF BIOSYNTHESIS

The desired CD's synthesis, to use starting materials of plants, or isolated phytochemicals can be classified into the following three categories top-down methods, bottom-up methods, from domestic resources.

### 22.2.1  TOP-DOWN METHODS

Top-down methods involve large precursor, which need to be gradually divided into small NPs (NPs). These methods face the problems of uniformity, which usually lead to the low desired product yield. In these methods, plants dried components, or their crude extracts could be used for the synthesis of required CDs. Different techniques could be further chosen for the related CDs (Fig. 22.1).

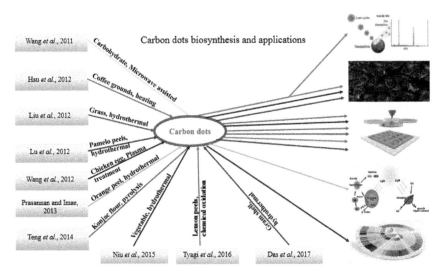

**FIGURE 22.1   (See color insert.)** Summary of some green approaches for carbon dots (CDs) synthesis and their application. The arrow color indicates the application of those synthesized CDs. The insets on the right hand side, from top to bottom: 1 = SALDI-MS, 2 = bio-imaging, 3 = chemical sensing, 4 = photocatalysis, and 5 = ink printing.

### 22.2.1.1  THERMAL PYROLYSIS

Thermal pyrolysis is the better option when we require CDs from the roots, leaves, stem, or branches of the plants, that is, herbs, shrubs, or trees, especially in the absence of any solvent. Dried ground or peeled plant materials will be further helpful in easy pyrolysis. As the plant materials are sensitive to temperature and usually at slightly high temperature decompose; hence, the lower temperature applied for longer times are the best option for CD's synthesis using this technique. Theoretically, the CDs obtained by this technique will be to some extent functionalized on their surface, that is, both in terms of their number and nature of functional groups, and will further require functionalization for their better solubility in water, and fluorescence to further use them as bio-imaging tools, or passivation by other molecules for conjugation with biomolecules. Moreover, as the surface of all the plant materials is mainly composed of cellulose, hence the large quantity of CDs will be obtained from the plant materials. Literature confirms the use of different plant sources for CD's synthesis by pyrolysis application. Wenbin Li and colleagues synthesized the CD from soyabean grounds through their thermal pyrolysis (Li et al., 2013b). Wei et al. (2013) reported the synthesis of blue fluorescence CDs with fair stability by the direct burning of paper. They also successfully applied these materials for bioimaging. Teng and their colleagues synthesized CDs from konjac flour which have similarity with pectin fiber, abundantly found inside the plant cell wall (Teng et al., 2014). The pyrolytic treatment of peanut shells was reported for fluorescent CDs synthesis by Xue et al. (2016). Natural chia seeds were reported by S. Solomon Jones and colleagues in order to synthesize CDs (Jones et al., 2017). Thermal treatment of rice grains for CDs synthesis is reported by Hemen Kalita and coworkers for CD's synthesis who also reported the quantum yield and photoluminescence for these materials (Kalita et al., 2016). Chinese paprika pyrolysis was carried out for CD's synthesis by Chen et al. (2017) who also monitored these materials for cell membrane crossing. Konwar et al. (2017), reported the use of tea as CD's precursor in their studies for graphene oxide reduction. Thermal pyrolysis of sago waste has been carried out for CD's synthesis by Xian Wen Tan and their team which they successfully employed for metal ion sensing (Tan et al., 2014).

## 22.2.1.2   HYDROTHERMAL TREATMENT

As principally, the plant materials are insoluble in water, hence this method will results usually in the low quantity of CD's synthesis. However, in the case of extracts, the high polar extract of plants, such as alcoholic or aqueous-alcoholic extracts is best option to be used in this method. The fluorescence, and antimicrobial potential, theoretically will be of medium range, while the synthetic yield will also be very low. Mewada et al. (2013) used *Trapa bispinosa* peels for CD's synthesis hydrothermally and obtained good yield, while their method involves crushing peels followed by centrifugation and refluxing at 90°C for 2 h. In another study, *Coffea liberica* (coffee) husks were used by Valle et al. (2017) to synthesis CDs through hydrothermal treatment, which they further applied in bio-imaging of water pathways in the rose plant. Liu et al. (2014) synthesized fluorescent CDs from bamboo leaves hydrothermally and applied these for detection of $Cu^{2+}$ ion. Lignin has been reported for the hydrothermal CD's synthesis by Chen and colleagues by using $H_2O_2$ as oxidizing agents (Chen et al., 2016). Shuyan Gao and their team synthesized nitrogen-doped CDs with their successful application for oxygen reduction reaction by the hydrothermal treatment of natural willow leaves (Gao et al., 2014). Wang et al. reported the CDs from coffee grounds through hydrazine-assisted hydrothermal route (Wang et al., 2016b). Liu and their team successfully synthesized CDs from grass leading to their successful application for the detection of $Cu^{2+}$ ions (Liu et al., 2012). Li et al. (2015a) in their synthesis of CDs for blood-related studies synthesized from a fiber named as α-cyclodextrin through its hydrothermal carbonization. In their hydrothermal synthesis of N-doped CDs, Edison and their team used L-ascorbic acid as carbon source along with β-alanine as –N source, carried their cytotoxicity studies and success-fully employed these CDs for bio-imaging purposes (Edison et al., 2016). *Hylocereus undatus* extract along with aqueous ammonia was employed for CD's synthesis by Arul and their team, whom also reported their cytotoxicity and biocompatibility (Arul et al., 2017). Ensafi et al. reported the synthesis of CDs through hydrothermal route and applied these for bio-imaging and sensing (Ensafi et al., 2017). Vandarkuzhali and associates synthesized CDs from the banana plant through their simple hydrothermal treatment followed by their application as biosensor and bio-imaging probes (Vandarkuzhalia et al., 2017). Jhonsi and thulasi reported the CD's synthesis from tamarind seeds by one-step hydrothermal process with stable and excitation depen-dent nature (Jhonsi and Thulasi, 2016). Wang and contemporaries reported the hydrothermal route for CD's synthesis from papaya powder as a natural carbon source and applied them for $Fe^{3+}$ and *Escherichia coli* sensing (Wang et al., 2016c). Rose-heart radish was used for CD's synthesis by Wen Liu and fellow workers, through hydrothermal route further applying these for $Fe^{+3}$ sensing and cell imaging (Liu et al., 2017a). In another study, a research team lead by Jumeng Wei reported the synthesis of CDs from corn flour through a simple one-pot hydrothermal route (Wei et al., 2014). *Ocimum sanctum* leaves as a carbon source for CD's synthesis was reported by Kumar and coworkers along with their application for Pb$^{2+}$ sensing and live cell imaging (Kumar et al., 2017). Hydrothermally, CDs were synthesized from *Punica granatum* (pomegranate) fruits by Kasibabu and team workers, with their applications for sensing bacterial (*Pseudomonas aeruginosa*) and fungal (*Fusarium avenaceum*) cells (Kasibabu et al., 2015). Pakchoi was hydrothermally converted to CDs by Niu and their squad (Niu et al., 2015). CDs were also synthesized from hydrothermal treatment of lemon peel wastes by Tyagi and group (Tyagi et al., 2016). Ginger juice was hydrothermally converted into CDs by Li and group fellows, exploring their potential inhibition against hepatocellular carcinoma (Li et al., 2014a). Wool and pig hairs were hydrothermally converted into CDs by Wang and their team followed by their successful applications against Cr (vi) ions (Wang et al., 2016d). Poushali Das and team synthesized CDs from gram shell and applied these for *E. coli* labeling (Das et al., 2017).

### 22.2.1.3  SOLVOTHERMAL TREATMENT

Solvothermal is the best method for the CD's synthesis from plant materials. In this method, the dried plant material, as well as the solvent extracts could be used for the synthesis of the desired CDs. Interestingly, this method can selectively be used for the synthesis of desired CDs in terms of their doping with different metals and nonmetals, as well as in view of their surface functionalization. Hence, it is the suitable method to obtain CDs with better fluorescence, good antimicrobial activity, and targeted tumor inhibition potential. Dang and their team synthesized blue fluorescent –N and –S CDs by heating caffeine with ammonium persulphate, the optical properties of which were further enhanced when further ammonia was added to this solution (Dang et al., 2018). Edible soyabean oil was solvothermally treated for CD's synthesis by Chen and Tseng (2017). Solvothermal large-scale synthesis of CDs from waste food and ethanol by Park and their coworkers along with their cytotoxic and bioimaging studies (Park et al., 2014).

### 22.2.1.4  FROM CARBON SOOT

Tripathi and Sarkar (2014), reported the synthesis of CDs from carbon soot, which they collected by burning mustard oil in the lamp, and after that washed the soot with toluene, alcohol, and acetone, followed by water for removing the impurities. The obtained CDs were further modified by treating with nitric acid.

### 22.2.1.5   HYDROGEN FLUORIDE (HF) SACCHARIFICATION

HF has been largely reported for the saccharification of wood, cellulose, and so forth, in both anhydrous and as well as liquid solution state. The resulting product, as reported by Hardt et al. (1982), usually gives solid lignin along with a water-soluble saccharide fraction. As these fractions, both of which contains fluorine, either adsorbed or bonded with other materials, and thus could lead to the CD's synthesis, possibly doped with fluorine-containing C–F bond, and will have important pharmaceutical applications.

### 22.2.1.6   ACIDIC OXIDATION

In natural products chemistry, usually, HF is used for the exfoliation of plant materials, such as roots and so forth. Chemical exfoliation could be applied for CD's synthesis from the plants.

### 22.2.1.7   CHEMICAL OXIDATION

Yan and coresearchers reported the synthesis of CDs by the chemical oxidation of starch (Yan et al., 2015). In another study, chemical oxidation of sugarcane bagasse by Thambiraj and Shankaran (2016), lead to the synthesis of fluorescent CDs with their applications in bio-sensing, drug delivery, and bio-imaging.

### 22.2.1.8   MECHANICAL EXFOLIATION

Mechanical exfoliation could be applied for the top-down synthesis of CDs, provided that plant materials should be already thermally converted to powdered form or should be oxidized by the action of acids.

### 22.2.1.9   MICROWAVE IRRADIATION

Rai and associates reported the synthesis of CDs from the hydrolytic microwave irradiation of lignosulfonate lignin (Rai et al., 2017). Dan Gu and collaborators reported the one-pot microwave treatment for N-doped CD's synthesis from lotus root and successfully applied these CDs for cell imaging and $Hg^{2+}$ detection (Gu et al., 2016). Wool was selected in their one-step microwave pyrolysis for CD's synthesis by Wang and fellow beings followed by their sensing application of glyphosate (Wang et al., 2016a). $Ag^+$ and $Au^{3+}$ CD's nanohybrids with Citrus limon extract were synthesized using hydrothermal microwave-assisted treatment by Sajid and their team (Sajid et al., 2016). Microwave-assisted treatment of algal blooms comprising of almost 10 plant species of Cyanophyceae, Chlorophyceae, Bacillariophyceae, and Euglenophyceae families after mixing with phosphoric acid resulted in the synthesis of solid fluorescent CDs which were further applied for cell viability assay for in vitro imaging (Ramanan et al., 2016). Fresh tomatoes were pyrolyzed through microwave treatment to obtain CDs having good fluorescence by Liu and their group in their studies for CD's UV emission and quantum confinement (Liu et al., 2017b). Yuan Jiao Feng and their team synthesized CDs from rose petal carbonization followed by their aqueous solution microwave irradiation leading to its successful application for tetracycline sensing (Feng et al., 2015). Wang and their side reported the synthesis of excitation dependent CDs from a variety of carbohydrates such glycerol, glycol, glucose, sucrose, and so forth. (Wang et al., 2011).

## 22.2.2   BOTTOM-UP METHODS

Bottom-up methods are those in which individual molecules are used as precursors, reacting with each other and form the bulky molecules of CDs. These methods are used for the CDs with more selectivity, more homogeneity, as well as better short and long-range order, and thus are comparatively expensive methods. In these methods, the building blocks are added onto the substrates to form the required nanostructures. These methods are purely termed as chemical methods, as these generally follow the synthetic routes used for different bond formation, through careful optimization for NP's synthesis. Phytochemicals, instead of whole plant components and plant extracts are suitable precursors for CDs production by these methods. Although individually each molecule will produce the related CDs, yet these methods can easily be classified on the basis of the different classes of compounds present in plants. Some of these are discussed herein.

### 22.2.2.1   ALKALOIDS

Literature confirms the better fluorescence of N-doped CDs compared to those without doping (Hola et al., 2017). As alkaloids have N-atom in their ring (Fig. 22.2), therefore these molecules could act as better precursors for N-doped CD's synthesis than other precursors. Dang et al. (2018), successfully prepared CDs from caffeine. In their well-established experiment, they report that caffeine alone or in presence of urea, there was no reaction or decomposition upon heating. However, when caffeine and urea in the presence of ammonium persulphate were heated, the subsequent CDs were synthesized. In case of alkaloids as CD's precursors, we can use purely isolated alkaloids (pyrolysis at high temperature, usually near to their melting points etc.), or the alkaloids mixture obtained after acidification/basification of plant extract (Verpoorte and Schripsema, 1994). Alkaloids having large molecules, could act as a whole for the synthesis of N-doped CDs, while in case of small molecules, these materials can be mixed with other molecules or solvents in which these materials act as surface or doping agents, while the other molecules provide the skeleton of carbon core. Alkaloids are one of the most important classes of compounds, usually isolated at higher polarity solvent systems from the plants. As noted, these heterocyclic phytochemicals contain –N in their ring, and thus can act as precursors for CD's synthesis, both doped intrinsically by –N atom, as well as containing –N functionalized side chains, in the form of $-NO_2$, $-NH_2$, and so forth. These synthesized CDs, compared to other non-doped CDs will show better fluorescence and sensing because of the nitrogen present in their structures. Some best alkaloids of medicinal importance, already reported can best act as precursors for CD's syntheses, for example, cocaine,

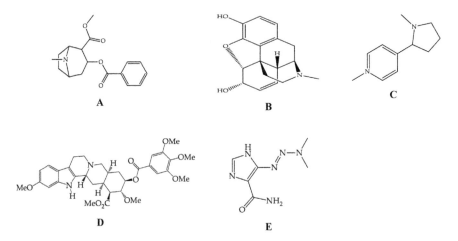

**FIGURE 22.2** Some important alkaloids (A) cocaine; (B) morphine; (C) nicotine; (D) reserpine, (E) dacarbazine.

nicotine, and so forth. The best synthetic methodology for CD's synthesis from this precursor is pyrolysis. The CDs obtained herein will have suitable applications for chemical sensing, along with others. Alkaloids having a higher number of –N atoms, such as dacarbazine and so forth, can be suitably used with other precursors acting as a source of carbon core skeleton. Recently, our group is working on the synthesis of CDs from the alkaloids based materials using different approaches involving thermal, hydrothermal and solvothermal methods.

## 22.2.2.2   FLAVONOIDS

A variety of flavonoids (Fig. 22.3), such as flavones, flavanones, flavonols, flavanonols, anthocyanidins, isoflavone, and so forth has been discovered and studied extensively for their biomedical applications. These classes of compounds have a slight difference in their structures, while basically contains the same skeleton.

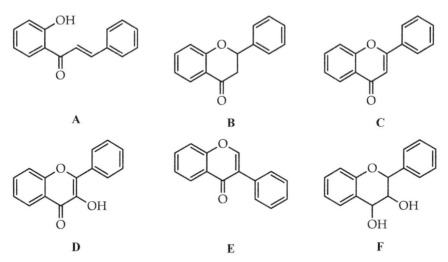

**FIGURE 22.3**   Some major classes of flavonoids (A) chalcone; (B) flavanone; (C) flavone; (D) flavanol; (E) isoflavone; (F) flavandiol.

From the literature, we did not find any reported method of these starting materials which has been applied in CD's synthesis. However, like many other reported chemicals, these phytochemicals can also be chosen as precursors for CD's synthesis. As usually, these contain a sufficient amount of carbon atoms in their skeleton along with oxygen, so these molecules could act individually as precursors for CD's synthesis with –COOH and –OH, and –O surface groups. Pyrolysis, solvothermal, and hydrothermal approaches could be utilized for CD's synthesis from flavonoids.

### 22.2.2.3  TERPENOIDS

Terpenoids represent a vast family of phytochemicals consisting of basic five-carbon isoprene units. Although few classes of terpenoids, consisting of simple isoprene units, consisting mainly of carbon and hydrogen atoms, are not the best choice to be employed for CD's synthesis. However, their treatment with other molecules could lead to some useful CD's synthesis. Apart from these simple molecules, there are other representative compounds from this family such as azadirachtin and artemisinin (Fig. 22.4), and so forth.

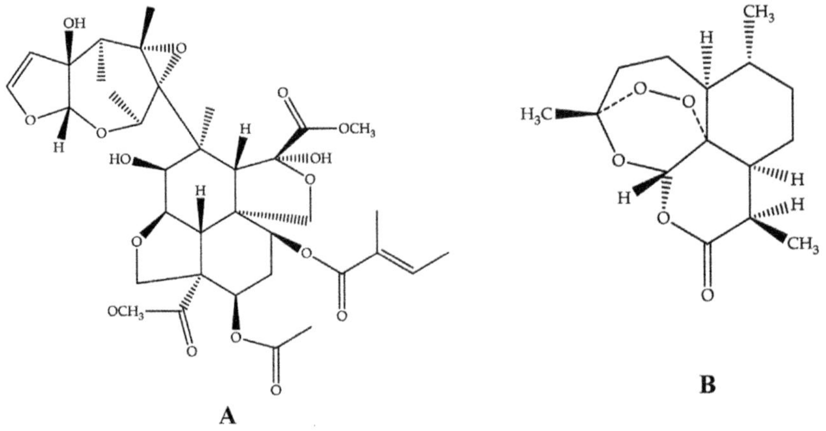

**FIGURE 22.4**    The chemical structure of (A) azadirachtin and (B) artemisinin.

### 22.2.2.4  STEROIDS

These can act as better precursors for carbogenic core synthesis of CDs, especially through thermal pyrolysis. However, for getting good quality CDs, these require hydrothermal or solvothermal synthetic methods along with other precursors for –O, –N and –S doping. Figure 22.5 is a typical steroid.

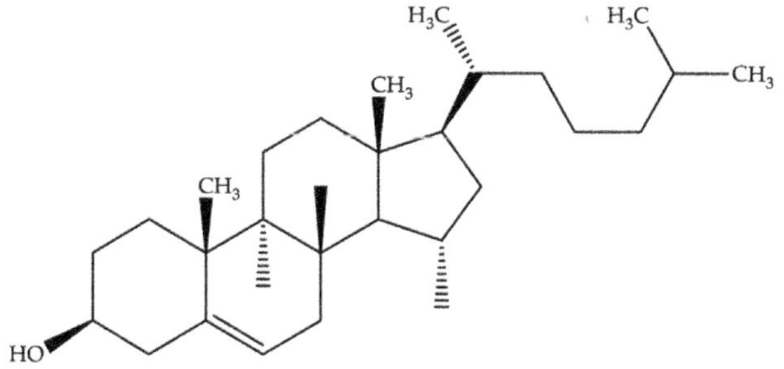

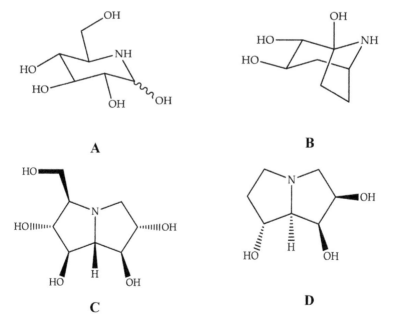

**FIGURE 22.5**   The chemical structure of cholesterol.

## 22.2.2.5   *GLYCOSIDES*

Glycosides (Fig. 22.6) are heavily modified sugar molecules, mainly consisting of –C, -H, and –O functional groups, thus can be used for simple CD's synthesis. However, some iminosugars, which have been reported for glycosidase inhibition (Zelli et al., 2016), contains –N in their ring, and hence could be applied for the synthesis of N-doped CDs.

**FIGURE 22.6**   Some glycosides (A) 1-deoxy-D-nojirimycin; (B) calystegin; (C) casuarine; (D) swainsonine.

## 22.2.2.6   NATURAL PHENOLS

Phenols are the natural antioxidant agents, found in a large number of plants. Phenols are a good source of –C, –H, and –O atoms, and are thus the best source for CDs with –COOH and –OH functional groups (Fig. 22.7). In case of their hydrothermal treatment, they could results in more oxidized CDs, while on mixing with solvents containing –N, or other precursors with –N atoms, can lead to the synthesis of N-doped CDs.

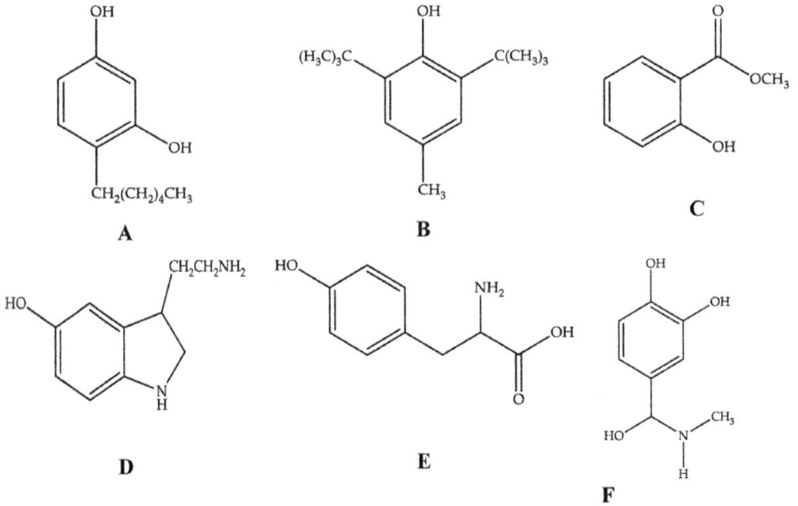

**FIGURE 22.7**   Examples of some phenols (A) 4-hexylresorcinol; (B) 2,6-di-tert-butyl-4-methylphenol; (C) methyl salicylate; (D) serotonin; (E) tyrosine (F) adrenalin.

## 22.2.2.7   CARBOHYDRATES (MONOSACCHARIDES AND POLYSACCHARIDES)

Carbohydrates are obviously abundant, readily available, and heterogeneity components of plants, requiring low carbonization temperature, cost-effective, having minimum cytotoxicity, and good water solubility. Hence, the better option to obtain CDs from these precursors is hydrothermal or solvothermal with a possible treatment of thermal carbonization. A number of monosaccharides, such as glucose, sucrose, as well as disaccharides such as fructose and so forth have been widely reported for CDs synthesis (Hill and Galan, 2017). Similarly, carbohydrates base polymers such as dextran and so forth have been utilized in CD's synthesis (Lim et al., 2015).

## 22.2.2.8  PHENAZINES

Phenazine $((C_6H_4)_2N_2)$ and its derivatives (Fig. 22.8) are present in large amount in plants, consisting basically of dibenzo annulated pyrazine ring. As it contains –N in its skeleton, hence its pyrolysis could lead to the synthesis of N-doped CDs. Its hydrothermal treatment will be more suitable, and solvothermal will be a suitable one to obtain CDs with better fluorescence.

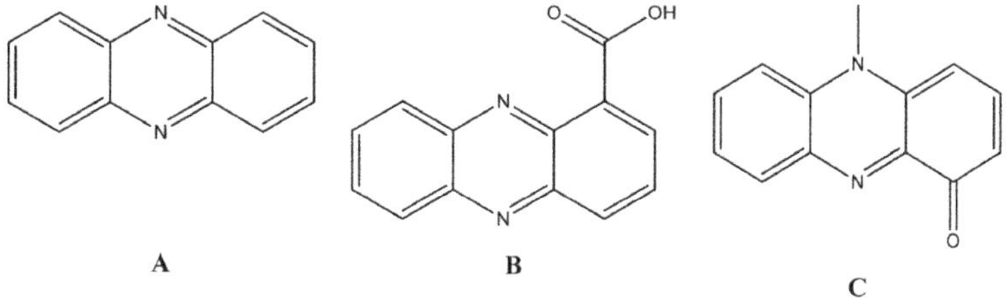

**FIGURE 22.8**  Phenazine and its derivatives (A) phenazine; (B) phenazine-1-carboxylic acid; (C) pyocyanin.

## 22.2.2.9  POLYKETIDES

Polyketides (Fig. 22.9) although are quite remarkable candidates for CDs synthesis, yet are already reported for their biological applications extensively and will not be favored for CD's synthesis of ordinary use.

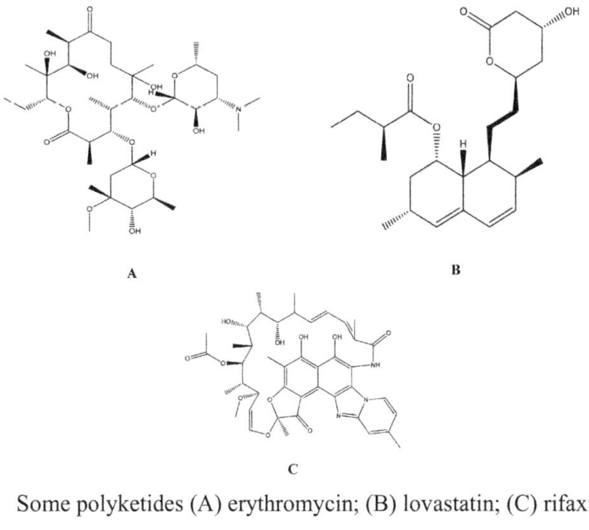

**FIGURE 22.9**  Some polyketides (A) erythromycin; (B) lovastatin; (C) rifaximin.

### 22.2.2.10   FATTY ACIDS

Plants contain a variety of fatty acids, basically shaped in their body as oils. A number of oils, such as soyabeans oil (Chen and Tseng, 2017) and mustard oil (Tripathi and Sarkar, 2014) has been reported for CD's synthesis. Being a raw material present in large quantity in plants, the fatty acids can be used easily for CD's synthesis. As actually composed of carboxylate groups and having liquid/oily physical state, these can be used as a starting material for simple CD's synthesis in individual form without adding any other precursor by thermal pyrolysis, while hydrothermal method will not be much fruitful due to immiscibility of oil with water, but on the other hand, solvothermal methods can be used for the synthesis of high-quality CDs using some good and miscible solvents.

### 22.2.3   FROM DOMESTIC RESOURCES

Sk et al. (2012) reported the extraction of alkaloids from readily available food caramels such as bread, sugar caramels, jiggery, biscuits, and corn flour. All these sources are carbohydrate-based and involve the heating of the source material which results in the conversion of the food carbohydrate material to CD's synthesis. In another report, Jiang and their team report the extraction of CDs from a routine drink of coffee with the successful application of these CDs in bio-imaging studies (Jiang et al., 2014).

### 22.3   APPLICATIONS

Owing to their simple chemistry, which usually leads to their low cytotoxicity, CDs are more suitable candidates for bio-applications. Herein, we more concisely discuss their bio-applications.

### 22.3.1   CHEMICAL SENSING

CDs have been reported for the detection of metal ions, known as chemical sensing. All the CDs, as discussed previously have the functional groups of –COOH, –OH, and so forth, which have the ability of coordinate bonding with metal ions, such as $Fe^{2+}$, $Fe^{3+}$, $Cu^+$, and $Cu^{2+}$ and so forth. Furthermore, the presence of doped N or S also facilitates their coordination with metal ions, making able CDs to behave as Lewis bases. Their applications for detection of heavy metals, for example, Cu, or Hg and so forth, is of much importance in biosciences, due to their hazardous effects on the environment and human health. The fluorescence of CDs results in substantial quenching after interactions with these ions, both in aqueous solutions as well as in live cells (Zhu et al., 2013; Li et al., 2014b; Yan et al., 2014).

## 22.3.2  BIO-SENSING

Researchers are eagerly interested to apply CDs as bio-imaging tools in biology, primarily due to their nontoxic and environmentally friendly nature in terms of their composition from carbon. These materials can be used for the bio-labeling of important cell organelles, such as for labeling and tracking proteins and so forth their mechanism of action is nearly the same, as described above in the sensing of metal ions. The CDs contain Lewis base atoms, or electron pair donor species, such as –N, or –O, which can readily form complexes with proteins and results in quenching of CD's fluorescence, resulting in confirming the site or the movement pattern of the resulting protein. To obtain better results from CDs, the surface of these materials are usually passivated by other molecules. In bio-imaging, the problems of bleaching, near infrared (IR), or IR range fluorescence, and stable long-standing emission are the main concerns of these materials application, which are primarily achieved by these materials passivation. Many reports are available regarding the synthesis of CDs, followed by their surface passivation and their successful applications in bio-imaging. A number of studies are available on their applications as biosensors (Qian et al., 2012; Ramanan et al., 2016).

## 22.3.3  NANOMEDICINE

Keeping in mind their low cell cytotoxicity, cheap synthetic precursors, and easy fabrication methods, CDs have wide applications in nano-medicines. These materials have their self-applications as anti-microbes, and also have applications as drug carriers inside the organism bodies. As drug carriers, their surfaces are loaded with different drugs, and through their better absorption, low size, and suitably loaded drug, these easily target the effector cells, thus preventing the loss of drug, and the less or no effects on other cells, while in most reports, these materials also enhanced the functionality of CDs (Park et al., 2014; Augustine et al., 2017).

## 22.3.4   PHOTOCATALYSIS

CDs in relation to graphene quantum dots, and in contrast to pure graphene have non zero band gap, primarily due to more electronegative atoms of –O, and thus act as best precursors for applications in photocatalysis (Xu et al., 2013). The photocatalytic property of CDs could be more improved by the doping of electronegative elements of –N, –S, and or –P. The simple carbon core skeleton and the presence of simple functional groups on the surface of CDs favor its application as photocatalysts for carrying out different reactions, because these two basic components of CDs can be easily engineered by introducing different doping agents as well as by introducing different functional groups on its surfaces (Gao et al., 2014). Photocatalysis is a green approach towards the synthesis of compounds with emphasis that this involves the use of sunlight for carrying out different reactions instead of using other energy sources such as heat and so forth, which not only cause financial burden but also causes the production of environmental dangerous species. Moreover, the sun as an ultimate source of energy causes the scientists to concentrate more on this source for carrying different reactions. However, extensive care is needed to carry out photocatalytic reactions as it has been reported that the high energy ultraviolet (UV) and short wavelength visible light may adversely damage CDs (Kumar et al., 2008).

## 22.4   PHOTOLUMINESCENCE

CDs are composed of $sp^2$ carbon atoms which in combination form can be classified into two categories. The main carbon core, which contains $sp^2$ carbon atoms, bonded in conjugation with each other, and their surface functional groups bonded to the main carbon core.

Literature favors the role of both the components in their PL (Baptista et al., 2015; Lim et al., 2015; Cayuela et al., 2016), which can be highlighted in the below lines.

## 22.4.1   CARBON CORE

The carbon atoms are both single and double bonded to their carbon skeleton having $sp^2$ hybridization. The presence of conjugations in their skeleton which also lead to their resonance in CDs structures leading to some extent the fluorescence of these nanomaterials, although some studies report their non-fluorescence nature (Cayuela et al., 2016), which is not to be accepted on a theoretical basis. Theoretically, the conjugated carbon atoms upon excitation undergo $\pi$------>$\pi^*$ transitions, which usually fall in the UV region of electromagnetic radiations ranging from 190–380 nm, appearing blue fluorescence. However, a non-photoluminescent observation for CDs having an only carbogenic core, or having the negligible influence of the surface functional groups, could be attributed theoretically to the counter effecting nature of graphite rings because as the CDs size increases their photoluminescence finally disappears.

The role of the main core in the photoluminescence is more prominent for CDs doped with other elements like –N, –O, –S, or –B and so forth. In such case, there also appear n------> $\pi^*$ transitions along with $\pi$------>$\pi^*$ transitions.

## 22.4.2  SURFACE GROUPS

The main advantage of CDs is their ability to be functionalized with a variety of functional groups, besides their nanosize and so forth; however, apart from a variety of other functional groups to be passivated on the surface of CDs after further modification of the surface, there are only a few types of functional groups such as –COOH, –OH, and C–O–C. These also undergo n------>$\pi^*$ transitions and $\pi$------>$\pi^*$ transitions, which cause the photoluminescence of CDs by itself or further enhance the photoluminescence of CDs.

## 22.5  CYTOTOXICITY

Owing to their simple structure, consisting of carbon core with few functional groups, these materials are regarded as less cytotoxic, and hence are prominently used for their applications in biological processes, such as bioimaging, bio-sensing, anti-cancerous studies, and so forth. Of the truth, little has been reported regarding their cytotoxic studies, especially in vivo, yet confirming their low cytotoxic nature.

## 22.5.1  CYTOTOXICITY IN ANIMALS

Below is given a short discussion about CD's cytotoxicity in different animals.

## 22.5.1.1   IN VITRO CYTOTOXICITY

CDs have intriguing optical and electro-optical properties attributed to their quantum confinement and edge effects (Dong et al., 2014). Their easy fabrication and dynamic chemical and optical inertness along with their fluorescent nature give them an extra weight in this huge family of carbon nanomaterials. As CDs have been reported for their bio-imaging, cell imaging, sensing, photovoltaic devices, and catalysis, hence their cytotoxic studies are also very important to be analyzed. In vitro hemocompatibility studies of fluorescent CDs were performed by Li and company showing that up to the concentration of ≤0.1 mg/mL they show a slight effects on blood components, while in higher doses it lead to adverse effects to blood, by impairing the structures and functions of red blood cells, interfering the fibrinogen microenvironment, and affecting blood coagulation (Li et al., 2015a). Li et al synthesized different types of PL CDs and studied their in vitro cytotoxicity studies against HeLa cell lines for imaging and antitumor cells targeting and drug delivery. The authors claim their low toxicity, even at the double concentrations used for bio-imaging (Li et al., 2010). Wang and Zhou claimed the synthesis of non-toxic fluorescent CDs, along with their imaging application for U87 glioma cancer cell lines (Wang and Zhou, 2014). The human breast cancer cell line MCF-7(ATCC) and human colon adenocarcinoma grade II cell line HT-29(ATCC) were used for the viability and labeling studies of ligand-functionalized CDs by Liu and colleagues. They found that CDs functionalized with long ligands are more cytotoxic than those with small ligands (Liu et al., 2015). Li and partners in their study report the effects of CDs on inhibiting the actions of human insulin. In their well-controlled study, they found that CDs in a concentration of 40 μg/mL are sufficient enough to inhibit the 0.2 mg/mL of human insulin from fibrillation at 65°C for 5 days, whereas insulin denatures in 3 h under the same conditions, thus confirming their low cytotoxicity (Li et al., 2015b). Yang et al. (2009), studied the in vitro effects of CDs comparatively with semiconductor quantum dots and found these less toxic in a concentration of 200 μg/mL against MCF-7 and HT-29 cells.

## 22.5.1.2   IN VIVO CYTOTOXICITY

Li and fellow workers reported the hemocompatibility of fluorescent CDs, confirming their concentration-dependent effects with the blood both in vivo. They found these CDs safe up to the concentrations of 50 mg/kg, as their in action for blood coagulation (Li et al., 2015a). In vivo assays of CDs were reported by Yang and fellows against male CD-1 mice (~25 g) exposed intravenously to CDs in a single injection of 200 µg carbon core-equivalent in 200 µl or three injections (with 4 h intervals between injections) of 333 µg in 333 µL each. Mice exposed to 0.9% NaCl aqueous solution were taken as the control group. The results confirm these materials as non-toxic to these species (Yang et al., 2009).

## 22.5.2   CYTOTOXICITY IN PLANTS

Many studies have been performed for the cytotoxicity of NPs against plants, although not more specifically regarding the CDs. The mechanism of NPs uptake in plants is generally considered as active transport, depending on the factors such as signaling, recycling, and regularization of the plasma membrane (Etxeberria et al., 2009). The general pathway, regarding the uptake of NPs by plants is through roots and their leaves. The roots absorb NPs from the soil, although in little concentration, also depending on their size and surface area and then transmit them to stem and leaves and so forth. All this passage within plant body occurs through the pores in the cell wall and cell membranes of the plant parts, hence this passage depends upon the size of NPs. In other words, large size NPs are less prone to be absorbed by the roots and then passed to other parts comparatively to small size CDs. Thus we can assume that the plant cell wall act as a semiperme-able membrane by allowing small size NPs while blocking large size NPs. The major accumulating organ of NPs is the tuber of the roots, followed by leaves and then stem. In small amounts, the NPs could also be absorbed by leaves, if these materials are already suspended in the air. For their passage, it has been proved that these materials are passed away within plant body through capillary action, and are accumulated within the empty spaces of plant cells. However, if their size is large enough, or they are no functional-ized well, then these materials can block the passage pathways of plants (Husen and Siddiqi 2014; Yadav, et al., 2014). Avanasi and team workers, in their report found that the NPs such as $C_{60}$ sorption in the soil depends upon the factors such as pH of the soil, clay content, and mineralogy and cation exchange capacity, thus alternatively the NPs uptake by plants is affected by the environment and nature of the soil containing the plants (Avanasi et al., 2014). A large variation has been observed in NPs effects on the different plants, as well as on the different components of the same plant, hence it is not concluding that what the exact effect on plants is. For example, the effect of Multi Wall Carbon Nano Tubes in a concentration of 2000 mg/L, the effect was a reduction in root length for Lettuce plant, while no effect on root was observed for Corn and Cucumber, and an increase was observed for ryegrass plant (Lin and Xing, 2007).

Tripathi and Sarkar (2014), in their studies of CDs effect on the growth of roots and shoots, found that soluble CDs greatly increase the growth rate of plant roots and shoots, both in dark and in the light. They also reported that the growth rate is higher in roots as compared to shoots. Husen and Siddiqi (2014), in their study, reveals that carbon nanotubes and fullerene $C_{60}$ increase the water retaining capacity, fruit production, and biomass production up to 118%. They also conclude that the seeds treated with fullerene $C_{60}$ bear more phytomedicine than the untreated seeds. In conclusion, they remark that both edible plants and vegetables increase their fruit and crop production after exposure to NPs. They also demonstrate that the non-functionalized NPs in some cases are toxic because they cannot readily form bonds with solvents, and thus cannot move freely. However, their cytotoxicity could be decreased if these materials are functionalized properly. As to date, the studies on the cytotoxicity of NPs are not enough which can lead us to a straightforward conclusion, however, regarding the applications of NPs against the cytotoxicity of plants, we should be extremely careful, and must keep these points in minds, the types of NPs we dealing with in terms of their physical and chemical properties, the dosage, type of plant, developmental stage, and time of treatment and so forth (Milewska-Hendel et al., 2016). Although it is considered in general, that crops and fruits production of plants could be increased by applying different NPs (Tripathi et al., 2017). However, regarding the future perspective, we could present that the objective of nanotechnology in crop and fruit production should emphasize to develop CDs having bactericidal, fungicidal, and uretic applications, which no doubt will be of great benefit as compared to their current available counterparts.

## 22.6  CHALLENGES AND FUTURE SCOPE

In this nascent field of biosynthetic CDs, we still have a long way to go. For instance, yet the fully established mechanism of these materials synthesis is not discovered, which require some well-designed experiments to be carried out. Another issue is of their classification, on which a high number of researchers are confusing between CDs and graphene quantum dots.

A better step could be to further study their synthesis systematically, and to explore their applications in other fields, with emphasis on material sciences. A high number of experiments to explore well step by step their synthesis, their properties and their applications in details are needed. On the other hand, the origin of their photoluminescence is also not explained well, as scientists have opposite views in terms of their experiments.

The top-down methods for biosynthesis are studied well to some extent, while the literature on the bottom up approaches from bioresources is almost nil. Moreover, among the top down methods, the hydrothermal methods have been implied largely, along with thermal and solvothermal, yet the other methods are not established well to date. Their cytotoxicity studies are also limited.

## 22.7   CONCLUSION

Compliant with the reputation of these petite discovered materials, one can easily conclude that CDs are among few of the best sources regarding materials and biology applications. Furthermore, it is also evident that plants and their phytochemicals have numerous resources, and indefinite advantages including the less hazardous impact on environment and organisms, and are thus more suitable candidates to be utilized for CDs synthesis. Surprisingly, this nascent field of research also achieved great successes as compared to other areas of science and technology. However, further optimization and more specific studies are required to explore well this field of CD's synthesis and their applications.

## ACKNOWLEDGMENT

Financial support from University of Chinese Academy of Sciences and The World Academy of Sciences under the CAS-TWAS President's fellowship program is highly acknowledged. I am also very thankful to the Editor Chukwuebuka Egbuna, to my research supervisor Professor Jian Ru Gong and my research instructor Assistant Professor Xing Qi as well as to Professor Zhang Guo Lin, Chengdu Institute of Biology for their good wishes and thoughtful insights during the accomplishment of this project.

## KEYWORDS

- **carbon dots**
- **biosynthesis**
- **top-down methods**
- **bottom-up methods**
- **cytotoxicity**

## REFERENCES

Arul, V.; Edison, T. N. J. I.; Lee, Y. R.; Sethuraman, M. G. Biological and Catalytic Applications of Green Synthesized Fluorescent N-Doped Carbon Dots Using Hylocereus Undatus. *J. Photochem. Photobiol. B* **2017**, *168,* 142–148.

Augustine, S.; Singh, J.; Srivastava, M.; Sharma, M.; Das, A. Malhotra, B. D. Recent Advances in Carbon Based Nanosystems for Cancer Theranostics. *Biomater. Sci.* **2017**, *5*(5), 901–952.

Avanasi, R.; Jackson, W. A.; Sherwin, B.; Mudge, J. F.; Anderson, T. A. C60 Fullerene Soil Sorption, Biodegradation, and Plant Uptake. *Environ. Sci. Technol.* **2014**, *48*(5), 2792–2797.

Baptista, F. R.; Belhout, S. A.; Giordani, S.; Quinn, S. J. Recent Developments in Carbon Nanomaterial Sensors. *Chem. Soc. Rev.* **2015**, *44*(13), 4433–4453.

Cayuela, A.; Soriano, M. L.; Carrillo-Carrion, C.; Valcarcel, M. Semiconductor and Carbon-Based Fluorescent Nanodots: The Need for Consistency. *Chem. Commun. (Cambridge).* **2016**, *52*(7), 1311–132.

Chen, T. H.; Tseng, W. L. Self-Assembly of Monodisperse Carbon Dots into High-Brightness Nanoaggregates for Cellular Uptake Imaging and Iron(III) Sensing. *Anal. Chem.* **2017**, *89*(21), 11348–11356.

Chen, X.; Zhang, W. Diamond Nanostructures for Drug Delivery, Bioimaging, and Biosensing. *Chem. Soc. Rev.* **2017**, *46*(3), 734–760.

Chen, X.; Wang, F.; Hyun, J. Y.; Wei, T.; Qiang, J.; Ren, X.; Shin, I.; Yoon, J. Recent Progress in the Development of Fluorescent, Luminescent and Colorimetric Probes for Detection of Reactive Oxygen and Nitrogen Species. *Chem. Soc. Rev.* **2016**, *45*(10), 2976–3016.

Chen, J.; Zhang, X.; Zhang, Y.; Wang, W.; Li, S.; Wang, Y.; Hu, M.; Liu, L.; Bi, H. Understanding the Capsanthin Tails in Regulating the Hydrophilic-Lipophilic Balance of Carbon Dots for a Rapid Crossing Cell Membrane. *Langmuir* **2017**, *33*(39), 10259–10270.

Chowdhury, D.; Gogoi, N.; Majumdar, G. Fluorescent Carbon Dots Obtained from Chitosan Gel. *RSC Adv.* **2012**, *2*(32), 12156.

Dang, D. K., Chandrasekaran, S.; Ngo, Y.-L. T.; Chung, J. S.; Kim, E. J.; Hur, S. H. One Pot Solid-State Synthesis of Highly Fluorescent N and S Co-Doped Carbon Dots and its use as Fluorescent Probe for Ag⁺ Detection in Aqueous Solution. *Sens. Actuators B* **2018**, *255,* 3284–3291.

Das, B.; Dadhich, P.; Pal, P.; Srivas, P. K.; Bankoti, K.; Dhara, S. Carbon Nanodots from Date Molasses: New Nanolights for the In Vitro Scavenging of Reactive Oxygen Species. *J. Mater. Chem. B* **2014**, *2*(39), 6839–6847.

Das, P.; Bose, M.; Ganguly, S.; Mondal, S.; Das, A. K.; Banerjee, S.; Das, N. C. Green Approach to Photoluminescent Carbon Dots for Imaging of Gram-Negative Bacteria *Escherichia coli. Nanotechnology* **2017,** *28*(19), 195501.

Dong, Y.; Lin, J.; Chen, Y.; Fu, F.; Chi, Y.; Chen, G. Graphene Quantum Dots, Graphene Oxide, Carbon Quantum Dots and Graphite Nanocrystals in Coals. *Nanoscale* **2014,** *6*(13), 7410–7415.

Edison, T. N.; Atchudan, R.; Sethuraman, M. G.; Shim, J. J.; Lee, Y. R. Microwave Assisted Green Synthesis of Fluorescent N-Doped Carbon Dots: Cytotoxicity and Bio-imaging Applications. *J Photochem Photobiol B.* **2016,** *161,* 154–161.

Ensafi, A. A.; Sefat, S. H.; Kazemifard, N.; Rezaei, B.; Moradi, F. A Novel One-Step and Green Synthesis of Highly Fluorescent Carbon Dots from Saffron for Cell Imaging and Sensing of Prilocaine. *Sens. Actuators B* **2017,** *253,* 451–460.

Etxeberria, E.; Gonzalez, P.; Pozueta, J. Evidence for Two Endocytic Transport Pathways in Plant Cells. *Plant Sci.* **2009,** *177*(4), 341–348.

Fan, Y. Z.; Zhang, Y.; Li, N.; Liu, S. G.; Liu, T.; Li, N. B.; Luo, H. Q. A Facile Synthesis of Water-Soluble Carbon Dots as a Label-Free Fluorescent Probe for Rapid, Selective and Sensitive Detection of Picric Acid. *Sens. Actuators B* **2017,** *240,* 949–955.

Feng, Y.; Zhong, D.; Miao, H.; Yang, X. Carbon Dots Derived from Rose Flowers for Tetracycline Sensing. *Talanta* **2015,** *140,* 128–133.

Gao, S.; Chen, Y.; Fan, H.; Wei, X.; Hu, C.; Wang, L.; Qu, L. A Green One-Arrow-Two-Hawks Strategy for Nitrogen-Doped Carbon Dots as Fluorescent Ink and Oxygen Reduction Electrocatalysts. *J. Mater. Chem. A* **2014,** *2*(18), 6320.

Gu, D.; Shang, S.; Yu, Q.; Shen, J. Green Synthesis of Nitrogen-Doped Carbon Dots from Lotus Root for Hg(II) Ions Detection and Cell Imaging. *Appl. Surf. Sci.* **2016,** *390,* 38–42.

Hardt, H.; Lamport, D. T. Hydrogen Fluoride Saccharification of Wood: Lignin Fluoride Content, Isolation of Alpha-D-Glucopyranosyl Fluoride and Posthydrolysis of Reversion Products. *Biotechnol. Bioeng.* **1982,** *24,* 903–918.

Hill, S.; Galan, M. C. Fluorescent Carbon Dots from Mono- and Polysaccharides: Synthesis, Properties and Applications. *Beilstein J. Org. Chem.* **2017,** *13,* 675–693.

Hola, K.; Sudolska, M.; Kalytchuk, S.; Nachtigallova, D.; Rogach, A. L.; Otyepka, M.; Zboril, R. Graphitic Nitrogen Triggers Red Fluorescence in Carbon Dots. *ACS Nano.* **2017,** *11*(12), 12402–12410.

Hsu, P.-C.; Shih, Z.-Y.; Lee, C.-H. Chang, H.-T. Synthesis and Analytical Applications of Photoluminescent Carbon Nanodots. *Green Chem.* **2012,** *14*(4), 917.

Huang, X.; Zhang, F.; Zhu, L.; Choi, K. Y.; Guo, N.; Guo, J.; Tackett, K.; Anilkumar, P.; Liu, G.; Quan, Q.; Choi, H. S.; Niu, G.; Sun, Y. P.; Lee, S.; Chen, X. Effect of Injection Routes on the Biodistribution, Clearance, and Tumor Uptake of Carbon Dots. *ACS Nano* **2013,** *7*(7), 5684–5693.

Huang, G.; Chen, X.; Wang, C.; Zheng, H.; Huang, Z.; Chen, D.; Xie, H. Photoluminescent Carbon Dots Derived from Sugarcane Molasses: Synthesis, Properties, and Applications. *RSC Adv.* **2017,** *7*(75), 47840–47847.

Husen, A.; Siddiqi, K. S. Carbon and Fullerene Nanomaterials in Plant System. *J. Nanobiotechnol.* **2014,** *12,* 16.

Jeong, C. J.; Roy, A. K.; Kim, S. H.; Lee, J. E.; Jeong, J. H.; In, I.; Park, S. Y. Fluorescent Carbon Nanoparticles Derived from Natural Materials of Mango Fruit for Bio-Imaging Probes. *Nanoscale* **2014,** *6*(24), 15196–15202.

Jhonsi, M. A.; Thulasi, S. A Novel Fluorescent Carbon Dots Derived from Tamarind. *Chem. Phys. Lett.* **2016,** *661,* 179–184.

Jiang, C.; Wu. H.; Song, X.; Ma, X.; Wang, J.; Tan, M. Presence of Photoluminescent Carbon Dots in Nescafes Original Instant Coffee: Applications to Bioimaging. *Talanta* **2014,** *127,* 68–74.

Jones, S. S.; Sahatiya, P.; Badhulika, S. One Step, High Yield Synthesis of Amphiphilic Carbon Quantum Dots Derived from Chia Seeds: A Solvatochromic Study. *New J. Chem.* **2017,** *41*(21), 13130–13139.

Kalita, H.; Mohapatra, J.; Pradhan, L.; Mitra, A.; Bahadur, D.; Aslam, M. Efficient Synthesis of Rice Based Graphene Quantum Dots and Their Fluorescent Properties. *RSC Adv.* **2016,** *6*(28), 23518–23524.

Kasibabu, B. S. B.; D'Souza, S. L.; Jha, S.; Singhal, R. K.; Basu, H.; Kailasa, S. K. One-Step Synthesis of Fluorescent Carbon Dots for Imaging Bacterial and Fungal Cells. *Anal. Methods* **2015,** *7*(6), 2373–2378.

Konwar, A.; Gogoi, N.; Majumdar, G.; Chowdhury, D. Green Chitosan-Carbon Dots Nanocomposite Hydrogel Film with Superior Properties. *Carbohydr. Polym.* **2015,** *115,* 238-245.

Konwar, A.; Baruah, U.; Deka, M. J.; Hussain, A. A.; Haque, S. R.; Pal, A. R.; Chowdhury, D. Tea-Carbon Dots-Reduced Graphene Oxide: An Efficient Conducting Coating Material for Fabrication of an E-Textile. *ACS Sustainable Chem. Eng.* **2017,** *5*(12), 11645–11651.

Kumar, A. S.; Ye, T.; Takami, T.; Yu, B. C.; Flatt, A. K.; Tour, J. M.; Weiss, P. S. Reversible Photo-Switching of Single Azobenzene Molecules in Controlled Nanoscale Environments. *Nano Lett.* **2008,** *8*(6), 1644–1648.

Kumar, A.; Chowdhuri, A. R.; Laha, D.; Mahto, T. K.; Karmakar, P.; Sahu, S. K. Green Synthesis of Carbon Dots from Ocimum Sanctum for Effective Fluorescent Sensing of $Pb^{2+}$ Ions and Live Cell Imaging. *Sens. Actuators B* **2017,** *242,* 679–686.

Li, Q.; Ohulchanskyy, T. Y.; Liu, R.; Koynov, K.; Wu, D.; Best, A.; Kumar, R.; Bonoiu, A.; Prasad, P. N. Photoluminescent Carbon Dots as Biocompatible Nanoprobes for Targeting Cancer Cells In Vitro. *J. Phys. Chem. C* **2010,** *114,* 12062–12068.

Li, L.; Wu, G.; Yang, G.; Peng, J.; Zhao, J.; Zhu, J. J. Focusing on Luminescent Graphene Quantum Dots: Current Status and Future Perspectives. *Nanoscale* **2013a,** *5*(10), 4015–4039.

Li, W.; Yue, Z.; Wang, C.; Zhang, W.; Liu, G. An Absolutely Green Approach to Fabricate Carbon Nanodots from Soya Bean Grounds. *RSC Adv.* **2013b,** *3*(43), 20662.

Li, C.-L.; Ou, C.-M.; Huang, C.-C.; Wu, W.-C.; Chen, Y.-P.; Lin, T.-E.; Ho, L.-C.; Wang, C.-W.; Shih, C.-C.; Zhou, H.-C.; Lee, Y.-C.; Tzeng, W.-F.; Chiou, T.-J.; Chu, S.-T.; Cang, J.; Chang, H.-T. Carbon Dots Prepared from Ginger Exhibiting Efficient Inhibition of Human Hepatocellular Carcinoma Cells. *J. Mater. Chem. B* **2014a,** *2*(28), 4564.

Li, F.; Liu, C.; Yang, J.; Wang, Z.; Liu, W.; Tian, F. Mg/N Double Doping Strategy to Fabricate Extremely High Luminescent Carbon Dots for Bioimaging. *RSC Adv.* **2014b,** *4*(7), 3201–3205.

Li, S.; Guo, Z.; Zhang, Y.; Xue, W.; Liu, Z. Blood Compatibility Evaluations of Fluorescent Carbon Dots. *ACS Appl. Mater. Interfaces* **2015a,** *7*(34), 19153–19162.

Li, S.; Wang, L.; Chusuei, C. C.; Suarez, V. M.; Blackwelder, P. L.; Micic, M.; Orbulescu, J.; Leblanc, R. M. Nontoxic Carbon Dots Potently Inhibit Human Insulin Fibrillation. *Chem. Mater.* **2015b,** *27*(5), 1764–1771.

Liang, Q.; Ma, W.; Shi, Y.; Li, Z.; Yang, X. Easy Synthesis of Highly Fluorescent Carbon Quantum Dots from Gelatin and Their Luminescent Properties and Applications. *Carbon* **2013,** *60,* 421–428.

Lim, S. Y.; Shen, W.; Gao, Z. Carbon Quantum Dots and Their Applications. *Chem. Soc. Rev.* **2015,** *44*(1), 362–381.

Lin, D.; Xing, B. Phytotoxicity of Nanoparticles: Inhibition of Seed Germination and Root Growth. *Environ. Pollut.* **2007,** *150*(2), 243–250.

Liu, S.; Tian, J.; Wang, L.; Zhang, Y.; Qin, X.; Luo, Y.; Asiri, A. M.; Al-Youbi, A. O.; Sun, X. Hydrothermal Treatment of Grass: a Low-Cost, Green Route to Nitrogen-Doped, Carbon-Rich, Photoluminescent Polymer Nanodots as an Effective Fluorescent Sensing Platform for Label-Free Detection of Cu(II) Ions. *Adv. Mater.* **2012,** *24*(15), 2037–2041.

Liu, Y.; Zhao, Y.; Zhang, Y. One-Step Green Synthesized Fluorescent Carbon Nanodots from Bamboo leaves for Copper(II) Ion Detection. *Sens. Actuators B* **2014,** *196,* 647–652.

Liu, J. H.; Cao, L.; LeCroy, G. E.; Wang, P.; Meziani, M. J.; Dong, Y.; Liu, Y.; Luo, P. G.; Sun, Y. P. Carbon "Quantum" Dots for Fluorescence Labeling of Cells. *ACS Appl. Mater. Interfaces* **2015,** *7*(34), 19439–19445.

Liu, W.; Diao, H.; Chang, H.; Wang, H.; Li, T.; Wei, W. Green Synthesis of Carbon Dots from Rose-Heart Radish and Application for $Fe^{3+}$ Detection and Cell Imaging. *Sens. Actuators B* **2017a,** *241,* 190–198.

Liu, W.; Li, C.; Sun, X.; Pan, W.; Yu, G.; Wang, J. Highly Crystalline Carbon Dots from Fresh Tomato: UV Emission and Quantum Confinement. *Nanotechnology* **2017b,** *28*(48), 485705.

Lu, W.; Qin, X.; Liu, S.; Chang, G.; Zhang, Y.; Luo, Y.; Asiri, A. M.; Al-Youbi, A. O.; Sun, X. Economical, Green Synthesis of Fluorescent Carbon Nanoparticles and Their Use as Probes for Sensitive and Selective Detection of Mercury(II) Ions. *Anal. Chem.* **2012,** *84*(12), 5351–5357.

Mewada, A.; Pandey, S.; Shinde, S.; Mishra, N.; Oza, G.; Thakur, M.; Sharon, M.; Sharon, M. Green Synthesis of Biocompatible Carbon Dots Using Aqueous Extract of *Trapa bispinosa* Peel. *Mater. Sci. Eng. C Mater. Biol. Appl.* **2013,** *33*(5), 2914–2917.

Milewska-Hendel, A.; Gawecki, R.; Zubko, M.; Stróż, D.; Kurczyńska, E. Diverse Influence of Nanoparticles on Plant Growth with a Particular Emphasis on Crop Plants. *Acta Agrobot.* **2016,** *69,* 4.

Niu, X.; Liu, G.; Li, L.; Fu, Z.; Xu, H.; Cui, F. Green and Economical Synthesis of Nitrogen-Doped Carbon Dots from Vegetables for Sensing and Imaging Applications. *RSC Adv.* **2015,** *5*(115), 95223–95229.

Park, S. Y.; Lee, H. U.; Park, E. S.; Lee, S. C.; Lee, J. W.; Jeong, S. W.; Kim, C. H.; Lee, Y. C.; Huh, Y. S.; Lee, J. Photoluminescent Green Carbon Nanodots from Food-Waste-Derived Sources: Large-Scale Synthesis, Properties, and Biomedical Applications. *ACS Appl. Mater. Interfaces* **2014,** *6*(5), 3365–3370.

Pinto, T. D. S.; Alves, L. A.; Cardozo, G. D. A.; Munhoz, V. H. O.; Verly, R. M.; Pereira, F. V.; Mesquita, J. P. D. Layer-by-Layer Self-Assembly for Carbon Dots/Chitosan-Based Multilayer: Morphology, Thickness and Molecular Interactions. *Mater. Chem. Phys.* **2017,** *186,* 81–89.

Prasannan, A.; Imae, T. One-Pot Synthesis of Fluorescent Carbon Dots from Orange Waste Peels. *Ind. Eng. Chem. Res.* **2013,** *52*(44), 15673–15678.

Qian, J.; Wang, D.; Cai, F. H.; Xi, W.; Peng, L.; Zhu, Z. F.; He, H.; Hu, M. L.; He, S. Observation of Multiphoton-Induced Fluorescence from Graphene Oxide Nanoparticles and Applications in In Vivo Functional Bioimaging. *Angew. Chem. Int. Ed. Engl.* **2012,** *51*(42), 10570–10575.

Rai, S.; Singh, B. K.; Bhartiya, P.; Singh, A.; Kumar, H.; Dutta, P. K.; Mehrotra, G. K. Lignin Derived Reduced Fluorescence Carbon Dots with Theranostic Approaches: Nano-Drug-Carrier and Bioimaging. *J. Lumin.* **2017,** *190,* 492–503.

Ramanan, V.; Thiyagarajan, S. K.; Raji, K.; Suresh, R.; Sekar, R.; Ramamurthy, P. Outright Green Synthesis of Fluorescent Carbon Dots from Eutrophic Algal Blooms for In Vitro Imaging. *ACS Sustainable Chem. Eng.* **2016,** *4*(9), 4724–4731.

Sajid, P. A.; Chetty, S. S.; Praneetha, S.; Murugan, A. V.; Kumar, Y.; Periyasamy, L. One-Pot Microwave-Assisted in situ Reduction of $Ag^+$ and $Au^{3+}$ ions by *Citrus limon* Extract and Their Carbon-Dots Based Nanohybrids: A Potential Nano-Bioprobe for Cancer Cellular Imaging. *RSC Adv.* **2016,** *6*(105), 103482–103490.

Sk, M. P.; Jaiswal, A.; Paul, A.; Ghosh, S. S.; Chattopadhyay, A. Presence of Amorphous Carbon Nanoparticles in Food Caramels. *Sci. Rep.* **2012,** *2,* 383.

Sun, D.; Ban, R.; Zhang, P.-H.; Wu, G.-H.; Zhang, J.-R.; Zhu, J.-J. Hair Fiber as a Precursor for Synthesizing of Sulfur- and Nitrogen-Co-Doped Carbon Dots with Tunable Luminescence Properties. *Carbon* **2013,** *64,* 424–434.

Tan, X. W.; Romainor, A. N. B.; Chin, S. F.; Ng, S. M. Carbon Dots Production via Pyrolysis of Sago Waste as Potential Probe for Metal Ions Sensing. *J. Anal. Appl. Pyrolysis* **2014,** *105,* 157–165.

Teng, X.; Ma, C.; Ge, C.; Yan, M.; Yang, J.; Zhang, Y.; Morais, P. C.; Bi, H. Green Synthesis of Nitrogen-Doped Carbon Dots from Konjac Flour with "off–on" Fluorescence by $Fe^{3+}$ and l-lysine for Bioimaging. *J. Mater. Chem. B* **2014,** *2*(29), 4631.

Thambiraj, S.; Shankaran D. R. Green Synthesis of Highly Fluorescent Carbon Quantum Dots from Sugarcane Bagasse Pulp. *Appl. Surf. Sci.* **2016,** *390,* 435–443.

Tripathi, S.; Sarkar, S. Influence of Water Soluble Carbon Dots on the Growth of Wheat Plant. *Appl. Nanosci.* **2014,** *5*(5), 609–616.

Tripathi, D. K.; Shweta; Singh, S.; Singh, S.; Pandey, R.; Singh, V. P.; Sharma, N. C.; Prasad, S. M.; Dubey, N. K.; Chauhan, D. K. An Overview on Manufactured Nanoparticles in Plants: Uptake, Translocation, Accumulation and Phytotoxicity. *Plant Physiol. Biochem.* **2017,** *110,* 2–12.

Tyagi, A.; Tripathi, K. M.; Singh, N.; Choudhary, S.; Gupta, R. K. Green Synthesis of Carbon Quantum Dots from *Lemon* Peel Waste: Applications in Sensing and Photocatalysis. *RSC Adv.* **2016,** *6*(76), 72423–72432.

Valle, F. J. R. D. D.; Tuppal, R. C. G.; Royes, F. D. De Los. Synthesis of Carbon Quantum Dots (CQDs) from *Coffea liberica. Pure and Applied Chemistry International Conference (PACCON).* **2017.**

Vandarkuzhalia, S. A. A.; Jeyalakshmi. V.; Sivaraman, G.; Singaravadivel, S.; Krishnamurthy, K. R.; Viswanathan, B. Highly Fluorescent Carbon Dots from Pseudo-Stem of Banana Plant: Applications as Nanosensor and Bio-Imaging Agents. *Sens. Actuators B* **2017,** *252,* 894–900.

Verpoorte, R.; Schripsema, J. Isolation, Identification, and Structure Elucidation of Alkaloids A General Overview. In *Alkaloids;* Linskens, H. F., Jackson, J. F., Eds.; Springer Berlin Heidelberg: Berlin, Heidelberg, 1994; pp 1–24.

Wang, J.; Wang, C.-F.; Chen, S. Amphiphilic Egg-Derived Carbon Dots: Rapid Plasma Fabrication, Pyrolysis Process, and Multicolor Printing Patterns. *Angew. Chem. Int. Ed.* **2012,** *51,* 9297–9301.

Wang, L.; Zhou, H. S. Green Synthesis of Luminescent Nitrogen-Doped Carbon Dots from Milk and its Imaging Application. *Anal. Chem.* **2014,** *86*(18), 8902–8905.

Wang, X.; Qu, K.; Xu, B.; Ren, J.; Qu, X. Microwave Assisted One-Step Green Synthesis of Cell-Permeable Multicolor Photoluminescent Carbon Dots Without Surface Passivation Reagents. *J. Mater. Chem.* **2011**, *21,* 2445–2450.

Wang, L.; Bi, Y.; Hou, J.; Li, H.; Xu, Y.; Wang, B.; Ding, H.; Ding, L. Facile, Green and Clean One-Step Synthesis of Carbon Dots from Wool: Application as a Sensor for Glyphosate Detection Based on the Inner Filter Effect. *Talanta* **2016a,** *160,* 268–275.

Wang, L.; Li, W.; Wu, B.; Li, Z.; Wang, S.; Liu, Y.; Pan. D.; Wu, M. Facile Synthesis of Fluorescent Graphene Quantum Dots from Coffee Grounds for Bioimaging and Sensing. *Chem. Eng. J.* **2016b,** *300,* 75–82.

Wang, N.; Wang, Y.; Guo, T.; Yang, T.; Chen, M.; Wang, J. Green Preparation of Carbon Dots with Papaya as Carbon Source for Effective Fluorescent Sensing of Iron (III) and *Escherichia coli. Biosens. Bioelectron.* **2016c,** *85,* 68–75.

Wang, S.; Niu, H.; He, S.; Cai, Y. One-Step Fabrication of High Quantum Yield Sulfur- and Nitrogen-Doped Carbon Dots for Sensitive and Selective Detection of Cr(vi). *RSC Adv.* **2016d,** *6*(109), 107717–107722.

Wei, J.; Shen, J.; Zhang, X.; Guo, S.; Pan, J.; Hou, X.; Zhang, H.; Wang, L.; Feng, B. Simple One-Step Synthesis of Water-Soluble Fluorescent Carbon Dots Derived from Paper Ash. *RSC Adv.* **2013,** *3*(32), 13119.

Wei, J.; Zhang, X.; Sheng, Y.; Shen, J.; Huang, P.; Guo, S.; Pan, J.; Feng, B. Dual Functional Carbon Dots Derived from Cornflour via a Simple One-Pot Hydrothermal Route. *Mater. Lett.* **2014,** *123,* 107–111.

Xu, X.; Ray, R.; Gu Y.; Ploehn, J. H.; Gearheart, L.; Raker, K.; Scrivens, W. A. Electrophoretic Analysis and Purification of Fluorescent Single-Walled Carbon Nanotube Fragments.*J. Am. Chem. Soc.* **2004,** *126,* 12736–12737.

Xu, Q.; Zhou, Q.; Hua, Z.; Xue, Q.; Zhang, C.; Wang, X.; Pan, D.; Xiao, M. Single-Particle Spectroscopic Measurements of Fluorescent Graphene Quantum Dots. *ACS Nano* **2013,** *7*(12), 10654–10661.

Xu, Y.; Tang, C.-J.; Huang, H.; Sun, C.-Q.; Zhang, Y.-K.; Ye, Q-F.; Wang, A.-J. Green Synthesis of Fluorescent Carbon Quantum Dots for Detection of $Hg^{2+}$. *Chin. J. Anal. Chem.* **2014,** *42*(9), 1252–1258.

Xue, M.; Zhan, Z.; Zou, M.; Zhang, L.; Zhao, S. Green Synthesis of Stable and Biocompatible Fluorescent Carbon Dots from Peanut Shells for Multicolor Living Cell Imaging. *New J. Chem.* **2016,** *40*(2), 1698–1703.

Yadav, T.; Mungray, A. A.; Mungray, A. K. Fabricated Nanoparticles: Current Status and Potential Phytotoxic Threats. *Rev. Environ. Contam. Toxicol.* **2014,** *230,* 83–110.

Yan, F.; Zou, Y.; Wang, M.; Mu, X.; Yang, N.; Chen, L. Highly Photoluminescent Carbon Dots-Based Fluorescent Chemosensors for Sensitive and Selective Detection of Mercury Ions and Application of Imaging in Living Cells. *Sens. Actuators B* **2014,** *192,* 488–495.

Yan, Z.; Shu, J.; Yu, Y.; Zhang, Z.; Liu, Z.; Chen, J. Preparation of Carbon Quantum Dots Based on Starch and Their Spectral Properties. *Luminescence* **2015,** *30*(4), 388–392.

Yang, S. T.; Wang, X.; Wang, H.; Lu, F.; Luo, P. G.; Cao, L.; Meziani, M. J.; Liu, J. H.; Liu, Y.; Chen, M.; Huang, Y.; Sun, Y. P. Carbon Dots as Nontoxic and High-Performance Fluorescence Imaging Agents. *J. Phys. Chem. C* **2009,** *113,* 18110–18114.

Zelli, R.; Bartolami, E.; Longevial, J.-F.; Bessin, Y.; Dumy, P.; Marra, A.; Ulrich, S. A Metal-Free Synthetic Approach to Peptide-Based Iminosugar Clusters as Novel Multivalent Glycosidase Inhibitors. *RSC Adv.* **2016,** *6*(3), 2210–2216.

Zhang, Z.; Zhang, J.; Chen, N.; Qu, L. Graphene Quantum Dots: An Emerging Material for Energy-Related Applications and Beyond. *Energy Environ. Sci.* **2012,** *5*(10), 8869.

Zhou, J.; Sheng, Z.; Han, H.; Zou, M.; Li, C. Facile Synthesis of Fluorescent Carbon Dots Using Watermelon Peel as a Carbon Source. *Mater. Lett.* **2012,** *66*(1), 222–224.

Zhu, S.; Meng, Q.; Wang, L.; Zhang, J.; Song, Y.; Jin, H.; Zhang, K.; Sun, H.; Wang, H.; Yang, B. Highly Photoluminescent Carbon Dots for Multicolor Patterning, Sensors, and Bioimaging. *Angew. Chem. Int. Ed. Engl.* **2013,** *52*(14), 3953–3957.

# PART IV
# Phytochemicals as Friends and Foes

# CHAPTER 23

# TOXIC PLANTS AND PHYTOCHEMICALS

CHUKWUEBUKA EGBUNA[1,*], ALAN THOMAS S.[2],
ONYEKA KINGSLEY NWOSU[3], OLUMAYOWA VINCENT ORIYOMI[4],
TOSKΛ L. KRYEZIU[5], SARAVANAN KALIYAPERUMAL[6], and
JONATHAN C. IFEMEJE[1]

[1]*Department of Biochemistry, Chukwuemeka Odumegwu Ojukwu University, Anambra State, Nigeria, Tel.: +2347039618485*

[2]*National Institute of Plant Science Technology, Mahatma Gandhi University, Kottayam, Kerala 686560, India*

[3]*National Biosafety Management Agency, Abuja, Nigeria*

[4]*Institute of Ecology and Environmental Studies, Obafemi Awolowo University, Ile-Ife, Osun State, Nigeria*

[5]*Department of Clinical Pharmacy, University of Pristina, Kosovo*

[6]*PG and Research Department of Zoology, Nehru Memorial College (Autnonomous), Puthanampatti-621007, Tiruchirappalli, India*

*Corresponding author. E-mail: egbuna.cg@coou.edu.ng; egbunachukwuebuka@gmail.com*
*ORCID: https://orcid.org/0000-0001-8382-0693*

## ABSTRACT

Poisonous plants produce toxic compounds which are capable of eliciting undesirable aftermath effects even death when in contact with living organisms such as plants, animal, and microorganism. These compounds include anticholinergic, severe gastrointestinal (GI) irritants, cardiac glycosides, central nervous system stimulants/hallucinogens, and cyanogens. Some of these toxic phytochemicals have also been harnessed for some therapeutic uses. This chapter focuses on plants that produce toxic phytochemicals which can be harnessed for destruction and implementation of death to mankind and livestock. Some of these plants include *Atropa belladonna, Brugmansia, Ricinus communis, Dieffenbachia* Schott, *Cicuta maculata, Hippomane mancinella, Aconitum,* and *Abrus precatorius.* These plant poisons can be classified into four toxidromes that include cytotoxic, cardiotoxic, gastrointestinal-hepatotoxic, and neurotoxic. The potentials of these phytochemicals to be readily harnessed for bioterrorism in a civilized society call for concern.

## 23.1   INTRODUCTION

Plants are the important sources of metabolites which have medicinal, industrial, and toxicological values. The major importance of plants have always been as a source of food for mankind, but the discovery of their pharmacological effects have enabled their efficient utilization by mankind as traditional medicine or the extraction of their pharmacological agents for drug formulation. The metabolite synthesized by these plants can either be a primary metabolite that enables the day to day sustenance of the plant or secondary metabolites that is needed in minimal quantity or large quantity for the protection or maintenance of the plant. Some of these phytochemicals synthesized by plant can be beneficial, while others can be extremely dangerous. These compounds include anticholinergic, severe gastrointestinal (GI) irritants, cardiac glycoside, central nervous system stimulant/ hallucinogens, and cyanogens (Holmes, 2000). The toxic nature of plants is one of the strategies in plant defense against herbivores (Rasmann and Agrawal, 2009).

The secondary metabolites present in poisonous plants produce multiple symptoms, called toxidromes (Holmes, 2000). These plant poisons can be classified into four toxidromes that include cytotoxic, cardiotoxic, gastrointestinal-hepatotoxic, and neurotoxic (Diaz, 2016). The toxic properties of secondary metabolites present in plants are now being exploited for the production of valuable medicines for the treatment of many diseases (McGaw et al., 2005). However, instead of their use for the goodness of mankind, some people are focusing on how to destroy the world by intentional release of toxins or biological entities into the environment (Khajja et al., 2011). Bioterrorism or biological warfare is the use of biological entities and deadly toxins for their intentional release into the environment. Many toxic plants (Fig. 23.1 and Table 23.1), have been implicated in one incidence or the other.

Plate 1: *Atropa belladonna*

Plate 2: *Brugmansia sp*

Plate 3: *Ricinus communis* L.

Plate 4: Seeds of *Ricinus communis* L.    Plate 5: *Dieffenbachia* Schott                Plate 6: *Cicuta maculate*

Plate 4: Seeds of *Ricinus communis* L.    Plate 5: *Dieffenbachia* Schott                Plate 6: *Cicuta maculate*

**FIGURE 23.1    (See color insert.)** Some poisonous plants.

TABLE 23.1  List of Toxic Plants and Their Effects.

| S/ No. | Scientific name (Common name) | Family | Parts of the plant | Toxic compound | Toxic effect/symptoms/ |
|---|---|---|---|---|---|
| 1. | *Atropa belladonna* (Deadly nightshade) | Nightshade family (Solanaceae) | Root | Tropane alkaloids, scopolamine, hyoscyamine, and atropine. | Poisoning in central (short-term memory loss) and peripheral nervous system (hypertension or hypotension) |
| 2. | *Brugmansia* (Angel's trumpets) | Solanaceae | Seeds, leaves and almost all parts | Alkaloids | Tachycardia, diarrhea, paralysis of smooth muscles, migraine headaches, visual and auditory hallucinations, mydriasis, rapid onset of cycloplegia, and death. |
| 3. | *Ricinus communis* L. (Castor bean) | Euphorbiaceae | Seeds | Ricin | Respiratory distress, diarrhea, and gastrointestinal (GI) hemorrhage |
| 4. | *Dieffenbachia* (Dumb cane or mother-in-law's tongue) | Araceae | Root | Hydrocyanic acid and Nonhistamine | Contact with the eyes cause irritation and may lead to loss of vision. |
| 5. | *Cicuta maculata* (spotted water hemlock or suicide root) | Apiaceae | Roots (bulbs) | Cicutoxin | Nausea, vomiting, trembling, tachycardia, mydriasis, coma, and blockage of the inhalation process. |
| 6. | *Hippomane mancinella* (Beach apple) | Euphorbiaceae | Whole plant | Hippoman A and B, phorbol, and physostigmine | Nausea, headache, and abdominal discomfort. Damage to the digestive tract and loss of vision. |
| 7. | *Aconitum* (Devils helmet, wolf's bane) | Ranunculaceae | Root | Aconitine-mesaconitine, hypaconitine, deoxyaconitine, and Pseudaconitin | GI disorders and fatal poisonings in excess. Ethanolic extract was found to have antioxidant, anti-inflammatory, and analgesic potential in rodents. |
| 8. | *Abrus precatorius* (Rosary pea, love pea, and Crab's eye) | Fabaceae | All parts, highly seed | Abrin, glycyrrhizin, aric acid, N-methyltryptophal | Vomiting, nausea, diarrhea, abdominal pain, and male antifertility. |

**TABLE 23.1** (Continued)

| S/ No. | Scientific name (Common name) | Family | Parts of the plant | Toxic compound | Toxic effect/symptoms/ |
|--------|-------------------------------|--------|---------------------|----------------|------------------------|
| 9. | *Actaea pachypoda* (Doll's-eyes, white baneberry) | Ranunculaceae | All parts especially the berries | Cardiogenic toxins | Sedative effect on human cardiac muscle tissue and cardiac arrest. |
| 10. | *Nerium Oleander* (Nerium or oleander) | Apocynaceae | All parts | Oleandrin and oleandrigenin, known as cardiac glycosides | The ingestion of the leaves causes deaths in human beings. It's one of the most poisonous garden plants. |
| 11. | *Ageratina altissima* (White snakeroot) | Asteraceae | All parts | Trementol | Credited with killing Abraham Lincoln's mother. Causes poisonous milk when cattle feed on it. |
| 12 | *Datura stramonium* (Jimsonweed or devil's snare or devil's trumpet or devil's cucumber) | Nightshade family | All parts | Tropane alkaloids atropine, scopolamine, hyoscyamine | Hallucinogenic effects and death. Used in traditional medicine to relieve asthma symptoms and as an analgesic during surgery |
| 13. | *Cerbera odollam* (Suicide tree) | Apocynaceae | Kernels | Cerberin, a potent toxin related to digoxin | Blocks the calcium ion channels in heart muscle, causing disruption of the heartbeat |
| 14. | *Colchicum autumnale* (Autumn crocus or naked ladies) | Colchicaceae | All parts | Colchicine | Poisoning resemble those of arsenic, and no antidote is known |
| 15. | *Convallaria majalis* (lily of the valley) | Asparagaceae | All parts | Contains 38 different cardiac glycosides (cardenolides) and saponin | Causes abdominal pain, vomiting, reduced heart rate, and skin rashes |
| 16. | *Dendrocnide moroides* (stinging brush, gympie gympie, the suicide plant) | Urticaceae | Hairs on leaves or twigs causes | Moroidin, a bicyclic octapeptide | The most toxic of the Australian species of stinging trees. Pains from stinging can persist for up to 2 years. |

**TABLE 23.1** *(Continued)*

| S/ No. | Scientific name (Common name) | Family | Parts of the plant | Toxic compound | Toxic effect/symptoms/ |
|---|---|---|---|---|---|
| 17. | *Taxus baccata* (English yew, European yew, or graveyard tree) | Taxaceae | All parts except its berries although seed is toxic | Alkaloid taxine | Accelerated heart rate, muscle tremors, convulsions, collapse, difficulty breathing, circulation impairment, and eventually cardiac arrest. Can reach 400–600 years of age. |
| 18. | *Hyacinthus orientalis* (Garden hyacinth) | Asparagaceae | All parts especially the bulb | Alkaloids | Very toxic if large quantities are consumed. |
| 19. | *Gelsemium sempervirens* (Carolina jasmine or evening trumpet flower) | Gelsemiaceae | All parts | Strychnine-related alkaloids gelsemine and gelseminine | Sap may cause skin irritation in sensitive individuals. Nectar is toxic to honeybees. |
| 20. | *Gloriosa superba* (Flame lily, climbing lily, or Tiger claw) | Colchicaceae | All parts | Colchicine, a toxic alkaloid | Can cause nausea, vomiting, numbness, tingling around the mouth, burning in the throat, and bloody diarrhea. |
| 21. | *Heracleum mantegazzia-num* (Giant hogweed) | Apiaceae | Leaves, roots, stems, flowers, and seeds | Linear derivatives of furanocoumarin | Phytophotodermatitis (severe skin inflammations). Phototoxic when the contacted skin is exposed to sunlight or to ultraviolet rays |
| 22. | *Bryophyllum delagoensis* (Mother of millions or devil's backbone) | Crassulaceae | All parts especially flowers | Bufadienolide cardiac glycosides | Can cause fatal poisoning. About 125 grazing cattle died in 1997 after ingestion. |
| 23. | *Lamprocapnos spectabilis* (Bleeding heart or lady-in-a-bath) | Papaveraceae | All parts | Isoquinoline-like alkaloids | Contact with the plant can cause skin irritation. |
| 24. | *Zigadenus glaberrimus* (Sandbog death camas) | Melanthiaceae | All parts | Zygacine an alkaloids | Poisonous and can cause death. |

**TABLE 23.1** (Continued)

| Sl No. | Scientific name (Common name) | Family | Parts of the plant | Toxic compound | Toxic effect/symptoms/ |
|---|---|---|---|---|---|
| 25. | *Wisteria sinensis* (Chinese wisteria) | Fabaceae | All parts | Wisterin, a toxic glycoside | Can cause nausea, vomiting, stomach pains, and diarrhea. |
| 26. | *Toxicodendron radicans* (Poison ivy) | Anacardiaceae | Sap | Urushiol | Urushiol-induced contact dermatitis, an itchy, irritating, and painful rash. Members of the genus *Toxicodendron* are poisonous too. |
| 27. | *Sanguinaria Canadensis* (Bloodroot or bloodwort or redroot) | Papaveraceae | Juice (red in color) | Sanguinarine, a benzylisoquinoline alkaloids | Sanguinarine kills animal cells by blocking the action of Na+/K+-ATPase transmembrane proteins |
| 28. | *Nicotiana glauca* (Tree tobacco) | Solanaceae | Leaves | Anabasine | Ingestion of the leaves can be fatal. |
| 29. | *Aristolochia clematitis* (Birthwort) | Aristolochiaceae | All parts | Aristolochic acid | Can cause renal failure. |
| 30. | *Anemone nemorosa* (Thimbleweed or smell fox) | Ranunculaceae | All parts | Protoanemonin | Can cause severe skin and GI irritation, nausea, and diarrhea. |

## 23.2   ATROPA BELLADONNA

*Atropa belladonna* (Fig. 23.1, **Plate 1**) is a perennial herbaceous plant in the nightshade family: Solanaceae, which includes tomatoes, potatoes, and aubergine. The name Atropa has a Greek origin from the goddess Atropos; while the name belladonna means 'beautiful lady' in Italian language (Rita and Animesh, 2011). Other researchers reported that *A. belladonna* takes the name from its lethal effects and its use in cosmetics (Rani and Prasad, 2013). The Greek fate about Atropa is mythological and it is connected to Goya Atropos; the goddess of death that cuts the thread of life. Some believed belladonna takes its origin in the renaissance because Italian' dames used the plant to improve color of skin and brightness of eyes. The substance atropine is extracted from Atropa plant which was used to prepare ointment called witches' whisper. Atropine has a charm used by poets, witches, writers, medical scientists, and alchemists. In middle ages, these plants were used in rituals to meet Satana because they have powers both hypnotics and aphrodisiacs. Fascinated by this substance, Will Enrich Peuckert investigated its use as a witchcraft recipe. He prepared and used the ointment on himself. He fell into a deep catalepsy and a 20 hsleep. During the period, he was tormented by horrible visions, monsters, infernal landscapes, diabolic beings, and satanic creatures.

### 23.2.1   TOXICITY EFFECTS OF ATROPA

Acute toxicity associated with *A. belladonna* is called anticholinergic toxidrome. The ingestion of 10 berries is toxic to an adult while ingestion of 3 berries is toxic to a child. The root of atropine belladonna is the most poisonous of the plant; this is closely followed by the leaves and flowers while the berries are the least toxic except to children. The risk of intoxication in children is often possibly confused with ingestion of other berries. Poisoning due to *A. belladonna* affects both central and peripheral nervous systems. Symptoms of belladonna poisoning in the central nervous system include confusion, disorientation, short-term memory loss, seizure, hallucinations, ataxia, psychosis, agitated delirium, coma, respiratory failure, or cardiovascular collapse. Symptoms in the peripheral system include dry mucous membranes, mydriasis with cycloplegia, hyperreflexia, flushed skin, reduced sound in the bowel or ileus, urinary retention, tachycardia, and hypertension or hypotension (Ulbricht et al., 2004). The anticholinergic toxidrome can be confused with post-traumatic brain damage and acute psychosis. These effects are dose-dependent. The therapeutic dose of atropine is 0.5–2.0 mg and its administration causes decrease in heart rate by vagal stimulation, dryness of mouth, inhibition of swearing, and mild dilation of pupils (Ulbricht et al., 2004). A dose of 2 mg causes anticholinergic toxidrome, 5 mg dose results in difficulty swallowing, headache, dry hot skin, and reduced intestinal peristalsis while 10 mg dose causes rapid pulse, hot dry scarlet skin, ataxia, mydriasis (iris obliterated), hallucination, coma, and delirium.

## 23.2.2 THERAPEUTIC USES OF ATROPA

Anticholinergic drugs like atropine are used in anaesthesia because they reduce the activity of salivatory glands and correct the vagal-induced bradycardia (AHA, 2005). Atropine is suitable for the cure of a lot of GI problems like abnormal intestinal motility, bile-stone, and irritate bowel syndrome. In oculistic, it is used because of its ability to induce mydriasis and reduce ocular accommodation since it stops cholinergic stimulation of muscle sphincter of iris and ciliary muscles. If atropine is administered locally, the ocular effect is considerable and accommodation returns normally after 7–12 days. In 2006, anticholinergic drugs were used also for initial treatment of early idiopathic Parkinson's disease (Rani and Prasad, 2013). Belladonna can be used to dilate stressed pupils in the eye. It is used in traditional treatments of assortment conditions like peptic ulcer, headache, menstrual symptoms, histaminic reactions, inflammation, and motion sickness. It is used to overcome bronchial spasms and whooping cough. It is used as an antidote for treating snakebites and as a gastric agent.

## 23.3 BRUGMANSIA

*Brugmansia* (Fig. 23.1, **Plate 2**) are poisonous flowering plants which belong to the family Solanaceae. Included in the family are potato, tobacco, tomatoes, and other deadly nightshades. Their large and fragrant flowers give them the common name "angel's trumpet." Often, the name *Brugmansia* is used to refer to *Datura,* a closely related genus. *Brugmansia* are pendulous woody trees or shrubs whose flowers are not erect and have no spines in their fruits. *Datura* species, on the other hand, are not pendulous but herbaceous with erect flowers and with their fruits mostly having spines. Currently, all the seven species of *Brugmansia* are extinct in the wild according to International Union for Conservation of Nature (IUCN) Red List compilation. The origin of *Brugmansia* was said to be subtropics of South America from Andes, Colombia, and Northern Chile to Southern Brazil. *Brugmansia* was also found in Africa, Costa Rica, Australia, and Asia. They are grown as ornamental plants and have become naturalized in isolated tropical areas around the globe.

## 23.3.1   TOXICITY EFFECTS OF BRUGMANSIA

Seeds, leaves, and all parts of *Brugmansia* are dangerously poisonous although with different degree of concentrations of alkaloids in the plant. The concentrations even vary with seasonal changes and hydration levels. Hence, it is almost impossible to determine a safe level of exposure to alkaloids of *Brugmansia*. It is noteworthy, that members of *Brugmansia* genera are highly toxic and fatal if ingested by humans, pets, and livestock. It is advisable to wear gloves when trimming the plant and avoid getting plant juices on the skin. The plant's tropane alkaloid was once stored in the pupal stage by adult butterfly, where it was used as a defense mechanism (botanical), making the plant less vulnerable to predator and pest attacks (Eich, 2008). Effects of *Brugmansia* ingestion include tachycardia, diarrhea, paralysis of smooth muscles, confusion, dry mouth, migraine headaches, visual and auditory hallucinations, mydriasis, rapid onset cycloplegia, and death (Wagstaff, 2008). *Brugmansia* have been described as terrifying rather than pleasurable due to their hallucinogenic effects. The author Christina Pratt, in An Encyclopedia of Shamanism, stated that *Brugmansia* induce powerful trances with violent and unpleasant effects, sickening after effects, and at times temporal insanity. Hallucinations resulting from *Brugmansia* ingestion are characterized by total loss of awareness, disconnection from reality, and amnesia. A reported case of this effect was the psychiatric and clinical neuroscience condition of a young man that amputated his own tongue and penis after taking a cup of *Brugmansia* tea. Some municipalities have prohibited the purchase, sale, and cultivation of *Brugmansia* plants. In the year 1994, 112 teenagers were admitted for treatment at different times in many hospitals in Florida after ingesting *Brugmansia* (Roberts and Wink, 1998). These reports suggest that *Brugmansia* could be a biological weapon of warfare particularly when applied to food ingested by humans.

## 23.3.2   THERAPEUTIC USES OF BRUGMANSIA

*Brugmansia* possess important alkaloids such as hyoscyamine, scopolamine, and atropine which have huge medicinal properties with antiasthmatic, spasmolytic, anticholinergic, narcotic, and anaesthetic effects. Many of these alkaloids and their equivalents are now being synthesized artificially (Bracci et al., 2013). *Brugmansia* have traditionally been applied by many South Americans in the preparation of medicines and as entheogen during religious or spiritual ceremonies (Harner, 1990). Medicines produced from *Brugmansia* have been used externally as ointment, poultice and tincture, or applied directly transdermally on skin. They have been used to treat rheumatism, aches and pains, headaches, dermatitis, orchitis, arthritis, infections, and as anti-inflammatory agents. *Brugmansia* are rarely used internally due to dangers of ingestion. They have been applied internally in the treatments of stomach and muscle pains, as sedative, as decongestant to induce vomiting and to expel worms and parasites (De Feo, 2004). Some parts of South America have administered *Brugmansia* in the treatment of unruly children; according to them, the ingestion prepares the children for their ancestors to directly admonish them in the spirit world thereby making them compliant or obedient. Others believed that when hardworking slaves and loved wives are dead, *Brugmansia* could be mixed with maize beer and tobacco leaves and applied on the dead bodies before they are buried to invoke the spirit of the master or husband who is alive to journey with them in the grave (De Feo, 2004).

## 23.4   RICINUS COMMUNIS L

*Ricinus communis* is a rapidly growing perennial shrub belonging to spurge family, Euphorbiaceae (Fig. 23.1, **Plate 3**). *R. communis* is commonly known as castor oil plant, castor bean and palma Christi or palm of Christ. It is a plant growing in the wild as well as cultivated as a crop throughout tropical and subtropical regions (Kumar, 2017). The cultivation of *R. communis* is for its seeds (Fig. 23.1, **Plate 4**) used for the extraction of castor oil which are of many purposes after proper detoxification procedures. Wide ranges of phytochemicals were isolated from this plant to date. Among them ricin, a toxin found in the seeds, was discovered by Hermann Stillmark in 1888 who observed that the ricin caused precipitation of serum proteins and agglutination of red blood cells (Roxas-Duncan and Smith, 2012a). Later studies revealed that this is due to the presence of less toxic and strong hemagglutinin ricin homologue called *R. communis* agglutinin ($RCA_{120}$), a tetrameric protein composed of two A-chains, which is 90% similar to the ricin A-chain, and two B-chains, which is 84% similar to the ricin B-chain (Roberts et al., 1985).

The action of ricin is its ability to inhibit the protein synthesis in all eukaryotic organisms by blocking the ribosomes through fatal modification of 28S rRNA (Lugnier et al., 1976). The ricin is a heterodimeric protein consists of a catalytic ricin toxin A (RTA) and a galactose-binding ricin toxin B (RTB) linked by a single disulfide bond. After the B chain binds to the cell surface the toxin enter into the cell through endocytosis and gradually the A chain separates from the B chain. The catalytic A chain cleaves a particular adenine residue present in the GAGA sequence of 28S rRNA and blocks the binding of elongation factors. The inability of elongation factors to bind ribosomes stops protein production. Thereby the cells starve for protein and die. Sometimes leads to the death of the organism itself (Bozza et al., 2015).

The lethal dose ($LD_{50}$) values were identified through various toxicity studies conducted in mice. $LD_{50}$ values obtained after giving the toxin through intraperitoneal and oral routes are 2.4–36 µg/kg and 21–30 mg/kg, respectively. The estimated $LD_{50}$ values for human is between 1–20 mg/kg of body weight and if the exposure of the toxin is through inhalation is much more toxic (1–10 µg/kg of body weight) (Sousa et al., 2017).

After exposure to ricin due to its toxicity, the affected person begins to show many clinical symptoms. Through different route the toxin gets inside the body and based on this symptom may vary. Inhalation of the toxin causes fever, cough, and respiratory distress, whereas ingestion leads to abdominal pain, diarrhea, soft dark feces, and GI hemorrhage. Both inhalation and ingestion lead to multi-organ failure (Sousa et al., 2017). Radioimmunoassay is more efficient than enzyme-linked immunosorbent assay (ELISA) which can detect ricin at a low amount. Nevertheless, the identification can be difficult due to quick binding of ricin and is metabolized before it is excreted (Kent, 2006). Unfortunately, there is no antidote available to treat the ricin poisoning (Wei-Gang et al., 2013; Noy-Porat et al., 2017).

Mainly in the field of monoclonal antibody (mAb) technology, many types of research are now actively going on for the production of antiricin (Wei-Gang et al., 2013). Recently, Wei-Gang et al., (2013) reported the production of effective mAb and tested in mice. The mAb, especially mAb D9 showed unexceptional results and 100% recovery of mice poisoned with 5 x $LD_{50}$ was observed. Noy-Porat et al. (2017) reported a tri-antibody-based cocktail was effective and has high survival rates when it was administrated to animals. From a clinical point of view, this knowledge would shed light on the development and production of the effective antidotes.

The highly toxic nature of ricin attracted the minds to use it for biological warfare (Olsnes, 2004). Many cases were reported from different parts of the world. The first plan to use ricin as a biological weapon was by United States, United Kingdom, and Canada in world war one and two. From large quantities of ricin, ricin-containing bombs (W bombs) were made during the world war two and in fact, it was never used in battle even it has surpassed the existing bioweapons of the time (Maman and Yehezkelli, 2005) (Roxas-Duncan and Smith, 2012a). The potential of ricin as bioweapon was studied by the Soviet Union during the cold war. Roxas-Duncan and Smith reported more than 20 incidents occurred from 1981 to 2011 (2012a, 2012b). Bozza et al. reported 20 incidents occurred during the period from 1981–2013 (2015). Few notable incidents involving the use of ricin for bioterrorist activities during the period from 1978–2017 were depicted in Table 23.2.

Many national security and health bodies treated and classified ricin as a threat agent. The US Centers for Diseases Control and Prevention classified ricin as a category B agent which considered it as moderately easy to disseminate and able to cause morbidity and low mortality. Likewise, the Chemical weapons convention monitored ricin as schedule one agent. Many countries, including the United States took decisions to prevent these types of attacks.

## 23.4.1   THERAPEUTIC USES OF RICIN

Indeed, the ricin is one of the potential phytochemical that has anticancer and lipolytic activities (Olsnes, 2004; Kumar, 2017). Many reported the possibilities of ricin to become a drug that can save a life instead of destroying life. Along with the use of nanotechnology, a potential way of developing a therapeutic drug from ricin for the treatment of cancer is in the not too distant future (Tyagi et al., 2015).

## 23.5   DIEFFENBACHIA

*Dieffenbachia* Schott (Fig. 23.1, **Plate 5**) is an herbaceous genus belonging to the family Araceae and commonly known as Dumb cane and Mother in law's tongue, which refers to the poisonous nature of temporary difficulties in speaking. *Dieffenbachia* is native to Mexico to tropical America and now distributed throughout tropics. Some species are commonly cultivated as ornamental plants and used as a houseplant due to their adaptability in shady places.

Most of the plant species belonging to the family Araceae are poisonous including the ornamental species such as *Dieffenbachia, Alocasia, Caladium,*

**TABLE 23.2** Few Notable Incidents Involving the Biological Toxin Ricin.

| Date | Location | Summary |
|---|---|---|
| September 1978 | London, UK | Georgi Markov, 49, a Bulgarian exile was murdered by Ken Alibek, working in association with Russia's biological weapons program, using an armed umbrella with ricin (Roxas-Duncan and Smith, 2012a). |
| August 1981 | Virginia, US | Boris Korczak, double agent, survived from the shot by an air gun with a pellet containing ricin (Roxas-Duncan and Smith, 2012a). |
| April 1991 | Alexandria, Minnesota | Minnesota Patriots Council members (Dennis Henderson, 37, Leroy Wheeler, 55, Richard Oelrich, 55, and Douglas Baker, 29) were convicted for planning to use ricin to kill local deputy sheriffs, IRS Agents, and U. S. Marshals. |
| November 1999 | Florida, US | James Kenneth Gluck in Tampa, Florida, threatened to assassinate the court officials Jefferson County, Colorado with ricin (Roxas-Duncan and Smith, 2012a). |
| October 2003 | South Carolina, US | A sealed container and an envelope with a threatening note found in a postal facility. The threatening note was if demands were unaccepted, they would release poison to the water supplies. Later, the investigation found that the ricin was present in the sealed container. |
| November 2011 | Toccoa; Gainesville, Georgia | Frederick Thomas, 73, Dan Roberts, 67, Crump, 71, and Ray, 58, were arrested for conspiracy against the government that included a plan to intentionally release ricin in five American cities such as Newark, Washington, Jacksonville, Atlanta, and New Orleans (Roxas-Duncan and Smith, 2012b). |
| April 2013 | Washington, DC, US | James Everett Dutschke Tupelo, Mississippi was arrested and punished for mailing envelopes containing ricin to President Barack Obama, Mississippi Senator Roger Wicker, and Lee County, Mississippi Judge Sadie Holland. |
| May 2013 | Washington, DC, US | Shannon Guess Richardson, 36, minor actress was arrested for sending letters containing ricin to Barack Obama, President of United States, former New York Mayor Michael Bloomberg, and the head of his gun-control group. On July 2017, she was sentenced to 18 years in prison (Bozza, et al., 2015). |
| December 2014 | New York, US | Cheng Le, 22, was arrested for possessing ricin and trying to resell for the use as a bioweapon. On March 2016, he was sentenced to 16 years in jail. |
| December 2017 | Shelburne, Canada | 70-year-old US woman tested the toxicity of ricin on retirement home residents (The Telegraph, 2017). |

*Philodendron*, and *Zantedeschia* (Barbi et al., 2011). The plant body of *Dieffenbachia* contains calcium oxalate crystals called raphides (Fig. 23.2), arranged in specialized crystal idioblasts (Prychid and Rudall, 1999). It was once believed that the raphides are the only root cause of the poisoning of *Dieffenbachia*. Unfortunately, raphides are not the primary cause of the toxicity but it helps (Arditti and Rodriguez, 1982). The mechanism for the cause of poisoning is not well understood. Walter reported that the toxicity is due to the combined actions of raphides, insoluble oxalate, and dumbain, a proteolytic enzyme named by Walter (1967). In addition, another toxic principle belonging to the cyanogenic glycosides which on hydrolysis yields hydrocyanic acid (Fig. 23.3) was found in *D. bausii* and *D. exotic*. Ladeira et al., found that a non-histamine compound is responsible for the toxicity and is unstable on vacuum drying at 50–60°C or heating to 100°C (1975).

**FIGURE 23.2**    Raphides of *dieffenbachia seguine* (Jacq.) schott (magnification 40X).

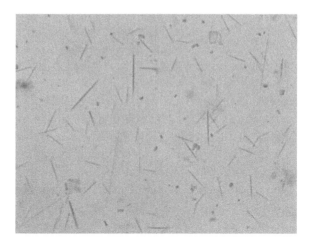

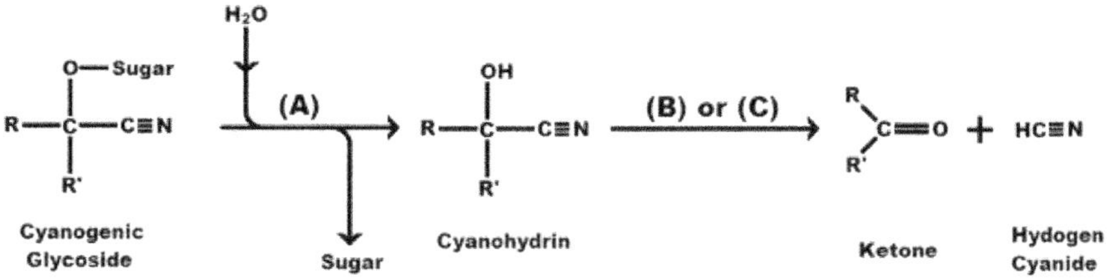

**FIGURE 23.3**    Pathway of hydrolysis of cyanogenic glycosides. R and R' represents various alkyl or aryl substituents. (A) Glycosidase, (B) Hydroxynitrile lyase, and (C) spontaneous.

Direct skin contact with the juice from the plant causes dermatitis with redness, pain, or itchiness. In contact with the eyes cause irritation and may lead to loss of vision (Reis, 2010; Botanical online, 2017). Thorough washing with water and examination of the eye is important after the exposure (Barceloux, 2008). To avoid more irritation thoroughly washes the affected body part with water (Sasseville, 1999; Botanical online, 2017). Painful oropharyngeal edema occurs when any part of the plant ingested (Altin et al., 2013). Altin et al. reported that a 70-year-old male patient poisoned by *Dieffenbachia* was applied to their clinic due to dysphagia, wounds on the lips and tongue, aphasia, sialorrhea, swelling on the face, and edema on the different parts of the mouth (2013). Sometimes it may lead to the death due to breathing difficulties if the patient failed to get any medical attention (Reis, 2010). Rinsing with a large amount of water gives relief to some extent and eliminates any small part of the plant present in the mouth (Botanical online, 2017). Generous fluid replacement is the most effective prophylactic treatment for those who ingested large amount, which prevents the calcium oxalate crystals deposition in different parts of the body. Treatment with sodium citrate for primary and secondary oxalosis leads to the inhibition of calcium oxalate crystallization. No specific biomarkers available for proper diagnosis (Barceloux, 2008).

Due to the toxic nature of this plant, it was used to torment many Jamaican slaves by rubbing the stem cuttings of *Dieffenbachia* in their mouth. Many experiments were carried out in concentration camp prisoners to study the efficiency of the juice, which could induce sterility in animals. However, the restricted growth of *Dieffenbachia seguine* to North America ended the experiments in Nazi concentration camps due to the unavailability of the plant in large quantities. Another incident was one in which a criminal rubbed *Dieffenbachia* inside the mouth of a witness to save himself from the punishment. Not many incidents were reported.

### 23.5.1   THERAPEUTIC USES OF DIEFFENBACHIA SPP.

However, medicinal uses of *Dieffenbachia spp.* were also reported. The grounded leaves of *D. seguine* were used in the treatment of rheumatism (Rashtra, 2008). *D. seguine* was used to induce sexual impotence among the male inhabitants of the Caribbean Islands as a method of contraception and its effects last for 28–48 h. Sliced roots of *Dieffenbachia* were used for gout treatment. In addition, *Dieffenbachia* had an industrial application in the process of sugar granulation (Arditti and Rodriguez, 1982).

## 23.6   *CICUTA MACULATA*

*Cicuta maculata* (the root of death) (Fig. 23.1, **Plate 6**), from the Apiaceae family is remarkable from the phytochemical viewpoint. Since prehistoric times, antique civilizations considered them as a considerable source of a toxic substance with lethal properties. Cases of poisonings from the consumption of contaminated livestock meat have been reported probably due to the fact that the livestock grazed on it. Because of these poisonings, this plant now has several names: "The root of death," "cowbane," "poisonous hemlock," "parsley," and so forth.

A large amount of these herbs has been discovered in the last few centuries, the main members: *C. bulbifera, C. douglasii, C. virosa* as well as *C. maculata*, with *C. maculata* containing four varieties: var. maculata, var. victorinii, v. bolanderi along with v. angustifolia. Their distinction is made only through the sequential isolation of the chromosomes where *C. maculata* is distinguished with 22 large somatic chromosomes. It is near to impossible to differentiate without the knowledge in genetic engineering by analyzing their chromosomal genotype.

These plants are native to the American zones so they are considered common and widely distributed, especially in the Illinois regions as a result of the high humidity in these countries. Recognized as a naturally occurring and particularly common plant in northern Canada, up to Alaska, these plants are found in wide and populated geographic areas. They usually prefer a brightening sunny weather with moderate humidity. Particularly, they can adapt to conditions of large droughts and loamy. Other places that they can be habited are in lakes and wet shores of the wetlands.

From the first point of view, these plants have a relatively long stalk, which grows up to 1–1.5 m. The leaves have a prolonged form, consisting of several leaves grouped and are at least 10 cm long in diameter. At the time of flowering, the appearance of the flowers is white color and constitutes of five petals, each with more sepals in a form that looks like an umbrella. Difficult to distinguish are the varieties through the free eye. It is almost impossible nevertheless but with the assistance of an electron microscope and the exercise of genetic engineering, scientists are able to distinguish those varieties. Meanwhile, there have been numerous poisonings because of the small difference with some nonpoisonous plants. The plants of this genus, in general, are distinguished for the identical blossoming time, from the month of May until September. The fruits undergo a developing phase from August until the end of September.

The forms of toxin penetration are diverse. Typically, penetration occurs through the skin or oral form and then into the stomach. Before, poisonings occur as a result of consumption of livestock, with the initial symptoms such as nausea, vomiting, and trembling, while the first forms of convulsions appeared in children. Later on, the symptoms go into more severe in the form of tachycardia, mydriasis, coma, and blockage of the inhalation process. In recent decades, the emergence of amnesia in poisoned persons has been confirmed.

Within the roots (bulbs), there is a very toxic substance causing this plant to take the pseudonym as the deadliest plant in the United States, it is called Cicutoxin, a molecule that actually has the structure of an unsaturated alcohol, namely an acetyl diol.

The lethal effect occurs according to a particular mechanism. When it enters the body, it causes damage to the central nervous system as a noncompetitive GABA receptor antagonist. The concentration of the toxin ranges from one variety to the other and to the genus as well. For example, the varieties of *C. maculata*: var. maculata are known for an average concentration of 1.01 mg/g of cicutoxin, the other is var. victorinii containing less, about 0.30–0.40 mg/g while var. bolanderi even less. The least poisonous variety: *C. virosa* contains about 0.01 mg/g. There have been also innumerable indirect poisonings from poultry origin by consumption of this birds that previously fed with Cicuta herbs, with fatal consequences for the persons in question. This is because the poultries contain specific enzymes that neutralize the poison, enzymes which are not present in humans.

The vegetations have caused many fatal deaths to millions of people from the confusion of this plant with other nonpoisonous plants. Moreover, after an extended investigation of the vegetation that has been consumed and the indications it presented, in the laboratory, Cicutoxin was discovered to be the main toxic compound by using the analytical methods of spectroscopy or HPLC techniques. Meanwhile, symptomatic as a primary therapy is preferable. This treatment consists of controlling of the seizures and respiratory tract management, which is also the major cause of the death of the poisoned from the collapse of the respiratory system in the form of asphyxia along with the damage of the cardiovascular system. In people over the age of 45, plenty of damages to the kidney are present in the form of nephritis, also hemodialysis is recommended to these people. Secondary therapy consists of by the administration of benzodiazepine class drugs as well as gastric lavage for the cleaning of toxins in the GI tract. In addition, tertiary therapy consists of the monitoring of rhabdomyolysis, low blood pressure, and metabolic acidosis.

## 23.7   HIPPOMANE MANCINELLA

*Hippomane mancinella* (Fig 23.1, **Plate 7**) is one of the deadliest vegetation of all time, so remarkable for this toxic characteristic, for it has also gained the record from Guinness World Records as the most toxic plants of all times. The terminology of this herb is diverse, the native calls it "the spring apple of death," while in the scientific terminology it is called *H. mancinella* of the Euphorbiaceae family. The word Hippomane came from the term "horse poison" while Mancinella means "small apple." Although the name may be slightly misleading, the whole plant is toxic and not just its characteristic fruit.

The genus of Hippomane consists of a large amount of variety of herbs, together numbering more than 5000 species, all around the world. The plant is mostly found in relatively tropical zones with a lot of rain for instance in areas such as the Equator, India, and also America, especially in the Caribbean and Florida, where people used these plants to make arrow to kill animals and also to use the quality wood of these herbs to make raw materials.

The herb has the appearance of a green apple plant and this has led to frequent confusion in the native cultures that live in the zones that these herbs live. In spring, the blossoming times begin with the flourishing of the flowers which have omnisexual ovaries (a characteristic of the genus). The fruits grow to several centimeters in diameter and look like in a form, shaped like an apple. The trunk reaches 2–3 m in weight and grows in the length of 13 m with grey colored barks. Countless antiquated civilizations consider it as a lethal plant by such a degree that only the interaction with the skin or the taste of the fruit can trigger blindness and other risks with serious consequences. It is so poisonous that even a minor bite of the small fruit very similar to the ordinary apple, it is enough to cause death. For this, its pseudonym came "little apple of death."

Initially, the first sign consists of nausea, headache, and abdominal discomfort, due to the damage to the digestive tract and ultimately is the loss of vision. Then the prognosis of keratoconjunctivitis dermatitis and lung damage diseases are caused by contact with the latex of the plant. A special case of poisoning with this plant has happened to the conquistador Juan Ponce de Leon, who was among the first European to have come into contact with this plant, during his expedition in the lands of Florida during the 16th century, a native habitat for cultivation and this plant. He was attacked with arrows from the natives of those areas that had been dyed with Manchineel latex, soon after he would die a few days later from those wounds. Last but not least, the appearance of benign and malignant tumors has been established due to ingesting by the patients. Poisoning can occur even in the name of the plant, especially during the raining conditions. Throughout that time, when the raindrops contact with the fruit, it causes it to release the toxic constituents in the form of blistering. As a consequence of this high risk, in tourist sites, it is desirable to be inscribed with a note nearby the vegetation to not stay close by to this plant at any moment. Actually, only one race of black iguana, which lives in Central America, is capable of surviving even after nourishing the plant for the single reason of resisting enzymes that neutralize the toxins.

It is also worth mentioning that scorching this plant is prohibited, for the reason that even the smoke can cause impaired vision and damage to the lungs if inhaled even in minor quantities. This plant comprises an enormous amount of deadly constituents and it is anticipated that numerous other to be sequestered in the forthcoming, it is worth mentioning that they mainly consist of terpene structure such as Hippoman A and B, phorbol, and also physostigmine and many other constituents in minor proportions.

Thus far, it is not identified why this plant generates so much poison, because of this, this plant now is endangered from extinction because humans have started to dispose of the plant to prevent potential poisonings. It is thought that the plant produces large quantities of toxins to make sure that its seeds are distributed safely even with water. Despite the fact they contain quite a lot of lethal effects, due to the worthy excellence as a carpeting material, societies have and still use it to construct diverse furniture's, leaving the raw material to stay in the sun to oxidize the toxic substances

In traditional medicine, they have been used as medical remedies for the treatment of edema while the distinguishing fruits are firstly dried in the sun and later on after the toxin has been evaporated, the fruits are used for patients with diuretic problems. So far, there are no remedies that can deactivate the lethal effects of this herb in humans, nevertheless, it has been shown success in practice, by cleaning the infected skin with soap and water and certainly to make a gastric scarring, so that the toxic substance cannot be absorbed into the organism and cause fatality.

## 23.8 ACONITUM

*Aconitum* (Fig. 23.1, **Plate 8**) is a perennial plant commonly known as Wolfsbane, Women's bane, Devils helmet, or Blue rocket. It is a member of the buttercup family which is also known as Family *Ranunculaceae*. They are extremely poisonous. It is currently found wild in parts of England and

Wales but can hardly be considered truly indigenous. *Aconitum* species contains a series of alkaloids exemplified by "Aconitine". About 14 have been identified including mesaconitine, hypaconitine, deoxyaconitine, and pseudaconitine among others. Pseudaconitine is found in the root and it is regarded as the most poisonous (ZIDBITS, 2011). It is a diterpene alkaloid. According to Fu et al. (2006), the *Aconitum* alkaloids may be divided into three subgroups – the first containing two ester bonds on the diterpene structure. This group activates voltage-dependent sodium channels and inhibit noradrenaline reuptake. This activation alongside excessive depolarization causes suppression of pain transmission. The second group blocks voltage-dependent sodium channels and have antiarrhythmic properties. They are monoesters and appear to be a competitive antagonist of the first group. The third group does not have an ester side chain and are less toxic than either of the other two groups.

Marked symptoms like GI (nausea, diarrhea, and vomiting) are noticed within a few minutes of oral administration of doses of *Aconitum*. It is then followed by the sensation of burning, numbness in mouth and face, tingling, and scorching in the abdomen (Extra Pharmacopoeia Martindale, 1958). After an hour, motor weakness and cutaneous sensation, fall in pulse and respiration until death occurs from asphyxia (Chisholm, 1911). However, some literature indicates that death may be from ventricular arrhythmias (Chan, 2009). Aconitine toxin is easily absorbed through the skin; therefore, poisoning can as well occur simply by harvesting the leaves without wearing gloves though severe toxicity will not be expected (Fu et al., 2006). In large doses, death occurs in as little as 2–6 h as only 20 mL of pseudoaconitine can kill an adult human (ZIDBIT, 2011). These alkaloids are a potent neurotoxin, and its poisonous mechanism is that it opens tetrodotoxin-sensitive sodium channels which increase the influx of sodium through the channels and interrupts repolarization. This certainly increases the excitability and promotes the ventricular dysrhythmias leading to asphyxia. Particularly, the enzyme acetylcholinesterase is moderately inhibited by pseudaconitine. This enzyme breaks down a neurotransmitter known as acetylcholine via hydrolysis. Inhibition of this enzyme effects a constant stimulation of the postsynaptic membrane by the neurotransmitter which it cannot cancel. This accumulation of acetylcholine may thus lead to the constant stimulation of the central nervous system, glands, and muscles.

The physiological antidotes are atropine, digitalis, and strophanthin which can be injected intravenously in maximal doses (Chan, 2009). However, emptying the stomach by tube or by a non-depressant emetic can serve as an early treatment. Other drugs like lidocaine, flecainide, procainamide, and barakol can be used for ventricular arrhythmias. Most of these drugs reduce the occurrence of aconitine-induced ventricular fibrillation and tachycardia probably due to the hindrance of intracellular sodium ion accumulation.

## 23.9   ABRUS PRECATORIUS

*Abrus precatorius* (Fig. 23.1, **Plate 9**) is a plant native to India and other tropical and subtropical areas of the world (Bhutia and Maiti, 2011). The plant is known by a variety of names including rosary pea, jequirity bean, love pea, cock's eyes, crab's eye, and so forth. It is of the *Fabaceae* family and the *Abrus* genus. The plant is seen growing wild throughout warm temperate tropical forests and is propagated through seeds. It is currently seen as an invasive wild plant in many regions especially in Northern America (Holm et al., 1979). When the plant grows to maturity, the roots become difficult to remove and the plants' ability to sucker, aggressive growth and hard-shelled seeds render an invasion very difficult to eradicate, thereby extremely challenging to avert reinfestation.

All parts of the *A. precatorius* are believed to be toxic with the highest concentrations found in the seed. As is often the case, they have an indigestible outer casing and there have been reports of them transiting through the digestive system without harm unless if chewed. It is toxic because of the presence of a toxin known as "Abrin" along with lower concentrations of glycyrrhizin, uric acid, and N-methyltryptophan (Reedman et al., 2008). Abrin is a dimer containing two protein subunits known as A and B. The two protein chains are linked through a disulfide bridge. The B chain allows the entry of the A chain into the cells by facilitating binding to cell surface receptors (transport proteins). Immediately the A chain enters into the cell, it prevents the synthesis of protein by inactivating the 60S subunit of the ribosome that eventually inhibits elongation factors; EF-1 and EF-2 causing cell death (Benson et al., 1975). One molecule of abrin is capable of inactivating up to 1,500 ribosomes per second (Benson et al., 1975). Abrin's structure is similar to insulin, ricin, botulinum, diphtheria, and cholera toxins (Shih and Goldfrank, 1998). The majority of poisoning cases involving abrin is the intake of the whole seed. Many of these cases often cause few or mild symptoms because the shell remains intact and there is limited toxin exposure. Wambebe and Amosun (1984) reported that the oral ingestion of the whole seed normally does not produce sickness since the shell guards the toxin against digestion. However, if chewed, ground, or crushed, poisoning occurs because abrin will be exposed to the GI tracts. The activity of abrin is lost on boiling and consequently, the seeds once cooked can be taken without any harmful effects.

When abrin is exposed to the digestive system, it is almost poorly absorbed but the toxin can produce GI symptoms with large exposures. These GI symptoms include vomiting, nausea, diarrhea, abdominal pain while other symptoms are a weakness, rapid and irregular pulse, vertigo, and rectal bleeding among others. When injected subcutaneously, the site of injection results to painful swelling and ecchymosis develops with the inflammation and necrosis. Generalized septicemia and hemolysis can as well occur; also convulsions and death may occur from cardiac paralysis with 3–4 days. Many studies have shown that symptoms of abrin ingestion are delayed from a few hours to 2–3 days and the medical course can take up to 10 days (Shih and Goldfrank. 1998). There is no identified or known toxicity level in human though sharp dose-lethality curve in animals has been observed (Arena, 1986).

Aggressive and immediate treatment is extremely necessary due to the delay between the ingestion and symptoms. Gastric emptying techniques like induced emesis, gastric lavage, and bowel irrigation, activated charcoal among others can serve as useful treatment processes. No precise antidote for abrin poisoning has been reported, however, correction of electrolyte abnormalities, intravenous fluids and proper handling of the stools and vomits are applied as supportive care measures pending when the diagnosis is confirmed. Reedman et al. (2008) reported of a suicide attempt of a 27 years old man that crushed the Abrus seed. In this case, the toxin exposure to the GI tract has led to severe diarrhea and vomiting. Though he self-administered activated charcoal few minutes of ingestion and delivered himself to the emergency department after deciding not to continue with the suicide, aggressive treatment for dehydration was administered in the emergency department and the outcome was good in contrast to previous similar cases that did not get immediate treatment and supportive care.

The major economic importance of *A. precatorius* seed was its use as beads and native jewelry due to its bright coloration and in percussion instruments. It was also used as a unit of measurement by Indians many years ago because the seed is very consistent with weight. Indians used it to weigh gold through a measure known as Ratti. Garaniya and Bapodia (2014) reported several compounds like trigonelline, xylose, choline, hypaphorine, inositol, isoflavones, abrine, and so forth, in the leaves of *Abrus precatorius*. These compounds may have suggested the biological and therapeutic importance of *A. precatorius* leaves as many studies reported it is antiemetic, used in treating epilepsy, tuberculosis, asthma, and trachoma.

## 23.10 CONCLUSION

Sometimes science becomes a weapon to destroy instead of becoming a tool to build. Bioterrorism is one of the examples, which employ biological entities to make biological weapons. Not only bacteria and viruses, toxins from the poisonous plants have a place in biological warfare. In addition, these toxins prove to have many pharmacological properties and can lead to the production of valuable drugs to cure various diseases.

## KEYWORDS

- **poisonous phytochemicals**
- **phytotoxins**
- **warfare weapons**
- **bioterrorism**
- **phytochemicals**

## REFERENCES

Altin, G.; Sanli, A.; Erdogan, B. A.; Paksoy, M.; Aydin, S.; Altintoprak, N. Severe Destruction of the Upper Respiratory Structures After Brief Exposure to a Dieffenbachia Plant. *J. Craniofacial Surg.* **2013,** *24*(3), 245–247.

American Heart Association (AHA). Management of Symptomatic Bradycardia and Tachycardia. Part 7.3, 2005. http://www.circulationaha.org.

Arditti, J.; Rodriguez, E. Dieffenbachia: Uses, Abuses and Toxic Constituents: A Review. *J. Ethnopharmacol.* **1982,** *5*(3), 293–302.

Arena, J. M. *Poisoning: Toxicology, Symptoms and Treatments;* Charles C Thomas: Springfield, IL, 1986; Vol. I, p 496.

Barbi, N S.; Lucchetti, L.; Pereira, N. A.; da. Silva, A. J. R. Chemistry of *Dieffenbachia picta.* In *Poisoning by Plants, Mycotoxins and Related Toxins;* Riet-Correa, F., Pfister, J., Schild, A. L., Wierenga, T. Eds.; CAB International: United Kingdom, 2011; pp 593–599.

Barceloux, D. G. Dieffenbachia and Other Oxalate-Containing House Plants. In *Medical Toxicology of Natural Substances*; Barceloux, D. G., Ed.; John Wiley & Sons: New Jersey, 2008, p 768.

Benson, S.; Olnes, S.; Pihl, A.; Skorve, J.; Abraham, A. K. On the Mechanism of Protein-Synthesis Inhibition by Abrin and Ricin. *Eur. J. Biochem.* **1975,** *59*, 573–580.

Bhutia, S. K.; Maiti, T. K. Crabs Eye (*Abrus precatorius*) Seed and Its Immunomodulatory and Antitumor Properties. In *Nuts and Seeds in Health and Disease Prevention;* Preedy, V., Watson, R., Patel, V., Eds.; Academic Press: USA, 2011, pp 409–405.

Botanical online. Toxicity of Dieffenbachia. https://www.botanicalonline.com/alcalo idesdiefenbaquiaangles.html (accessed Dec 11, 2017).

Bozza, W.P.; Tolleson, W. H.; Rivera Rosado, L. A.; Zhang, B. Ricin Detection: Tracking Active Toxin. *Biotech. Adv.* **2015**, *33*(1), 117–123.

Bracci, A.; Daza-Losada, M.; Aguilar, M.; Feo, V.; Miñarro, J.; Rodríguez-Arias, M. A Methanol Extract of *Brugmansia arborea* Affects the Reinforcing and Motor Effects of Morphine and Cocaine in Mice. *Evidence-Based Complementary and Alternative Medicine;* 2013, Article ID 482976, 1-7. http://dx.doi.org/10.1155/2013/482976.

Chan, T. Y. Aconitine Poisoning. *Clin. Toxicol.* **2009**, *47*(4), 279–285.

Chisholm, H. Aconitine. *Encyclopaedia Britannica*, 11th ed.; Cambridge University Press: USA, 1911; Vol. 1, pp 151–152.

De Feo, V. The Ritual Use of Brugmansia Species in Traditional Andean Medicine in Northern Peru. *Econ. Bot.* **2004**, *58*(1), 221–229.

Diaz, J. H. Poisoning by Herbs and Plants: Rapid Toxidromic Classification and Diagnosis. *Wild. Environ. Medi.* **2016**, *27*(1), 136–152.

Eich, E. *Solanaceae and Convolvulaceae - Secondary Metabolites*; Springer-Verlag, Berlin Heidelberg, Germany. 2008, pp 157–158.

*Extra Pharmacopoeia Martindale.* 24th Edn.; The Pharmaceutical Press: London, 1958; Vol. 1, p 38.

Fu, M.; Wu, M.; Qiao, Y.; Wang, Z. Toxicological Mechanism of Aconitum Alkaloids. *Pharmazie* **2006**, *61*, 735–741.

Garaniya, N.; Bapodra, A. Ethnobotanical and Phytopharmacological Potential of *Abrus precatorius* L: A Review. *Asian. Pac. J. Trop. Biomed.* **2014**, *4*(1), 527–534.

Harner, M. J. *The Way of the Shaman;* Harper and Row: New York, 1990, Vol. 8, p 171.

Holm, L.P.; Juan, V.V.; James, P.P.; Donald, L. *A Geographical Atlas of the World Weeds*; John Wiley and Sons: New York, 1979; Vol. 2, p 391.

Holmes, C.H. Poisonous Plants: Perils in Nature. *J. Pharm. Pract.* **2000**, *13*(2), 125–129.

Kent, K. J. Ricin. In *Handbook of Bioterrorism and Disaster Medicine;* Antosia, R. E.,Cahill, J. D., Eds; Springer: Germany, 2006; pp 142–146.

Khajja, B. S.; Sharma, M.; Singh, R.; Mathur, G.K. Forensic Study of Indian Toxicological Plants as Botanical Weapon (BW): A Review. *J. Environ. Anal. Toxicol.* **2011**, *1*, 112. doi: 10.4172/2161-0525.1000112.

Kumar, M. A Review on Phytochemical Constituents and Pharmacological Activities of *Ricinus communis* L. Plant. *Int. J. Pharmacogn. Phytochem. Res.* **2017**, *9*(4), 466 472.

Lugnier, A. A. J.; Dirheimer, G.; Madjar, J. J.; Reboud J. P.; Gordon, J.; Howard, G. A. Action of Ricin from *Ricinus communis* L. Seeds on Eukaryote Ribosomal Proteins. *FEBS Lett.* **1976**, *67*(3), 343–347.

Maman, M.; Yehezkelli, Y. Ricin: A Possible, Noninfectious Biological Weapon. In *Bioterrorism and Infectious Agents: A New Dilemma for the 21st Century;*Fong I. W., Alibek, K. Eds.; Emerging Infectious Diseases of the 21st Century book series; Springer Science & Business Media: New York, 2005, pp 205–216.

McGaw, L. J.; Eloff, J. N.; Meyer, J. J. M. Screening of 16 Poisonous Plants for Antibacterial, Anthelmintic and Cytotoxic Activity in Vitro. *S. Afr. J. Bot.* **2005**, *71*(3), 302–306.

Noy-Porat, T.; Alcalay, R.; Epstein, E.; Sabo, T.; Kronman, C.; Mazor, O. Extended Therapeutic Window for Post-Exposure Treatment of Ricin Intoxication Conferred by the use of High-Affinity Antibodies. *Toxicon* **2017**, *127*, 100–105.

Olsnes, S. The History of Ricin, Abrin and Related Toxins. *Toxicon* **2004**, *44*, 361–370.

Prychid, C. J.; Rudall, P. J. Calcium Oxalate Crystals in Monocotyledons: A Review of their Structure and Systematics. *Ann. Bot.* **1999,** *84*(6), 725–739.

Rani, A.; Prasad, M. P. Studies on the Organogenesis of *Atropa Belladonna* in in-Vitro Conditions. *Int. J. Biotechnol. Bioeng. Res.* **2013,** *4*(5), 457-464.

Rasmann, S.; Agrawal, A. A. Plant Defense Against Herbivory: Progress in Identifying Synergism, Redundancy, and Antagonism Between Resistance Traits. *Curr. Opin. Plant Biol.* **2009,** *12*(4), 473–478.

Reedman, L.; Shih, R. D.; Hung, O. Survival after an Intentional Ingestion of Crushed Abrus Seeds. *West. J. Emerg. Med.* **2008,** *9*(3), 157–159.

Reis, V. M. S. dos. Dermatoses provocadas por plantas (fitodermatoses). *An. Bras. Dermatol.* **2010,** *85*, 479–489.

Rita, P.; Animesh, D. K. An Updated Overview on Atropa *belladonna* L. *Int J. Pharm.* **2011,** *2*(11), 11–17.

Roberts, M. F.; Wink, M. *Alkaloids: Biochemistry, Ecology, and Medicinal Applications;* Plenum Press: New York, 1998, p 28.

Roxas-Duncan, V. I.; Smith, L. A. Ricin Perspective in Bioterrorism. In. *Bioterrorism;* Morse, S., Ed.; InTech: London, 2012a; pp. 133–158.

Roxas-Duncan V. I.; Smith, L. A. Of Beans and Beads: Ricin and Abrin in Bioterrorism and Biocrime. *J. Bioterrorism Biodef.* **2012b,** *S2,*002. DOI: 10.4172/2157-2526.S2-002.

Sasseville, D. Phytodermatitis. *J. Cutaneous Med. Surg.* **1999,** *3*(5), 263–279.

Shih, R. D.; Goldfrank, L. R. Plants. In *Goldfrank's Toxicological Emergecies;* Goldfrank, L. R., Flomenbaum, N. E., Lewin, N. A., Weisman, R. S., Howland, M. A., Hoffman, R. S., Eds.; Appleton and Lange: East Norwalk Connecticut, 1998; Vol. XVI.

Sousa, N. L.; Cabral, G. B.; Vieira, P. M.; Baldoni, A. B.; Aragão, F. J. L. Bio-Detoxification of Ricin in Castor Bean (*Ricinus communis* L.) Seeds. *Sci. Rep.* **2017,** *7*(1), 15385.

The Telegragh. FBI arrests 70-year-old US woman who tested homemade ricin on fellow retirement home residents. 2017. Available: https://www.telegraph.co.uk/news/2017/12/02/70-year-old-us-woman-tested-ricin-neighbours-fbi/ (accessed Jan 18, 2018).

Tyagi, N.; Tyagi, M.; Pachauri, M.; Ghosh, P. C. Potential Therapeutic Applications of Plant Toxin-Ricin in Cancer: Challenges and Advances. *Tumor Biol.* **2015,** *36*, 8239.

Ulbricht, C.; Basch, E.; Hammerness, P.; Vora, M.; Wylie, J.; Woods, J. An Evidence-Based Systematic Review of Belladonna by the Natural Standard Research Collaboration. Natural Standard Review *J. Herb. Pharmacother.* **2004,** *4*, 61–90.

Wagstaff, D. J. *International Poisonous Plants Checklist: An Evidence-Based Reference;* CRC Press, 2008, p 69.

Wambebe, C.; Amosun, S. L. Some Neuromuscular Effects of the Crude Extracts of the Leaves of *Abrus precatorius. J. Ethnopharmacol.* **1984,** *11*, 49–58.

Walter, W.G. Dieffenbachia Toxicity. *JAMA.* **1967,** *201*(2), 140–141.

Wei-Gang, H.; Junfei, Y.; Chau, D.; Chen, H. C.; Lillico, D.; Justin, Y.; Negrych L. M.; Cherwonogrodzky, J. W. Conformation-Dependent High-Affinity Potent Ricin-Neutralizing Monoclonal Antibodies. *BioMed. Res. Int.* **2013,** Article ID 471346. DOI:10.1155/2013/471346.

ZIDBITS. The Top 10 Deadliest Plants. July 2011. https://www.google.com.ng/amp/s/zidbits.com/2011/07/the-top-10-deadliest-plants/amp/ (accessed Oct 19, 2017).

# CHAPTER 24

# PHYTOCHEMICALS AS PROOXIDANTS

ANDREW G. MTEWA

*Department of Chemistry, Institute of Technology, Malawi University of Science and Technology, Malawi*

*Pharmbiotechnology and Traditional Medicine Center of Excellence, Mbarara University of Science and Technology, Uganda*

*E-mail: amtewa@must.ac.mw; andrewmtewa@yahoo.com; Tel.: +265 999643272/+256 794547811*

*\*ORCID: https://orcid.org/0000-0003-2618-7451*

## ABSTRACT

Prooxidants refers to biological active compounds capable of inducing oxidative stress by either enhancing the generation of ROS or by inhibiting antioxidant systems. Phytochemicals are commonly known for their positive identity but least is known about their prooxidant characteristics. For example, the well-known vitamin C with a powerful antioxidant function has since been implicated as prooxidants because they reduce metallic ions which lead to the generation of free radicals through the Fenton reaction. Little research has been conducted on these characteristics of phytochemicals. Prooxidant characteristics of phytochemicals potentially lead to oxidative stress manifested in degenerative disorders among other ailments, but they are also important in the healing and/prevention of the same particularly when in the right equilibrium with antioxidants in the right timing and environments. This chapter has included some selected mechanisms and methods in the determination of prooxidant activities and system balance for phytochemicals.

## 24.1  INTRODUCTION

Phytochemicals should be understood to be a group of various compounds mostly acknowledged for their botanical species bioactivities. They include compounds known to be essential nutrients essential for normal physiological body functions, phytotoxins, which are toxic to body cells and systems (Shibamoto and Bjeldanes, 2009). Antinutrients which interfere with the normal nutrient absorption processes, antioxidants, which are known to be responsible for fighting reactive oxygen species (ROS) and prooxidants, which facilitate the presence and functions of ROS. The discovery of the positive effects of phytochemicals has seen the media, the supplements and nutrition industries and the vegetarian community is promoting and lobbying for the promotion of antioxidant supplements and natural botanical foods to counteract the adverse effects of oxidative stress. Too much of the presence of antioxidants, which may be sourced from phytochemicals from natural fruits and vegetables or supplementation with synthetic antioxidants can be a life-threatening hazard in some cases. There is little or almost no interest in consumers to explore and understand the negative contribution of phytochemicals to bioactivities. Recently, some science researchers have found interest to dig deeper into the prooxidative potential and bioactivities that the natural botanic wonder compounds have so to inform the public in general and further research. Prooxidants can be defined as oxidative stress-inducing chemicals by generating reactive radical species (Reactive oxygen and reactive nitrogen species [RNS]) or by inhibiting the activities of antioxidant species.

In ginkgo extracts, the flavonoids which exist primarily as glycosylated derivatives of quercetin and kaempferol have been shown to be extremely effective free radical scavengers (Huang, 1999; Blumenthal et al., 2000). In another development, quercetin has also been reported both in vitro and in vivo studies to have prooxidant activities just as several tea polyphenols (Geetha et al., 2005; Kapetanovic, 2013).

## 24.2  KNOWLEDGE GAP IN PHYTOCHEMICALS AS PROOXIDANTS

The common understanding about phytochemicals in herbal debates and arguments is the general consensus that they are present in most plants including fruits and vegetables which are beneficial to our health with no adverse effects (Mtewa, 2017). Most people who take herbs for their phytochemicals have full belief in the remediation or are too desperate to think otherwise, they take almost anything someone advises them without any questions. The consequence of the perception that phytotherapy is all safe is that many people are dying from unknown bioactivities of phytochemicals that are often overlooked or neglected right by the scientist on the bench to the policymaker. In as much as phytotherapy is proven to be effective, it must be through research as well to look at the holistic approach in vitro, together with the dynamic mechanisms in tissue and cells.

There are sparing or unclear disseminated pieces of information from research done on phytochemicals as being one of the potential culprits in the pathogenesis of most of the terminal diseases in the modern world. Due to the lack of serious attention, we may be having cases that are taking thousands or more lives from the activities of phytochemicals as prooxidants. There is a need for robust research and sensitization on the awareness of this ugly side of phytotherapy to better understand them and inform research and policy.

## 24.3   OXIDATIVE STRESS

Oxidative stress ensues when reactive oxygen and other radical species' generation exceeds their removal by antioxidant molecules. However, it should be noted that oxidative stress is a usual phenomenon in the body and under normal circumstances, the important physiological intracellular intensities of ROS, RNS, or other reactive species are maintained at acceptably low levels by various enzymatic systems participating in the in vivo redox homeostasis (Rahal et al., 2014).

Research has shown that oxidative stress remains a prerequisite mechanism of either the beginning or the progression of several disorders which includes cardiovascular diseases (CVDs), neurological diseases and disorders, obesity, cerebrovascular diseases, cancer, rheumatoid arthritis, aging, and myocardial ischemia (Boskabadi et al., 2013). Apart from environmental factors, some food activities and dietary habits are known to cause oxidative stress by means of different radical reaction mechanisms. These radicals, commonly known as free radicals, are molecules that have one or more unpaired electrons in their outer most shell and they include hydroxyl ($OH^{\cdot}$), nitric oxide ($NO^{\cdot}$), and superoxide($O_2^{-}$) also known as a hyperoxide (Aruoma, 1996).

Free radicals have the capacity of damaging protein, DNA, and lipid structural integrities leading to oxidative stress (Sies and DeGroot, 1992). Oxidative stress can generally be regarded as a disparity between the prooxidant and antioxidant systems in the body. Depending on the type, concentration and level of RNS and/or ROS, tissue-antioxidant status, exposure-duration, exposure to free radicals and their metabolites leads to different levels of increased proliferation, cell cycle interruption, and necrosis and apoptosis (Halliwell, 2008).

### 24.3.1 CONTRIBUTION OF PHYTOCHEMICALS TO OXIDATIVE STRESS

Research indicates that prooxidant phytochemicals contribute to oxidative stress. Stone et al. showed that some phytochemicals from dietary products contribute to colorectal cancer (CRC) pathogenesis (2014). Polyphenols in tea have been reported to be contributing to oxidative stress both in human and animal models (Kapetanovic, 2013). Epigallocatechin gallate (EGCG), for example, was reported to demonstrate prooxidant activities, yield hydrogen peroxide, and ROSs, ultimately leading to oxidative stress and in vitro cytotoxicity (Yamamoto et al., 2003). One of the known mechanisms of undergoing EGCG is protein carbonylation (Ishii et al., 2010). These prooxidant mechanisms may impact on redox-dependent cellular actions and lead to toxicity. Some phytochemicals are capable of undergoing chlorination reactions under favorable environments (Zhu and Fan, 2011). The chlorinated products, which could be phenolic or quinones are mostly carcinogenic and can cause oxidative stress. Chelation of flavonoids with metals like copper, iron, and other heavy metals facilitates their conversion to species that can easily react and show prooxidant activities (Rahal et al., 2014). Sometimes, the auto-oxidation properties of phytochemicals may end up with the development of prooxidant characteristics. The underlying antioxidant activities of quercetin, for example, may also participate in prooxidant reactions due to the auto-oxidation of quercetin that yields hydroxyl radicals within the cell and may be responsible for it's cytotoxic, mutagenic and/or biocidal effects (Geetha et al., 2005).

### 24.3.2 REMEDIATION FROM PROOXIDANT ACTIVITIES

To control the activities of prooxidants, antioxidants can be added to out-of-hand prooxidant systems. Some of the commonly known antioxidants used in the counter activity are butylated hydroxyanisole, ascorbic acid, butylated hydroxytoluene, propyl gallate, and sodium bisulfite.

## 24.3.2.1   ANTIOXIDANT-PROOXIDANT BALANCE

The antioxidant capacity and prooxidant potential of biological systems in the body are usually in an approximate equilibrium, the balance of which is an important key to maintaining cellular homeostasis. Some phytochemical derivatives are capable of existing as prooxidants and antioxidants. For example, it is complicated to tell the role of ascorbate, a mineral salt of ascorbic acid, as a cellular protectant. Phytochemicals can exist as a prooxidant due to one or both of the following two possibilities;

1. The ability to reduce transitional metal ions in order to enable participation in Fenton chemistry reactions.
2. Formation of ROS from their (phytochemical) metal catalyzed auto-oxidation.

On the other hand, antioxidant activities may occur due to one, two, or all of the three possibilities;

1. The ability to regenerate phenolic antioxidants that had been converted to other forms such as α-tocopherol.
2. The ability to shift the redox balance of transitional metal redox couples to reduce their natural preference of participating in prooxidant reactions.
3. The ability of directly reacting with free radicals or non-radical oxidants and engage the oxidants which ultimately give less reactive products

As reported in the literature by Gulcin et al. (2004) and Mtewa et al. (2017), exogenous factors such as ionization radiation, certain pollutants, solvents, and pesticides as well as endogenous factors such as stimulated polymorph nuclear leukocytes and macrophages can cause prooxidant-antioxidant imbalance. Prooxidant-antioxidant imbalance should be considered a serious life-threatening issue because the continued production of toxic species may cause oxidative stress and endothelial tissue injury (Boskabadi et al., 2013).

## 24.4   BENEFITS OF PROOXIDANT PHYTOCHEMICALS

Though not clear, currently, there is a scientific debate on which one between antioxidants and prooxidants has an ultimate influence in curing cancer. One school of thought suggests that antioxidant activities are solely responsible for cancer cell death by reducing the amount of ROSs which facilitates proliferation. Similarly, there exists some scientific evidence that prooxidants have given satisfactory results in chemotherapy due to their selective nature between normal and malignant cells. DeRoose and Duthie reported the positive roles dietary phytochemicals have in the maintenance of health and resilience to oxidative stress (2015). Prooxidants, having the ability to induce oxidative stress and selectively effectively so on cancerous cells, they eventually cause cell death to the cancers without or with little damage to the neighboring normal cells. This is one reason that makes cancer treatment so complex; its complex metabolomics: the activities of antioxidants and prooxidants are not standard in every single patient, it depends on several other issues including nature of the particular antioxidant and/or the prooxidant chemicals involved (including metal reducing potential, chelating behavior, and solubility), the stage of tumorigenesis and the microenvironment around the affected cells (including pH). The same treatment with a phytochemical antioxidant or prooxidant in one patient may give different results in another, that's why others suggest that the treatment of cancer should be given as isolated cases to individuals and adopt what suits them.

## 24.5   INFLUENCE OF METALS ON THE PROOXIDATIVE PROPERTIES

Transitional metals are important entities around phytochemicals for their prooxidant activities. Some of the popular and well-known antioxidant flavonoids have been reported to act as prooxidant when a transition metal is available (Halliwell, 2008). In vivo studies have shown phytochemicals to be mutagenic in such environments (Sahu and Gray, 1993; Sahu and Gray, 1994). These phytochemical metal-initiated prooxidant activities depend on their structures and the number of free -OH substitutions on the structures as observed in flavonoids (Guohua et al., 1997). The more the hydroxyl substitution groups available on a molecule, the more resilient the prooxidant activities.

Copper has been reported to be capable of reacting with flavonoids to give a prooxidant flavonoid derivative (Rahal et al., 2014) and has proven to be a good catalyst to the making of prooxidants in phytochemicals (Fukumoto and Mazza, 2000). Prooxidant activities of phenolics in the presence of copper (II) form radical hydroxyl in a series of reactions adapted (Akyuz et al., 2013) and outlined as follows:

Initial phenolic oxidation reaction produces a semiquinone radical, PhO;

$$PhOH + Cu^{2+} \rightarrow PhO^{\bullet} + Cu^{+} + H^{+}$$

This reacts with oxygen molecules to generate an autocatalytic oxygen species that in turn oxidizes the parent polyphenolic compound to give a semiquinone again and a hydrogen peroxide.

$$PhO^{\bullet} + O_2 \rightarrow PhO^{+} + O_2^{\bullet-}$$

$$PhOH + O_2^{\bullet-} + H^{+} \rightarrow PhO^{\bullet} + H_2O_2$$

Hydrogen peroxide can also be made through the disproportionation of the autocatalytic oxygen species.

$$2O_2^{\bullet-} + 2H^{+} \rightarrow H_2O_2 + O_2$$

Copper ions that were formed essentially from the initial reactions partly bind to proteins that contain thiol functional groups.

$$Protein\text{-}SH + Cu^{+} \rightarrow Protein\text{-}S\text{-} + Cu^{I} + H^{+}$$

The copper bound proteins may react with hydrogen peroxide to give hydroxide radicals through a reaction in the order of the Fenton-reaction.

$$Protein\text{-}S\text{-}Cu^{I} H_2O_2 \rightarrow Protein\text{-}S\text{-}Cu^{II} + {}^{\bullet}OH + OH^{-}$$

which eventually leads to the degeneration and damage of proteins, observable as cancer or other mentioned ailments.

$$Protein\text{-}S\text{-}Cu^{II} + {}^{\bullet}OH \rightarrow Protein_{(Damaged)} + Cu^{2+}$$

## 24.6 METHODS IN PROOXIDATION

### 24.6.1 DETERMINATION OF PROOXIDANT ACTIVITY

Generally, the same methods used in the establishment of antioxidant activities can be used to draw prooxidant activities of phytochemicals. However, there are several other assays that are in use in the determination of prooxidant activities of phytochemical compounds. The three methods generally outlined in this chapter can be employed on plant extracts or pure plant isolates including specific phytochemicals. The three methods are as discussed and used by in literature (Akyuz et al., 2013).

### 24.6.1.1 THE CUPRIC REDUCING ANTIOXIDANT CAPACITY (CUPRAC) ASSAY

The CUPRAC assay is considered as the recommended assay for the determination of prooxidant activities of polyphenolic compounds. Phosphate buffer (PH 7.4) is prepared and (1.0 mL) is put in a test tube. The following are then added to the same test tube: Polyphenolic compound, which may also be the methanolic (80%, v/v) test plant extract (1.0 mL, 1.0 mm), Cupper (II) solution (1.0 mL, 1.0 mm), egg white solution (1.0 mL), and distilled water (1.0 mL). Instead of water, rosmarinic acid or caffeic acid (1.0 mL, 1.0 mM) may be used for standard addition. This gives the final mixture at 5.0 mL total volume. Vortex, this mixture for 5 min and keep at room temperature for 25 min, precisely. At the end of the 25 min, add TCA solution (1.0 mL, 10%) to the incubated solution and precipitate proteins.

The prepared mixture is centrifuged for 10 min at 4000 rpm, and the upper phase which is liquid is decanted and the precipitate is washed (three times) with distilled water. Neocuproine (Nc) solution (1.0 mL) is added to the precipitate and 1.0 mL of CUPRAC buffer ($NH_4Ac$) is also added, then stirred and diluted to 4.1 mL distilled water. The moisture is filtered off using a filter paper and absorbance is recorded (450 nm) against CUPRAC reagent as blank. Prooxidant activity is determined from the difference between the values of absorbance recorded in the absence and the presence of the plant extract or polyphenolic sample.

### 24.6.1.2 ATOMIC ABSORPTION SPECTROSCOPY (AAS) ASSAY METHOD

In this method, everything is prepared and added as in the CUPRAC assay but differs at the point of adding the CUPRAC buffer. In this method, CUPRAC buffer is replaced with protein dissolution buffer which is added to the incubation product and then the copper passed into the solution is determined by Atomic Absorption Spectroscopy (AAS).

## 24.6.1.3   PROTEIN CARBONYL DETECTION ASSAY

This method measures absorbance (370 nm) of dinitrophenylhydrazones (DNP) which is generated from the reaction of 2,4-dinitrophenylhydrazine (DNPH) with the carbonyl group that emerges from the oxidation of proteins. The following solutions are added to a test tube: Phosphate buffer (1.0 mL, PH 7.4), Polyphenolic compound, which may be a methanolic (80%, v/v) plant extract or pure phytochemical isolate (1.0 mL, 1.0 mm), egg white solution (1.0 mL), distilled water (2.0 mL), and copper (II) compound solution (1.0 mL, 1.0 mm). The final mixture is stirred and incubated for 30 min in a water bath (37°C). At the end of the 30 min, DNPH solution (1.0 mL, 10.0 mm) is added and then incubated for 30 min (room temperature). Trichloroacetic acid (TCA) solution (1.0 mL, 10%) is added to the incubation solution and proteins get precipitated. The mixture is centrifuged for 10 min at 4000 rpm and the upper phase (liquid) is decanted. The precipitate is washed (3 times, with distilled water). Add protein dissolution buffer (1.0 mL) to the precipitate and dilute with distilled water to 4.1 mL. In a separate test tube, all the incubation solutions are added. The final mixture of 6.0 mL total volume is stirred but not incubated. DNPH and TCA solutions are added and the method proceeds with subsequent procedures as above. The test tube with the final mixture is filtered and absorbance (370 nm) is recorded against distilled water. The prooxidant activity is determined using the difference between values of absorbance recorded in the absence and presence of phenolic compounds, plant extracts, or pure isolates. When dealing with plant extracts, measure with compensation for the initial absorbance ($A_{370}$) values that were not resulting from the oxidation of protein.

## 24.6.2   DETERMINATION OF PROOXIDANT – ANTIOXIDANT IMBALANCE

As discussed already, the prooxidant and antioxidant activity balance is essential for normal functioning of body cells and systems. Therefore, it is imperative to focus on studies that indicate the position of the balance for leads to drug modifications through structure-activity-relationship (SAR).

## 24.6.2.1   PROOXIDANT-ANTIOXIDANT BALANCE (PAB) ASSAY (THE GENERAL PRINCIPLE)

The method described here is in vivo, as carried out by Boskabadi et al. (2013) which is a validated modification of a method as used by Alamdari et al. (2007). The modification was sought because researchers are yet to have a consensus on the methods for measuring oxidative stress in vivo. It is based on two different redox reactions, occurring simultaneously. In the enzymatic reaction, TMB (3,3',5,5'-tetramethylbenzidine), the chromogen, is oxidized to a colored cation by peroxides and in the chemical reaction, the TMB cation (colored) is reduced to another compound (colorless) by antioxidants. At this point, the absorbance is compared with the photometric absorbance of a series of standard solutions that are prepared by combining different proportions (0–100%) of 250 mmol/L hydrogen peroxide, as well as a characteristic of hydro-peroxides which is a pointer to total oxidant status, with uric acid (3 mmol/L) as a representative of the antioxidant capacity (in 10 mmol/L NaOH). It should be emphasized that uric acid and hydrogen peroxide do not interact with one another, nor do they neutralize the activities of each other. The said photometric comparison is carried out using an ELISA reader. Samples (blood) are collected in the morning into plain serum tubes after an overnight fast and are let clot between 30 and 60 min. With exclusion criteria of hemolyzed samples, serum is obtained by centrifuging at 2500 rpm (room temperature) and stored at $-20°C$ for analyses that follow.

- **PAB Assay (Step-by-step method)**

TMB (60 mg) is dissolved in dimethyl sulfoxide (DMSO, 10 mL). In the preparation of the cations of TMB, TMB/DMSO solution (400 mL) is added to acetate buffer (0.05 mol/L buffer, pH 4.5, 20 mL), and then fresh chloramine-T (100 mmol/L, 70 mL) solution is added to this 20 mL buffer. The solution is well mixed and incubated for 2 h (room temperature) in the dark. Peroxidase enzyme solution (25 units) is added to 20 mL of TMB cation solution, dispensed in 1 mL and stored at $-20°C$. To prepare the TMB solution, TMB/DMSO (200 mL) is added to 10 mL of acetate buffer (0.05 mol/L buffer, pH 5.8) and the working solution is prepared by mixing 1 mL TMB cation with 10 mL of TMB solution. The working solution is incubated for 2 min at room temperature in the dark and must be used immediately. About 10 mL of each of the sample, blank (distilled water) and the standard is mixed with the working solution (200 mL) in each well of a

96-well plate, which is then incubated in the dark at 37°C for 12 min. At the end of incubation, hydrochloric acid (2 N, 100 mL) is added to each well and the optical density (OD) is measured in an ELISA reader (450 nm) with reference wavelengths of 570 or 620 nm. A standard curve from the values is plotted in relation to the standard samples. PAB values are expressed in an arbitrary HK unit, which is the hydrogen peroxide percentage in the calibration mixture (standard solution) multiplied by 6. The values of the unknown test samples are then calculated based on the values got from the standard curve derived above. For the determination of the precision of the modified PAB method, the *"inter "* and *"intra "* assay coefficient of variation (CV %) is determined.

## KEYWORDS

- **prooxidants**
- **antioxidant**
- **phytochemicals**
- **reactive oxygen species**
- **oxidative stress**

## REFERENCES

Akyuz, E.; Ozyurek, M.; Guclu, K., Novel Prooxidant Activity Assay for Polyphenols, Vitamins C and E Using Modified CUPRAC Method. *Talanta* **2013**, *115*(2013), 583–589.

Alamdari, D. H.; Paletas, K.; Pegiou, T.; Sarigianni, M.; Befani, C.; Koliakos, G. A Novel Assay for the Evaluation of the Prooxidant-Antioxidant Balance, Before and After Antioxidant Vitamin Administration in Type II Diabetes Patients. *Clin. Biochem.* **2007**, *40*(3–4), 248–54.

Aruoma, O. I. Assessment of Potential Prooxidant and Antioxidant Actions. *J. Am. Oil. Chem. Soc.* **1996**, *73*(12), 1617–1625.

Blumenthal, M.; Goldberg, A.; Brinckmann, J. HerbalMedicine: Expanded Commission E Monographs. *Integr. Med. Comm.* **2000**, *133*(6), 487.

Boskabadi, H.; Moeini, M.; Tara, F.; Tavallaie, S.; Saber, H.; Nejati, R.; Hosseini, G.; Mostafavi-Toroghi, H.; Ferns, G. A. A.; Ghayour-Mobarhan, M., Determination of Prooxidant-Antioxidant Balance During Uncomplicated Pregnancy Using Rapid Assay. *J. Med. Biochem.* **2013**, *32*(13), 227–232.

DeRoose, B.; Duthie, G. G. Role of Dietary Prooxidants in the Maintenance of Health and Resilience to Oxidative Stress. *Mol. Nutr. Food. Res.* **2015**, *59*(7), 1229–1248.

Fukumoto, L. R.; Mazza, G. Assessing Antioxidant and Prooxidant Activities of Phenolic Compounds. *J. Agric. Food Chem.* **2000,** *48*(8), 3597–3604.

Geetha, T.; Malhotra, V.; Chopra, K.; Kaur, I. P. Antimutagenic and Antioxidant/Prooxidant Activity of Quercetin. *Ind. J. Exp. Biol.* **2005,** *43*, 61–67.

Gulcin, I.; Kufrevioglu, O. I.; Oktay, M.; Buyukokuroglu, M. E. Antioxidant, Microbial, Antiulcer and Analgesic Activities of Nettle (*Urtica dioica* L.). *J. Ethnopharmacol.* **2004,** *90*, 205–215.

Guohua, G.; Emin, S.; Ronald, L. Antioxidant and Prooxidant Behavior of Flavonoids: Structure-Activity Relationships. *Free Rad. Bio. Med.* **1997,** *22*(5), 749–760.

Halliwell, B. Are Polyphenols Antioxidants or Pro-Oxidants? What Do We Learn from Cell Culture and in Vivo Studies? *Arch. Biochem. Biophys.* **2008,** *476*(2), 107–112.

Huang, K. C. *Pharmacology of Chinese Herbs;* CRC Press: Boca Raton, Fla, USA, 1999.

Ishii, T.; Minoda, K.; Bae, M. J.; Mori, T.; Uekusa, Y.; Ichikawa, T.; Aihara, Y.; Furuta, T.; Wakimoto, T.; Kan, T.; Nakayama, T. Binding Affinity of Tea Catechins for HSA; Characterization by High Performance Affinity Chromatography with Immobilized Albumin Column. *Mol. Nutr. Food. Res.* **2010,** *54*(6), 816–822.

Kapetanovic, I. Toxicity of Green Tea Polyphenols. In *Tea in Health and Disease Prevention;* Preedy, V. R., Ed.; Academic Press: Florida, 2013; pp 1435–1447.

Mtewa, A. G. K. Antibacterial Potency Stability, pH and Phytochemistry of Some Malawian Ready-to-Serve Aqueous Herbal Formulations used Against Enteric Diseases. *Int. J. Herb. Med.* **2017,** *5*(3), 1–05.

Mtewa A. G. K.; Biswick T. T.; Mwatseteza J. F. Antioxidant Potential and Total Phenolic Changes in Malawian Herbal Formulations Stored in Clay Pots and Plastic Bottles. *J. Free Rad. Antiox.* **2017,** *144,* 479–494.

Rahal, A.; Kumar, A.; Singh, V.; Yadav, B.; Tiwari, R.; Chakraborty, S.; Dhama, K. Oxidative Stress, Prooxidants, and Antioxidants: The Interplay. *Biomed Res. Int.* [online] **2014.** https://www.ncbi.nlm.nih.gov/pubmed/24587990. DOI: 10.1155/2014/761264.

Sahu, S. C.; Gray, G. C. Interactions of Flavonoids, Trace Metals, and Oxygen: Nuclear DNA Damage and Lipid Peroxidation Induced by Myricetin. *Cancer Lett.* **1993,** 70(1–2), 73–79.

Sahu, S. C.; Gray, G. C. Kaempferol-Induced Nuclear DNA Damage and Lipid Peroxidation. *Cancer Lett.* **1994,**85(2), 159–164.

Shibamoto, T.; Bjeldanes, L. *Introduction to Food Toxicology;* 2nd ed.; Academic Press: USA, 2009; Vol. 2, p 124.

Sies, H.; De Groot, H. Role of Reactive Oxygen Species in Cell Toxicity. *Toxicol. Lett.***1992,** *64–65,* 547–551.

Stone, W. L.; Krishnan, K.; Campbell, S. E.; Palau, V. E. The Role of Antioxidants and Pro-Oxidants in Colon Cancer. *World J. Gastrointest. Oncol.* **2014,** *6*(3), 55–66.

Yamamoto, T.; Hsu, S.; Lewis, J.; Wataha, J.; Dickinson, D.; Singh, B.; Bollag, W. B.; Lockwood, P.; Ueta, E.; Osaki, T.; Schuster, G. Green Tea Polyphenol Causes Differential Oxidative Environments in Tumor Versus Normal Epithelial Cells. *J. Pharmacol. Exp. Ther.* **2003,** *307*, 230–236.

Zhu, B. Z.; Fan, R. Metal Independent Pathways of Chlorinated Phenol/Quinone Toxicity. *Adv. Mol. Toxicol.* **2011,** *5,* 1–43.

# CHAPTER 25

# PHYTOCHEMICALS AS ANTINUTRIENTS

CHUKWUEBUKA EGBUNA

*Department of Biochemistry, Chukwuemeka Odumegwu Ojukwu University, Uli, Anambra State, Nigeria, Tel.: +2347039618485*

*E-mail: egbuna.cg@coou.edu.ng; egbunachukwuebuka@gmail.com*

*ORCID: https://orcid.org/0000-0001-8382-0693*

## ABSTRACT

Antinutrients are components of food that acts to reduce nutrient intake, digestion, absorption, and utilization. They are present in many food substances such as legumes, nuts, pulses, seeds, whole grains, alcoholic drinks, carbonated drinks, coffee and tea, pharmaceutical drugs, tobacco and plants of the nightshade family. Antinutrients could be natural or synthetic and can be beneficial to some extent especially when in moderate amount. The major antinutrients considered in this chapter are phytate, lectins, tannins, protease inhibitors, saponins, and oxalates while others are listed.

## 25.1 INTRODUCTION

Antinutrients are natural or synthetic compounds found in a variety of foods especially grains, beans, legumes, and nuts that interfere with the absorption of vitamins, minerals, and other nutrients (Table 25.1). Some phytochemicals have the potentials to be beneficial for the overall maintenance of the health of an organism in terms of serving as antioxidants, antimicrobial, anti-inflammatory or as a drug, or lead compound for the discovery of new drugs. But in high concentration, some phytochemicals can act as an antinutrient (Ifemeje et al., 2014). This has been a cause for concern since current recommendations suggest the increased consumption of plants such as grains, fruits, and vegetables in order to prevent or possibly cure cardiovascular

**TABLE 25.1** Antinutrients, Sources, and Effects.

| Sr./ No. | Antinutrients | Sources | Antinutritional effects |
|---|---|---|---|
| 1. | Phytic acid/ phytate | Grains, legumes, nuts, oilseeds | Interferes with the absorption of minerals such as Ca, Cu, Fe, Mn, and Zn. Inhibits certain essential digestive enzymes called amylase, trypsin, and pepsin |
| 2. | Oxalates | Sesame seeds, soybeans, peanuts, black and brown millets, spinach, sweet potatoes, cocoa powder. | Binds divalent minerals, for example, $Ca^{2+}$ and prevent its absorption in the human body |
| 3 | Tannins | Tea, coffee, red wine, berries, fruit juice, nuts, for example, hazelnuts, walnuts | Enzyme inhibitor and prevent adequate digestion which can cause protein deficiency and gastrointestinal problems |
| 4. | Lectins | Legumes, wheat, nuts, corn, potatoes, tomatoes, eggplant | Reduce nutrient absorption and can cause indigestion, bloating, and gas for many people |
| 5. | Saponins | Quinoa, legumes such as soybeans and chickpeas | Affect the gastrointestinal lining, contributing to the leaky gut syndrome and autoimmune disorders. |
| 6. | Trypsin Inhibitors | Grain-containing products, including cereals, porridge, bread and even baby foods. | Trypsin and chymotrypsin inhibitors. Inhibitors cause a Th1-driven immune response, activation of TLR4, and cause intestinal inflammation |
| 7. | Lipase inhibitors | *Panax japonicas* (Japanese Ginseng) *Platycodon grandiflorus*. Generally, plants that contain saponins, polyphenols, terpenes. Microbial compounds: lipstatin (marketed as orlistat), panclicins, valilactone | Interfere with enzymes, such as human pancreatic lipase and could be beneficial to prevent obesity and related diseases |
| 8. | Amylase inhibitors | Alpha-amylase inhibitor, a protein from the microbe *Streptomyces tendae*. Wheat or white kidney bean (*Phaseolus vulgaris*) | Prevent the action of enzymes that break the glycosidic bonds of starches and other complex carbohydrates, preventing the release of simple sugars and absorption by the body. It may aid weight loss, a beneficial prospect |
| 9. | Isoflavones | Soybeans, red clover, green tea, split peas, pigeon peas, peanuts, chickpeas | Classified as phytoestrogens and considered endocrine disruptors |

**TABLE 25.1** *(Continued)*

| Sr./ No. | Antinutrients | Sources | Antinutritional effects |
|---|---|---|---|
| 10. | Solanine | Nightshade vegetables such as eggplant, peppers, and tomatoes. | Can cause "poisoning" and symptoms such as nausea, diarrhea, vomiting, stomach cramps, burning of the throat, headaches, and dizziness |
| 11. | Glucosinolates | Broccoli, brussel sprouts, cabbage, and cauliflower | Prevent the uptake of iodine, affecting the function of the thyroid and thus are considered goitrogens |
| 12. | Chaconine | Found in corn and plants of the Solanaceae family, for example, potatoes | Causes digestive issues, especially when uncooked and eaten in high amounts |
| 13. | Alcohol | Fermented carbohydrates, such as sugars or starches. Beer, brandy, gin, rum, wine | Slowing down the processes related to the digestion of protein, carbohydrates, and fat. Alcohol also forces expulsion of zinc and inhibits absorption of the B complex of vitamins |
| 14. | Flavonoids | Fruits and vegetables, berries, apple, tomato | At very high concentration can chelate metals such as iron and zinc and reduce the absorption of these nutrients. Moreover, inhibit digestive enzymes and may also precipitate proteins |
| 15. | Avidin | Egg whites contain high levels of avidin, a protein that binds to vitamin B7 (biotin) and stops it being absorbed | Deficiency of vitamin B7 may result in dry scaly skin, fatigue, loss of appetite, nausea, vomiting, mental depression as well as tongue inflammation and high cholesterol |
| 17. | Sulfites | Cider, white wine, and dried fruit | Sulfites can also cause edema, anaphylaxis, and rhinitis |
| 18. | Gluten | Proteins found in wheat, rye, barley, and triticale | Gluten can cause inflammation and digestive problems. Those with celiac disease experience an immune response when exposed to gluten |

disease, diabetes, and cancer (Shahidi, 1997). Although these concentration-dependent effects may be manipulated in such a way that advantage is taken from their health-related benefits so that management of chronic diseases becomes possible (Shahidi, 1997). The most important antinutrients are phytate, tannins, protease inhibitors, calcium oxalate, and lectins.

## 25.2   PHYTOCHEMICALS AND ANTINUTRIENTS

### 25.2.1   PHYTATE

Phytic acid, also known as phytate or inositol hexakisphosphate (IP6), is the principal storage form of phosphorus in many plant tissues, especially bran and seeds. Phytic acid (Fig. 25.1), with its highly negatively charged structure, is a very reactive compound and particularly attracts positively charged ions such as those of zinc and calcium. Phytate is not digestible to humans or nonruminant animals, this is because these animals lack the digestive enzyme phytase required to remove phosphate from the inositol in the phytate molecule. On the other hand, ruminants readily digest phytate because of the phytase produced by rumen microorganisms.

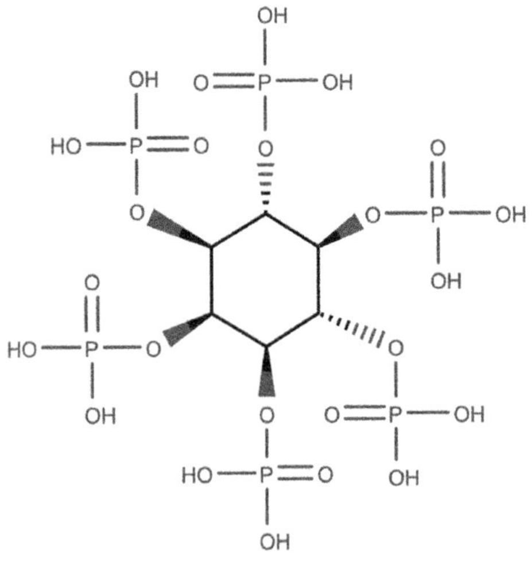

**FIGURE 25.1**   Structure of phytic acid.

According to Egbuna and Ifemeje (2015), an excessive amount of phytic acid in the diet can form insoluble complexes with multi-charged metals, such as copper, zinc, calcium, and iron which results to a deficit in the absorption of some dietary minerals and leads to mineral deficiencies. Phytic acid has been reported to interact with other compounds by the formation of ternary complexes of phytic acid with protein and carbohydrate (starch) which reduces their bioavailability and digestion (Makkar et al., 2007). Although home food preparation methods can degrade phytic acid. Simply cooking the food will reduce the phytic acid to some degree. More effective methods are soaking in an acid medium, lactic acid fermentation, and sprouting. There exist some beneficial aspects of phytate. For instance, it has been reported to lower blood glucose response to starchy foods, reduction of plasma cholesterol, and levels of triacylglycerols.

## 25.2.2   LECTINS

Lectins are carbohydrate-binding proteins, macromolecules that are highly specific for sugar moieties. They are often referred to as a nutrient that needs to be avoided due to their ability to selectively bind to carbohydrates (or the carbohydrate portion of certain proteins). Foods with high concentrations of lectins, such as beans, cereal grains, seeds, nuts, seafood, the nightshade family of vegetables (i.e., tomatoes, eggplant, peppers, and potatoes), may be harmful if consumed in excess or in uncooked or improperly cooked form. Adverse effects may include nutritional deficiencies and immune (allergic) reactions. Possibly, most effects of lectins are due to gastrointestinal distress through interaction of the lectins with the gut epithelial cells which has been reported to create a leaky gut. Many beneficial functions of lectins in animals exist. They help in the regulation of cell adhesion, glycoprotein synthesis, and the control of protein levels in the blood, and so forth. Other uses of lectins are in affinity chromatography for purifying glycoproteins and its use as biochemical weapon.

## 25.2.3   PHENOLIC COMPOUNDS

Phenolic compounds found in foods generally contribute to their astringency and may also reduce the availability of certain minerals such as zinc (Shahidi, 1997). During thermal processing, phenolic compounds can undergo oxidation to quinones, which may combine with amino acids, thus making them nutritionally unavailable (Shahidi, 1997). Phenolic compounds include flavonoids, isoflavonoids, and tocopherols, among others.

Tannin or tannic acid is an astringent, a polyphenolic biomolecule that can bind to proteins, amino acids, alkaloids and other organic compounds and possibly cause their precipitation. Moreover, tannins may reduce cell wall digestibility by binding bacterial enzymes and form indigestible complexes with cell wall carbohydrates.

Flavonoids (not often considered as antinutrient) are polyphenol with great antioxidant functions. At very high concentration, flavonoids can chelate metals such as iron and zinc and reduce the absorption of these nutrients. They also inhibit digestive enzymes and may also precipitate proteins. In one experiment, flavonoids were found to be strong topoisomerase inhibitors and induce deoxyribonucleic acid mutations in the MLL gene, which are common findings in neonatal acute leukemia (Thirman et al., 1993; Strick et al., 2000).

### 25.2.4   PROTEASE INHIBITORS

As the name implies, protease inhibitors prevent the functions of protease, the digestive enzymes that break down proteins in the digestive tract. They are found in legumes (particularly soybeans) and grains. It has been reported that protease inhibitors from soybeans are associated with a decrease in the development of chemically induced cancers. Owing to their particular protein nature, protease inhibitors can be easily denatured by heat treatment.

### 25.2.5   SAPONINS

Saponins are soap-like substances characterized by their ability to create foam in water. They cause a bitter or astringent taste in plants. Saponins can bind to various nutrients inhibiting the body ability to absorb them. They have also been found to inhibit digestive enzymes (trypsin and chymotrypsin), interfering with thyroid function and causing hemolysis. The amphiphilic structure of saponins allows them to latch on to cholesterol molecules attached to the surface of intestinal cells. Once they attach, they stimulate a reaction that creates pores in the cell surface, producing increased permeability (i.e., leaky gut) and allowing substances to enter the bloodstream. Their ability to attach to cholesterol can be beneficial in reducing cholesterol. Substances high in saponins are soybeans, potatoes, aloe vera, ash gourd, baobab fruit, chickpeas, corn silk, and horse chestnuts contain saponins. Saponins can be helpful in preventing colon cancer by binding primary bile acids which prevent primary bile acids from forming secondary bile acids that cause colon cancer. Figure 25.2 shows solanine, a typical saponin.

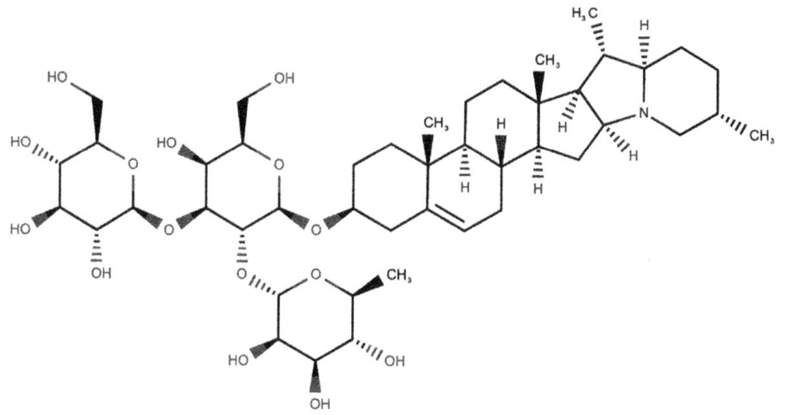

**FIGURE 25.2**   A typical saponin: solanine.

## 25.2.6   OXALATE

Oxalate (IUPAC: ethanedioate) is the dianion with the formula $C_2O_4^{2-}$, also written $(COO)_2^{2-}$ (Fig. 25.3). In the body, oxalic acid combines with divalent metallic cations such as calcium ($Ca^{2+}$) and iron ($Fe^{2+}$) to form crystals of the corresponding oxalates which are then excreted in urine as minute crystals. Oxalate crystals can be razor sharp and may cause damage to various tissues. Apart from food sources, oxalate can be formed in the body through cadmium catalyzed transformation of vitamin C into oxalic acid. Cadmium can result from smoking heavily, ingesting substances contaminated with Cd, or from industrial exposure to Cd.

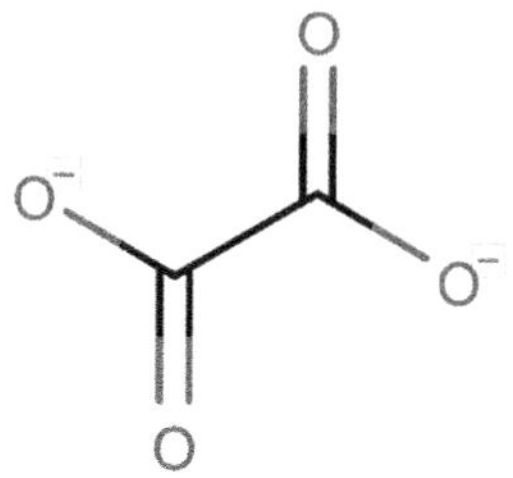

**FIGURE 25.3**   Structure of oxalate.

## 25.3   CONCLUSION

Antinutrients can be harmful especially when in excess and can be beneficial as well in moderate amount. Although many traditional methods of food preparation such as fermentation, cooking, and malting increase the nutritive quality of plant foods by reducing certain antinutrients such as phytic acid, polyphenols, and oxalic acid. For instance, soaking foods high in antinutrients can significantly reduce lectin and phytate content

## KEYWORDS

- antinutrients
- phytate
- oxalate
- lectins
- saponins

## REFERENCES

Egbuna, C.; Ifemeje, J. C. Biological Functions and Anti-Nutritional Effects of Phytochemicals in Living System. *IOSR J. Pharm. Biol. Sci. (IOSR-JPBS)* **2015**, *10*(2), 10–19.

Ifemeje, J. C.; Egbuna, C.; Eziokwudiaso, J. O.; Ezebuo, F. C. Determination of the Anti-Nutrient Composition of Ocimum Gratissimum, Corchorus Olitorius, Murraya Koenigii Spreng and Cucurbita Maxima. *Int. J. Innovation Sci. Res.* **2014**, *2*(3), 127–133.

Makkar, H. P. S.; Siddhuraju, S.; Becker, K. *Plant Secondary Metabolites; Methods in Molecular Biology;* Springer: New York, 2007.

Shahidi, F. *Antinutrients and Phytochemicals in Food;* ACS Symposium Series; American Chemical Society: Washington, DC, 1997.

Strick, R.; Strissel, P. L.; Borgers, S.; Smith, S. L.; Rowley, J. D. Dietary Bioflavonoids Induce Cleavage in the MLL Gene and may Contribute to Infant Leukemia. *Proc. Natl. Acad. Sci. U.S.A.* **2000**, *9*(9), 4790–4795.

Thirman, M. J.; Gill, H. J.; Burnett, R. C., Mbangkollo, D.; McCabe, N. R.; Kobayashi, H.; et al. Rearrangement of the MLL Gene in Acute Lymphoblastic and Acute Myeloid Leukemias with 11q23 Chromosomal Translocations. *N. Engl. J. Med.* **1993**, *329*(13), 909–914.

# INDEX

**C**

# W

# X

# Y

# Z

Printed and bound by CPI Group (UK) Ltd, Croydon, CR0 4YY

23/10/2024

01777703-0019